MW01517857

Wind Power Generation and Wind Turbine Design

WITPRESS

WIT Press publishes leading books in Science and Technology.
Visit our website for the current list of titles.
www.witpress.com

WIT eLibrary

Home of the Transactions of the Wessex Institute, the WIT electronic-library provides the
international scientific community with immediate and permanent access to individual
papers presented at WIT conferences. Visit the WIT eLibrary at
http://library.witpress.com

Wind Power Generation and Wind Turbine Design

Edited by:

Wei Tong

Kollmorgen Corp., USA

WITPRESS Southampton, Boston

Edited by: **Wei Tong**, *Kollmorgen Corp., USA*

Published by

WIT Press
Ashurst Lodge, Ashurst, Southampton, SO40 7AA, UK
Tel: 44 (0) 238 029 3223; Fax: 44 (0) 238 029 2853
E-Mail: witpress@witpress.com
http://www.witpress.com

For USA, Canada and Mexico

WIT Press
25 Bridge Street, Billerica, MA 01821, USA
Tel: 978 667 5841; Fax: 978 667 7582
E-Mail: infousa@witpress.com
http://www.witpress.com

British Library Cataloguing-in-Publication Data
A Catalogue record for this book is available
from the British Library

ISBN: 978-1-84564-205-1

Library of Congress Catalog Card Number: 2009943185

*The texts of the papers in this volume were set
individually by the authors or under their supervision.*

Contents

CHAPTER 2
Wind resource and site assessment **49**
Wiebke Langreder

CHAPTER 3
Aerodynamics and aeroelastics of wind turbines................................... 89
Alois P. Schaffarczyk

CHAPTER 4
Structural dynamics of wind turbines... 121
Spyros G. Voutsinas

CHAPTER 17
Design of support structures for offshore wind turbines 559
J. van der Tempel, N.F.B. Diepeveen, D.J. Cerda Salzmann & W.E. de Vries

CHAPTER 20
Wind turbine noise measurements and abatement methods 641
Panagiota Pantazopoulou

CHAPTER 21
Wind energy storage technologies 661
Martin Leahy, David Connolly & Noel Buckley

Preface

Along with the fast rising energy demand in the 21st century and the growing recognition of global warming and environmental pollution, energy supply has become an integral and cross-cutting element of economies of many countries. To respond to the climate and energy challenges, more and more countries have prioritized renewable and sustainable energy sources such as wind, solar, hydropower, biomass, geothermal, etc., as the replacements for fossil fuels.

Wind is a clean, inexhaustible, and an environmentally friendly energy source that can provide an alternative to fossil fuels to help improve air quality, reduce greenhouse gases and diversify the global electricity supply. Wind power is the fastest-growing alternative energy segment on a percentage basis with capacity doubling every three years. Today, wind power is flourishing in Europe, North America, and some developing countries such as China and India. In 2009, over 37 GW of new wind capacity were installed all over the world, bringing the total wind capacity to 158 GW. It is believed that wind power will play a more active role as the world moves towards a sustainable energy in the next several decades.

The object of this book is to provide engineers and researchers in the wind power industry, national laboratories, and universities with comprehensive, up-to-date, and advanced design techniques and practical approaches. The topics addressed in this book involve the major concerns in wind power generation and wind turbine design. An attempt has been made to include more recent developments in innovative wind technologies, particularly from large wind turbine OEMs. This book is a useful and timely contribution to the wind energy community as a resource for engineers and researchers. It is also suitable to serve as a textbook for a one- or two-semester course at the graduate or undergraduate levels, with the use of all or partial chapters.

To assist readers in developing an appreciation of wind energy and modern wind turbines, this book is organized into four parts. Part 1 consists of five chapters, covering the basics of wind power generation. Chapter 1 provides overviews of the history of wind energy applications, fundamentals of wind energy and basic knowledge of modern wind turbines. Chapter 2 describes how to make wind resource assessment, which is the most important step for determining initial feasibility in a wind project. The assessment may pass through several stages such as initial site identification, detailed site characterizations, site suitability, and energy yield and losses. As a necessary tool for modeling the loads of wind turbines and designing rotor blades, the detail review of aerodynamics, including analytical theories and experiments, are presented in Chapter 3. Chapter 4 provides an overview of the frontline research on structural dynamics of wind turbines, aiming at assessing the integrity and reliability of the complete construction against varying external loading over the targeted lifetime. Chapter 5 discusses the issues related to wind turbine acoustics, which remains one of the challenges facing the wind power industry today.

Part 2 comprises seven chapters, addressing design techniques and developments of various wind turbines. One of the remarkable trends in the wind power industry is that the size and power output from an individual wind turbines have being continuously increasing since 1980s. As the mainstream of the wind power market, multi-megawatts wind turbines today are extensively built in wind farms all over the world. Chapter 6 presents the detail designing methodologies, techniques, and processes of these large wind turbines. While larger wind turbines play a critical role in on-grid wind power generation, small wind turbines are widely used in residential houses, hybrid systems, and other individual remote applications, either on-grid or off-grid, as described in Chapter 7. Chapter 8 summarises the principles of operation and the historical development of the main types of vertical-axis wind turbines. Due to some significant advantages, vertical-axis turbines will coexists with horizontal-axis turbines for a long time. The innovative turbine techniques are addressed in Chapter 9 for the direct drive superconducting wind generators and in Chapter 10 for the tandem wind rotors. To fully utilize the wind resource on the earth, offshore wind turbine techniques have been remarkably developed since the mid of 1980s. Chapter 11 highlights the challenges for the offshore wind industry, irrespective of geographical locations. To shed new light on small wind turbines, Chapter 12 focuses on updated state-of-the-art technologies, delivering advanced small wind turbines to the global wind market with lower cost and higher reliability.

Part 3 contains five chapters, involving designs and analyses of primary wind turbine components. As one of the most key components in a wind turbine, the rotor blades strongly impact the turbine performance and efficiency. As shown in Chapter 13, the structural design of turbine blades is a complicated process that requires know-how of materials, modeling and testing methods. In Chapter 14, the implementation of the smart rotor concept is addressed, in which the aerodynamics along the blade is controlled and the dynamic loads and modes are dampened. Chapter 15 explains the gear design criteria and offers solutions to the various gear design problems. Chapter 16 involves the design and analysis of wind turbine towers. In pace with the increases in rotor diameter and tower height for large wind turbines, it becomes more important to ensure the serviceability and survivability of towers.

For offshore wind turbines, the design of support structures is described in Chapter 17. In this chapter, the extensive overviews of the different foundation types, as well as their fabrications and installations, are provided.

Part 4 includes four chapters, dealing with other important issues in wind power generation. The subject of Chapter 18 is to describe approaches to determine the wind power curves, which are used to estimate the power performing characteristics of wind turbines. Cooling of wind turbines is another challenge for the turbine designers because it strongly impacts on the turbine performance. Various cooling techniques for wind turbines are reviewed and evaluated in Chapter 19. As a complement of Chapter 5, Chapter 20 focuses on engineering approaches in noise measurements and noise abatement methods. In Chapter 21, almost all up-to-the date available wind energy storage techniques are reviewed and analyzed, in view of their applications, costs, advantages, disadvantages, and prospects.

To comprehensively reflect the wind technology developments and the tendencies in wind power generation all over the world, the contributors of the book are engaged in industries, national laboratories and universities at Australia, China, Denmark, Germany, Greece, Ireland, Japan, Sweden, The Netherlands, UK, and USA.

I gratefully acknowledge all contributors for their efforts and dedications in preparing their chapters. The book has benefited from a large number of reviewers all over the world. With their constructive comments and advice, the quality of the book has been greatly enhanced. Finally, special thanks go to Isabelle Strafford and Elizabeth Cherry at WIT Press for their efficient work for publishing this book.

Wei Tong
Radford, Virginia, USA, 2010

List of Contributors

Stephan Barth
ForWind – Center for Wind Energy Research of
   the Universities of Oldenburg, Bremen and
   Hannover
D-26129 Oldenburg
Germany
Email: stephan.barth@forwind.de

Biswajit Basu
School of Engineering
Trinity College Dublin
Dublin 2
Ireland
Email: basub@tcd.ie

Harald Bersee
Faculty of Aerospace Engineering
Delft University of Technology
Kluyverweg 1
2628 CN Delft
The Netherlands
Email: H.E.N.Bersee@tudelft.nl

Sumit Bose
Global Research Center
General Electric Company
Niskayuna, NY 12309
USA
Email: bose@ge.com

Kim Branner
Wind Energy Division
Risø National Laboratory for Sustainable Energy
DK-4000 Roskilde
Denmark
Email: kibr@risoe.dtu.dk

Povl Brøndsted
Materials Research Division
Risø National Laboratory for Sustainable Energy
DK-4000 Roskilde
Denmark
Email: pobr@risoe.dtu.dk

Denis Noel Buckley
The Charles Parsons Initiative
Department of Physics
University of Limerick
Castletroy, Limerick
Ireland
Email: noel.buckley@ul.ie

David Connolly
The Charles Parsons Initiative
Department of Physics
University of Limerick
Castletroy, Limerick
Ireland
Email: david.connolly@ul.ie

Paul Cooper
School of Mechanical, Materials and
    Mechatronic Engineering
University of Wollongong
Wollongong, NSW 2522
Australia
Email: pcooper@uow.edu.au

Niels F. B. Diepeveen
Department of Offshore Engineering
Delft University of Technology
2628 CN Delft
The Netherlands
Email: n.f.b.diepeveen@tudelft.nl

Laszlo Fuchs
Division of Fluid Mechanics
Lund University
S-22100 Lund
Sweden
Email: Laszlo.Fuchs@energy.lth.se

Ray Hicks
Ray Hicks Ltd
Llangammarch Wells, Powys
LD4 4BS
UK
Email: raymondhicks@btinternet.com

John W. Holmes
Materials Research Division
Risø National Laboratory for Sustainable Energy
DK-4000 Roskilde
Denmark
Email: jwho@risoe.dtu.dk

Anton W. Hulskamp
Faculty of Aerospace Engineering
Delft University of Technology
Kluyverweg 1
2629 HS Delft
The Netherlands
Email: A.W.Hulskamp@tudelft.nl

Yanlong Jiang
Department of Man-Machine and Environment
    Engineering
Nanjing University of Aeronautics and
    Astronautics
Nanjing 210016
China
Email: jiang-yanlong@nuaa.edu.cn

Toshiaki Kanemoto
Department of Mechanical and Control
    Engineering
Kyushu Institute of Technology
1-1Sensui, Tobata,
Kitakyushu, Fukuoka, 804-8550
Japan
Email: kanemoto@mech.kyutech.ac.jp

Koichi Kubo
Graduate School of Engineering
Kyushu Institute of Technology
1-1 Sensui, Tobata,
Kitakyushu, Fukuoka, 804-8550
Japan
Email: h584104k@tobata.isc.kyutech.ac.jp

Wiebke Langreder
Wind&Site, Suzlon Energy A/S
DK 8000 Århus C
Denmark
Email: wiebke.langreder@suzlon.com

Martin John Leahy
The Charles Parsons Initiative
Department of Physics
University of Limerick
Castletroy, Limerick
Ireland
Email: martin.leahy@ul.ie

Clive Lewis
Converteam UK Ltd
Rugby, Warwickshire
CV21 1BU
UK
Email: clive.lewis@converteam.com

Hikary Matsumiya
Hikarywind Lab., Ltd
5-23-4 Seijo, Setagaya-ku
Tokyo 157-0066
Japan
Email: Hikaruwind@aol.com

Patrick Milan
ForWind – Center for Wind Energy Research of
    the Universities of Oldenburg, Bremen and
    Hannover
D-26129 Oldenburg
Germany
Email: patrick.milan@uni-oldenburg.de

Panagiota Pantazopoulou
BRE
Bucknalls Lane
Watford, Hertfordshire WD25 9XX
UK
Email: PantazopoulouP@bre.co.uk

Joachim Peinke
ForWind – Center for Wind Energy Research of
    the Universities of Oldenburg, Bremen and
    Hannover
D-26129 Oldenburg
Germany
Email: joachim.peinke@forwind.de

David J. Cerda Salzmann
Department of Offshore Engineering
Delft University of Technology
2628 CN Delft
The Netherlands
Email: d.j.cerdasalzmann@tudelft.nl

Alois P. Schaffarczyk
Center of Excellence for Wind Energy
    (CEWind)
Kiel University of Applied Sciences
Grenzstrasse 3
D-24149 Kiel
Germany
Email: alois.schaffarczyk@fh-kiel.de

Bent F. Sørensen
Materials Research Division
Risø National Laboratory for Sustainable Energy
DK-4000 Roskilde
Denmark
Email: bent.soerensen@risoe.dk

Lawrence S. Staudt
Center for Renewable Energy
Dundalk Institute of Technology
Dundalk, County Louth
Ireland
Email: Larry.Staudt@dkit.ie

Robert-Zoltan Szasz
Department of Energy Sciences
Lund University
P.O. Box 118
221 00 Lund
Sweden
Email: Robert-Zoltan.Szasz@energy.lth.se

Jan van der Tempel
Department of Offshore Engineering
Delft University of Technology
2628 CN Delft
The Netherlands
Email: j.vandertempel@tudelft.nl

Wei Tong
Kollmorgen Corp.
201 W. Rock Road
Radford, VA 24141
USA
Email: wei.tong@kollmorgen.com

Spyros G. Voutsinas
School of Mechanical Engineering
National Technical University of Athens
15780 Zografou
Athens, Greece
Email: spyros@fluid.mech.ntua.gr

W. E. de Vries
Department of Offshore Engineering
Delft University of Technology
2628 CN Delft
The Netherlands
Email: w.e.devries@tudelft.nl

Matthias Wächter
ForWind – Center for Wind Energy Research of
    the Universities of Oldenburg, Bremen and
    Hannover
D-26129 Oldenburg
Germany
Email: matthias.waechter@uni-oldenburg.de

Lawrence D. Willey
Energy Wind
General Electric Company
300 Garlington Road
Greensville, SC 29602
USA
Email: lawrence.willey@ge.com
    lwilley@clipperwind.com (present)

Danian Zheng
Infrastructure Energy
General Electric Company
300 Garlington Road
Greenville, SC 29615
USA
Email: danian.zheng@ge.com

PART I

BASICS IN WIND POWER GENERATION

CHAPTER 1

Fundamentals of wind energy

Wei Tong
Kollmorgen Corporation, Virginia, USA.

The rising concerns over global warming, environmental pollution, and energy security have increased interest in developing renewable and environmentally friendly energy sources such as wind, solar, hydropower, geothermal, hydrogen, and biomass as the replacements for fossil fuels. Wind energy can provide suitable solutions to the global climate change and energy crisis. The utilization of wind power essentially eliminates emissions of CO_2, SO_2, NO_x and other harmful wastes as in traditional coal-fuel power plants or radioactive wastes in nuclear power plants. By further diversifying the energy supply, wind energy dramatically reduces the dependence on fossil fuels that are subject to price and supply instability, thus strengthening global energy security. During the recent three decades, tremendous growth in wind power has been seen all over the world. In 2009, the global annual installed wind generation capacity reached a record-breaking 37 GW, bringing the world total wind capacity to 158 GW. As the most promising renewable, clean, and reliable energy source, wind power is highly expected to take a much higher portion in power generation in the coming decades.

The purpose of this chapter is to acquaint the reader with the fundamentals of wind energy and modern wind turbine design, as well as some insights concerning wind power generation.

1 Wind energy

Wind energy is a converted form of solar energy which is produced by the nuclear fusion of hydrogen (H) into helium (He) in its core. The H $\rightarrow$ He fusion process creates heat and electromagnetic radiation streams out from the sun into space in all directions. Though only a small portion of solar radiation is intercepted by the earth, it provides almost all of earth's energy needs.

Wind energy represents a mainstream energy source of new power generation and an important player in the world's energy market. As a leading energy technology, wind power's technical maturity and speed of deployment is acknowledged, along with the fact that there is no practical upper limit to the percentage of wind that can be integrated into the electricity system [1]. It has been estimated that the total solar power received by the earth is approximately 1.8×10^{11} MW. Of this solar input, only 2% (i.e. 3.6×10^9 MW) is converted into wind energy and about 35% of wind energy is dissipated within 1000 m of the earth's surface [2]. Therefore, the available wind power that can be converted into other forms of energy is approximately 1.26×10^9 MW. Because this value represents 20 times the rate of the present global energy consumption, wind energy in principle could meet entire energy needs of the world.

Compared with traditional energy sources, wind energy has a number of benefits and advantages. Unlike fossil fuels that emit harmful gases and nuclear power that generates radioactive wastes, wind power is a clean and environmentally friendly energy source. As an inexhaustible and free energy source, it is available and plentiful in most regions of the earth. In addition, more extensive use of wind power would help reduce the demands for fossil fuels, which may run out sometime in this century, according to their present consumptions. Furthermore, the cost per kWh of wind power is much lower than that of solar power [3].

Thus, as the most promising energy source, wind energy is believed to play a critical role in global power supply in the 21st century.

2 Wind generation

Wind results from the movement of air due to atmospheric pressure gradients. Wind flows from regions of higher pressure to regions of lower pressure. The larger the atmospheric pressure gradient, the higher the wind speed and thus, the greater the wind power that can be captured from the wind by means of wind energy-converting machinery.

The generation and movement of wind are complicated due to a number of factors. Among them, the most important factors are uneven solar heating, the Coriolis effect due to the earth's self-rotation, and local geographical conditions.

2.1 Uneven solar heating

Among all factors affecting the wind generation, the uneven solar radiation on the earth's surface is the most important and critical one. The unevenness of the solar radiation can be attributed to four reasons.

First, the earth is a sphere revolving around the sun in the same plane as its equator. Because the surface of the earth is perpendicular to the path of the sunrays at the equator but parallel to the sunrays at the poles, the equator receives the greatest amount of energy per unit area, with energy dropping off toward the poles. Due to the spatial uneven heating on the earth, it forms a temperature gradient from the equator to the poles and a pressure gradient from the poles to the equator. Thus, hot air with lower air density at the equator rises up to the high atmosphere and moves towards the poles and cold air with higher density flows from the poles towards the equator along the earth's surface. Without considering the earth's self-rotation and the rotation-induced Coriolis force, the air circulation at each hemisphere forms a single cell, defined as the meridional circulation.

Second, the earth's self-rotating axis has a tilt of about 23.5° with respect to its ecliptic plane. It is the tilt of the earth's axis during the revolution around the sun that results in cyclic uneven heating, causing the yearly cycle of seasonal weather changes.

Third, the earth's surface is covered with different types of materials such as vegetation, rock, sand, water, ice/snow, etc. Each of these materials has different reflecting and absorbing rates to solar radiation, leading to high temperature on some areas (e.g. deserts) and low temperature on others (e.g. iced lakes), even at the same latitudes.

The fourth reason for uneven heating of solar radiation is due to the earth's topographic surface. There are a large number of mountains, valleys, hills, etc. on the earth, resulting in different solar radiation on the sunny and shady sides.

2.2 Coriolis force

The earth's self-rotation is another important factor to affect wind direction and speed. The Coriolis force, which is generated from the earth's self-rotation, deflects the direction of atmospheric movements. In the north atmosphere wind is deflected to the right and in the south atmosphere to the left. The Coriolis force depends on the earth's latitude; it is zero at the equator and reaches maximum values at the poles. In addition, the amount of deflection on wind also depends on the wind speed; slowly blowing wind is deflected only a small amount, while stronger wind deflected more.

In large-scale atmospheric movements, the combination of the pressure gradient due to the uneven solar radiation and the Coriolis force due to the earth's self-rotation causes the single meridional cell to break up into three convectional cells in each hemisphere: the Hadley cell, the Ferrel cell, and the Polar cell (Fig. 1). Each cell has its own characteristic circulation pattern.

In the Northern Hemisphere, the Hadley cell circulation lies between the equator and north latitude 30°, dominating tropical and sub-tropical climates. The hot air rises at the equator and flows toward the North Pole in the upper atmosphere. This moving air is deflected by Coriolis force to create the northeast trade winds. At approximately north latitude 30°, Coriolis force becomes so strong to balance the pressure gradient force. As a result, the winds are defected to the west. The air accumulated at the upper atmosphere forms the subtropical high-pressure belt and thus sinks back to the earth's surface, splitting into two components: one returns to the equator to close the loop of the Hadley cell; another moves along the earth's surface toward North Pole to form the Ferrel Cell circulation, which lies between north latitude 30° and 60°. The air circulates toward the North Pole along the earth's surface until it collides with the cold air flowing from the North Pole at approximately north latitude 60°. Under the influence of Coriolis force, the moving air in this zone is deflected to produce westerlies. The Polar cell circulation lies between the North Pole and north latitude 60°. The cold air sinks down at the

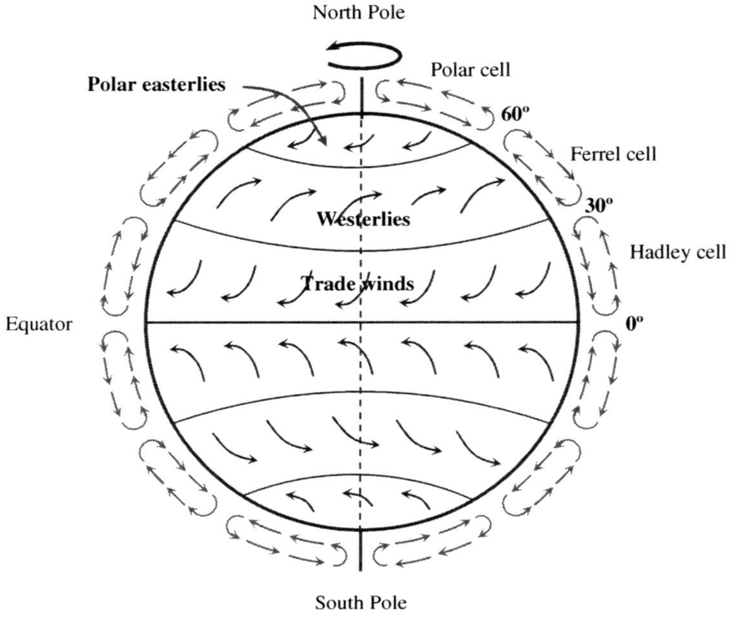

Figure 1: Idealized atmospheric circulations.

North Pole and flows along the earth's surface toward the equator. Near north latitude 60°, the Coriolis effect becomes significant to force the airflow to southwest.

2.3 Local geography

The roughness on the earth's surface is a result of both natural geography and
manmade structures. Frictional drag and obstructions near the earth's surface generally retard with wind speed and induce a phenomenon known as wind shear. The
rate at which wind speed increases with height varies on the basis of local conditions of the topography, terrain, and climate, with the greatest rates of increases
observed over the roughest terrain. A reliable approximation is that wind speed
increases about 10% with each doubling of height [4].

In addition, some special geographic structures can strongly enhance the wind
intensity. For instance, wind that blows through mountain passes can form mountain jets with high speeds.

3 History of wind energy applications

The use of wind energy can be traced back thousands of years to many ancient
civilizations. The ancient human histories have revealed that wind energy was
discovered and used independently at several sites of the earth.

3.1 Sailing

As early as about 4000 B.C., the ancient Chinese were the first to attach sails to their primitive rafts [5]. From the oracle bone inscription, the ancient Chinese scripted on turtle shells in Shang Dynasty (1600 B.C.–1046 B.C.), the ancient Chinese character "ᕼ" (i.e., "ᕾ", sail - in ancient Chinese) often appeared. In Han Dynasty (220 B.C.–200 A.D.), Chinese junks were developed and used as ocean-going vessels. As recorded in a book wrote in the third century [6], there were multi-mast, multi-sail junks sailing in the South Sea, capable of carrying 700 people with 260 tons of cargo. Two ancient Chinese junks are shown in Figure 2. Figure 2(a) is a two-mast Chinese junk ship for shipping grain, quoted from the famous encyclopedic science and technology book *Exploitation of the works of nature* [7]. Figure 2(b) illustrates a wheel boat [8] in Song Dynasty (960–1279). It mentioned in [9] that this type of wheel boats was used during the war between Song and Jin Dynasty (1115–1234).

Approximately at 3400 BC, the ancient Egyptians launched their first sailing vessels initially to sail on the Nile River, and later along the coasts of the Mediterranean [5]. Around 1250 BC, Egyptians built fairly sophisticated ships to sail on the Red Sea [9]. The wind-powered ships had dominated water transport for a long time until the invention of steam engines in the 19th century.

3.2 Wind in metal smelting processes

About 300 BC, ancient Sinhalese had taken advantage of the strong monsoon winds to provide furnaces with sufficient air for raising the temperatures inside furnaces in excess of 1100°C in iron smelting processes. This technique was capable of producing high-carbon steel [10].

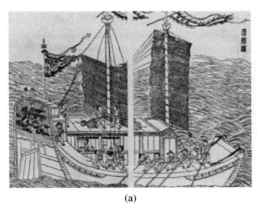 

(a)   (b)

Figure 2: Ancient Chinese junks (ships): (a) two-mast junk ship [7]; (b) wheel boat [8].

The double acting piston bellows was invented in China and was widely used in metallurgy in the fourth century BC [11]. It was the capacity of this type of bellows to deliver continuous blasts of air into furnaces to raise high enough temperatures for smelting iron. In such a way, ancient Chinese could once cast several tons of iron.

3.3 Windmills

China has long history of using windmills. The unearthed mural paintings from the tombs of the late Eastern Han Dynasty (25–220 AD) at Sandaohao, Liaoyang City, have shown the exquisite images of windmills, evidencing the use of windmills in China for at least approximately 1800 years [12].

The practical vertical axis windmills were built in Sistan (eastern Persia) for grain grinding and water pumping, as recorded by a Persian geographer in the ninth century [13].

The horizontal axis windmills were invented in northwestern Europe in 1180s [14]. The earlier windmills typically featured four blades and mounted on central posts – known as Post mill. Later, several types of windmills, e.g. Smock mill, Dutch mill, and Fan mill, had been developed in the Netherlands and Denmark, based on the improvements on Post mill.

The horizontal axis windmills have become dominant in Europe and North America for many centuries due to their higher operation efficiency and technical advantages over vertical axis windmills.

3.4 Wind turbines

Unlike windmills which are used directly to do work such as water pumping or grain grinding, wind turbines are used to convert wind energy to electricity. The first automatically operated wind turbine in the world was designed and built by Charles Brush in 1888. This wind turbine was equipped with 144 cedar blades having a rotating diameter of 17 m. It generated a peak power of 12 kW to charge batteries that supply DC current to lamps and electric motors [5].

As a pioneering design for modern wind turbines, the Gedser wind turbine was built in Denmark in the mid 1950s [15]. Today, modern wind turbines in wind farms have typically three blades, operating at relative high wind speeds for the power output up to several megawatts.

3.5 Kites

Kites were invented in China as early as the fifth or fourth centuries BC [11]. A famous Chinese ancient legalist Han Fei-Zi (280–232 BC) mentioned in his book that an ancient philosopher Mo Ze (479–381 BC) spent three years to make a kite with wood but failed after one-day flight [16].

4 Wind energy characteristics

Wind energy is a special form of kinetic energy in air as it flows. Wind energy can be either converted into electrical energy by power converting machines or directly used for pumping water, sailing ships, or grinding gain.

4.1 Wind power

Kinetic energy exists whenever an object of a given mass is in motion with a translational or rotational speed. When air is in motion, the kinetic energy in moving air can be determined as

$$E_k = \tfrac{1}{2} m \bar{u}^2 \tag{1}$$

where m is the air mass and $\bar{u}$ is the mean wind speed over a suitable time period. The wind power can be obtained by differentiating the kinetic energy in wind with respect to time, i.e.:

$$P_w = \frac{dE_k}{dt} = \frac{1}{2} \dot{m} \bar{u}^2 \tag{2}$$

However, only a small portion of wind power can be converted into electrical power. When wind passes through a wind turbine and drives blades to rotate, the corresponding wind mass flowrate is

$$\dot{m} = \rho A \bar{u} \tag{3}$$

where ρ is the air density and A is the swept area of blades, as shown in Fig. 3. Substituting (3) into (2), the available power in wind P_w can be expressed as

$$P_w = \tfrac{1}{2} \rho A \bar{u}^3 \tag{4}$$

An examination of eqn (4) reveals that in order to obtain a higher wind power, it requires a higher wind speed, a longer length of blades for gaining a larger swept area, and a higher air density. Because the wind power output is proportional to the cubic power of the mean wind speed, a small variation in wind speed can result in a large change in wind power.

4.1.1 Blade swept area

As shown in Fig. 3, the blade swept area can be calculated from the formula:

$$A = \pi \left[(l+r)^2 - r^2 \right] = \pi l (l + 2r) \tag{5}$$

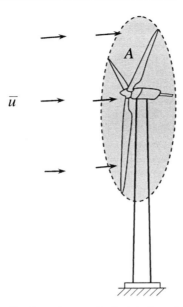

Figure 3: Swept area of wind turbine blades.

where l is the length of wind blades and r is the radius of the hub. Thus, by doubling the length of wind blades, the swept area can be increased by the factor up to 4. When $l \gg 2r$, $A \approx \pi\, l^2$.

4.1.2 Air density

Another important parameter that directly affects the wind power generation is the density of air, which can be calculated from the equation of state:

$$\rho = \frac{p}{RT} \tag{6}$$

where p is the local air pressure, R is the gas constant (287 J/kg-K for air), and T is the local air temperature in K.

The hydrostatic equation states that whenever there is no vertical motion, the difference in pressure between two heights is caused by the mass of the air layer:

$$dp = -\rho g\, dz \tag{7}$$

where g is the acceleration of gravity. Combining eqns (6) and (7), yields

$$\frac{dp}{p} = -\frac{g}{RT} dz \tag{8}$$

The acceleration of gravity g decreases with the height above the earth's surface z:

$$g = g_0 \left(1 - \frac{4z}{D} \right) \tag{9}$$

where g_0 is the acceleration of gravity at the ground and D is the diameter of the earth. However, for the acceleration of gravity g, the variation in height can be ignored because D is much larger than $4z$.

In addition, temperature is inversely proportional to the height. Assume that $dT/dz = c$, it can be derived that

$$p = p_0 \left(\frac{T}{T_0}\right)^{-g/cR} \tag{10}$$

where p_0 and T_0 are the air pressure and temperature at the ground, respectively.

Combining eqns (6) and (10), it gives

$$\rho = \rho_0 \left(\frac{T}{T_0}\right)^{-(g/cR+1)} = \rho_0 \left(1 + \frac{cz}{T_0}\right)^{-(g/cR+1)} \tag{11}$$

This equation indicates that the density of air decreases nonlinearly with the height above the sea level.

4.1.3 Wind power density

Wind power density is a comprehensive index in evaluating the wind resource at a particular site. It is the available wind power in airflow through a perpendicular cross-sectional unit area in a unit time period. The classes of wind power density at two standard wind measurement heights are listed in Table 1.

Some of wind resource assessments utilize 50 m towers with sensors installed at intermediate levels (10 m, 20 m, etc.). For large-scale wind plants, class rating of 4 or higher is preferred.

Table 1: Classes of wind power density [17].

Wind power class	10 m height		50 m height	
	Wind power density (W/m^2)	Mean wind speed (m/s)	Wind power density (W/m^2)	Mean wind speed (m/s)
1	<100	<4.4	<200	<5.6
2	100–150	4.4–5.1	200–300	5.6–6.4
3	150–200	5.1–5.6	300–400	6.4–7.0
4	200–250	5.6–6.0	400–500	7.0–7.5
5	250–300	6.0–6.4	500–600	7.5–8.0
6	300–350	6.4–7.0	600–800	8.0–8.8
7	>400	>7.0	>800	>8.8

4.2 Wind characteristics

Wind varies with the geographical locations, time of day, season, and height above the earth's surface, weather, and local landforms. The understanding of the wind characteristics will help optimize wind turbine design, develop wind measuring techniques, and select wind farm sites.

4.2.1 Wind speed

Wind speed is one of the most critical characteristics in wind power generation. In fact, wind speed varies in both time and space, determined by many factors such as geographic and weather conditions. Because wind speed is a random parameter, measured wind speed data are usually dealt with using statistical methods.

The diurnal variations of average wind speeds are often described by sine waves. As an example, the diurnal variations of hourly wind speed values, which are the average values calculated based on the data between 1970 and 1984, at Dhahran, Saudi Arabia have shown the wavy pattern [18]. The wind speeds are higher in daytime and the maximum speed occurs at about 3 p.m., indicating that the daytime wind speed is proportional to the strength of sunlight. George et al. [19] reported that wind speed at Lubbock, TX is near constant during dark hours, and follows a curvilinear pattern during daylight hours. Later, George et al. [20] have demonstrated that diurnal wind patterns at five locations in the Great Plains follow a pattern similar to that observed in [19].

Based on the wind speed data for the period 1970–2003 from up to 66 onshore sites around UK, Sinden [21] has concluded that monthly average wind speed is inversely propositional to the monthly average temperature, i.e. it is higher in the winter and lower in the summer. The maximum wind speed occurs in January and the minimum in August. Hassanm and Hill have reported that the month-to-month variation of mean wind speed values over the period of 1970–1984 at Dhahran, Saudi Arabia has shown the wavy pattern [13]. However, because the variation in temperature at Dhahran is small over the whole year, there is no a clear correlation between wind speed and temperatures.

The year-to-year variation of yearly mean wind speeds depends highly on selected locations and thus there is no common correlation to predict it. For instance, except for several years, the annual mean wind speeds decrease all the way from 1970 to 1983 at Dhahran, Saudi Arabia [18]. In UK, this variation displays in a more fluctuated matter for the period 1970–2003 [21]. Similarly, a significant variation in the annual mean wind speed over 20-year period (1978–1998) is reported in [22], with maximum and minimum values ranging from less than 7.8 to nearly 9.2 m/s. The long-term wind data (1978–2007) obtained from automated synoptic observation system of meteorological observatories were analyzed and reported by Ko et al. [23]. The results show that fluctuation in yearly average wind speed occurs at the observed sites; it tends to slightly decrease at Jeju Island, while the other two sites have random trends.

4.2.2 Weibull distribution

The variation in wind speed at a particular site can be best described using the Weibull distribution function [24], which illustrates the probability of different mean wind speeds occurring at the site during a period of time. The probability density function of a Weibull random variable $\bar{u}$ is:

$$f\left(\bar{u},k,\lambda\right)=\begin{cases}\dfrac{k}{\lambda}\left(\dfrac{\bar{u}}{\lambda}\right)^{k-1}\exp\left(-\left(\dfrac{\bar{u}}{\lambda}\right)^{k}\right) & \bar{u}\geq 0\\[2ex] 0 & \bar{u}<0\end{cases} \tag{12}$$

where λ is the scale factor which is closely related to the mean wind speed and k is the shape factor which is a measurement of the width of the distribution. These two parameters can be determined from the statistical analysis of measured wind speed data at the site [25]. It has been reported that Weibull distribution can give good fits to observed wind speed data [26]. As an example, the Weibull distributions for various mean wind speeds are displayed in Fig. 4.

4.2.3 Wind turbulence

Wind turbulence is the fluctuation in wind speed in short time scales, especially for the horizontal velocity component. The wind speed $u(t)$ at any instant time t can be considered as having two components: the mean wind speed $\bar{u}$ and the instantaneous speed fluctuation $u'(t)$, i.e.:

$$u\left(t\right)=\bar{u}+u'\left(t\right) \tag{13}$$

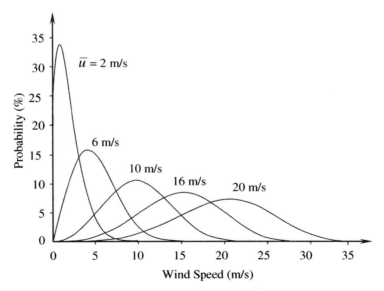

Figure 4: Weibull distributions for various mean wind speeds.

Wind turbulence has a strong impact on the power output fluctuation of wind turbine. Heavy turbulence may generate large dynamic fatigue loads acting on the turbine and thus reduce the expected turbine lifetime or result in turbine failure.

In selection of wind farm sites, the knowledge of wind turbulence intensity is crucial for the stability of wind power production. The wind turbulence intensity I is defined as the ratio of the standard deviation σ_u to the mean wind velocity $\bar{u}$:

$$I = \frac{\sigma_u}{\bar{u}} \qquad (14)$$

where both σ_u and $\bar{u}$ are measured at the same point and averaged over the same period of time.

4.2.4 Wind gust

Wind gust refers to a phenomenon that a wind blasts with a sudden increase in wind speed in a relatively small interval of time. In case of sudden turbulent gusts, wind speed, turbulence, and wind shear may change drastically. Reducing rotor imbalance while maintaining the power output of wind turbine generator constant during such sudden turbulent gusts calls for relatively rapid changes of the pitch angle of the blades. However, there is typically a time lag between the occurrence of a turbulent gust and the actual pitching of the blades based upon dynamics of the pitch control actuator and the large inertia of the mechanical components. As a result, load imbalances and generator speed, and hence oscillations in the turbine components may increase considerably during such turbulent gusts, and may exceed the maximum prescribed power output level [27]. Moreover, sudden turbulent gusts may also significantly increase tower fore-aft and side-to-side bending moments due to increase in the effect of wind shear.

To ensure safe operation of wind farms, wind gust predictions are highly desired. Several different gust prediction methods have been proposed. Contrary to most techniques used in operational weather forecasting, Brasseur [29] developed a new wind gust prediction method based on physical consideration. In another study [30], it reported that using a gust factor, which is defined as peak gust over the mean wind speed, could well forecast wind gust speeds. These results are in agreement with previous work by other investigators [31].

4.2.5 Wind direction

Wind direction is one of the wind characteristics. Statistical data of wind directions over a long period of time is very important in the site selection of wind farm and the layout of wind turbines in the wind farm.

The wind rose diagram is a useful tool of analyzing wind data that are related to wind directions at a particular location over a specific time period (year, season, month, week, etc.). This circular diagram displays the relative frequency of wind directions in 8 or 16 principal directions. As an example shown in Fig. 5, there are 16 radial lines in the wind rose diagram, with 22.5° apart from each other. The length of each line is proportional to the frequency of wind direction. The frequency

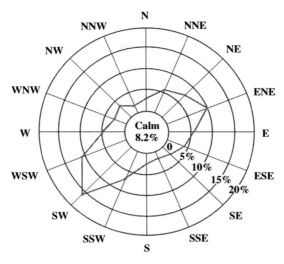

Figure 5: Wind rose diagram for wind directions.

of calm or near calm air is given as a number in the central circle. Some wind rose diagrams may also contain the information of wind speeds.

4.2.6 Wind shear

Wind shear is a meteorological phenomenon in which wind increases with the height above the ground. The effect of height on the wind speed is mainly due to roughness on the earth's surface and can be estimated using the Hellmann power equation that relates wind speeds at two different heights [33]:

$$u(z) = u(z_0)\left(\frac{z}{z_0}\right)^a \tag{15}$$

where z is the height above the earth's surface, z_0 is the reference height for which wind speed $u(z_0)$ is known, and a is the wind shear coefficient. In practice, a depends on a number of factors, including the roughness of the surrounding landscape, height, time of day, season, and locations. The wind shear coefficient is generally lower in daytime and higher at night. Empirical results indicate that wind shear often follows the "1/7 power law" (i.e. $a = 1/7$). The values of wind shear coefficient for different surface roughness are provided in [34].

Because the power output of wind turbine strongly depends on the wind speed at the hub height, modern wind turbines are built at the height greater than 80 m, for capturing more wind energy and lowering cost per unit power output.

5 Modern wind turbines

A modern wind turbine is an energy-converting machine to convert the kinetic energy of wind into mechanical energy and in turn into electrical energy. In the recent three decades, remarkable advances in wind turbine design have been achieved along with modern technological developments. It has been estimated that advances in aerodynamics, structural dynamics, and micrometeorology may contribute to a 5% annual increase in the energy yield of wind turbines [35].

Various wind turbine concepts have been developed and built for maximizing the wind energy output, minimizing the turbine cost, and increasing the turbine efficiency and reliability.

5.1 Wind turbine classification

Wind turbines can be classified according to the turbine generator configuration, airflow path relatively to the turbine rotor, turbine capacity, the generator-driving pattern, the power supply mode, and the location of turbine installation.

5.1.1 Horizontal-axis and vertical-axis wind turbines

When considering the configuration of the rotating axis of rotor blades, modern wind turbines can be classified into the horizontal-axis and vertical-axis turbines. Most commercial wind turbines today belong to the horizontal-axis type, in which the rotating axis of blades is parallel to the wind stream. The advantages of this type of wind turbines include the high turbine efficiency, high power density, low cut-in wind speeds, and low cost per unit power output.

Several typical vertical-axis wind turbines are shown in Fig. 6. The blades of the vertical-axis wind turbines rotate with respect to their vertical axes that are perpendicular to the ground. A significant advantage of vertical-axis wind turbine is that the turbine can accept wind from any direction and thus no yaw control is needed. Since the wind generator, gearbox, and other main turbine components can be set up on the ground, it greatly simplifies the wind tower design and construction, and consequently reduces the turbine cost. However, the vertical-axis wind turbines must use an external energy source to rotate the blades during initialization. Because the axis of the wind turbine is supported only on one end at the ground, its maximum practical height is thus limited. Due to the lower wind power efficiency, vertical-axis wind turbines today make up only a small percentage of wind turbines.

5.1.2 Upwind and downwind wind turbines

Based on the configuration of the wind rotor with respect to the wind flowing direction, the horizontal-axis wind turbines can be further classified as upwind and downwind wind turbines. The majority of horizontal-axis wind turbines being used today are upwind turbines, in which the wind rotors face the wind. The main advantage of upwind designs is to avoid the distortion of the flow field as the wind passes though the wind tower and nacelle.

For a downwind turbine, wind blows first through the nacelle and tower and then the rotor blades. This configuration enables the rotor blades to be made more flexible without considering tower strike. However, because of the influence of the distorted unstable wakes behind the tower and nacelle, the wind power output

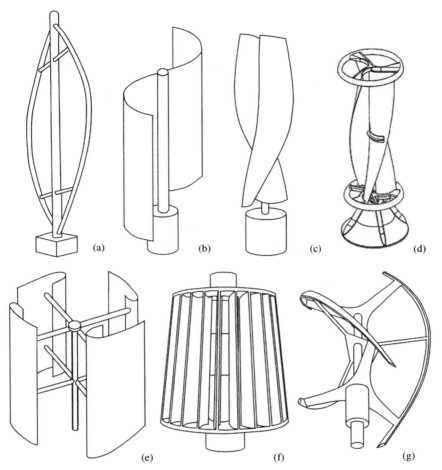

Figure 6: Several typical types of vertical-axis wind turbines: (a) Darrius; (b) Savonius; (c) Solarwind™ [36]; (d) Helical [37]; (e) Noguchi [38]; (f) Maglev [39]; (g) Cochrane [40].

generated from a downwind turbine fluctuates greatly. In addition, the unstable flow field may result in more aerodynamic losses and introduce more fatigue loads on the turbine. Furthermore, the blades in a downwind wind turbine may produce higher impulsive or thumping noise.

5.1.3 Wind turbine capacity

Wind turbines can be divided into a number of broad categories in view of their rated capacities: micro, small, medium, large, and ultra-large wind turbines. Though a restricted definition of micro wind turbines is not available, it is accepted that a turbine with the rated power less than several kilowatts can be categorized as micro wind turbine [41]. Micro wind turbines are especially suitable in locations where the electrical grid is unavailable. They can be used on a per-structure basis, such as street lighting, water pumping, and residents at remote areas, particularly in developing countries. Because micro wind turbines need relatively low cut-in speeds at start-up and operate in moderate wind speeds, they can be extensively installed in most areas around the world for fully utilizing wind resources and greatly enhancing wind power generation availability.

Small wind turbines usually refer to the turbines with the output power less than 100 kW [42]. Small wind turbines have been extensively used at residential houses, farms, and other individual remote applications such as water pumping stations, telecom sites, etc., in rural regions. Distributed small wind turbines can increase electricity supply in the regions while delaying or avoiding the need to increase the capacity of transmission lines.

The most common wind turbines have medium sizes with power ratings from 100 kW to 1 MW. This type of wind turbines can be used either on-grid or off-grid systems for village power, hybrid systems, distributed power, wind power plants, etc.

Megawatt wind turbines up to 10 MW may be classified as large wind turbines. In recent years, multi-megawatt wind turbines have become the mainstream of the international wind power market. Most wind farms presently use megawatt wind turbines, especially in offshore wind farms.

Ultra-large wind turbines are referred to wind turbines with the capacity more than 10 MW. This type of wind turbine is still in the earlier stages of research and development.

5.1.4 Direct drive and geared drive wind turbines

According to the drivetrain condition in a wind generator system, wind turbines can be classified as either direct drive or geared drive groups. To increase the generator rotor rotating speed to gain a higher power output, a regular geared drive wind turbine typically uses a multi-stage gearbox to take the rotational speed from the low-speed shaft of the blade rotor and transform it into a fast rotation on the high-speed shaft of the generator rotor. The advantages of geared generator systems include lower cost and smaller size and weight. However, utilization of a gearbox can significantly lower wind turbine reliability and increase turbine noise level and mechanical losses.

By eliminating the multi-stage gearbox from a generator system, the generator shaft is directly connected to the blade rotor. Therefore, the direct-drive concept is more superior in terms of energy efficiency, reliability, and design simplicity.

5.1.5 On-grid and off-grid wind turbines

Wind turbines can be used for either on-grid or off-grid applications. Most medium-size and almost all large-size wind turbines are used in grid tied applications. One of the obvious advantages for on-grid wind turbine systems is that there is no energy storage problem.

   As the contrast, most of small wind turbines are off-grid for residential homes, farms, telecommunications, and other applications. However, as an intermittent power source, wind power produced from off-grid wind turbines may change dramatically over a short period of time with little warning. Consequently, off-grid wind turbines are usually used in connection with batteries, diesel generators, and photovoltaic systems for improving the stability of wind power supply.

5.1.6 Onshore and offshore wind turbines

Onshore wind turbines have a long history on its development. There are a number of advantages of onshore turbines, including lower cost of foundations, easier integration with the electrical-grid network, lower cost in tower building and turbine installation, and more convenient access for operation and maintenance.

   Offshore wind turbines have developed faster than onshore since the 1990s due to the excellent offshore wind resource, in terms of wind power intensity and continuity. A wind turbine installed offshore can make higher power output and operate more hours each year compared with the same turbine installed onshore. In addition, environmental restrictions are more lax at offshore sites than at onshore sites. For instance, turbine noise is no long an issue for offshore wind turbines.

5.2 Wind turbine configuration

Most of the modern large wind turbines are horizontal-axis turbines with typically three blades. As shown in Fig. 7, a wind turbine is comprised of a nacelle, which

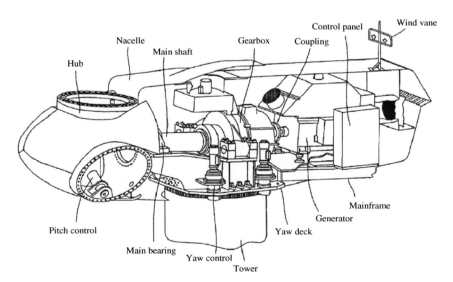

Figure 7: A horizontal-axis wind turbine configuration. Courtesy of the US Patent & Trademark Office.

is positioned on the top of a wind tower, housing the most turbine components inside. Three blades (not shown) mounted on the rotor hub, which is connected via the main shaft to the gearbox. The rotor of the wind generator is connected to the output shaft of the gearbox. Thus, the slow rotating speed of the rotor hub is increased to a desired high rotating speed of the generator rotor.

Using the pitch control system, each blade is pitched individually to optimize the angle of attack of the blade for allowing a higher energy capture in normal operation and for protecting the turbine components (blade, tower, etc.) from damaging in emergency situations. With the feedback information such as measured instantaneous wind direction and speed from the wind vane, the yaw control system provides the yaw orientation control for ensuring the turbine constantly against the wind.

5.3 Wind power parameters

5.3.1 Power coefficient

The conversion of wind energy to electrical energy involves primarily two stages: in the first stage, kinetic energy in wind is converted into mechanical energy to drive the shaft of a wind generator. The critical converting devices in this stage are wind blades. For maximizing the capture of wind energy, wind blades need to be carefully designed.

The power coefficient C_p deals with the converting efficiency in the first stage, defined as the ratio of the actually captured mechanical power by blades to the available power in wind:

$$C_p = \frac{P_{me,out}}{P_w} = \frac{P_{me,out}}{(1/2)\rho A \bar{u}^3} \tag{16}$$

Because there are various aerodynamic losses in wind turbine systems, for instance, blade-tip, blade-root, profile, and wake rotation losses, etc., the real power coefficient C_p is much lower than its theoretical limit, usually ranging from 30 to 45%.

5.3.2 Total power conversion coefficient and effective power output

In the second stage, mechanical energy captured by wind blades is further converted into electrical energy via wind generators. In this stage, the converting efficiency is determined by several parameters

- Gearbox efficiency η_{gear} – The power losses in a gearbox can be classified as load-dependent and no-load power losses. The load-dependent losses consist of gear tooth friction and bearing losses and no-load losses consist of oil churning, windage, and shaft seal losses. The planetary gearboxes, which are widely used in wind turbines, have higher power transmission efficiencies over traditional gearboxes.

- Generator efficiency η_{gen} – It is related to all electrical and mechanical losses in a wind generator, such as copper, iron, load, windage, friction, and other miscellaneous losses.
- Electric efficiency η_{ele} – It encompasses all combined electric power losses in the converter, switches, controls, and cables.

Therefore, the total power conversion efficiency from wind to electricity η_t is the production of these parameters, i.e.:

$$\eta_t = C_p \eta_{gear} \eta_{gen} \eta_{ele} \tag{17}$$

The effective power output from a wind turbine to feed into a grid becomes

$$P_{eff} = C_p \eta_{gear} \eta_{gen} \eta_{ele} P_w = \eta_t P_w = \tfrac{1}{2}(\eta_t \rho A \bar{u}^3) \tag{18}$$

5.3.3 Lanchester–Betz limit

The theoretical maximum efficiency of an ideal wind turbomachine was derived by Lanchester [43] in 1915 and Betz [44] in 1920. It was revealed that no wind turbomachines could convert more than 16/27 (59.26%) of the kinetic energy of wind into mechanical energy. This is known as Lanchester–Betz limit (or Lanchester–Betz law) today.

As shown in Fig. 8, $\bar{u}_1$ and $\bar{u}_4$ are mean velocities far upstream and downstream from the wind turbine; $\bar{u}_2$ and $\bar{u}_3$ are mean velocities just in front and back of the wind rotating blades, respectively. By assuming that there is no change in the air velocity right across the wind blades (i.e. $\bar{u}_2 = \bar{u}_3$) and the pressures far upstream and downstream from the wind turbine are equal to the static pressure of the undisturbed airflow (i.e. $p_1 = p_4 = p$), it can be derived that

$$P_2 - P_3 = \tfrac{1}{2}\rho(\bar{u}_1^2 - \bar{u}_4^2) \tag{19}$$

and

$$\bar{u}_2 = \bar{u}_3 = \tfrac{1}{2}(\bar{u}_1 + \bar{u}_4) \tag{20}$$

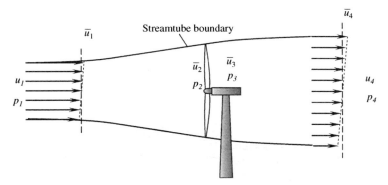

Figure 8: Airflow through a wind turbine.

Thus, the power output of mechanical energy captured by wind turbine blades is

$$P_{\text{me,out}} = \tfrac{1}{2} \rho A \bar{u}_2 (\bar{u}_1^2 - \bar{u}_4^2) = \tfrac{1}{2} \rho A \bar{u}_1^3 \, 4a(1-a)^2 \tag{21}$$

where a is the axial induction factor, defined as

$$a = \frac{\bar{u}_1 - \bar{u}_2}{\bar{u}_1} \tag{22}$$

Substitute eqn (21) into (16) (where $\bar{u}_1 = \bar{u}$), yields

$$C_{\text{p}} = 4a(1-a)^2 \tag{23}$$

This indicates that the power coefficient is only a function of the axial induction factor a. It is easy to derive that the maximum power coefficient reaches its maximum value of 16/27 when $a = 1/3$ (see Fig. 9).

5.3.4 Power curve

As can be seen from eqn (18), the effective electrical power output from a wind turbine P_{eff} is directly proportional to the available wind power P_{w} and the total effective wind turbine efficiency η_{t}.

The power curve of a wind turbine displays the power output (either the real electrical power output or the percentage of the rated power) of the turbine as a function of the mean wind speed. Power curves are usually determined from the field measurements. As shown in Fig. 10, the wind turbine starts to produce usable power at a low wind speed, defined as the cut-in speed. The power output increases continuously with the increase of the wind speed until reaching a saturated point, to which

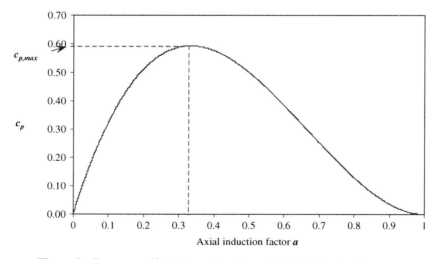

Figure 9: Power coefficient as a function of axial induction factor a.

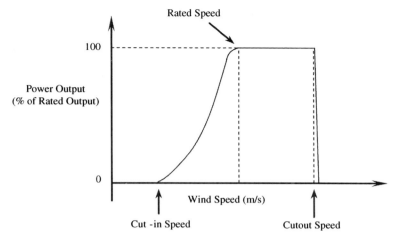

Figure 10: Typical wind turbine power curve.

the power output reaches its maximum value, defined as the rated power output. Correspondingly, the speed at this point is defined as the rated speed. At the rated speed, more increase in the wind speed will not increase the power output due to the activation of the power control. When the wind speed becomes too large to potentially damage the wind turbine, the wind turbine needs to shut down immediately to avoid damaging the wind turbine. This wind speed is defined as the cut-out speed. Thus, the cut-in and cut-out speeds have defined the operating limits of the wind turbine.

There are a number of methods available for forecasting the wind turbine power performance curves. Based on statistical tools, a comparison of five different methods has been performed by Cabezon *et al.* [45]. The best results were obtained when the fuzzy logic tool and tuning over the transfer functions were applied for wind turbines. More recently, based on a stochastic model for the power conversion process, Gottschall and Peinke [46] proposed a dynamic method for estimating the power performance curves and the dynamic approach has verified to be more accurate than the common IEC standard [47]. A novel method, based on the stochastic differential equations of diffusive Markov processes, was developed to characterize wind turbine power performance directly from high-frequency fluctuating measurements [48].

5.3.5 Tip speed ratio

The tip speed ratio is an extremely important factor in wind turbine design, which is defined as the ratio of the tangential speed at the blade tip to the actual wind speed, i.e.:

$$\lambda = \frac{(l+r)\omega}{\bar{u}} \tag{24}$$

where l is the length of the blade, r is the radius of the hub, and ω is the angular speed of blades.

If the blade angular speed ω is too small, most of the wind may pass undisturbed though the blade swept area making little useful work on the blades. On the contrary, if ω is too large, the fast rotating blades may block the wind flow reducing the power extraction. Therefore, there exists an optimal angular speed at which the maximum power extraction is achieved. For a wind turbine with n blades, the optimal angular speed can be approximately determined as [49]:

$$\omega_{opt} \approx \frac{2\pi}{n} \frac{\bar{u}}{L} \tag{25}$$

where L is the length of the strongly disturbed air stream upwind and downwind of the rotor.

Substituting eqn (25) into (24), the optimal tip speed ratio becomes

$$\lambda_{opt} \approx \frac{2\pi}{n} \left(\frac{l+r}{L} \right) \tag{26}$$

Empirically, the ratio $(l + r)/L$ is equal to about 2. Thus, for three-blade wind turbines (i.e. $n = 3$), $\lambda_{opt} \approx 4\pi/3$.

If the aerofoil blade is designed with care, the optimal tip speed ratio may be about 25–30% higher than the calculated optimal values above. Therefore, a wind turbine with three blades would have an optimal tip speed ratio [49]:

$$\lambda_{opt} = \frac{4\pi}{3}(1.25 \sim 1.30) \approx 5.24 \sim 5.45 \tag{27}$$

5.3.6 Wind turbine capacity factor

Due to the intermittent nature of wind, wind turbines do not make power all the time. Thus, a capacity factor of a wind turbine is used to provide a measure of the wind turbine's actual power output in a given period (e.g. a year) divided by its power output if the turbine has operated the entire time. A reasonable capacity factor would be 0.25–0.30 and a very good capacity factor would be around 0.40 [50]. In fact, wind turbine capacity factor is very sensitive to the average wind speed.

5.4 Wind turbine controls

Wind turbine control systems continue to play important roles for ensuring wind turbine reliable and safe operation and to optimize wind energy capture. The main control systems in a modern wind turbine include pitch control, stall control (passive and active), yaw control, and others.

Under high wind speed conditions, the power output from a wind turbine may exceed its rated value. Thus, power control is required to control the power output within allowable fluctuations for avoiding turbine damage and stabilizing the power output. There are two primary control strategies in the power control: pitch control and stall control. The wind turbine power control system is used to control the power output within allowable fluctuations.

5.4.1 Pitch control

The pitch control system is a vital part of the modern wind turbine. This is because the pitch control system not only continually regulates the wind turbine's blade pitch angle to enhance the efficiency of wind energy conversion and power generation stability, but also serves as the security system in case of high wind speeds or emergency situations. It requires that even in the event of grid power failure, the rotor blades can be still driven into their feathered positions by using either the power of backup batteries or capacitors [51] or mechanical energy storage devices [52].

Early techniques of active blade pitch control applied hydraulic actuators to control all blades together. However, these collective pitch control techniques could not completely satisfy all requirements of blade pitch angle regulation, especially for MW wind turbines with the increase in blade length and hub height. This is because wind is highly turbulent flow and the wind speed is proportional to the height from the ground. Therefore, each blade experiences different loads at different rotation positions. As a result, more superior individual blade pitch control techniques have been developed and implemented, allowing control of asymmetric aerodynamic loads on the blades, as well as structural loads in the non-rotating frame such as tower side-side bending. In such a control system, each blade is equipped with its own pitch actuator, sensors and controller.

In today's wind power industry, there are primarily two types of blade pitch control systems: hydraulic controlled and electric controlled systems. As shown in Fig. 11, the hydraulic pitch control system uses a hydraulic actuator to drive the blade rotating with respect to its axial centreline. The most significant advantages of hydraulic pitch control system include its large driving power, lack of a gearbox, and robust backup power. Due to these advantages, hydraulic pitch control systems historically dominate wind turbine control in Europe and North America for many years.

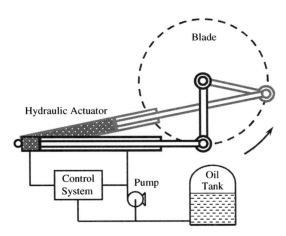

Figure 11: Hydraulic pitch control system.

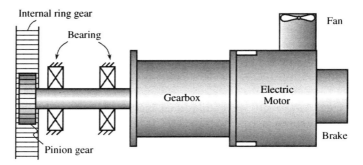

Figure 12: Electric pitch control system.

The electric pitch control systems have been developed alternatively with the hydraulic systems. This type of control system has a higher efficiency than that of hydraulic controlled systems (which is usually less than 55%) and avoids the risk of environmental pollution due to hydraulic fluid being split or leaked.

In an electric pitch control system as shown in Fig. 12, the motor connects to a gearbox to lower the motor speed to a desired control speed. A drive pinion gear engages with an internal ring gear, which is rigidly attached to the roof of the rotor blade. Alternatively, some wind turbine manufacturers use the belt-drive structure adjusting the pitch angle. The use of electric motors can raise the responsiveness rate and sensitivity of blade pitch control. To enhance operation reliability, the use of redundant pitch control systems was proposed to be equipped in large wind turbines [53].

5.4.2 Stall control

Besides pitch control, stall control is another approach for controlling and protecting wind turbines. The concept of stall control is that the power is regulated through stalling the blades after rated speed is achieved.

Stall control can be further divided into passive and active control approaches. Passive stall control is basically used in wind turbines in which the blades are bolted to the hub at a fixed installing angle. In a passive stall-regulated wind turbine, the power regulation relies on the aerodynamic features of blades. In low and moderate wind speeds, the turbine operates near maximum efficiency. At high wind speeds, the turbine is automatically controlled by means of stalled blades to limit the rotational speed and power output, protecting the turbine from excessive wind speeds.

Compared with pitch control, a passive stall control system has a simple structure and avoids using a complex control system, leading to high reliability of the control system. In addition, the power fluctuations are lower for stall-regulated turbines. However, this control method has some disadvantages, such as lower efficiency, the requirement of external equipment at the turbine start, larger dynamic loads acting on the blades, nacelle, and tower, dependence on reliable brakes for the operation safety. Therefore, this control technique has been primarily used for small and medium wind turbines. Since the capacity of wind turbines has entered the multi-megawatt power range in recent years, pitch control has become dominant in the wind power market.

The active stall control technique has been developed for large wind turbines. An active stall wind turbine has stalling blades together with a blade pitch system. Since the blades at high wind speeds are turned towards stall, in the opposite direction as with pitch-control systems, this control method is also referred to as negative pitch control. Compared with passive stall control, active control provides more accurate control on the power output and maintains the rated power at high wind speeds. However, with the addition of the pitch-control mechanism, the active stall control mode increases the turbine cost and decreases operation reliability.

With megawatt wind turbines becoming the mainstream in the wind power industry from the late 1990s, pitch control is more favorable than stall control. It has been reported that the number of pitch-regulated turbines is four times higher than that of stall-regulated turbines and the trend is going to continue in coming decades [41].

5.4.3 Yaw control
In order to maximize the wind power output and minimize the asymmetric loads acting on the rotor blades and the tower, a horizontal-axis wind turbine must be oriented with rotor against the wind by using an active yaw control system. Like wind pitch systems, yaw systems can be driven either electrically or hydraulically. Generally, hydraulic yaw systems were used in the earlier time of the wind turbine development [54]. In modern wind turbines, yaw control is done by electric motors. The yaw control system usually consists of an electrical motor with a speed reducing gearbox, a bull gear which is fixed to the tower, a wind vane to gain the information about wind direction, a yaw deck, and a brake to lock the turbine securely in yaw when the required position is reached. For a large wind turbine with high driving loads, the yaw control system may use two or more yaw motors to work together for driving a heavy nacelle (see Figure 7).

In practice, the yaw error signals obtained from the wind vane are used to calculate the average yaw angle in a short interval. When this average yaw angle exceeds the preset threshold, the yaw motor is activated to align the turbine with the wind direction. Thus, with heavily filtered wind direction measurements, the actions of yaw control are rather limited and slow.

5.4.4 Other control approaches
In the early time of wind turbine design, ailerons were once used to control the power output. This method involves placing moveable flaps on the trailing edge of rotor blades [55]. The ailerons change the lift and drag characteristics of the blades and eventually change the rotor torque, which enable to regulate rotor speed and rotor power output. However, this method was less successful and was soon abandoned.

Another possibility is to yaw the rotor partly out of the wind to decrease power. This technique of yaw control is in practice used only for tiny wind turbines (>1 kW) [56].

6 Challenges in wind power generation

While wind power generation offers numerous benefits and advantages over conventional power generation, there are also some challenges and problems need to be seriously addressed. The wide range of challenges and problems, from long-term environmental influences to thermal management of wind turbines, must be carefully considered in response to the rapid growth of wind power generation.

6.1 Environmental impacts

Modern wind farms today may contain a large number of large-size wind turbines. Therefore, their impacts on the environment cannot be ignored. One of the impacts is that poorly sited wind energy facilities may block bird migration routes and hurt or kill birds.

Though blade rotation speeds are rather low for large wind turbines at their normal operation, the tangential speeds at the blade tips could be higher than 70 m/s. At such high speeds, birds flying through the blade sweeping areas may be easily hurt or killed by colliding with blades. It has been reported by the US National Academy of Science that wind turbines may kill up to 40,000 birds per year in US [57]. Though this number is much smaller than the 80 million birds killed by cars each year, it is important to evaluate the long-term influence on local geography, seasonal bird abundance and the species at risk. To reduce the bird death, using bird scares to drive birds away from wind farms has been considered. A more recent study has revealed that fossil-fuelled power stations appear to pose a much greater threat to avian wildlife than wind and nuclear power technologies [58].

Today, this problem becomes less important. Before building a wind farm, a series of environmental assessments have to be completed to avoid bird migration routes and to minimize other environmental impacts. Once the wind farm is built, further monitoring takes place to better understand the ongoing relationship between birds and the wind farm [59].

Building wind farms will change the character of local landscape. Modern large wind turbines are more than 100 m tall and thus can be seen at a far distance. In practice, the visual effect for local residents is a significant consideration and is always scrutinized for wind projects. To minimize the visual effect, wind turbines usually use neutral colors such as light grey or off-white [60]. Strategies to minimize visual effects involve the spacing, design, and uniformity of turbines, markings or lighting, roads and service buildings [61]. There are a number of analytical tools available to assist understanding and testing of the effect of wind farms on visual amenity.

6.2 Wind turbine noise

With the extensive build up of wind power plants and the population growth all over the world, the influence of wind turbine noise to the nearby residents becomes a problem not to be neglected. Wind turbine noise consists of aerodynamic noise from rotating blades and mechanical vibration noise from gearboxes and generators. For a modern large wind turbine, aerodynamic noise from the blades is considered to be the dominant noise source.

A detailed review of available wind turbine noise standards, regulations, and guidelines in Europe, North America, and Australia was made by Ramakrishnan [62]. Though the noise limits vary significantly country to country, the approximate noise level at nighttimes in most European countries and Canada ranges from 35 to 40 dBA.

There are two components in aerodynamic noise: (1) airfoil self noise, that is, the noise produced by the blade in an undisturbed inflow and is caused by the interaction in the boundary layer with the blade trailing edge; and (2) inflow turbulence noise which is caused by the interaction of upstream atmospheric turbulence with the blade and depends on the atmospheric conditions. Both airfoil self noise and inflow turbulence noise mechanisms are dependent on a number of parameters such as wind speed, angle of attack, radiation direction, and airfoil shape.

There are a number of techniques for reducing aerodynamic noise produced by wind turbine blades. One of them is to use serrated blades at their trailing edges. It can improve blade aerodynamic characteristics and reduce the noise induced by Karman vortex street [63]. Another is to use turbulence generating means, placed on the leeward surface side and at the outer section of the blade, to reduce noise [66]. In a recent US patent application, it has reported that with an anti-noise device at the blade trailing edge, it allows altering the characteristics of the boundary layer and therefore modifies emitted noise [67].

The field measurements of GE wind turbines have shown that the use of the optimized blades and the serrated blades can reduce average overall noise by 0.5 and 3.2 dBA, respectively [68]. In a field test of a 2.3 MW wind turbine, the overall noise level reduction provided by blade serrations is over 6 dBA for at least two frequencies [69].

6.3 Integration of wind power into grid

Wind is a highly intermittent energy source for causing overall fluctuation in wind power generation. Electricity generated from wind turbines strongly depends on the local weather and geographic conditions that can fluctuate a great deal more than with some renewable energy sources such as hydropower.

With the increasing share of wind energy in the global power market, a large amount of wind power is integrated into existing grids. Thus, the expected growth in wind power could soon exceed the current capability of grids with today's technology. To prepare this situation in advance, the influence of intermittent wind power on the grid stability and system security must be properly addressed.

The impacts of wind power to a power grid depend on the level of wind power penetration, grid size and generation mix of electricity in the grid. Undoubtedly, there is no problem for low wind power penetration in a large power grid. However, integrating large utility-scale wind power presents unique challenges. These challenges call into questions such as: How to ensure system controllability? How to manage new kinds of variability and uncertainty [70]? The detailed analysis regarding the impacts of wind power on power systems can be found in [71].

6.4 Thermal management of wind turbines

Large wind turbines are usually installed far away from urban areas and often operate under severe climate conditions, thus experiencing large variations in environmental temperatures. As a consequence, there is a need for a wind turbine to have a robust thermal control system for maintaining temperature levels inside the nacelle within specified limits.

During turbine operation, heat is generated from electric/electronic devices and rotating mechanical components (e.g. gearboxes and bearings) as a result of various power losses. For ensuring safe and reliable operation and preventing failure of the turbine, heat generated in the wind turbine must be dissipated efficiently.

Wind turbine cooling includes:

- Wind generator cooling
- Electronic and electric equipment cooling
- Gearbox cooling
- Other components/subsystems cooling

New cooling techniques have continuously been innovated in all cooling modes. A method was proposed to utilize incoming wind to cool the wind turbine. This wind assisted cooling system sucks in wind flow from an air inlet port on the top of the nacelle, fills the received airflow into the generator and finally exhausts at the front of the nacelle [72]. Some large wind generators use water or oil cooling for dealing with high thermal loads [73]. While the turbine benefits high cooling efficiency, it also suffers lower reliability and higher cost for adding such a complex cooling system.

The main challenge for electronic devices in a wind turbine is that they must withstand a wide range of ambient temperatures, usually from –40 to +55°C. In addition, they must be protected from dusts and moisture, as well as electrical shocks from lightning. There are several cooling modes in electronic cooling, including passive or active air cooling, forced single- or multi-phase liquid cooling, and phase change cooling. Under high ambient temperature conditions, a cooling or ventilation system is necessary to prevent overheating of electronic devices.

In cold climates, heating may be required for

- Warming up the lubrication oil in gearboxes
- Heating blades and hub to prevent them from icing over
- Raising the temperature inside the control cabinets toward a desired temperature range to prevent electronic devices from malfunctioning

6.5 Wind energy storage

Today developing advanced, cost-effective storage technologies of electric energy still remains a challenge, which may limit the widespread application of wind energy. The research and development (R&D) of new energy storage systems are highly desired to meet cyclical energy demands and stabilize power output, especially for large-scale wind farms.

The technologies for wind energy storage have been developed over several decades to convert wind energy into various forms of energy, including:

- Electrochemical energy in batteries and super capacitors
- Magnetic energy in superconducting magnetic energy storage (SMES)
- Kinetic energy in rotating flywheels
- Potential energy in pumped water at higher altitudes
- Mechanical energy in compressed air in vast geologic vaults
- Hydrogen energy by decomposing water

Among these techniques, the most popular method is to use batteries. However, there are some drawbacks to regular batteries, such as cost, short lifetime, corrosion, and disposal concerns [74]. Research and development of innovative batteries are underway. It has reported that lithium-ion battery technology is projected to provide stationary electrical energy solutions to enable the effective use in renewable energy sources. It is expected that safe and reliable lithium-ion batteries will soon be connected to solar cells and wind turbines [75]. Sodium-sulfur battery is another promising candidate for energy storage [76]. This type of batteries is preferably used to store renewable energy such as wind, sunlight, and geothermal heat [77]. The detailed review of electrical energy storage can be found in [78].

6.6 Wind turbine lifetime

Modern wind turbines are designed for the lifetime of 20–30 years. A critical challenge facing turbine manufacturers and wind power plants is how to achieve the lifetime goals while at the same time minimize the costs of maintenance and repair. However, improving the operational reliability and extending the lifetime of wind turbines are very difficult tasks for a number of reasons:

- Wind turbines have to be exposed to various hostile conditions such as extreme temperatures, wind speed fluctuations, humidity, dust, solar radiation, lightning, salinity and frequent onslaughts of rain, hail, snow, ice, and sandstorms.
- A modern wind turbine consists of a large number of components and systems; each of them has its own lifetime. According to the Cannikin law, failure must first occur in the component or system with the shortest lifetime.

- A wind turbine is subjected to a large variety of dynamic loads due to wind fluctuations in speed and direction and numerous starts and stops of the system. Some primary parts or components have to withstand heavy fatigue loads [79].
- Advanced high-strength, fatigue-resistant materials are vital to some key components in modern large wind turbines due to the continuous increase in blade length, hub height, and turbine weight.
- As a complex engineering system, a wind turbine must be designed at the system level rather than part/component level as a common practice in some turbine manufacturers.

6.7 Cost of electricity from wind power

Although the wind power industry appears to be booming in recent years worldwide, achieving continuous cost reduction in wind power generation continues to be a challenge and a key focus for the wind industry.

Wind power is characterized by low variable costs and relatively high fixed costs. The main factors governing wind power economics are [80]:

- Investment costs, including wind turbines, foundations, and grid connection
- Operation and maintenance (O&M) costs, including regular maintenance, repairs, insurance, spare parts, and administration
- Wind turbine's electricity production cost, which highly depends on the wind turbine capacity, wind farm size, and average wind speed at the chosen site
- Wind turbine lifetime
- Discount rate

Among these, the most important factors are the wind turbines' electricity production and their investment costs. The trends towards lager wind turbines and larger wind farms help reduce both investment and O&M costs per kilowatt-hour (kWh) produced.

Though the price of electricity from wind has fallen approximately 90% over the last 30 years because of the developments of wind technology, it is still more expensive than those from coal or natural gas. It has been predicted by Electric Power Research Institute that even for plants coming online in 2015, wind energy would cost nearly one-third more than coal and about 14% more than natural gas [81]. This is the greatest obstacle for wind power to increase its share in the electric power market. A recent study [82] indicates that wind energy in US today still depends on federal tax incentives to compete with fossil fuel prices, and technology progress could dominate future cost competitiveness.

The global financial and economic crisis, which started from early 2008, has dramatically altered the pace of wind development. With reduced power consumption, the prices of fossil fuels (e.g. coal and natural gas) have greatly decreased, putting even more pressures on the wind power industry to continuously drive down wind power costs for staying competitive in the present challenging economic times.

7 Trends in wind turbine developments and wind power generation

Wind turbine technology has been developed by continuously optimizing turbine design, improving turbine performance, and enhancing overall turbine efficiency. There have been several generations of development and improvement in wind turbine technology, concentrated on blades, generators, direct drive techniques, pitch and yaw control systems, and so on. To provide more electrical energy from wind technology in the next several decades, it requires

- Developing innovative techniques
- Decreasing wind turbine costs through technology advancement
- Optimizing manufacturing processes and enhancing manufacturing operations
- Improving wind turbine performance and efficiency
- Reducing operating and maintenance costs
- Expanding wind turbine production capacities

The current major trends in the development of wind turbines are towards higher power, higher efficiency and reliability, and lower cost per kilowatt machines.

7.1 High-power, large-capacity wind turbine

One of the significant developments in wind turbine designing and manufacturing in recent years is the increase in the wind turbine capacity of individual wind turbines. From machines of just 25 kW two decades ago, the commercial range of modern wind turbines sold today is typically 1–6 MW. At the same time, 7–10 MW wind turbines are underway in some larger wind turbine OEMs. With this trend, innovative techniques have been developed and new materials have been adopted for optimizing the wind turbine performance and minimizing the operation and manufacturing costs. Enercon has installed the present world's largest wind turbine E-126 in Germany and is in the process of installing more units in Belgium. The E-126 turbine is rated at 6 MW with the rotor diameter of 126 m [83]. Clipper Windpower has announced that it is planning to build a 7.5 MW offshore wind turbine [84].

However, while high-powered wind turbines enable to increase wind power output per unit and lower the cost per kWh, there are some significant challenges facing wind turbine engineers:

a. Failure rates of wind turbines depend not only on turbines' operational age but also their rated power. High-power, large-size wind turbines have shown significant higher annual failure rates due to the primary failures of the control system, drivetrain, and electronic/electrical components. Because most of mega-watt wind turbines were usually among the first models installed, they show high early failure rates that decrease slightly throughout their years of operation [85].

b. Wind velocity is proportional to the height from the earth's surface. With the continuously increasing blade length of large wind turbines, the differences of the dynamic wind loads between the rotating blades become significantly large, resulting in a large resultant unbalanced fatigue load on the turbine blades, and a resultant unbalanced torsional moment on the main shaft, and in turn, on the wind tower.

c. During wind turbine's operation, a minimum clearance must be maintained between the blade tips and the wind tower. Therefore, high blade stiffness is required to avoid the collision between the blades and the tower. In practice, the maximum blade length is constrained by required stiffness and stresses of blades.

d. Large wind turbines become more susceptible to variations in wind speed and intensity across the swept area.

e. Transportation and installation of long-length blades remain challenges to the wind power industry. The length of a blade for a 4.5–5 MW wind turbine ranges 50–70 m. It is very difficult to ship such long blades through current highways and installed on the top of 120–160 m wind towers.

f. The tower strength is another consideration. For a given survivable wind speed, the mass of a wind turbine is approximately proportional to the cube of its blade length and the output power is proportional to the square of it blade length. Typically, the mass of a 4.5–5 MW is of 200–500 tons. It was reported that doubling the tower height generally requires doubling the diameter as well, increasing the amount of material by a factor of 8 [86].

To ensure the sustainability of the increase in power output and turbine size, all these challenges must be carefully and effectively addressed.

7.2 Offshore wind turbine

With several decades of experience with onshore wind technology, offshore wind technology has presently become the focus of the wind power industry. Due to the lower resistance, wind speeds over offshore sea level are typically 20% higher than those over nearby lands. Thus, according to the wind power law, the offshore wind power can capture much more power than the onshore one. This indicates that an offshore wind turbine may gain a higher capacity factor than that of its land-based counterpart. In addition, because the offshore wind speeds are relatively uniform with the lower variations and turbulence, it enables the offshore wind turbines to simplify the control systems and reduces blade and turbine wears.

Sweden installed the first offshore wind turbine in 1990, with the unit capacity of 220 kW. Denmark built its first demonstration offshore wind turbines in 1991, which consists of 11 units, with the unit capacity of 450 kW. With the developments of offshore wind technology in the next several years, offshore wind turbines entered the stage of industrial production in 2001. Today, high capacity wind turbines focus on the offshore application. In 2009, nearly 600 MW offshore wind power were added and connected to electric grids, basically by European countries, bringing the total accumulative installed offshore wind power capacity to more than 2,000 MW. It is expected that in 2010 ten additional European

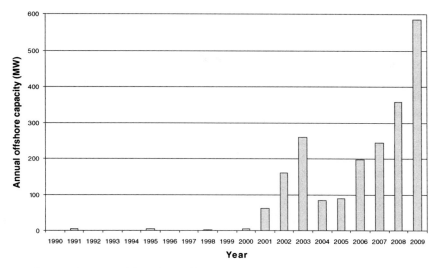

Figure 13: Annual installed offshore wind power capacity (all but 2009
data from [1]).

offshore wind farms will be completed to add approximately 1,000 MW online,
which represents a growth rate of 75% compared to 2009. The related foundation
technologies for offshore applications are also being developed for the erection of
higher capacity wind turbines. The annual installed offshore wind power capacities from 1990 to 2009 are shown in Fig. 13.

Offshore wind turbines are installed in seawater, with greater risks of structural
corrosion, particularly under conditions of high wave, sea salt splashing and low
temperature. To avoid or at least delay the corrosion to protect wind turbines, a
number of techniques have been developed. One of them involves the use of electrochemical reactions to prevent the steel corrosion, known as cathodic protection.
This is done using a small negative voltage applied to metal. A new method, called
impressed current cathodic protection, was invented by Brown and Hefner [87].
This method includes providing an impressed current anode electrochemically
coupled to the wind turbine support structure to the impressed current anode to
operate the impressed current anode.

In dealing with corrosion control, coating technology plays an important role in
the wind power industry. The adoption of high performance of multi-coating systems has achieved satisfactory results to prevent external and internal corrosions of
wind turbines. The coating materials used today include thermally sprayed metal
(zinc and aluminum), siloxane, acrylic, epoxy, polyurethane, etc. It is expected to
develop inorganic hybrid materials (such as polysiloxane) as super durable finishes
to meet more aggressive environments [88].

7.3 Direct drive wind turbine

In a direct drive wind turbine, rotor blades directly drive the rotor of the wind
generator. By eliminating the multi-stage gearbox, which is one of the most

easily damaged components in a MW wind turbine, the number of rotating parts is greatly reduced and the turbine structure is considerably simplified. As a result, it significantly increases the reliability and efficiency of the wind turbine and reduces turbine noise and maintenance costs. Since direct drive wind generators operate at relative low speeds, it reduces the wear and tear of the generator. In order to identify suitable generator concepts for director drive wind turbines, Bang et al. have compared various direct drive generator systems and concluded that direct drive permanent-magnet synchronous machines are more superior in terms of the energy yield, reliability and maintenance costs [89].

Though the concept of direct drive wind turbine has been proposed for a long time, the modern direct drive techniques have become available until recent three decades. Presently, direct drive wind turbines have been manufactured by large wind OEMs. Siemens installed two innovative 3.6 MW direct drive wind turbines at a site in west Denmark in 2008. The feasibility of building 10 MW direct-drive wind turbine was investigated by Polinder et al. [90].

However, direct drive wind turbines have some disadvantages in terms of the cost, size, and mass, making them difficult in manufacturing, shipping, and installing. Without a gearbox, the rotor diameter in a direct drive generator must be made larger enough to maintain a relative high rotating speed at the air gap. A lot of structural material must be added to keep the stator and rotor in place for maintaining the air gap. Therefore, the direct drive wind turbine has a larger size and a higher weight. For instance, the Siemens 3.6 MW direct drive wind turbine has a total weight of 265 tons (nacelle 165 tons and rotor 100 tons), as compared with 235 tons for a 3.6 MW geared turbine [91]. This requires the higher strength for the turbine tower. According to Bang et al. [89], the cost of a 3 MW direct drive PM synchronous generator system could be 35% higher than that of a 3 MW induction generator system with three stage gearbox. To make the direct drive wind turbines more attractive to the wind market, all these disadvantages must be solved.

7.4 High efficient blade

Rotor blade design can be split into structural and aerodynamic design. During normal wind turbine operation, rotor blades have to withstand enormous dynamic loads. The bending moment due to the gravity load results in up to 10^8 load cycle alternations within the turbine lifetime. In addition, there are stochastic alternating loads caused by wind turbulence and the effects of ageing of the materials due to the weather [92]. As wind turbines become larger and larger, the length, size, and weight of blades increase accordingly. For instance, the blade diameter in a large wind turbine could be longer than 100 m, which is higher than the wingspan and length of Boeing 747-400 at 64.4 and 70.7 m, respectively. There is no doubt that these blades require extremely high fatigue strength.

In the blade structural design, one indicator is the blade weight/swept area ratio (or swept area density in some references). It is highly desired to minimize this ratio while satisfying the blade strength requirements. Most blades used today are made from composite materials such as glass-fibre epoxy, carbon epoxy, fibre-reinforced plastic, etc. With epoxy resin/glass-fibre material, the weight/swept area ratio of 1–1.5 kg/m^2 can be achieved up to a rotor diameter of 62 m [93]. With the trend toward long-length and larger-size blades, high-strength, fatigue-resistant materials such as metallic materials need to be considered.

The aerodynamic design of wind turbine blades is important as it determines the wind energy capture. With advanced CFD tools, the shape of aerodynamic profile and dimensions of blade can be preliminarily determined and the blade optimization can be achieved via field tests. As a good example, a study of aerodynamic and structural design for wind turbines larger than 5 MW was reported by Hillmer *et al.* [94].

There are a number of improvements achieved in the rotor blade design. A new type of wind turbine blades, named "STAR" (Sweep Twist Adaptive Rotor) blades, was specially designed for low-wind-speed regions. The test results have shown that the STAR blades can shed 20% of the root moment via tip twist of about 3° and yield 5–10% annual energy capture than the regular blades [95].

Researchers at Purdue University and Sandia National Laboratory have developed an innovative technique that uses sensors and computational software to constantly monitor forces exerted on wind turbine blades. The data is fed into an active control system that precisely adjusts the shape of rotor blades to respond to changing winds. The technique could also help improve turbine reliability by providing critical real-time information to the control system for preventing damage to blades from high winds [96]. Recent aerodynamic research has revealed that an increase of the aerodynamic efficiency of a wind turbine rotor may be achieved by extending the turbine blades to very close to the wind turbine nacelle [97].

7.5 Floating wind turbine

Dr. Sclavounos at MIT is among the first to develop the concept of floating wind turbines in deep water. He and his team in 2004 integrated a wind turbine with a floater. According to their analysis, the floater-mounted turbines could work in water depths of up to 200 m [98].

The world's first commercial-scale floating wind turbine has being constructed in deep water far from land [100]. The turbine is mounted on a floating turbine platform on the sea surface and anchored to the seabed with three strong chains. Changing the length of the chains could allow the turbine to operate in water depths between 50 and 300 m, enough to take it far out into the deep ocean. By comparing with existing offshore wind turbines, floating wind turbine is more economic in the installation and shipping. Electricity would be sent ashore using undersea cables.

Sway, a Norwegian company, plans to launch its prototype of floating wind turbines in 2010. The turbine is to be mounted on an elongated floating mast, connected to the seabed by a metal tube. The turbine mast is designed to sway with wind and waves, and can lean at an angle of up to 15° [101].

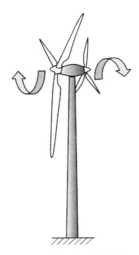

Figure 14: Illustration of a dual-blade set wind turbine.

7.6 Wind turbine with contra-rotating rotors

Coaxial contra-rotating propellers have been widely applied to aircrafts, marines, ships, and torpedoes for improving the propulsion efficiency and offsetting system reactive torques. The efficiency of contra-rotating propellers was found 6–16% higher than that of normal propellers [102]. Late, the contra-rotating concept has been introduced into the wind turbine design. The contra-rotating type of wind turbines has usually two sets of blades that rotate in opposite directions. These two sets of blades can be arranged either one behind another at the turbine front, or separately at the turbine front and rear (see Fig. 14).

Schönball [103] is one of the first to describe a mechanism which composes of two contra-rotating rotors. In the generator, a rotor is driven by one wind wheel and a regular stationary stator driven by another wind wheel. Owing to the opposed rotation of two wind wheels, the relative speed between the two rotors may be doubled and thus the efficiency is correspondingly increased, compared with a conventional generator having one rotor. Similarly, McCombs [104] developed a wind turbine equipped with two sets of blades that propel the rotor and stator directly. More recently, Wachinski [105] proposed a drive device for an improved windmill composed of two counter-rotative sets of blades. In 2005, with two granted US patents [106], Kowintec Inc. successfully built up a 100 kW and 1 MW contra-rotating wind turbine prototypes. The test results have shown that with the dual-blade and dual-rotor, the power production efficiency of the turbine increases 24% and the system cost decreases 20–30%, by comparing with a conventional single rotor unit [108].

The benefits of using contra-rotating wind turbines are:

- Enhancing wind energy capturing capability – A wind turbine with a contra-rotating system can capture more wind energy to convert into electricity, by comparing with a single set of blades. In a contra-rotating win turbine, each set of blades contributes independently to the total power output. Because the length of each set of blades can be made differently, the wind turbine may produce electricity at a lower cut-in speed by the set of blades with a shorter blade length.

- Achieving higher power density – With the increase in power output and the moderate increase in wind turbine volume, the power density becomes higher.
- Reducing wind turbine cost per kWh – Though the total cost of this type of wind turbines increases due to the addition of extra parts, the cost per kWh will decrease for the large increase in power output.
- Increasing wind turbine operation reliability – Some types of contra-rotating wind turbines eliminate gearboxes to simplify the design and increase the turbine operation reliability.

However, this type of wind turbine also has some drawbacks. The main negative effect is the wake vortices created by the first set of blades which can substantially lower the performance of the second set of blades. In fact, the wake vortices will enhance the wind turbulence intensity to strengthen unbalanced dynamic loads on the second set of blades, decreasing the mean wind velocity and the power captured by the second set of blades.

7.7 Drivetrain

In a geared wind turbine, the term "drivetrain" usually encompasses all rotating parts, from the rotor hub, the main gearbox, to the generator. The main gearbox is one of the most important components in the wind turbine, for increasing the slow rotation speed of the blades to the desired high speed of the generator's rotor. It is also the most expensive component in the turbine and can easily fail before reaching the intended life. With the increase in the turbine size and capacity in the last decade, the gearbox has been subjected to even greater loads and stresses. There are significant challenges presented to gearbox designers and manufacturers.

Wind turbines typically use planetary gears to divide torque along three paths and reduce individual loads on each gear. However, torsional loads twist gears out of alignment, and slight dimensional variation in gearbox components, indicate that planetary gears do not equally share the load. Misaligned gears, shock loads, and uneven forces lead to highly localized stress and eventually fracture along the gear edges. To solve these problems, an innovative type of gearbox has been developed (Fig. 15), using the Integrated Flex-pin Bearing (IFB) to equalize gear loads, eliminate misalignment, and dramatically improve wind turbine reliability. This novel design increases torque capacity of planetary gears up to 50% [109].

Based on the failure analysis of gearboxes, the Gearbox Reliability Collaborative initiated at US National Renewable Energy Laboratory (NREL) provides a fresh approach toward better gearboxes that combines the resources of key members of

Figure 15: Flex-drive planetary gearbox. Courtesy of [109].

the supply chain to investigate design-level root causes of field problems and solutions that will lead to higher gearbox reliability [111].

7.8 Integration of wind and other energy sources

One of the notable characteristics in the wind power generation is its uncertainty due to the sudden change in both wind speed and direction, especially for off-grid wind power generation systems. Therefore, the power output from wind turbines fluctuates from time to time. When wind turbines are connected to a small or isolated grid, the power output from other generators must be varied in response to these variations and fluctuation in order to keep system frequency and voltage within predefined limits. For this purpose, it is beneficial to integrate wind and other complementary energy sources to form hybrid power systems for assuring the stability and reliability of power supply and reducing the requirement for the wind energy storage.

7.8.1 Wind–solar hybrid system

Both wind and solar energy are highly intermittent electricity generation sources. Time intervals within which fluctuations occur span multiple temporal scales, from seconds to years. These fluctuations can be subdivided into periodic fluctuations (diurnal or annual fluctuations) and non-periodic fluctuations related to the weather change.

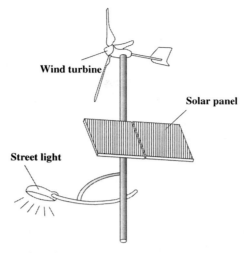

Figure 16: Wind–solar hybrid system for street lights.

Wind and solar energy are complementary to each other in time sequence and regions. In the summer, sunlight is intensive and the sunshine duration is long but there is less wind. In the winter, when less sunlight is available, wind becomes strong. During a day, the sunshine is strong while wind is weak. After sunset, the wind is strengthened due to large temperature changes near the earth's surface. It has been reported that the effects of complementarity are more dramatic in certain periods and locations at Serbia [112]. The analyses and test data of wind–solar hybrid power systems have shown that the optimum combination of the wind–solar hybrid system lies between 0.70 and 0.75 of solar energy to load ratio with the minimized life cycle cost. For all load demands, the leveled energy cost for the wind–solar hybrid system is always lower than that of standalone solar or wind system [113].

Because the major operating time for wind and solar systems occurs at different periods of time, wind–solar hybrid power systems can ensure the reliability of electricity supply. The applications of wind–solar hybrid systems ranges extensively from residential houses to municipal and industrial facilities, either grid-connected or standalone. For instance, as an independent power supply source, wind–solar hybrid systems have been widely used in China for street lighting (see Fig. 16). The world's largest wind–solar power test base, integrating wind power, photovoltaic power and energy storage, is being constructed at Zhangbei, China. The project will have an installed capacity to generate 300 MW of wind power, 100 MW of solar power and 75 MW of chemical energy storage [114].

7.8.2 Wind–hydro hybrid system
Hydropower generation is to convert potential energy in water into electrical energy by means of hydropower generators. As a renewable and clean energy

source, hydropower accounts for the dominant portion of electricity generated from all renewable sources.

In many locations of the world, hydropower is complementary with wind power, while the seasonal wind power distribution is higher in winter and spring but lower in summer and fall, hydropower is lower in the dry seasons (winter and spring) but higher in the wet seasons (summer and fall). Thus, the integration of wind and hydropower systems can provide significant technical, economic, and systematic benefits for both systems. Taking a reservoir as a means of energy regulation, "green" electricity can be produced with wind–hydro hybrid systems.

7.8.3 Wind–hydrogen system

Hydrogen is an energy carrier and can be produced from a variety of resources such as water, fossil fuels, and biomass. As a fuel with a high energy density, hydrogen can be stored, transported and then converted into electricity by means of fuel cells at end users. It is widely recognized that wind power, solar power and other renewable energy power generation systems can be integrated with the electrolysis hydrogen production system to produce hydrogen fuel. The largest wind-to-hydrogen power system in the UK has been applied to a building that is fuelled solely by wind and "green" hydrogen power with the developed hydrogen mini-grid system technology. In this system, electricity generated from a wind turbine is mainly used to provide to the building and excess electricity is used to produce hydrogen using a state-of-the-art high-pressure alkaline electrolyser [115].

7.8.4 Wind–diesel power generation system

Wind power can be combined with power produced by diesel engine-generator systems to provide a stable supply of electricity. In response to the variations in wind power generation and electricity consumption, diesel generator sets may operate intermittently to reduce the consumption of the fuel. It was reported that a viable wind–diesel stand-alone system can operate with an estimated 50–80% fuel saving compared to power supply from diesel generation alone [116].

Wind–diesel hybrid power systems have been studied since 1995 in the US. Till now, many new techniques have been developed and a large number of wind–diesel power generation systems have been installed all over the world [117]. According to the proportion of wind use in the system, three different types of wind–diesel systems can be distinguished: low, medium, and high penetration wind–diesel systems. Presently, low penetration systems are used at the commercial level, whereas solutions for high penetration wind–diesel systems are at the demonstration level. The technology trends include the development of robust and proven control strategies [41].

References

[1] Zervos, A., Wind power as a mainstream energy source. *Proc. of the 2009 European Wind Energy Conf.*, Marseille, March 2009.

[2] Nelson, V., *Wind Energy – Renewable Energy and the Environment*, CRC Press, 2009.

[3] El-Ali, A., Moubayed, N. & Outbib, R., Comparison between solar and wind energy in Lebanon. *Proc. of 9th Int. Conf. on Electrical Power Quality and Utilisation*, Barcelona, 2007.

[4] Gipe, P., *Wind Energy Comes of Age*, John Wiley & Sons, 1995.

[5] Vestas, Discover the unique power of the wind, http://www.vestas.com/en/ modern-energy/experience-the-wind/wind-power-through-the-ages.aspx.

[6] Wan, Zhen, *Strange Things of the South* (南州异物志), wrote in Wu of Three Kingdoms period (220–280).

[7] Song, Yingxing, *Exploitation of the Works of Nature* (天工开物), 1637.

[8] Yu, Changhui, *Important Issues in Coast Defense* (防海辑要), 1842.

[9] Wells, H.G., *The Outline of History*, The Macmillan Company, New York, 1921.

[10] Juleff, G., An ancient wind-power iron smelting technology in Sri Lanka. *Nature*, **379(6560)**, pp. 60–63, 1996.

[11] Temple, R., *The Genius of China: 3,000 Years of Science, Discovery, and Invention*, Simon & Schuster, 1986.

[12] Windmill dreams, http://www.china.org.cn/english/NM-e/151624.htm

[13] Hassanm, Y.H. & Hill, D.R., *Islamic Technology: An Illustrated History*, Cambridge University Press, 1986.

[14] Drachmann, A.G., Heron's windmill, *Centaurus*, **7**, pp. 145–151, 1961.

[15] Krohn, S., Danish wind turbines: an industrial success story, 2002, http:// www.talentfactory.dk/media(483,1033)/Danish_Wind_Turbine_Industry,_ an_industrial_succes_story.pdf

[16] Han, Fei, *Han Fei Zi* (韩非子), wrote during 280 B.C.–232 B.C.

[17] American wind energy association. Basic principles of wind resource evaluation, http://www.awea.org/faq/basicwr.html

[18] Siddiqi, A.H., Khan, S. & Rehman, S., Wind speed simulation using wavelets. *American Journal of Applied Sciences*, **2(2)**, pp. 557–564, 2005.

[19] George, J.M., Peterson, R.E., Lee, J.A. & Wilson, G.R., Modeling wind and relative humidity effects on air quality. *Int. Specialty Conf. on Aerosols and Atmospheric Optics: Radiative Balance and Visual Air Quality*, Snowbird, Utah, 1994.

[20] George, J.M., Wilson, G.R. & Vining, R.C., Modeling hourly and daily wind and relative humidity. *Int. Conf. on Air Pollution from Agricultural Operations*, pp. 183–190, Ames, Iowa, 1996.

[21] Sinden, G., Characteristics of the UK wind resource: long-term patterns and relationship to electricity demand. *Energy Policy*, **35(1)**, pp. 112–127, 2007.

[22] European Wind Energy Association. Wind energy – the facts, part I: Technology, The annual variability of wind speed, 2009.

[23] Ko, K.N., Kim, K.B., Kang, M.J., Oh, H.S. & Huh, J.C., Variation in wind characteristics for 30 years on Jeju Island, Korea. *Proc. of the 2009 European Wind Energy Conf.*, Marseille, March 2009.

[24] Weibull, W., A statistical distribution function of wide applicability. *ASME Journal of Applied Mechanics*, **18(3)**, pp. 239–297, 1951.

[25] Ulgen, K. & Hepbasli, A., Determination of Weibull parameters for wind energy analysis of Izmir, Turkey. *International Journal of Energy Research*, **26(6)**, pp. 495–506, 2002.

[26] Yilmaz, V. & Celik, H.E., A statistical approach to estimate the wind speed distribution: the case of Gelibolu region. *Doğuş Üniversitesi Dergisi*, **9(1)**, pp. 122–132, 2008.

[27] Barbu, C. & Vyas, P., System and method for loads reduction in a horizontal-axis wind turbine using upwind information, US patent application, 20,090,047,116, 2009.

[28] Spudić, V., Marić, M. & Perić, N., Neural networks based prediction of wind gusts, *European Wind Energy Conf.*, Marseille, France, 2009.

[29] Brasseur, O., Development and application of a physical approach to estimating wind gusts. *Monthly Weather Review*, **129(1)**, pp. 5–25, 2001.

[30] Cook, K.R. & Gruenbacher, B., Assessment of methodologies to forecast wind gust speed, National Weather Service Weather Forecast Office, 2008, http://www.crh.noaa.gov/ict/?n=windgust

[31] Mitsuta, Y. & Tsukamoto, O., Studies on spatial structure of wind gust. *Journal of Applied Meteorology*, **28(11)**, pp. 1155–1160, 1989.

[32] Paulsen, B.M. & Schroeder, J.L., An examination of tropical and extratropical gust factors and the associated wind speed histograms. *Journal of Applied Meteorology*, **44(2)**, pp. 270–280, 2005.

[33] Vanek, F.M. & Albright, L.D., *Energy Systems Engineering: Evaluation & Implementation*, McGraw-Hill, 2008.

[34] Gipe, P., *Wind Power: Renewable Energy for Home, Farm, and Business*, Chelsea Green Publishing Company, 2004.

[35] Shikha, S., Bhatti, T.S. & Kothari, D.P., Aspect of technology development of wind turbines. *Journal of Energy Engineering*, **129(3)**, pp. 81–95, 2003.

[36] Blueenegy, Solarwind™ turbine, http://www.bluenergyusa.com

[37] Naskali, P.H., MacLean A., Gray, N.C.C., Lewis, J.H. & Newall, A.P., Helical wind turbine, US Patent 7,344,353, 2005.

[38] Noguchi, T., Windmill for wind power generation, US Patent 7,084,523, 2006.

[39] Thomas, J., Colossal magnetic levitation wind turbine proposed, Science & Technology, July 27, 2007, http://www.treehugger.com/files/2007/07/colossal_magnet.php

[40] Cochrane, R.C., Vertical-axis wind turbine with LED display, GB patent 2415750, 2006.

[41] European Wind Energy Association. Wind energy – the facts, part I: Technology, 2009, http://www.wind-energy-the-facts.org/en

[42] Doran, A., Classification of wind turbines, http://www.articlesbase.com/electronics-articles/classification-of-wind-turbines-429469.html

[43] Lanchester, F.W., A contribution to the theory of propulsion and the screw propeller. *Trans. Institution of Naval Architects*, Vol. XXX, p. 330, March 25, 1915.

[44] Betz, A., Das maximum der theoretisch möglichen ausnützung des windes durch windmotoren. *Zeitschrift für das gesamte Turbinewesen*, pp. 307–309, September 20, 1920.

[45] Cabezon, D., Marti, I., Isidro, M.J.S. & Perez, I., Comparison of methods for power curve modelling, Global Windpower 2004, Chicago, Illinois, 2004.

[46] Gottschall, J. & Peinke, J., How to improve the estimation of power curves for wind turbines. *Environmental Research Letters*, **3**(1), paper No. 015005, 2008.

[47] IEC 61400-12-1. Power performance measurements of grid connected wind turbines, 2005.

[48] Anahua, E., Barth, St. & Peinke, J., Markovian power curves for wind turbine. *Wind Energy*, **11**(3), pp. 219–232, 2007.

[49] Ragheb, M., Optimal rotor tip speed ratio, 2009, https://netfiles.uiuc. edu/mragheb/www/NPRE%20475%20Wind%20Power%20Systems/ Optimal%20Rotor%20Tip%20Speed%20Ratio.pdf

[50] Renewable Energy Research Laboratory/University of Massachusetts at Amherst. Wind power: capacity factor, intermittency, and what happens when the wind doesn't blow? http://www.ceere.org/rerl/about_wind/ RERL_Fact_Sheet_2a_Capacity_Factor.pdf

[51] Brandt, K. & Zeumer, J., Wind power plant comprising a rotor blade adjusting device, US Patent 7,256,509, 2007.

[52] Setec. Pitch-system including pitch-master and pitch-drive, http://www. setec-dresden.com/index.php?id=22

[53] Weitkamp, R., Lütze, H., Riesberg, Riesberg, A. & Anemüller, J., Means for adjusting the rotor blade of a wind power plant rotor, US Patent 6,783,326, 2004.

[54] Hildingsson, S. & Westin, T., Yawing system for adjusting a wind turbine into a required wind direction by turbine the turbine about a yawing axle, US Patent 5,990,568, 1999.

[55] Kaźmiekowski, M.P., Krishnan, R. & Blaabjerg, F., *Control in Power Electronics*, Academic Press, 2002.

[56] Danish Wind Industry Association. Power control of wind turbines, http:// www.windpower.org/en/tour/wtrb/powerreg.htm

[57] National Academy of Sciences. Environmental impacts of wind-energy projects, National Research Council, 2007.

[58] Sovacool, B.K., Contextualizing avian mortality: a preliminary appraisal of bird and bat fatalities from wind, fossil-fuel, and nuclear electricity. *Energy Policy*, **37**, pp. 2241–2248, 2009.

[59] Canadian Wind Energy Association. Birds, bats and wind energy, http:// www.canwea.ca/images/uploads/File/NRCan_-_Fact_Sheets/6_wildlife.pdf

[60] Wind farms & visual aesthetics, http://www.countrysideenergyco-op.ca/ files/cec_flyer_windfarm_visual_aesthetics_20060713a_w.pdf

[61] Gipe, P., Design as if people matter: aesthetic guidelines for the wind industry, http://www.wind-works.org/articles/design.html

[62] Ramakrishnan, R., Wind turbine facilities noise issues, Alolos Engineering Corporation, http://www.ene.gov.on.ca/envision/env_reg/er/documents/2008/Noise%20Report.pdf

[63] Vijgen, P.M.H.W., Howard, F.G., Bushnell, D.M. & Holmes, B.J., Serrated trailing edges for improving lift and drag characteristics of lifting surface, US patent 5,088,665, 1992.

[64] Dassen, A.G.M. & Hagg, F., Wind turbine, US patent 5,533,865, 1996.

[65] Shibara, M., Furukawa, T., Hayashi, Y. & Kata, E., Wind turbine provided with nacelle, US patent 6,830,436, 2004.

[66] Godsk, K. & Nielsen, T., A pitch controlled wind turbine blade, a wind turbine and use hereof, WO/2006/122574, 2006.

[67] Gil, A.M. & Rueda, L.M.G., Wind turbine blade with anti-noise devices, US patent application 20,080,298,967, 2008.

[68] Oerlemans, S., Fisher, M., Maeder, T. & Kögler, K., Reduction of wind turbine noise using optimized airfoils and trailing-edge serrations. *AIAA Journal*, **47(6)**, 2009.

[69] Kögler, K., Herr, S. & Fisher, M., Wind turbine blades with trailing edge serrations, US patent application, US20090074585, 2009.

[70] Jones, L.E., Integrating variable renewable generation in utility operations. *Utility/T&D*, Vol. 14.04, April 2009.

[71] European Wind Energy Association. Wind energy – the facts, part II: Grid integration, 2009, http://www.wind-energy-the-facts.org/en

[72] Bagepalli, B., Barnes, G.R., Gadre A.D., Jansen, P.L., Bouchard Jr., C.G., Jarczynski, E.D. & Garg, J., Wind turbine generators having wind assisted cooling systems and cooling methods, US Patent 7,427,814, 2008.

[73] Fischer, T. & Vilsbøll, N., Offshore wind turbine with liquid-cooling, US Patent 6,520737, 2003.

[74] University of Colorado at Boulder. Electrical energy storage, http://www.colorado.edu/engineering/energystorage/intro.html

[75] Daniel, C., Materials and processing for lithium-ion batteries. *JOM*, **60(9)**, pp. 43–48, 2008.

[76] Wen, Z.Y., Cao, J.D., Gu, Z.H., Xu, X.H., Zhang, F.L. & Lin, Z.X., Research on sodium sulphur battery for energy storage. *Solid State Ionics*, **179**, 2008.

[77] Tamakoshi, T. & Atsumi, S., Operational guidance device of sodium-sulphur battery, US Patent Application 20,080,206,626, 2008.

[78] US Department of Energy, Office of Science. Basic research needs for electrical energy storage, 2007, http://www.er.doe.gov/bes/reports/files/EES_rpt.pdf

[79] Kensche, C.W., Fatigue of components for wind turbines. *International Journal of Fatigue*, **28(10)**, pp. 1363–1374, 2006.

[80] EWEA, Wind power economics, http://www.ewea.org/fileadmin/ewea_documents/documents/press_releases/factsheet_economy2.pdf

[81] Wald, M.L., Cost works against alternative and renewable energy sources in time of recession, *The New York Times*, March 28, 2009, http://www.nytimes.com/2009/03/29/business/energy-environment/29renew.html

[82] Logan, J. & Kaplan, S.M., Wind power in the United States: technology, economic, and policy issues, 2008, http://www.fas.org/sgp/crs/misc/RL34546.pdf

[83] Enercon, Installing giant windmills, Windblatt Magazine, Issue 4, 2008.

[84] Sievert, T., World's largest offshore wind energy turbine to be developed in North England, http://www.windfair.net/press/4254.html

[85] Echavarria, E., Hahn, B. & van Bussel, G.J.W., Reliability of wind turbine technology through time. *Journal of Solar Energy Engineering*, **130(3)**, 2008.

[86] PESWiki, PowerPedia: Wind turbine, http://peswiki.com/index.php/PowerPedia:Wind_Turbine

[87] Brown, D.A. & Hefner, R.E., Corrosion protection for wind turbine unites in a marine environment, US Patent 7,230,347, 2007.

[88] Thick, J., Offshore corrosion protection of wind farms, http://www.2004ewec.info/files/23_1400_jamesthick_01.pdf

[89] Bang, D., Polinder, H., Shrestha, G. & Ferreira, J.A., Review of generator systems for direct-drive wind turbines. *Proc. of the 2008 European Wind Energy Conf.*, Brussels, April 2008.

[90] Polinder, H., Bang, D., van Rooij, R.P.J.O.M., McDonald, A.S. & Mueller, M.A., 10MW wind turbine direct drive generator design with pitch and active stall control. *Proc. of International Electric Machines & Drives Conf.*, IEEE International, 2007.

[91] De Vries, E., Siemens tests its direct drive: will direct drive bring down the cost of energy from large-scale turbines? *Renewable Energy World Magazine*, **11(5)**, 2008.

[92] Hau, E., *Wind Turbines: Fundamentals, Technologies, Application, Economics*, 2nd edition, Springer, 2005.

[93] Scherer, R., Blade design aspects. *Renewable Energy*, **16(1)**, pp. 1272–1277, Elsevier Science, 1999.

[94] Hillmer, H., Borstelmann, T., Schaffarczyk, P.A. & Dannenberg, L., Aerodynamic and structural design of multiMW wind turbine blades beyond 5MW. *Journal of Physics: Conference Series* **75**, 2007.

[95] Sandia National Laboratories. More efficient wind turbine blade designed, http://www.sciencedaily.com/releases/2007/03/070319180042.htm

[96] Purdue University. 'Smart turbine blades' to improve wind power, http://esciencenews.com/articles/2009/05/01/smart.turbine.blades.improve.wind.power

[97] Engström, S., Hernnäs, B., Parkegren, C. & Waernulf, S., Development of NewGen – a new type of direct-drive generator. *Proc. of Nordic Wind Power Conf.*, Espoo, Finland, May 2006.

[98] Lee, K., Sclavounos, P.D. & Wayman, E.N., Floating wind turbines. *20th Workshop on Water Waves and Floating Bodies*, Spitsbergen, Norway, 2005.

[99] Sclavounos, P.D., Tracy, C. & Lee, S., Floating offshore wind turbines: responses in a seastate, Pareto optimal designs and economic assessment, MIT.

[100] Adam, D., Floating wind turbines poised to harness ocean winds, http://www.guardian.co.uk/environment/2008/jul/16/windpower. renewableenergy

[101] Sway, Principles of concept, http://sway.no/index.php?id=16

[102] Vanderover, J.S. & Visser, K.D., Analysis of a contra-rotating propeller driven transport aircraft, ftp://ftp.clarkson.edu/.depts/mae/public_html/papers/vanderover.pdf

[103] Schönball, W., Electrical generator arrangement, US Patent 3,974,396, 1976.

[104] McCombs, J.C., Machine for converting wind energy to electrical energy, US Patent 5,506,453, 1996.

[105] Wachinski, A., Drive device for a windmill provided with two counter-rotative propellers, US Patent 7,384,239, 2008.

[106] Shin, C. & Hur, M.C., Over-drive gear device, US Patent 5,222,924, 1993.

[107] Shin, C., Multi-unit rotor blade system integrated wind turbine, US Patent 5,876,181, 1999.

[108] Kowintec Corp., http://www.kowintec.com/english/products/main.htm

[109] Korane, K.J., Flexible gears bolster wind-turbine reliability. *Machine Design*, **79(15)**, pp. 24–28, 2007.

[110] Fox, G., Epicyclic gear system, US Patent 7,056,259, 2006.

[111] Musial, W., Butterfield, S. & McNiff, B., Improve wind turbine gearbox reliability. *2007 European Wind Energy Conf.*, Milan, Italy, 2007.

[112] Gburčik, P., Gburčik, V., Gavrilov, M., Srdanovič, V. & Mastilović, S., Complementary regimes of solar and wind energy in Serbia. *Geographica Pannonica*, **10**, pp. 22–25, 2006.

[113] Muralikrishna, M. & Lakshminarayana, V., Hybrid (solar and wind) energy systems for rural electrification. *ARPN Journal of Engineering and Applied Science*, **3(5)**, pp. 50–58, 2008.

[114] Wang, M.H., Project of wind-solar energy storage system locating at Zhangbei, *People's Daily – Overseas Edition* (in Chinese), June 30, 2009.

[115] Ragan, S., UK's first hydrogen powered building, March 2, 2009, http://www.v1energy.com/articles/features/217-uks-first-hydrogen-powered-building

[116] Danvest Energy. Wind-diesel instructions, http://www.danvest.com/filesfordownload/wind-diesel.pdf

[117] Wikipedia, Wind-diesel hybrid power systems, http://en.wikipedia.org/wiki/Wind-Diesel_Hybrid_Power_Systems

CHAPTER 2

Wind resource and site assessment

Wiebke Langreder
Wind & Site, Suzlon Energy, Århus, Denmark.

Wind farm projects require intensive work prior to the finalizing of a project. The wind resource is one of the most important factors for the financial viability of a wind farm project. Wind maps representing the best estimate of the wind resource across a large area have been produced for a wide range of scales, from global down to local government regions. They do not substitute for wind measurements – rather they serve to focus investigations and indicate where on-site measurements would be merited. This chapter explains how wind resource can be assessed. The steps in this process are explained in detail, starting with initial site identification. A range of aspects concerning wind speed measurements is then covered including the choice of sensors, explaining the importance of proper mounting and calibration, long-term corrections, and data analysis. The difficulty of extrapolating the measured wind speed vertically and horizontally is demonstrated, leading to the need for flow models and their proper use. Basic rules for developing a layout are explained. Having analysed wind data and prepared a layout, the next step is energy yield calculation. The chapter ends by exploring various aspects of site suitability.

1 Initial site identification

The wind resource is one of the most critical aspects to be assessed when planning a wind farm. Different approaches on how to obtain information on the wind climate are possible. In most countries where wind energy is used extensively, some form of general information about the wind is available. This information could consist of wind maps showing colour coded wind speed or energy at a specific height. These are often based on meso-scale models and in the ideal case, are validated with ground-based stations. The quality of these maps varies widely and depends on the amount and precision of information that the model has been fed with, the validation process and the resolution of the model.

Wind atlases are normally produced with Wind Atlas Analysis and Application Program (WAsP, a micro-scale model, see Section 4.3) or combined models which involve the use of both meso- and micro-scale models, and are presented as a collection of wind statistics. The usefulness of these wind statistics depends very much on the distance between the target site and the stations, the input data they are based on, the site as well as on the complexity of the area, both regarding roughness and orography. Typically the main source of information for wind atlases is meteorological stations with measurements performed at a height of 10 m. Meso-scale models additionally use re-analysis data (see Section 3.1.1). Care has to be taken since the main purpose of these data is to deliver a basis for general weather models, which have a much smaller need for high precision wind measurements than wind energy. Thus the quality of wind atlases is not sufficient to replace on-site measurements [1].

Nature itself frequently gives reasonable indications of wind resources. Particularly flagged trees and bushes can indicate a promising wind climate and can give valuable information on the prevailing wind direction.

A very good source of information for a first estimate of the wind regime is production data from nearby wind farms, if available.

No other step in the process of wind farm development has such significance to the financial success as the correct assessment of the wind regime at the future turbine location. Because of the cubic relationship between wind speed and energy content in the wind, the prediction of energy output is extremely sensitive to the wind speed and requires every possible attention.

2 Wind speed measurements

2.1 Introduction

The measured wind climate is the main input for the flow models, by which you extrapolate the spot measurement vertically and horizontally to evaluate the energy distribution across the site. Such a resource map is the basis for an optimised layout. The number and height of the measurement masts should be adjusted to the complexity of the terrain as with increasing complexity, the capability of flow models to correctly predict the spatial variation of the wind decreases. The more complex the site, the more and the higher masts have to be installed to ensure a reasonable prediction of the wind resource.

Unfortunately wind measurements are frequently neglected. Very often the measurement height is insufficient for the complexity of the site, the number of masts is insufficient for the size of the site, the measurement period is too short, the instruments are not calibrated, the mounting is sub-standard or the mast is not maintained. It cannot be stressed enough that the most expensive part when measuring wind is the loss of data. Any wind resource assessment requires a minimum measurement period of one complete year in order to avoid seasonal biases. If instrumentation fails due to lightning strike, icing, vandalism or other reasons and the failure is not spotted rapidly, the lost data will falsify the results and as a consequence the measurement period has to start all over again. Otherwise the increased uncertainty might jeopardise the feasibility of the whole project.

2.2 Instruments

2.2.1 General

Wind speed measurements put a very high demand on the instrumentation because the energy density is proportional to the cube of the mean wind speed. Furthermore, the instruments used must be robust and reliably accumulate data over extended periods of unattended operation. The power consumption should be low so that they can operate off the grid.

Most on-site wind measurements are carried out using the traditional cup anemometer. The behaviour of these instruments is fairly well understood and the sources of error are well known. In general, the sources of error in anemometry include the effects of the tower, boom and other mounting arrangements, the anemometer design and its response to turbulent and non-horizontal flow characteristics, and the calibration procedure. Evidently, proper maintenance of the anemometer is also important. In some cases, problems arise due to icing of the sensor, or corrosion of the anemometer at sites close to the sea. The current version of the internationally used standard for power curve measurements, the IEC standard 61400-12-1 [2], only permits the use of cup anemometry for power curve measurements. The same requirements for accuracy are valid for wind resource measurements. Therefore it is advisable to use also these instruments for wind resource assessment.

Solid state wind sensors (e.g. sonics) have until recently not been used extensively for wind energy purposes, mainly because of their high cost and a higher power consumption. These have a number of advantages over mechanical anemometers and can further provide measurements of turbulence, air temperature, and atmospheric stability. However, they also introduce new sources of error which are less known, and the overall accuracy of sonic anemometry is lower than for high-quality cup anemometry [3].

Recently, remote sensing devices based either on sound (Sodar) or on laser (Lidar) have made an entry into the market. Their clear merit is that they replace a mast which can have practical advantages. However, they often require more substantial power supplies which bring other reliability and deployment issues. Also more intensive maintenance is required since the mean time between failures does not allow unattended measurements for periods required for wind resource assessment. While the precision of a Lidar seems to be superior to the Sodar, and often comparable to cup anemometry [4], both instruments suffer at the moment from short-comings in complex terrain due to the fact that the wind speed sampling takes place over a volume, and not at a point.

Remote sensing technologies are currently evolving very rapidly and it is expected they will have a significant role to play in the future.

2.2.2 Cup anemometer

The cup anemometer is a drag device and consists typically of three cups each mounted on one end of a horizontal arm, which in turn are mounted at equal angles to each other on a vertical shaft. A cup anemometer turns in the wind because the drag coefficient of the open face cup is greater than the drag coefficient of the smooth surface of the back. The air flow past the cups in any horizontal direction turns the cups in a manner that is proportional to the wind speed. Therefore, counting the turns of the cups over a set time period produces the average wind speed for a wide range of speeds.

Despite the simple geometry of an anemometer its measurement behaviour depends on a number of different factors. One of the most dominant factors is the so-called angular response, which describes what components of the wind vector are measured [3]. A so-called vector anemometer measures all three components of the wind vector, the longitudinal, lateral and vertical component. Thus this type of anemometer measures independently of the inflow angle and is less sensitive to mounting errors, terrain inclination and/or thermal effects. However, for power curve measurements the instrument must have a cosine response thus measuring only the horizontal component of the wind [2]. Since for energy yield calculations the measurement behaviour of the anemometer used for the power curve and used for resource assessment should be as similar as possible, it is advisable to also use an anemometer with a cosine response for resource assessment. One of the key arguments for using such an instrument for power curve measurements is that the wind turbine utilises only the horizontal component. This is, however, a very simplified approach as, particularly for large rotors, three-dimensional effects along the blades leads to a utilisation of energy from the vertical component. Care has to be taken when using a cosine response anemometer as it is sensitive to mounting errors.

One of the most relevant dynamic response specifications is the so-called overspeeding. Due mainly to the aerodynamic characteristics of the cups, the anemometer tends to accelerate faster than it decelerates, leading to an over-estimate of wind speed particularly in the middle wind speed range.

Another dynamic response specification is the response length or distance constant, which is related to the inertia of the cup anemometer. The dynamic response can be described as a first order equation. When a step change of wind speed from U to $U + \Delta u$ hits the anemometer it will react with some delay of exponential shape. The distance constant, i.e. the column of air corresponding to 63% recovery time for a step change in wind speed, should preferably be a few meters or less. Different methods to determine the response length are described in [3].

2.2.3 Ultrasonic anemometer

Ultrasonic or sonic anemometers use ultrasonic waves for measuring wind speed and, depending on the geometry, the wind direction. They measure the wind speed based on the time of flight between pairs of transducers. Depending on the number of pairs of transducers, either one-, two- or three-dimensional flow can be measured. The travelling time forth and back between the transducers is different because in one direction the wind speed component along the path is added to the

sound speed and subtracted from the other direction. If the distance of the transducers is given with s and the velocity of sound with c then the travelling times
can be expressed as

$$t_1 = \frac{s}{c+u} \quad \text{and} \quad t_2 = \frac{s}{c-u} \tag{1}$$

These equations can be re-arranged to eliminate c and to express the wind speed
u as a function of t_1, t_2 and s. The sole dependency on the path length is advantageous, as the speed of sound depends on air density and humidity:

$$u = \frac{s}{2}\left(\frac{1}{t_1} - \frac{1}{t_2}\right) \tag{2}$$

It can be seen that once u is known, c can be calculated and from c the temperature can be inferred (slightly contaminated with humidity, this is known as the
"sound virtual temperature"). The spatial resolution is determined by the path
length between the transducers, which is typically 10–20 cm. Due to the very fine
temporal resolution of 20 Hz or better the sonic anemometer is very well suited for
measurements of turbulence with much better temporal and spatial resolution than
cup anemometry.

The measurement of different components of the wind, the lack of moving parts,
and the high temporal resolutions make the ultrasonic anemometer a very attractive wind speed measurement device. The major concern, inherent in sonic anemometry, is the fact that the probe head itself distorts the flow – the effect of which
can only be evaluated in detail by a comprehensive wind tunnel investigation. The
transducer shadow effect is a particularly simple case of flow distortion and a well-
known source of error in sonics with horizontal sound paths. Less well known are
the errors associated with inaccuracies in probe head geometry and the temperature sensitivity of the sound transducers. The measurement is very sensitive to
small variations in the geometry, either due to temperature variations and/or
mechanical vibrations due to wind. Finally, specific details in the design of a given
probe head may give rise to wind speed-dependent errors.

2.2.4 Propeller anemometer

A propeller anemometer typically has four helicoid-shaped blades. This propeller
can either be mounted in conjunction with a wind vane or in a fixed two- or three-
dimensional arrangement (Fig. 1). While a cup anemometer responds to the differential drag force, both drag and lift forces act to turn the propeller anemometer.
Similar to a cup anemometer the response of the propeller anemometer to slow
speed variations is linear above the starting threshold.

Propeller anemometers have an angular response that deviates from cosine. In fact
the wind speed measured is somewhat less than the horizontal component [5]. If a
propeller is used in conjunction with a vane the propeller is in theory on average
oriented into the wind and thus the angular response is not so relevant. However,
the vane often shows an over-critical damping which leads to misalignment and

Figure 1: Propeller anemometer in fixed three-dimensional arrangement.

thus to an under-estimate of the wind speed. If propellers are mounted in a fixed arrangement the under-estimate of the wind speed is even more significant as the axis of the propeller is not aligned with the wind direction.

2.2.5 Remote sensing

An alternative to mast-mounted anemometry are ground-based remote sensing systems. Two systems have found some acceptance in the wind energy community: Sodar and Lidar. Both the Sodar (SOund Detection And Ranging) and the Lidar (LIght Detection And Ranging) use remote sensing techniques based respectively on sound and light emission, in combination with the Doppler effect. The signal emitted by the Sodar is scattered by temperature fluctuations while the signal emitted by a Lidar is scattered by aerosols. In contrast to the very small measurement volume of a cup anemometer, both remote sensing devices measure large volumes, which change with height. Both types require significantly more power than a cup anemometer making the use of a generator necessary (for the majority of models) if no grid is available.

2.2.5.1 Sodar

Different types of Sodars are available with different arrangements of the loudspeaker and receiver. Most commonly the sound pulse generated by a loudspeaker array can be tilted by electronically steering the array to different directions (phased array Sodar). The combination of three beams, one in the vertical direction and

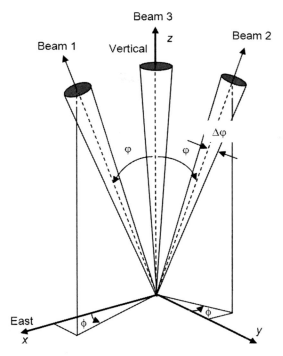

Figure 2: The principle of a three-beam phased array Sodar [7].

two others, tilted to the vertical and perpendicular to each other's planes produce the three-dimensional velocity field (Fig. 2).

One of the drawbacks of a Sodar is its dependency on temperature fluctuations which in turn means that a Sodar is at a disadvantage under neutral atmospheric conditions, which are related to high wind speed situations [8]. Another type of Sodar, the bistatic, reacts on velocity inhomogeneities rather than temperature fluctuations and does not exhibit this problem. However, there is no commercially available bistatic Sodar and the principle requires separate, spatially separated transmitter and receiver, which makes it less practical than conventional Sodar.

The signal-to-noise ratio of the Sodar is also known to deteriorate with height, resulting in a reduced number of valid signal returns. Thus a measured profile is difficult to interpret as the profile might be based on a different number of measurements for each height.

At the same time, due to the tilt of the beams, the measurement takes place in three non-overlapping volumes. Assuming a typical tilt angle of 17°, the distance between the tilted volumes and the distance between the vertical and a tilted volume is presented in Table 1 for a number of altitudes. As a consequence, the interpretation of a measured profile becomes difficult because the measurement volume changes with height, especially in complex terrain.

Since the Sodar is an acoustic system, the presence of noise sources can influence the Sodar's function. The source of noise could be rain but any other noise,

Table 1: The distance between the three beams at a number of heights [9].

Altitude (m)	Distance (m)	
	Tilt–vertical	Tilt–tilt
40	11.7	16.5
80	23.4	33.1
120	35.1	49.6
160	46.8	66.2
200	58.5	82.7

for example from animals, can have an adverse effect. A particularly critical issue is the increased background noise due to high wind speeds [9]. False echoes from the Sodar's enclosure or nearby obstacles can also lead to a falsified signal. Other parameters which may influence the Sodar measurement are errors in the vertical alignment of the instrument, temperature changes at the antenna and, especially for three-beam Sodars, changes in wind direction [8].

The measurement accuracy of Sodar systems cannot match that of cup anemometry and unless high acoustical powers are used, their availability falls in high wind speeds. A hybrid system comprising a moderately tall mast (say 40 m) and a relatively low power (30–100 W electrical) Sodar has many attractive features – the high absolute accuracy and high availability of the cup anemometer complements the less accurate but highly relevant vertical resolution obtained from the Sodar [10].

2.2.5.2. Lidar

Until recently, making wind measurements using Lidars was prohibitively expensive and essentially limited to the aerospace and military domain. Most limitations were swept aside by the emergence of coherent lasers at wavelengths compliant with fibre optic components (so-called 'fibre lasers'). Since light can be much more precisely focused and spreads in the atmosphere much less than sound, Lidar systems have an inherently higher accuracy and better signal to noise ratio than a Sodar. The Lidar works by focusing at a specific distance and it measures the scattering from aerosols that takes place within the focal volume.

The operation of the Lidar is influenced by atmospheric conditions (e.g. fog, density of particles in the air). Lack of particles influences its response, sometimes prohibiting measurement while fog can severely attenuate the beam before it reaches the measurement height. Rain also reduces the Lidar's ability to measure, as scattering from the falling droplets can result in errors in the wind speed, particularly its vertical components. Other parameters that influence the measurement are, as in the case of the Sodar: errors in the vertical alignment of the instrument and uncertainties in the focusing height. The Lidar,

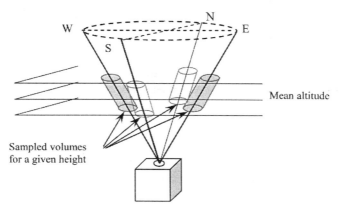

Figure 3: Working principle of a pulsed Lidar.

being an optical instrument, is also susceptible to influences from the presence of dirt on the output window; hence there is a need for a robust cleaning device of the Lidar window [9].

Two working principles of fibre-base Lidars are in use: one uses a continuous-wave system with height discrimination achieved by varying focus. The laser light is emitted through a constantly rotating prism giving a deflection of 30° from the vertical. This Lidar system scans a laser beam about a vertical axis from the ground, intercepting the wind on a 360° circumference. By adjusting the laser focus, winds may be sampled at a range of heights above ground level.

The other system uses a pulsed signal with a fixed focus. It has a 30° prism to deflect the beam from the vertical but here the prism does not rotate continuously. Instead, the prism remains stationary whilst the Lidar sends a stream of pulses in a given direction, recording the backscatter in a number of range gates (fixed time delays) triggered by the end of each pulse [10] (see Fig. 3).

Unlike pulsed systems, continuous-wave Lidar systems do not inherently 'know' the height from which backscatter is being received. A sensibly uniform vertical profile of aerosol concentration has to be assumed in which case the backscattered energy is from the focused volume. The obtained radial wind speed distribution in this case is dominated by the signal from the set focus distance. The assumption of vertical aerosol homogeneity unfortunately fails completely in the fairly common case of low level clouds (under 1500 m). Here, the relatively huge backscatter from the cloud base can be detected even though the cloud is far above the focus distance. The resulting Doppler spectrum has two peaks – one corresponding to the radial speed at the focused height and a second corresponding to the (usually) higher speed of the cloud base. Unless corrected for, this will introduce a bias to the wind speed measurement. For this reason, the continuous-wave Lidar has a cloud-correction algorithm that identifies the second peak and rejects it from the

Table 2: Beam half-length versus focal distance of a continuous-wave Lidar [11].

Altitude (m)	Beam half-length, L (m)
40	2.5
60	6
100	16
200	65

spectrum. An extra second scan with a near-collimated output beam is inserted into the height cycle. The spectra thus measured are used to remove the influence of the clouds at the desired measuring heights [10].

The two systems have fundamental differences concerning the measurement volume. The continuous-wave system adjusts the focus so winds may be sampled at a range of heights above ground level. The backscattered signal comes mainly from the region close to the beam focus, where the signal intensity is at its maximum. While the width of the laser beam increases in proportion to focus height, its probe length increases non-linearly (roughly the square of height; Table 2). The vertical measuring depth of the pulsed system depends on the pulse length and is constant with sensing range [10]. A continuous-wave system can measure wind speed at heights from less than 10 m up to a maximum of about 200 m. Pulsed Lidars are typically blinded during emission of the pulse and this restricts their minimum range to about 40 m, with the maximum range usually limited only by signal-to-noise considerations and hence dependent on conditions.

Like the Sodar, both Lidar systems rely on the assumption of horizontally homogeneous flow. In complex terrain this assumption is violated, increasingly so as the terrain complexity increases. There are indications that errors of 5–10% in the mean speed are not uncommon [12–14]. Only a multiple Lidar system, in which units are separated along a suitably long baseline, could eliminate this inherent error as explained above.

In general, care has to be taken when performing short-term measurements with remote sensing devices. The vertical profile varies significantly with different atmospheric stabilities. Thus the measurement campaign using remote sensing should, similarly to cup anemometry, be a minimum of 1 year.

Currently work is in progress for a Best Practice Guideline for the use of remote sensing.

2.3 Calibration

As explained in Section 2.2.2, the turns of an anemometer are transformed into a wind speed measurement by a linear function. The scale and offset of this transfer function are determined by wind tunnel calibration of the anemometer. Strict requirements concerning the wind tunnel test are specified in [2]. Please note that

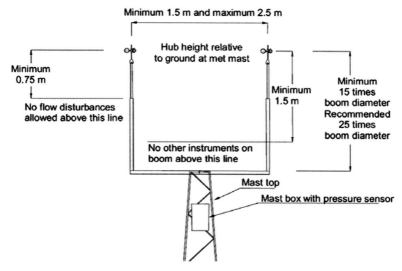

Figure 4: Example top mounted anemometers [2].

such strict calibration procedures are only in place for cup anemometers. Highest quality calibrations are ensured when calibrating in wind tunnels that have been accredited by MEASNET. MEASNET members participate regularly in a round robin test to guarantee interchangeability of the results, which has increased the quality of the calibration significantly. It should be kept in mind that even calibrations according to highest standard bear an uncertainty of around 1–2%.

Currently there are no calibration standards available for sonics and remote sensing devices.

2.4 Mounting

Accurate wind speed measurements are only possible with appropriate mounting of the cup anemometry on the meteorological mast. In particular, the anemometer shall be located such that the flow distortion due to the mast and the side booms is minimised. The least flow distortion is found by mounting the anemometer on top of the mast at a sufficient distance to the structure. Other instruments, aviation lighting, and the lightning protection should be mounted in such a way that interference with the anemometer is avoided. Figure 4 shows a possible top mounting arrangement.

Boom-mounted anemometers are influenced by flow distortion of both the mast and the boom. Flow distortion due to the mounting boom should be kept below 0.5% and flow distortion due to the mast should be kept below 1%. If the anemometer is mounted on a tubular side boom, this can be achieved by mounting the anemometer 15 times the boom diameter above the boom. The level of flow distortion due the mast depends on the type of mast and the direction the anemometer is facing with respect to the mast geometry and the main wind direction.

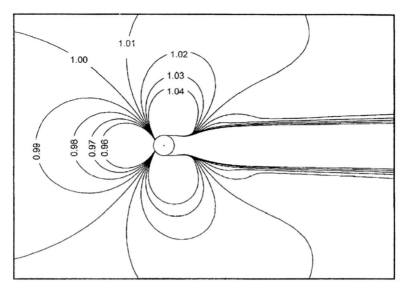

Figure 5: Iso-speed plot of local flow speed around a cylindrical mast, normalised by free-field wind speed (from left); analysis by two-dimensional
Navier-Stokes computations [3].

Figure 5 shows an example of flow distortion around a tubular mast. It can be seen
that there is a deceleration of the flow upwind to the mast, acceleration around it and
a wake behind it. The least disturbance can be seen to occur if facing the wind at 45°.

The flow distortion around a lattice mast is somewhat more complicated to
determine. Additionally to the orientation of the wind and the distance of the anemometer to the centre of the mast it also depends on the solidity of the mast and
the drag. Figure 6 shows an example of flow distortion. Again a deceleration in
front of the mast can be observed while there is acceleration at the flanks. Minimum distortion is achieved when the anemometer is placed at an angle of 60°.
More details on how to determine the flow distortion can be found in [3] or [2].

In general, mounting of the anemometer at the same height as the top of the
mast should be avoided since the flow distortion around the top of the mast is
highly complex and cannot be corrected for.

2.5 Measurement period and averaging time

The energy yield is typically calculated referring to the annual mean wind speed
of the site. Unfortunately the annual average wind speed varies significantly.
Depending on the local climate the annual averages of wind speed might vary
around ±15% from one year to the next. To reduce the uncertainty of the inter-
annual variability it is strongly recommended to perform a long-term correction
of the measured data (see Section 3.1). On a monthly scale, the variations of the

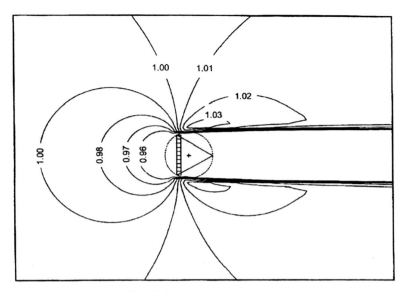

Figure 6: Iso-speed plot of local flow speed around a triangular lattice mast with a C_T of 0.5 normalised by free-field wind speed (from left); analysis by two-dimensional Navier-Stokes computations and actuator disk theory [3].

wind speed are much more significant, reaching – depending on the local climate – variations of ±50%. Thus it is of crucial importance to measure complete years since the monthly variations are too large to be corrected and, as a consequence, partial year data suffers from a seasonal bias.

Wind data is normally sampled for 10 min, the logger then calculates the averages and the standard deviations, the latter being necessary for determining the turbulence intensity. The averaging period of 10 min allows the direct use of the data for load calculations as they also refer to 10-min averages. Great care has to be taken when the averaging period is shorter than 10 min as the standard deviation and thus the turbulence intensity will be lower in comparison to 10-min averaging period. This can lead to under-estimated turbulence-induced loads.

3 Data analysis

3.1 Long-term correction

3.1.1 Introduction
Variability is an intrinsic feature of climate. Precautions are necessary as the weather changes from year to year and between consecutive decades. The data which form the basis of any wind resource study cover a limited period of time, which in many cases is less than 5 years. The question therefore arises: to what extent is that period representative for the longer-term climate. A study of climatic

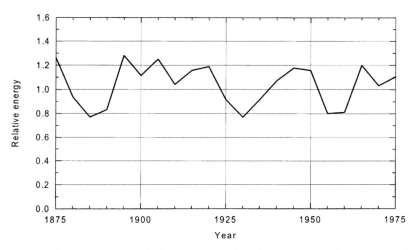

Figure 7: Mean energy in wind for consecutive 5-year periods based on a time series from Hesselø, Denmark, 1873–1982 [17].

variability in Northern Europe shows that variations in wind energy of up to 30% can be expected from one decade to another (see Fig. 7). In another study [15] it was found from an analysis of the expected power output for a 45 m high wind turbine over a 22-year period that the inter-annual variation in power corresponds to a mean relative standard deviation of approximately 13%. For the proper assessment of the economics of wind power utilisation, such variability must obviously be borne in mind [16].

Any short-term measurement should therefore be correlated with long-term data covering ideally 30 years from a representative station. Unfortunately it is very difficult in most cases to obtain reliable, consistent long-term data since weather stations are frequently subject to changes in environment by, for example, growing vegetation or building activity. Therefore – and particularly in very remote areas – the use of re-analysis data might be a feasible alternative. Most commonly used is NCEP/NCAR data from NCEP (National Centre for Environmental Prediction) and NCAR (National Center for Atmospheric Research) in the US. These institutes continuously perform a global re-analysis of weather data. The objective is to produce homogeneous data sets covering a decade or more of weather analysis with the same data assimilation systems. This means that data from synoptic weather stations, radio sondes, pilot balloons, aircraft, ships, buoys and satellites are collected, controlled, gridded, and prepared for initialisation of a global numerical weather prediction model. By running past-periods on a global scale, the data is re-analysed in a globally consistent manner which enables it to be used in some situations for long-term correlation purposes.

The advantages are that the period of available re-analysed data can be much longer than the period of available on-site data, the data is neither subject to icing nor seasonal influences from vegetation and finally there should be no long-term equipment trends. However the quality of re-analysis data still depends on the quality of the input data, with the result that in sparsely instrumented regions, the re-analysis data can still suffer from deficiencies in individual instrumentation and data coverage. The same care should be applied to the use of re-analysis data, as with ground-based data. The NCEP/NCAR data is available in the form of pressure and surface wind data in a 2.5° grid corresponding to a spacing of approximately 250 km. The data consists of values of wind speed and direction for four instantaneous values per day (every 6 hours).

The statistical method for long-term correcting data is called Measure-Correlate-Predict or MCP. This method is based on the assumption that the short- and long-term data sets are correlated. This correlation can be established in different ways depending on the data quality and the comparability of the two wind climates.

3.1.2 Regression method

If one mast is correlated with a second on-site mast, a linear regression either omnidirectional or by wind direction sectors (typically 30°) might be best suited. For the concurrent period, the wind speeds are plotted versus each other and a linear regression based on the least-square fit is established. This relationship is used to extend the shorter data set with synthetic data based on the longer data set. The regression coefficient is a measure of the quality of the correlation. R^2 should not be less than 70%. The same method might be appropriate for a short-term measurement on-site and a reference station in some distance if the orography and the wind roses are closely related. If the wind roses vary, care has to be taken since the wind rose of the reference station will be transferred to the site by applying a linear regression, which can have a significant impact on the layout as well as the energy yield. Another inherent problem of this methodology is the decreasing temporal correlation between the site and the reference station with increasing distance. The introduction of averaging of the two data sets can improve the correlation.

3.1.3 Energy index method

Rather than transposing the wind distribution from the reference station the energy index method determines a correction factor for the short-term data. For the concurrent period, the energy level of the reference data set is determined and compared with the long-term energy. The resulting ratio is then applied as correction to the short-term on-site data set. The correlation is best proven comparing monthly mean wind speeds of the two data sets.

This method has the main advantage that the on-site measured wind rose is not altered. The energy index method is particularly suited for NCEP/NCAR data since the low temporal resolution of NCEP/NCAR prohibits the use of the more detailed regression method. However, care has to be taken since NCEP/NCAR data represents only geostrophic wind and does not reflect local wind climates like wind tunnel effects across a mountain pass or thermal effects.

3.2 Weibull distribution

It is very important for the wind industry to be able to relatively simply describe the wind regime on site. Turbine designers need the information to optimise the design of their turbines, so as to minimise generating costs. Turbine investors need the information to estimate their income from electricity generation.

One way to condense the information of a measured time series is a histogram. The wind speeds are sorted into wind speed bins. The bin width is typically 1 m/s. The histogram provides information how often the wind is blowing for each wind speed bin.

The histogram for a typical site can be presented using the Weibull distribution expressing the frequency distribution of the wind speed in a compact form. The two-parameter Weibull distribution is described mathematically as

$$f(u) = \frac{k}{A}\left(\frac{u}{A}\right)^{k-1} \exp\left(-\left(\frac{u}{A}\right)^k\right) \tag{3}$$

where $f(u)$ is the frequency of occurrence of wind speed u. The scaling factor A is a measure for the wind speed while the shape factor k describes the shape of the distribution. The cumulative Weibull distribution $F(u)$ gives the probability of the wind speed exceeding the value v and is given by the simple expression:

$$F(u) = \exp\left(-\left(\frac{u}{A}\right)^k\right) \tag{4}$$

Following graph shows a group of Weibull distributions with a constant mean wind speed of 8 m/s but varying k factor. Note that high wind speeds become more probable with a low k factor.

The Weibull distribution can degenerate into two special distributions, namely for $k = 1$ the exponential distribution and $k = 2$ the Rayleigh distribution. Since observed wind data exhibits frequency distributions which are often well described by a Rayleigh distribution, this one-parameter distribution is sometimes used by wind turbine manufacturers for calculation of standard performance figures for their machines. Inspection of the k parameter shows that, especially for Northern European climates, the values for k are indeed close to 2.0.

On a global scale, the k factor varies significantly depending upon local climate conditions, the landscape, and its surface (Fig. 8). A low k factor (<1.8) is typical for wind climates with a high content of thermal winds. A high k factor (>2.5) is representative for very constant wind climates, for example trade winds. Both Weibull A and k parameters are dependent on the height and are increasing up to 100 m above ground (Fig. 9). Above 100 m the k parameter decreases.

The Weibull distribution is a probability density distribution. The median of the distribution corresponds to the wind speed that cuts the area into half. This means that half the time it will be blowing less than the median wind speed, the other half

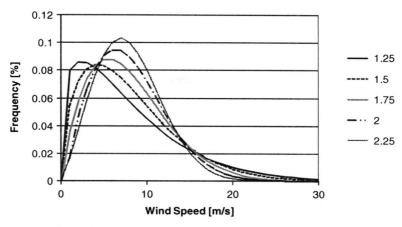

Figure 8: Weibull distributions for constant mean wind speed (8 m/s) and varying k factors.

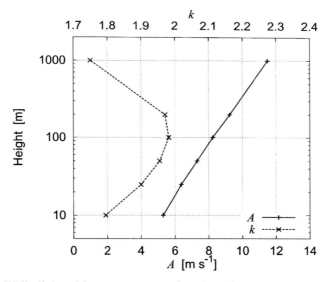

Figure 9: Weibull A and k parameters as a function of height for a roughness class 2 [16].

it will be blowing faster. The mean wind speed is the average of the distribution. The wind speed with the highest frequency is called the modal value.

Many different methods can be used for fitting the two Weibull parameters to a histogram. Since in general the observed histograms will show deviations, a fitting procedure must be selected which focuses on the wind speed range relevant to the application. Here the emphasis should be on the energy containing part of the spectrum. Thus a moment fitting method is normally used with focus on medium to high wind speeds but not extreme wind speeds. The requirements for determining the two Weibull parameters are that the total wind energy in the fitted Weibull distribution and the observed distribution are equal. Additionally the frequencies of occurrence of the wind speeds higher than the observed average wind speed have to be the same for both distributions. The combination of these two requirements leads to an equation in k only, which can be solved by a standard root-finding algorithm.

As a rough approximation, the relationship between the Weibull parameter A and k and the mean wind speed can be described as follows:

$$\overline{U} \approx A\left(0.568 + \frac{0.434}{k}\right)^{1/k} \tag{5}$$

4 Spatial extrapolation

4.1 Introduction

Analyses of wind speed measurements lead to a detailed description of the wind climate at one point. In order to calculate the energy yield of the whole wind field, the results from the measurements have to be extrapolated horizontally to cover the wind farm area, and vertically if the measurements were not performed at hub height. Normally for the spatial extrapolation, computer models are used which are primed with wind speed data measured on site.

In the following section, parameters affecting the vertical extrapolation are described. This is followed by a description of the concept of flow models, and particularly WAsP as the most commonly used model in the industry.

4.2 Vertical extrapolation

4.2.1 Introduction
The planetary boundary layer (PBL), also known as the atmospheric boundary layer (ABL), is the lowest part of the atmosphere and its behaviour is directly influenced by its contact with the planetary surface. Above the PBL is the "free atmosphere" where the wind is approximately geostrophic (parallel to the isobars) while within the PBL the wind is affected by surface drag and turns across the isobars. The free atmosphere is usually non-turbulent, or only intermittently turbulent. The surface layer is the lowest part of the ABL. Its height is normally taken as around 10% of the ABL height but varies significantly during seasons and day times.

The change of wind speed with height is described by the vertical wind profile, which is the key for extrapolation of the measured wind speed to hub height. Generally the wind speed increases with increasing height above the ground.

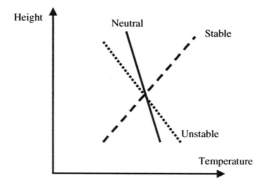

Figure 10: Vertical temperature profile: changes of temperature with height.

The vertical wind profile in the surface layer can be described by a number of simplified assumptions.

The profile depends, next to roughness and orography, on the vertical temperature profile, which is also referred to as atmospheric stability. Three general cases can be categorised (Fig. 10). In neutral conditions, the temperature profile is adiabatic meaning that there is equilibrium between cooling/heating and expansion/contraction with no vertical exchange of heat energy. The temperature decreases with around 1°C per 100 m in this situation. Neutral conditions are typical for high wind speeds. The vertical wind profile depends only on roughness and orography.

In unstable conditions, the temperature decreases with height faster than in the neutral case. This is typically the case during summer time where the ground is heated. As a result of the heating the air close to ground starts rising since the air density in the higher layers is lower (convective conditions). As a consequence, a vertical exchange of momentum is established leading to a higher level of turbulence. The vertical wind shear is generally small in these situations due to the heavy mixing.

In stable conditions, which are typical for winter or night time, the air close to the ground is cooler than the layers above. The higher air density with increasing height suppresses all vertical exchange of momentum. Thus turbulence is suppressed. The wind shear however can be significant since there is little vertical exchange. During these conditions large wind direction gradients can occur.

4.2.2 Influence of roughness

In neutral conditions and flat terrain with uniform roughness the vertical profile can be described analytically by the power law:

$$\frac{u(h_1)}{u(h_2)} = \left(\frac{h_1}{h_2}\right)^{a} \tag{6}$$

$u(h_1)$ and $u(h_2)$ are the wind speeds at heights h_1 and h_2. This is merely an engineering approximation where the wind shear exponent α is a function of height z, surface roughness, atmospheric stability and orography. Therefore, a measured wind shear exponent is only valid for the specific measurement heights and location and should thus never be used for vertical extrapolation of the wind speed, which is unfortunately very often done leading to erroneous results.

More helpful is the logarithmic law (log law) which in flat terrain and neutral conditions expresses the change of wind speed as a function of surface roughness:

$$u(h) = \frac{u_*}{\kappa} \ln\left(\frac{h}{z_0}\right)$$

(7)

The wind speed u depends on the friction velocity u_*, the height above ground h, the roughness length z_0 and the Kármán constant κ, which equals 0.4. Applying the above equation for two different heights and knowing the roughness length z_0 allows the extrapolation of the wind speed to a different height:

$$u_2\left(h_2\right) = u_1\left(h_1\right)\frac{\ln(h_2 / z_0)}{\ln(h_1 / z_0)}$$

(8)

The surface roughness length describes the roughness characteristics of the terrain. It is formally the height (in m) at which the wind speed becomes zero when the logarithmic wind profile is extrapolated to zero wind speed. A corresponding system uses roughness classes. A few examples of roughness lengths and their corresponding classes are given in Table 3.

In offshore conditions the roughness length varies with the wave condition, which in turn is a function of wind speed, wind direction, fetch, wave heights and length. However, recent surveys have shown that the vertical profile offshore is heavily influenced by the effect of atmospheric stability [18], because the roughness length offshore will in most cases be several orders of magnitude smaller than onshore.

Table 3: Example surface roughness length and class.

Cover	$z0$ (m)	$z0$ as roughness class
Offshore	0.0002	0
Open terrain, grass, few isolated obstacles	0.03	1
Low crops, occasional large obstacles	0.10	2
High crops, scattered obstacles	0.25	2.7
Parkland, bushes, numerous obstacles	0.50	3.2
Regular large obstacle coverage (suburb, forest)	0.5–1.0	3.2–3.7

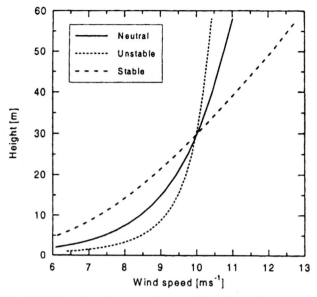

Figure 11: Wind profiles for neutral, unstable and stable conditions [16].

4.2.3 Influence of atmospheric stability

The logarithmic law presented above can be expanded to take atmospheric stability into account:

$$u(h) = \frac{u^*}{\kappa} \left(\ln\left(\frac{h}{z_0}\right) - \Psi \right) \qquad (9)$$

Ψ is a stability-dependent function, which is positive for unstable conditions and negative for stable conditions. The wind speed gradient is diminished in unstable conditions (heating of the surface, increased vertical mixing) and increased during stable conditions (cooling of the surface, suppressed vertical mixing). Figure 11 shows an example of the effect of atmospheric stability when extrapolating a measured wind speed at 30 m to different heights.

4.2.4 Influence of orography

The term orography refers to the description of the height variations of the terrain. While in flat terrain the roughness is the most dominant parameter, in hilly or mountainous terrain the shape of the terrain itself has the biggest impact on the profile.

Over hill or mountain tops the flow will be generally accelerated (Fig. 12). As a consequence the logarithmic wind profile will be distorted: both steeper and then less steep depending on height. The degree of distortion depends on the steepness

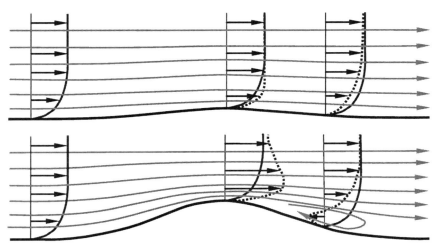

Figure 12: Effect of topography on the vertical wind speed profile gentle hill (top), steep slope (bottom).

of the terrain, on the surface roughness and the stability. In very steep terrain the flow across the terrain might become detached and form a zone of turbulent separation. As a rule of thumb this phenomena is likely to happen in terrain steeper than 30% corresponding to a 17° slope. The location and dimensions of the separation zone depend on the slope and its curvature as well as roughness and stability. In cases of separation, the wind speed profile might show areas with negative vertical gradient, where the wind speed is decreasing with height.

4.2.5 Influence of obstacles
Sheltering of the anemometer by nearby obstacles such as buildings leads to a distortion of the vertical profile. The effect of the obstacles depends on their dimensions, position and porosity.

Figure 13 sketches out the reduction of wind speed behind an infinite long two-dimensional obstacle. The hatched area relates to the area around the obstacle which is highly dependent on the actual geometry of the obstacle. The flow in this area can only be described by more advanced numerical models such as computational fluid dynamic (CFD) models.

4.3 Flow models

4.3.1 General
Due to the complexity of the vertical extrapolation described above, the prediction of the variation of wind speed with height is usually calculated by a computer model, which is specifically designed to facilitate accurate predictions of wind farm energy. These models also estimate the energy variation over the site area

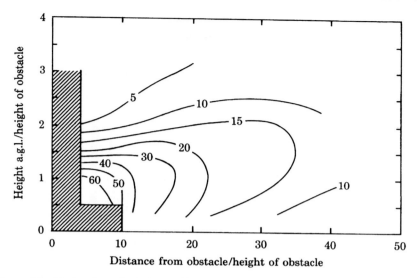

Figure 13: Reduction of wind speed in percent due to shelter by a two-dimensional obstacle [17].

and using separate physics, the wake interaction between the wind turbines. The use of such tools allows the energy production of different layouts, turbine types and hub heights to be rapidly established once the model has been set up. Site flow calculations are commonly undertaken using the WAsP model, which has been widely used within the industry over the past decades.

However in the last few years use of CFD codes is increasing, although CFD tools are typically used in addition to and not instead of more simple tools, to investigate specific flow phenomena at more complex sites. CFD tools must be used with care, as the results are quite sensitive to modelling assumptions and to the skill of the user of the code. Typical use of CFD tools is, firstly, to give another estimate of the local acceleration effects at the sites, and secondly, to identify hot spots, in other words areas where the wind conditions are particularly difficult for wind turbines. In particular, such tools are starting to be used to assist in the micro-siting of wind turbines on more complex sites [1]. Often it is the only means by which turbulence and shear across the site can be estimated.

4.3.2 WAsP
The challenge is to take a topographical map and the long-term corrected wind climate at a known point and use this information to calculate the long-term wind speed at all points on the map. WAsP is accomplishing this task using a double vertical and horizontal extrapolation. The idea behind this is quite simple. The local measurements on site are cleaned from local effects like obstacles, roughness and orography to calculate the geostrophic wind climate. Having determined the

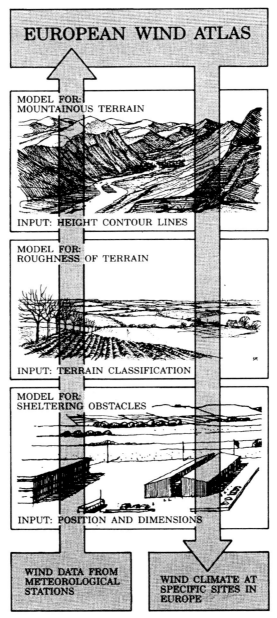

Figure 14: Principle of WAsP [17].

geostrophic wind climate this way in one position, the effect of the local obstacles, roughness and orography of the wind turbines location can be calculated for a different position assuming that the geostrophic wind climate is the same at this position. This principle is indicated in Fig. 14.

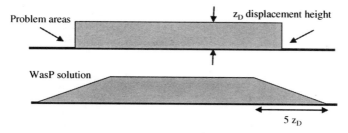

Figure 15: Modelling forest in WAsP: displacement height [19].

As explained above the vertical profile also depends on the atmospheric stability. Even at moderate wind speeds, deviations from the logarithmic profile occur when the height exceeds a few tens of meters. Deviations are caused by the effect of buoyancy forces in the turbulence dynamics; the surface roughness is no longer the only relevant surface characteristic, but has to be supplemented by parameters describing the surface heat flux. With cooling at night, turbulence is lessened, causing the wind profile to increase more rapidly with height; conversely, daytime heating causes increased turbulence and a wind profile more constant with height (see Section 4.2.3).

In order to take into account the effects of the varying surface heat flux without the need to model each individual wind profile, a simplified procedure was adopted in WAsP which only requires the climatological average and root mean square of the annual and daily variations of surface heat flux. This procedure introduces the degree of 'contamination' by stability effects to the logarithmic wind profile when conditions at different heights and surfaces are calculated [16].

It is important to appreciate that as the distance of the turbines from the meteorological mast increases, the uncertainty in the prediction also increases. This increase in uncertainty is typically more rapid in complex terrain than in simple terrain. When developing a site the increased uncertainty should be reflected in the number of measurement masts on site and the measurement height. As a rule of thumb the measurement height should be minimum 2/3 of the planned hub height.

A great challenge is the modelling of forests using WAsP. To model the wind speed correctly in WAsP a so-called displacement height must be introduced together with a very high roughness [19]. The displacement height is an artificial increase of terrain height for the area covered by forest. It should be around 2/3 of the tree height depending on the tree's density and the shape of the canopy.

At the edge of the forest the displacement height should taper off linearly out to a distance of five times the displacement height (Fig. 15). The displacement height shall correct for the speed up of the wind as the forest to some extent acts like an artificial hill leading to accelerated flow across the forest. The roughness length to be applied for forested areas should be in the order of 0.4 to more than 1 m. The increased roughness will lead to a wind profile exhibiting a higher shear when modelling the forest [19]. This approximation is only valid in simple terrain.

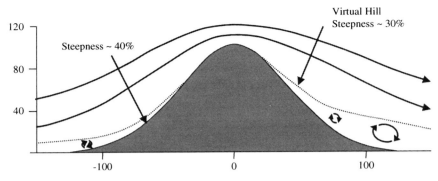

Figure 16: Effect of a steep hill – flow separation [21].

Table 4: Strengths and weaknesses of WAsP.

Strength	Weakness
Easy to use	Valid only for near-neutral conditions
Cheap and fast	Problems in complex terrain with flow separation
Validated, limitations are known and can be dealt with	Valid only for surface layer

One of the key simplifications of WAsP allowing this double extrapolation is the linearisation of the Navier-Stokes equations. This simplification is the main reason for WAsP's limitation in complex terrain where the terrain slope exceeds 30° and flow separation is likely to occur. Several studies have shown that WAsP tends to make prediction errors in complex terrain [20]. Figure 16 shows that due to the separation zones the stream lines follow the shape of a virtual hill with a somewhat reduced steepness. However, WAsP uses the real hill steepness to calculate the speed-up and is thus over-predicting the wind speed.

A correction of this model bias is possible using the so-called Ruggedness Index (RIX) which is defined as the percentage of the area around an object that has steepness above 30% (corresponds to 17° slope), thus the WAsP model assumptions are violated (Table 4). Figure 17 shows the relationship between expected wind speed error and the difference in complexity between the position of the measurement mast and the position of the future wind turbine. If the reference site (measurement site) is less rugged and the predicted sites (WTGs) are very rugged, the difference ΔRIX will be positive, and thus according to several studies an over-prediction of the wind speed can be expected. If the reference site (measurement site) is more rugged and the predicted sites (WTGs) are less rugged, the difference ΔRIX will be negative, and thus according to several studies an under-prediction of the wind speed can be expected. The model bias will vary from site to site. It must be emphasised that the ΔRIX description is a very much simplified description of the complexity variations in the terrain and that more data analysis is needed like that shown in Fig. 17.

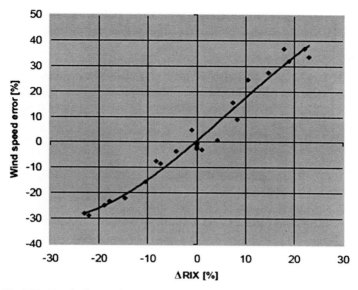

Figure 17: WAsP wind speed prediction error as a function of difference in ruggedness indices between the predicted and the predictor site [20].

5 Siting and site suitability

5.1 General

During the process of siting (also referred to as micro-siting), the locations of the future wind turbines are determined. Apart from the wind resource this process is driven by a number of other factors such as technical risks, environmental impact, planning restrictions and infrastructure costs.

Technical risks can very often be mitigated by adjusting the layout to suit the site-specific conditions. High turbulence as a main driver of fatigue loads can be avoided by maintaining sufficient distances between the turbines, keeping clear of forests and other turbulence-inducing terrain features like cliffs. Steep terrain slopes should be avoided to reduce stresses on the yaw system and blades and thereby improve the energy output.

5.2 Turbulence

5.2.1 Ambient turbulence

The turbulent variations of the wind speed are typically expressed in terms of the standard deviation σ_u of velocity fluctuations. This is measured over a 10-min period and normalised by the average wind speed $\bar{U}$ and is called turbulence intensity I_u:

$$I_u = \frac{\sigma_u}{\bar{U}} \tag{10}$$

The variation in this ratio is caused by a large natural variability, but also to some extent because it is sensitive to the averaging time and the frequency response of the sensor used.

Two natural sources of turbulence can be identified: thermal and mechanical. Mechanical turbulence is caused by vertical wind shear and depends on the surface roughness z_0. The mechanically caused turbulence intensity at a height h in neutral conditions, in flat terrain and infinite uniform roughness z_0 can be described as

$$I_u = \frac{1}{\ln(h / z_0)} \qquad (11)$$

This equation shows an expected decrease of turbulence intensity with increasing height above ground level.

Thermal turbulence is caused by convection and depends mainly on the temperature difference between ground and air. In unstable conditions, with strong heating of the ground, the turbulence intensity can reach very large values. In stable conditions, with very little vertical exchange of momentum, the turbulence is generally very low. The impact of atmospheric stability is considerable in low to moderate wind speeds.

The turbulence intensity varies with wind speed. It is highest at low wind speeds and shows an asymptotic behaviour towards a constant value at higher wind speeds (Fig. 18). Typical values of I_u for neutral conditions in different terrains at typical hub heights of around 80 m are listed in Table 5.

Forests cause a particularly high ambient turbulence and require special attention (Table 5). While special precautions allow estimating the mean wind speed in or near forest with standard flow models, the turbulence variations can only be modelled with more advanced models. Different concepts are available for more advanced CFD codes. One method to model forest is the simulation via an aerodynamic drag

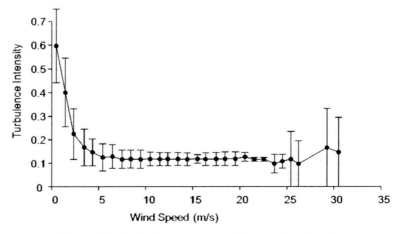

Figure 18: Turbulence versus wind speed (onshore).

Table 5: Typical hub height turbulence intensities for different land covers.

Land cover	Typical I_u (%)
Offshore	8
Open grassland	10
Farming land with wind breaks	13
Forests	20 or more

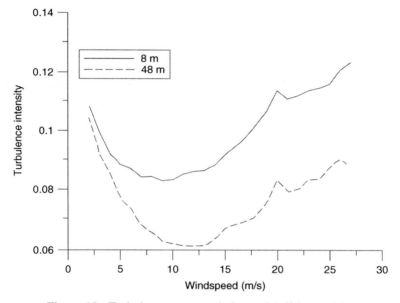

Figure 19: Turbulence versus wind speed (offshore) [23].

term in the momentum equations, parameterised as a function of the tree height and leaf density. The turbulence model might also be changed to simulate the increased turbulence.

As a rule of thumb measured data indicates that the turbulence intensity created by the forest is significant within a range of five times the forest height vertically and 500 m downstream from the forest edge in a horizontal direction. Outside these boundaries the ambient turbulence intensity is rapidly approaching normal values [22].

In or near forest, the mechanically generated turbulence intensity increases in high wind speeds due to the increasing movement of the canopy. A similar phenomenon can be observed in offshore conditions where increasing waves lead to increased turbulence in high wind speeds (Fig. 19). The increasing turbulence with increasing wind speed is of great importance when calculating the extreme gusts for a site.

Table 6: Definition of IEC classes [25].

	Class				
	I	II	III	IV	S
U_{ref} (m/s)	50	42.5	37.5	30	Site-specific
U_{ave} (m/s)	10	8.5	7.5	5	

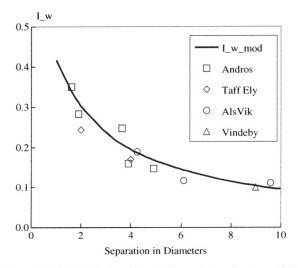

Figure 20: Wake-induced turbulence intensity I_{wake} [24].

5.2.2 Wake-induced turbulence

The wake of a wind turbine is characterised by the velocity deficit in comparison to the free flow and an increased level of turbulence. While the velocity deficit leads to a reduced energy output (Fig. 23) the turbulence in the wake leads to increased mechanical loads. The wake-induced turbulence I_{wake} is of a different nature than the ambient turbulence I_0 as the length scales are different. The wind turbine has to withstand both types of turbulence added up depending on its class (Table 6):

$$I_{wind\,farm} = \sqrt{I_0^2 + I_{wake}^2} \qquad (12)$$

Figure 20 shows the empirically measured, maximum wake turbulence as a function of distance between two wind turbines expressed in rotor diameters. As a rule of thumb wind turbines should not be closer than the equivalent of 5 rotor diameters in the main wind direction. Perpendicular to the main wind direction, a minimum distance of 3 rotor diameters should be maintained.

Turbulence particularly affects the blade root flap-wise as well as leading to varying torsion in the main shaft which is fed through the gearbox into the generator. Additionally turbulence causes thrust loads in the tower. The international

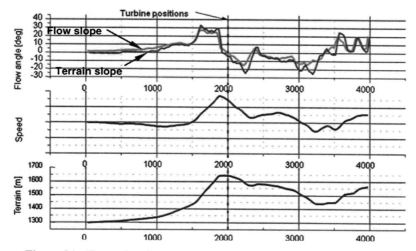

Figure 21: Example of speed-up and inflow angle across a mountain.

design standard IEC [25] assumes a maximum level of total turbulence of 18% for a wind turbine designed for high turbulence.

5.3 Flow inclination

A layout should aim for a high energy yield. Since the flow is accelerated across hills one idea could be to place the wind turbines at the point with the highest acceleration which is typically at the location of greatest curvature of the slope of the terrain. The unwanted side effect of such a location is that the flow might not be horizontal as it follows the shape of the mountain, thus the inflow angle is not zero. The askew wind vector leads to lower production as the angular response of the wind turbine can be simplified as a cosine and thus the horizontal component of the wind mainly contributes to the energy generation.

Additionally the wind turbine will be subject to higher loads when being exposed to large inflow angles. The fatigue loads on the blades will increase since the angle of attack changes during one rotation. Furthermore, the bending loads of the rotating parts of the drive train are increased, and finally the yaw drive will be subject to extra loads due to the uneven loading of the rotor. Thus a good layout avoids these locations. Figure 21 shows an example where the turbine position has been moved back from the edge of the cliff to a position with a much lower inflow angle (and slightly lower wind speed).

Information about the flow inclination can be obtained using ultrasonic or propeller anemometers with the restrictions described earlier. However, the obtained data only reflects a single location. Only flow models are capable to evaluate the flow inclination for each wind turbine location.

Care has to be taken, though when moving too far back from cliff edges, as this area is prone to separation leading not only to increased turbulence but in the worst

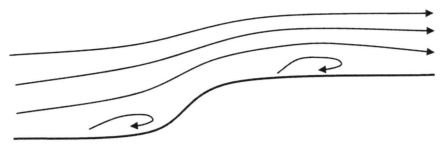

Figure 22: An example of flow separation over a cliff.

case to reverse flow imposing the potential of serious damage to the wind turbine (Fig. 22). The vertical and horizontal extent of such unsuitable areas can be estimated using CFD.

The international design standard IEC [25] assumes an inflow angle of maximum 8° for load calculations.

5.4 Vertical wind speed gradient

The loads on the rotor depend on the wind speed difference between the bottom and the top of the rotor. The gradient is normally expressed by the wind shear exponent a from the power law (eqn (6)). Most load cases assume a gradient of 0.2 between the bottom and the top of the rotor [25]. Note that in some cases, loads will be larger for very small or negative a.

The gradient causes changing loads of the blades as the angle of attack changes with each rotation. Thus the gradient adds to the fatigue loads of the blade roots. Furthermore the rotating parts of the drive train are stressed. The gradient is affected by four different phenomena:

- *Terrain slope.* The logarithmic profile can be heavily distorted by terrain slopes. While in flat terrain the wind speed increases with height, steep slopes might lead to a decrease with height. This is particularly likely for sites where the flow separates and does not follow the shape of the terrain anymore (Fig. 12). As a consequence the wind speed exponent might exceed the design limit in some sections of the rotor.
- *Roughness/obstacles.* If wind turbines are located closely behind obstacles like for example a forest the vertical wind speed profile might be again heavily distorted and areas of the rotor are exposed to large gradients. The degree of deformation of the profile depends not only on the geometry of the obstacle but also on its porosity.
- *Layout.* As explained above, the wake of a wind turbine is a conical area behind the rotor with increased turbulence and reduced wind speed since the rotor of the wind turbine has extracted kinetic energy from the flow. Figure 23 shows the effect of the wake on the profile, where the dotted line represents the free flow and the straight line the profile 5.3 rotor diameter behind the wind turbine.

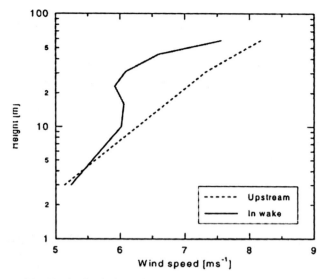

Figure 23: Vertical wind profile in front and behind a wind turbine [16].

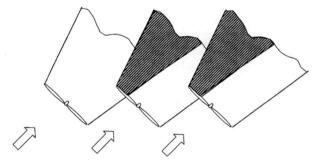

Figure 24: Horizontal gradient due to part wake operation.

It can clearly be seen that some areas of the deformed profile show very large gradients. If the wind turbines operate in part wake situations (Fig. 24), the wind turbines behind the front row will not only be exposed to vertical gradients but also to horizontal gradients which add significantly to the loads.

- *Atmospheric stability.* As explained above, the vertical wind speed profile depends on the vertical temperature profile. With increasing heating of the ground (unstable conditions) the turbulence increases and therefore the profile is being "smoothed" out. This leads to a steep wind speed profile characterised by very little increase of wind speed with height. Stable conditions in contrast are characterised by a very flat profile with significant wind speed gradients (see Fig. 11). Depending on the prevailing atmospheric stability of the site the gradient of 0.2 assumed for most load cases can be exceeded. Additionally in stable conditions significant wind direction shears are common.

An optimised layout can avoid excessive gradients related to the first three phenomena by staying clear from steep slopes, obstacles and allow a sufficient spacing between the wind turbines.

The gradient is often determined by measurements. Care has to be taken when the measurement height is lower than hub height since the wind shear exponent is a function of height, thus a gradient measured at lower height is not representative for higher heights. Be also aware of the fact that because the measured wind speed differences can be relatively small, the resulting measurement uncertainties can become quite significant. The preferred method in these cases is to model the wind speed profile using flow models.

6 Site classification

6.1 Introduction

When designing wind turbines a set of assumptions describing the wind climate on-site is made. As the robustness of a wind turbine is directly related to the costs of the machine, a system has become common of grouping sites into four different categories to allow cost optimisation of the wind turbines. The site class depends on the mean wind speed and extreme wind speed at hub height referred to as IEC classes [25]. Furthermore three different turbulence classes have been introduced, for low, medium and high turbulence sites. The term extreme wind or U_{ref} is used for the maximum 10-min average wind speed with a recurrence of 50 years at hub height. Please note that U_{ref} is not related to the mean wind speed.

6.2 Extreme winds

Selecting a suitable turbine for a site requires knowledge about the expected extreme wind in the form of the maximum 10-min average wind speed at hub height with a recurrence period of 50 years (U_{ref} according to the IEC 61400-1 [25]). While local building codes frequently generalise the amplitude of extreme events across large areas, on-site measured wind data allows for a more precise site-specific prediction of the expected 50-year event. A number of methods offer the possibility to estimate the on-site 50-year maximum 10-min wind speed from available shorter term on-site data. The IEC 61400-1 does not prescribe any preferred method for this purpose though.

Under certain assumptions, a Gumbel analysis [26] can be applied to predict the 50-year extreme wind speed on the basis of the measured on-site maximum wind speeds. If these assumptions are correct, then the measured maximum wind speeds should resemble a straight line in the Gumbel plot.

Different methodologies are available to extract extreme events from a short-term data set and further to fit the Gumbel distribution to these extremes. The parameters describing the Gumbel distribution are generally determined from the linear fit in a Gumbel plot. Hereby the so-called reduced variant (transformed probability of non-exceedence) is plotted versus wind speed. The reduced variant (probability) can be expressed in different ways leading to different plotting positions. Furthermore, a variety of fitting options are available leading to a vast number of different results [27]. However, all methods have one problem in common. The resulting 50-year estimate is highly correlated to the highest measured wind speed event of the time series used for the analysis [28].

The European Wind Turbine Standard, EWTS [29] offers an option to estimate the extreme wind based on the wind speed distribution on site rather than a measured time series. It suggests a link between the shape of the wind distribution and the extreme wind. The EWTS relates a ratio determined by the Weibull k factor to the annual average wind speed to estimate the 50-year extreme wind speed. This factor is 5 for $k = 1.75$, <5 for higher values of k and >5 for lower k values (i.e. high values for distributions with long tails, and low values for distributions with lower frequency of high wind speeds). The extreme wind is calculated as this factor times the yearly averaged wind speed.

The extreme 3-s gust or also called design wind speed of the wind turbine (U_{e50}) is a function of the U_{ref} and the turbulence intensity. The extreme gust is estimated using the relationship, where the turbulence intensity I_{ext} is a high wind speed turbulence value estimated from the measured data:

$$U_{e50} = U_{ref}(1 + 2.8 I_{ext}) \qquad (13)$$

This relationship is based on experimental data as well as theoretical work (Fig. 25). As mentioned in Section 5.2.1 care has to be taken in offshore and forest situations as the turbulence intensity does not show asymptotic behaviour but increases with increasing wind speed. It is thus much more difficult to estimate I_{ext} under these conditions.

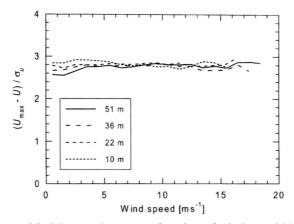

Figure 25: Measured gust as a function of wind speed [16].

7 Energy yield and losses

7.1 Single wind turbine

The power production varies with the wind speed that strikes the rotor. The wind speed at hub height is normally used as a reference for the power response of the wind turbine. Knowing the power curve of a wind turbine $P(u)$, the mean power production can be estimated using the probability density function of the wind speed at hub height $f(u)$, which is typically expressed as a Weibull distribution (see eqn (3)):

$$P = \int_0^\infty f(u)P(u)\,du = \int_0^\infty \frac{k}{A}\left(\frac{u}{A}\right)^{k-1} \exp\left(-\left(\frac{u}{A}\right)^k\right)P(u)\,du \qquad (14)$$

This integral cannot be computed analytically and thus has to be solved numerically. For this purpose the power curve is divided into a sufficient number of linear sections, typically for 0.5 m/s steps. The power output can now be calculated by summing up the produced energy for each wind speed bin.

Care has to be taken as the power of the wind is proportional to the air density. A power curve normally refers to an air density of 1.225 kg/m^3 which corresponds to a temperature of 15°C at sea level. A higher elevation and/or a warmer site will lead to a lower air density and thus to a lower energy output.

Special considerations should be given to the fact that a number of site-specific parameters influence the power curve and thus the energy yield most significantly turbulence and wind shear [30].

Another way of stating the annual energy output from a wind turbine is to look at the capacity factor for the turbine in its particular location. By capacity factor we mean its actual annual energy output divided by the theoretical maximum output if the machine were running at its rated (maximum) power during all of the 8760 h of the year. The capacity factors may theoretically vary from 0 to 100%, but in practice they will mostly be around 30–40%.

7.2 Wake and other losses

As mentioned in Section 5.2.2 and shown in Fig. 23 the wake of a wind turbine is characterised by a velocity deficit. This will lead to a reduced efficiency of any wind turbine operating in its wake. The so-called park loss is dependent on the thrust curve of the rotor, on the wake decay constant (which in turn is a function of the ambient turbulence intensity), and the distance between the turbines. High ambient turbulence intensity will lead to an increased mixing between the wake and the surrounding undisturbed flow. As a consequence the opening angle of the wake increases, thus resulting in smaller park losses than low turbulence situations.

Two models are commonly used in the industry, the N.O. Jensen model and the Eddy-Viscosity model of Ainslie [32]. The N.O. Jensen model is a simple, single wake kinematic model, in terms of an initial velocity deficit and a wake decay constant. It is based on the assumption that the wake right behind the wind turbine has a starting diameter equal to the rotor diameter and is linearly expanding as a function of the downwind distance. The Ainslie model is based on a numerical solution of the Navier-Stokes equations with an eddy viscosity closure in cylindrical coordinates. The eddy viscosity is described by the turbulent mixing due to the induced turbulence, generated within the shear layer of the wake, and the ambient turbulence [33].

Other sources of losses could be related to availability, electrical losses, high wind hysteresis, environmental conditions like icing and blade degradation and curtailments, of which wind sector management is the most common one. As explained earlier turbine loading is influenced by the wake effects from nearby machines. For some wind farms with particularly close spacing, it may be necessary to shut down certain wind turbines for certain wind directions. Typically this is done in a wind climate with a very dominant and narrow main wind direction.

7.3 Uncertainty

Uncertainty analysis is an important part of any assessment of the long-term energy production of a wind farm. The most important contributors to the uncertainty are:

- on-site wind speed measurements
- long-term correction
- flow modelling
- performance of the wind turbine
- park losses

Behind each point a vast number of aspects have to be considered and quantified including anemometer calibration, mounting effects, inter-annual variability, quality of the long-term data, vertical and horizontal extrapolation, quality of the topographic input, power curve uncertainty, wake model, etc. Some inspiration can be found in [34].

References

[1] European Wind Energy Association. *Wind Energy – The Facts*. ISBN-978-1-84407-710-6, 2009.
[2] IEC 61400-12-1. Wind Turbines Part 12-1: *Power performance measurements of electricity producing wind turbines*, 2005.
[3] IEA. Recommended practices for wind turbine testing, Part 11: *Wind speed measurement and use of cup anemometry*, 1999.
[4] Lindelöw, P., Courtney, M., Mortensen, N.G. & Wagner, R., Are Lidars good enough? Accuracy of AEP predictions in flat terrain generated from measurements by conically scanning wind sensing Lidars. *Proc. of European Wind Energy Conf.*, Marseille, France, 2009.
[5] Brock, F.V. & Richardson, S.J., *Meteorological Measurement Systems*, Oxford University Press, 2001.

 [6] Monna, W.A.A., Comparative investigation of dynamic properties of some propeller vanes. WR 78-11, KNMI, The Netherlands, 1978.
 [7] Bradley, S., Antoniou, I., von Hünerbein, S., Kindler, D., de Noord, M. & Jørgensen, H., Sodar calibration procedure, Final reporting WP3, EU WISE project NNE5-2001-297, University of Salford, 2005.
 [8] Antoniou, I., Jørgensen, H.E., Ormel, F., Bradley, S., von Hünerbein, S., Emeis, S. & Warmbier, G., On the theory of Sodar measurement techniques. Risø-R-1410, Risø National Laboratory, Roskilde, Denmark, 2003.
 [9] Antoniou, I., Jørgensen, H.E., Bradley, S.G., von Hunerbein, S., Cutler, N., Kindler, D., de Noord, M. & Warmbier, G., The profiler inter-comparison experiment (PIE). *Proc. of European Wind Energy Conf.*, London, UK, 2004.
[10] Courtney, M., Wagner, R. & Lindelöw, P., Commercial Lidar profilers for wind energy. A comparative guide. *Proc. of European Wind Energy Conf.*, Brussels, Belgium, 2008.
[11] Antoniou, I., Jørgensen, H.E., Mikkelsen, T., Pedersen, T.F., Warmbier, G. & Smith, D., Comparison of wind speed and power curve measurements using cup anemometer, a Lidar and a Sodar. *Proc. of European Wind Energy Conf.*, London, UK, 2004.
[12] Foussekis, D. & Georgakopoulos, T., Investigating wind flow properties in complex terrain using 3 Lidars and a meteorological mast. *Proc. of European Wind Energy Conf.*, Marseille, France, 2009.
[13] Bingöl, F., Mann, J. & Foussekis, D., Lidar performance in complex terrain modelled by **WAsP Engineering**. *European Wind Energy Conf.*, Marseille, France, 2009.
[14] Dupont, E., Lefranc, Y. & Sécolier, C., A Sodar campaign in complex terrain for data quality evaluation and methodological investigations. *Proc. of European Wind Energy Conf.*, Marseille, France, 2009.
[15] Petersen, E.L., Troen, I., Frandsen, S. & Hedegaard, K., Wind atlas for Denmark – A rational method for wind energy siting. Risø-R-428, Risø National Laboratory, Roskilde, Denmark, 1981.
[16] Petersen, E.L., Mortensen, N.G., Landberg, L., Højstrup, J. & Frank, H.P., Wind power meteorology. Risø-R-1206, Risø National Laboratory, Roskilde, Denmark, 1997.
[17] Troen, I. & Petersen, E.L., European wind atlas, ISBN 87-550-1482-8, Risø National Laboratory, Roskilde, Denmark, 1989.
[18] Lange, B., Modelling the marine boundary layer for offshore wind power utilization, Ph.D. thesis, University Oldenburg, Germany, 2002.
[19] Dellwik, E., Jensen, N.O. & Landberg, L., Wind and forests – general recommendations for using WasP, 2004, http://www.bwea.com/pdf/trees/Risoe.pdf
[20] Mortensen, N.G., Bowen, A.J. & Antoniou, I., Improving WAsP predictions in (too) complex terrain. *Proc. of European Wind Energy Conf.*, Athens, Greece, 2006.
[21] Wood, N., The onset of flow separation in neutral, turbulent flow over hills. *Boundary Layer Meteorology*, **76**, pp. 137–164, 1995.

[22] Pedersen, H.S. & Langreder, W., Forest-added turbulence: a parametric study on turbulence. *Journal of Physics*, Conference Series **75** – 012062, 2007.

[23] Barthelmie, R., Hansen, O.F., Enevoldsen, K., Motta, M., Pryor, S., Højstrup, J., Frandsen, S., Larsen, S. & Sanderhoff, P., Ten years of measurements of offshore wind farms – what have we learnt and where are uncertainties? The Science of Making Torque from Wind, Delft, 2004.

[24] Frandsen, S. & Thøgersen, M., Integrated fatigue loading for wind turbines in wind farms by combining ambient turbulence and wakes. *Wind Engineering*, **23**(6), pp. 327–339, 1999.

[25] IEC 61400-1. Wind Turbines Part 1: *Design requirements*, 2005.

[26] Gumbel, E.J., *Statistics of Extremes*, Columbia University Press, 1958.

[27] Palutikof, J.P., Brabson, B.B., Lister, D.H. & Adcock, S.T., A review of methods to calculate extreme wind speeds. *Meteorological Applications*, **6**, pp. 119–132, 1999.

[28] Langreder, W., Højstrup, J. & Svenningsen, L., Extreme wind estimates with modest uncertainty – a contradiction? *Proc. of Windpower Conf.*, Chicago, 2009.

[29] Dekker, J.W.M. & Pierik, J.T.G., *European Wind Turbine Standard II*. ECN-C-99-073, The Netherlands, 1999.

[30] Langreder, W., Kaiser, K., Hohlen, H. & Højstrup, J., Turbulence corrections for power curves. *Proc. of European Wind Energy Conf.*, London, UK, 2004.

[31] Antoniou, I. & Pedersen, S.M., Influence of turbulence, wind shear and low level jets on the power curve and the AEP of a wind turbine. *Proc. of European Wind Energy Conf.*, Marseille, France, 2009.

[32] Ainslie, J.F., Calculating the flowfield in the wake of wind turbines. *Journal of Wind Engineering and Industrial Aerodynamics*, **27**, pp. 213–224, 1988.

[33] Zigras, D. & Mönnich, K., Farm efficiencies in large wind farms. *Proc. of Windpower Conf.*, Los Angeles, USA, 2007.

[34] Albers, A., Uncertainty analysis and optimisation of energy yield predictions as basis for risk evaluation of wind farm projects. *Proc. of European Wind Energy Conf.*, Madrid, Spain, 2003.

CHAPTER 3

Aerodynamics and aeroelastics of wind turbines

Alois P. Schaffarczyk

University of Applied Sciences Kiel, Kiel, Germany.

Aerodynamics and aeroelastics of wind turbines are presented. First, the basic results of analytical, numerical and experimental work are reviewed, then the impact on commercial systems is discussed. A short section on non-standard wind turbines is finally included.

1 Introduction

Aerodynamics is a necessary tool for modeling the loads and power output of a wind turbine. Unlike other related applications such as ship propellers [5] and helicopters [6], there is no comprehensive and up-to-date presentation of this important subject. The reader is given a short introduction to current knowledge. A readable review of, especially, the German efforts during the 1950s and 1960s was given by Hütter [33]. Hansen and Butterfield [26] and Hansen *et al.* [25] present more up-to-date reviews.

It is assumed that the inflow velocity is more or less stationary, thereby omitting turbulence as rapid variations above 1 Hz and also neglecting diurnal variation. These distinctions are meant to be in the spirit of standard regulations, as for example given by the IEC (International Electrotechnical Commission) or Germanischer Lloyd (GL). Therefore no presentation of wake aerodynamics is found in this chapter. The interested reader may find a review of these items in [64]. This rest of this chapter is divided into seven sections.

Section 2 gives an account of analytical theories developed largely before the emergence of digital computers, beginning with the global momentum theories of Rankine [43] and Froude [23]. Several developments and extensions of these have emerged only recently. Section 3 introduces the most important development of the late 20th century: computational fluid dynamics (CFD). Therefore the reader should be familiar with the basics of fluid dynamics [2,3] and viscous fluid flow [4], together with some of the basics of CFD. Section 4 is devoted to experimental work on wind turbines. Two distinct branches can be identified: so-called free-field experiments, carried out in the open air and those performed under controlled inflow conditions within wind tunnels. Considerations are restricted to NASA-AMES *blind comparison experiment* and the European MEXICO (Measurements and Experiments in controlled conditions) project, performed in Europe's largest wind tunnel in the Netherlands. After describing aeroelastics in Sections 5 and part 6 in general, the impact of this elaborate scientific work on commercial wind turbines is presented. This is a somewhat difficult and delicate task, as most of this work and even the results are confidential. In practice, this means that only public-domain work and non-standard turbines will be considered. Section 8 concludes this discussion with a summary and an outlook for future developments after the aerodynamics of some unconventional turbines are presented in Section 7.

2 Analytical theories

The first work which provided a simple complete model of the global flow around a wind turbine is the so-called actuator disk (AD) theory. It was first developed by Rankine and Froude to describe the flow around ship propellers. Figure 1 shows an overview of all possible flow states which can occur. Recently [58] it was possible to reproduce all flow states observed by Glauert from a numerical full-field AD model.

The main idea is the introduction of a slipstream (Fig. 2) behind the rotor. Energy is extracted by decelerating the inflow v_1 to v_2 at the rotor and v_3 far downstream.

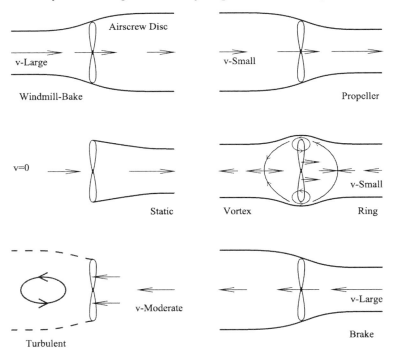

Figure 1: Flow states of propellers and wind turbines [12].

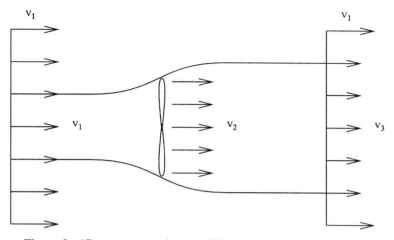

Figure 2: 1D momentum theory of Froude, Lanchester and Betz.

Applying the equations of conservation of mass, energy (Bernoulli's equation) and momentum:

$$v_2 = \frac{1}{2}(v_1 + v_3),$$

(1)

(Froude's law) and

$$C_P = 4a(1 - a^2),$$

(2)

are obtained. C_P is the non-dimensional power

$$C_P := \frac{P}{(\rho/2)Av_1^3},$$

(3)

$$A = \frac{\pi}{4}D^2.$$

(4)

A turbine of diameter D has a swept area of $A = (\pi/4)D^2$. The most important parameter in eqn 2 is the so-called *axial interference factor* [12] often also called velocity induction $a = v_2/v_1$. Differentiating eqn 2 with respect to a one can easily show that maximum energy is extracted when $a = 1/3$ and $C_P = C_P^{Betz} = 16/27 = 0.59$. This law was found independently by Lanchester in 1915 and Betz in 1925. Using the same arguments the main force on the turbine, the thrust in the wind direction:

$$C_T = \frac{T}{(\rho/2)Av_1^2}.$$

(5)

is seen to be $c_T(a) = 4a(1-a)$ at Betz' value of $a_{Betz} = 1/3$ $c_T(a=1/3) = 8/9 \sim 0.9$. . This shows that a wind turbine is heavily loaded at the optimum condition.

The following limitations apply to the theory:

- it is implicitly assumed that there is **no** slipstream as there are no radial components
- calculation of the full details of the slipstream expansion cannot be performed as the theory does not consider the radial velocity component
- the axisymmetric disk is assumed to be infinitesimally thin.

As already discussed by Betz [9] and further by Loth McCoy [70] in the context of a double AD for vertical axis wind turbines there is a possibility to *beat* Betz to some extent (roughly to 0.64 for a double AD). A recent discussion for *beating Betz* with general devices was given by Jamieson [34].

Wind turbines are rotating machines, and a very important dimensionless number is the tip speed ratio (TSR) defined as

$$\lambda := \frac{\Omega R}{v_{wind}}, \tag{6}$$

where Ω is the angular velocity of the turbine.

Figure 3 gives a graph of various c_P against λ. Apart from its own data, data from the classical literature, for example [14,18], was also included. Two items are

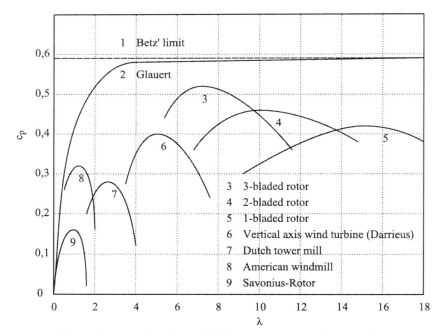

Figure 3: c_P as function of TSR, various types of wind turbines.

of particular interest. Firstly, all turbines have a maximum c_P and two kinds of turbine can be distinguished, those with a maximum c_P around 6–10 (so-called fast running machines) and those with maximum below 4 (so-called slow running machines). Secondly the curve must tend towards $c_P = 0$ when $\lambda = 0$. This was discussed by Glauert [12] and stems from the fact that in addition to the axial induction a, a second a' has to be introduced, where a' is defined by

$$a' = \frac{\omega}{2\Omega}.$$
(7)

Here ω is the local angular velocity of the flow. It comes from Newton's third law as applied to angular momentum. If dr is an increment of radius, the torque is now

$$dM = 4r^3 v_1(1-a)a' \, dr$$
(8)

and the total power becomes

$$c_P = 8\lambda^2 \int_0^1 a'(1-a)\left(\frac{r}{R}\right)^3 d\left(\frac{r}{R}\right).$$
(9)

The two parameters a and a' have to be optimized. A third equation, the so-called orthogonality condition of Glauert [12] is the condition

$$a'(1+a')x^2 = a(1-a)$$
(10)

with $x := \omega r/v_1$ being the local TSR. Figure 4 gives a sketch of the arrangement of the velocity vectors. Optimization results in the condition

$$a' = \frac{1-3a}{1+4a}$$
(11)

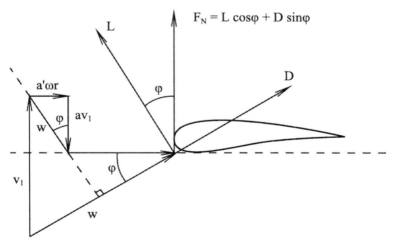

Figure 4: Velocity and force triangles.

which shows that $1/4 < a < 1/3$ must hold. These losses as compared to Betz' ideal limit are called *swirl* losses. Figure 5 shows the quantitative dependence. Two other mechanisms have to be introduced, to model all effects shown in Fig. 3. They are the so-called tip-losses and profile-drag losses. An AD was defined as a compact disk, formally having infinitely many blades. To estimate the effect of a finite number of blades, the two models are used. One is based on conformal mapping of the flow around a stack of plates to that of a rotor with a finite number of blades (given by Prandtl [42]) and the second is based on the theory of propeller flow of Goldstein [24]. A recent investigation has been made by Sørensen and Okulov [39,40]. Usually a reduction factor F is introduced to account for the decreasing forces on the blade towards the tip:

$$F = \frac{2}{\pi} arccos \, \exp\{-f\}, \qquad (12)$$

with

$$f = \frac{B}{2} \frac{R-r}{r} \sqrt{1+\lambda^2} . \qquad (13)$$

λ as expressed by eqn 6.

Comparison with measurements by Shen *et al.* [55] resulted in a new empirical tip-loss model for use in AD and CFD simulations. Recently Sharpe [53] has revised the arguments of Glauert and extended them slightly. Mikkelsen *et al.* [38] applied his numerical AD method to investigate this effect. His findings were that

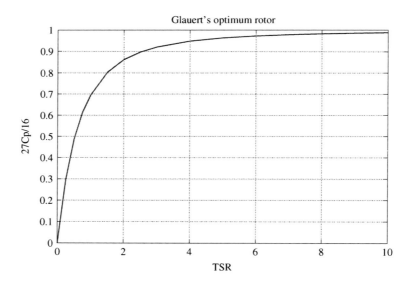

Figure 5: c_P/c_P^{Betz} as a function of TSR.

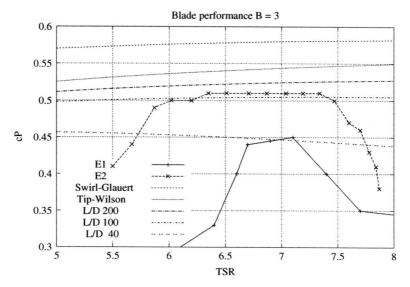

Figure 6: Comparison of swirl drag and profile-drag losses against measured values of actual turbines.

despite the fact that the otherwise neglected pressure decrease in the near wake gives higher c_P in the inboard section, Betz' limit is valid globally.

When considering drag losses, one has to imagine that a rotor blade can be regarded as an aerodynamic device experiencing two forces drag (in flow direction) and lift (orthogonal to that). The lift force and part of the drag force (per unit span) are due to the pressure around the airfoil. Figure 7 illustrates this. The c_P is defined as

$$c_P = \frac{p}{(\rho/2)v^2} \qquad (14)$$

where c is the chord of the airfoil.

Here:

$$c_D = \frac{D}{(\rho/2)v^2 c \cdot 1}, \qquad (15)$$

and

$$c_L = \frac{L}{(\rho/2)v^2 c \cdot 1} \qquad (16)$$

are defined. It can be shown that lift gives rise to no loss as it is perpendicular to the flow, which is not the case for drag. A measure of efficiency for profiles is

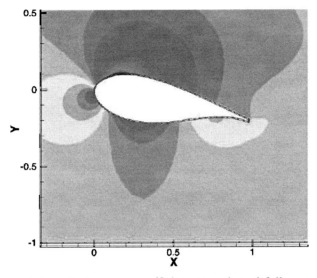

Figure 7: Pressure coefficient around an airfoil.

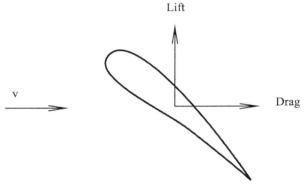

Figure 8: Lift and drag around an airfoil; v is a sample inflow velocity and is not
to be confused with a relative velocity.

defined as the lift-to-drag-ratio L/D. Usually this number is around 100. In total
Figs. 10 and 11 are obtained. Compared to Fig. 3 the quantitative influence of drag
and finite bladenumber on c_P are presented. In addition Fig. 6 shows the improvement of state-of-the art commercial wind turbines of one manufacturer over a span
of 20 years. Clearly one can see that high-performance airfoils have to be used to
reach for a c_P^{max} in the order of 0.52. It is clear that $B = 1$ is very special, and one
might assume that no such turbines were manufactured. This in fact was not the
case. In the late 1980s the German company MBB manufactured the so-called
Monopteros (see Fig. 9), a single-blade turbine. To demonstrate the big differences
data from Rohrbach et al. [46] is included.

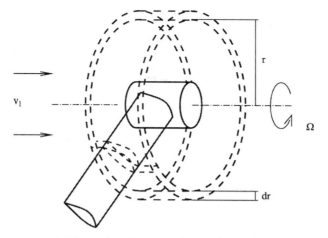

Figure 9: Blade elements of a rotor.

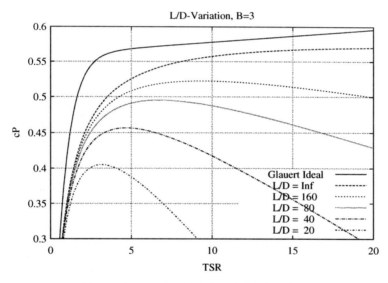

Figure 10: c_P via TSR: $B = 3$, L/D varies.

Sørensen and Okulov [39,40] recently formulated a vortex theory for these types of rotors (see Fig. 11).

To sum up: at the present time a limit of $c_P = 0.52$ seems to have been reached by modern turbines, which is only possible if specially designed airfoils are used. A c_P of 0.45 (typical for turbines in the early 1990s) is obtained using old profiles from the aerospace industry (see Figs 3 and 6).

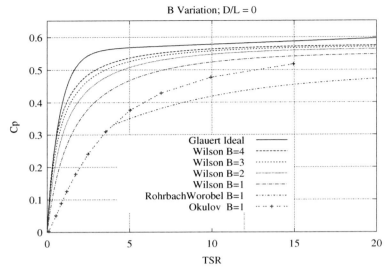

Figure 11: c_P via TSR: $D = 0$, bladenumber varies.

Figure 12: Single bladed wind turbine Monopteros by MBB, Photo: Schaffarczyk.

2.1 Blade element theories

Blade element theory (see Fig. 9) divides the rotor into several finite length sections of Δr (or dr mentioned above). Then forces applied on these annuli are compared to those from airfoil theory. Implicitly it is assumed that there is no mutual interference of the sections. With reference to Fig. 9 and denoting with $d\dot{m}$ the mass flow through the disk and with v_t the tangential velocity, a thrust

$$dT = d\dot{m}(v_1 - v_3), \tag{17}$$

$$dT = 4a(1+a)2\pi r \, dr \, \frac{\rho}{2} v_1^2 \tag{18}$$

and torque

$$dQ = d\dot{m} \, v_t r, \tag{19}$$

$$dQ = 2\pi r \, dr \, \rho v_2 \, 2\omega r^2, \tag{20}$$

$$dQ = 4\pi r^3 \rho v_1 \, \Omega(1-a)a'. \tag{21}$$

increment is obtained.

Now comparison with the forces resulting from the airfoil sections is performed. From Fig. 8 the force coefficient in the inflow direction is

$$C_N = C_L \cdot cos(a) + C_D \cdot sin(a). \tag{22}$$

Here the flow angle φ can be computed from the velocity triangle.

$$tan(\varphi) = \frac{(1-a)v_1}{(1+a')\omega r}. \tag{23}$$

The angle of attack and the flow angle is related via the twist angle ϑ:

$$a = \varphi - \vartheta. \tag{24}$$

The square of the velocity is

$$w^2 = ((1-a)v_1)^2 + (\omega(1+a'))^2. \tag{25}$$

All relevant data is then calculated using an interaction scheme. It is to be noted that in the inner part of a blade, values of greater than 0.5 and even 0.52 are observed. Then the simple momentum theoretical value $c_T(a) = 4(1-a) \cdot a$ is no longer valid. An empirical extension for $0.5 < a < 1.0$ must be used. The deviation starts at $a \approx 0.3$ and gives $c_T(a = 1)$ of approximately 2.

Many engineering codes, such as the PROP Code of Walker and co-workers [18] follow this approach. These codes rely heavily on measured aerodynamic data for airfoils. In the early days of wind energy, airfoils from ordinary airplanes were used. Since then, special airfoils for wind turbines have been developed, mainly at Stuttgart (FX-Series), Delft (DU profiles) and Risø. Today, not only power optimization but also load reduction has to be included into profile and blade design. Another important issue is a phenomenon called stall delay within rotating boundary layers. Since its first observation by Himmelskamp [30] in 1945, it has become evident that much behavior cannot be explained without the phenomenon, which is also important for swept flow. For 3D boundary layers the ECN model [45]:

$$f = 3\left(\frac{c}{r}\right)^2 \tag{26}$$

with f defined as $C_{L,3D} = C_{L,2D} + f(2\pi\alpha - C_{L,2D})$ is often used.

2.2 Optimum blade shape

Neglecting drag all relevant forces can be derived from lift via (compare to Fig. 9)

$$\frac{dT}{dr} = Bc\frac{\rho\omega^2}{2}C_L \cdot \cos\varphi \qquad (27)$$

and torque

$$\frac{dQ}{dr} = Bc\frac{\rho\omega^2}{2}C_L \cdot \sin\varphi \cdot r. \qquad (28)$$

Adding to eqn (10) the momentum balance for thrust, equation (17), and torque equation (19) solution for a and a' is possible. The following final equations are obtained:

$$\frac{a}{(1-a)^2} = \frac{\sigma C_L \cdot \cos\varphi}{4\sin^2(\varphi)}, \qquad (29)$$

$$\frac{a'}{1+a'} = \frac{\sigma C_L}{4\cos\varphi} \qquad (30)$$

With $\sigma = BC/2\pi r$ solution for $c \cdot C_L$ which is proportional to the circulation:

$$\frac{Bc\omega C_L}{2\pi v_1} = \lambda_r \sigma C_L = \frac{4\sin\varphi(2\cos\varphi - 1)}{1+2\cos\varphi}, \qquad (31)$$

can be performed. Division by $r/R = \lambda_r/\lambda_R$ results in the desired ratio c/r against r/R (Fig. 13). Together with Glauert's theory the more extended approach of Wilson [18] and De Vries [17] , which includes also the lift to drag ratio and tip losses:

$$\frac{(1-aF)aF}{(1-a)^2} = \frac{\sigma C_L \cdot \cos\varphi}{4\sin^2(\varphi)}, \qquad (32)$$

$$\frac{a'F}{1+a'} = \frac{\sigma C_L}{4\cos\varphi}. \qquad (33)$$

with F from eqn 12 is obtained. It can be seen that for $C_L / C_D = 100$ and $B = 3$, there are only small differences from the Glauert theory. Recently [40] it was found that the widely used optimization approach by Betz may be overcome by the older one of Joukowsky [36] thereby stating that a constantly loaded rotor with a finite number of blades may be superior to that with a Betz-type load distribution.

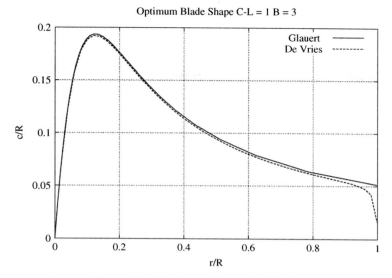

Figure 13: Optimum shape of a blade.

3 Numerical CFD methods applied to wind turbine flow

As a complement to analytical theories and experiments, CFD provides a third approach in developing applied methods. In its purest form, only the differential equations of Navier and Stokes (NS):

$$\nabla \cdot v = 0, \tag{34}$$

$$\rho \frac{Dv}{Dt} = f - \nabla p + \mu \Delta v = 0 \tag{35}$$

together with suitable boundary conditions and a description of the blade geometry is used. Unfortunately this ambitious goal cannot be reached at the present time. The main obstacle is the emergence of turbulence at higher Reynolds number (RN):

$$Re = \frac{vL}{v} \tag{36}$$

calculated from kinematic viscosity ($v = 1.5 - 10^{-5}$ m^2/s^{-1} for air) and a typical length L and a typical velocity v. Present wind turbines have a RN of several million based on blade chord. Only Direct Numerical Simulation (DNS) solves the (NS) without any further modeling. At the present time (2009), only airfoil calculations up to RN of a few thousands have been carried out at the price of months of CPU time and terrabytes of data. After Large Eddy Simulation (LES), the next stage of simplification is Reynolds averaged Navier Stokes equations (RANS). An ensemble average, which is assumed to be a time average by an ergodic hypothesis is carried out for all flow quantities.

Turbulence then emerges as a never ending hierarchy of higher correlations which has to be truncated by the so-called closure assumption. In wind-turbine applications at the present time the $k - \omega$ shear stress transport (SST) extension of Menter isfrequently used. k is the turbulent kinetic energy (per unit mass) and ω is a local frequency scale. Unfortunately all these empirical turbulence models describe only fully developed turbulent flow and are not able to resolve the transitional region from laminar to turbulent flow. There are good reasons to believe that parts of the blade must be laminar because otherwise the large L2D ratio cannot be achieved for c_P of the order of 0.5.

At UAS Kiel from 2000–2003 a program [65] comparing 2D-CFD simulation with measurements was carried out. A 30% thick airfoil from Delft University was chosen: DU-W-300-mod. For including transitional flow properties to e^N- method of Stock was included in DLR's structured CFD-Code FLOWer. The main findings of this project were as follows:

- Mesh generation with mostly orthogonal grids is very important. Therefore a hyperbolic type of generating equation, namely

$$x_\xi x_\eta + y_\xi y_\eta = 0, \tag{37}$$

$$x_\xi y_\eta - y_\xi y_\eta = \Delta A. \tag{38}$$

was chosen. For details see [62].

- The e^N method has to be parameterized with an N (usually between 6 and 9) which is related to a surrounding turbulence intensity by Mack's correlation. Comparison with wind tunnel measurements was difficult because the turbulence intensity was not known exactly.
- Prediction of c_L^{max}, meaning computation of flow separation (stall) was also difficult, because the flow started to become unsteady.
- Transition points were predicted correctly as long as only laminar separation or Tollmien Schlichting (TS) instabilities triggered transition.
- In addition, drag effects for example from Carborundun or Zig-Zag band were difficult to predict [28].

The situation becomes more complicated when whole blades are investigated. Correct prediction of overall power curves is seen to be very difficult. The picture is uneven [37] but parametric studies such as for wings including winglets [35] give valuable insight into the flow field and comparative changes. This is especially important when discussing performance enhancements.

Much better results can be achieved when no turbulence has to be included in the discussion. This is always the case, when basic studies have been performed

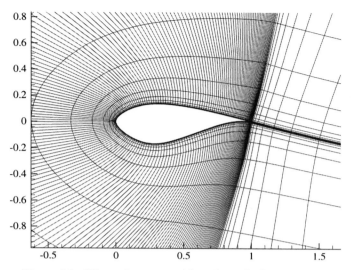

Figure 14: 2D mesh generated by a hyperbolic generator.

which cannot easily be checked by hand calculations. An example of such methods are the modification of the AD and actuator line method initiated by J. Sørensen and further developed by R. Mikkelsen. Another promising field of application is site-assessment and wind-farm optimization. The second torque conference at Copenhagen in 2007 gives an impressive report of the progress which has been made in this field. One has to bear in mind that the usual problems with turbulence modeling are the same as in other fields mentioned above.

4 Experiments

4.1 Field rotor aerodynamics

From 1992 to 1997, the International Energy Agency (IEA) conducted large-scale comparisons of measurements on research-type wind turbines. The project was termed Annex 14 [52] and was designed to

- validate and develop design codes,
- investigate design principles for stall controlled turbines.

Five turbines from ECN in the Netherlands, IC/RAC in the UK, NREL in the USA, RISO in Denmark and DUT in the Netherlands were considered. The diameter ranged from 10 to 27 m. The profiles used were quiet diverse, ranging from the NACA 44xx and NACA 632xx to the NLF 0416 and the S809. One objective was to investigate the correlation of measured performance and 2D-profile data from wind tunnel measurements.

As a result many problems with consistent interpretation of all the data were identified. Especially the definition of a local angle of attack was very difficult. This of course is also the case when 3D-CFD investigations are discussed. Several approaches have been tried, one being a correlation between the location of the stagnation pressure line and AOA. A second is the so-called inversed BEM, where c_N and c_T are correlated to the local flow angle. One may object that both methods rely on 2D assumptions for connecting forces and velocities so they may fail when 3D effects are strong. A somewhat detailed review is given by van Rooij [45] and Shen *et al.* [54].

It may be useful to note that a new program subsequent to the above-mentioned experiments will be started in early 2009 which will be undertaken on the research wind turbine E30 at the University of Applied Sciences Flensburg. During this experiment, using an aerodynamic glove, shear stresses which indicate the state of the boundary layer (laminar, transitional or fully developed turbulent) will be measured for the first time [51].

4.2 Chinese-Swedish wind tunnel investigations

In a joint effort of Chinese and Swedish aerodynamicists, a 4.75-m diameter wind turbine was investigated in the late 1980s [21,44].

4.3 NREL unsteady aerodynamic experiments in the NASA AMES-wind tunnel

As a result of the rather disappointing findings from open air experiments a big effort was mounted by NREL, the US National Renewable Energy Laboratory. A 10-m diameter two-bladed rotor was put into the world's largest wind-tunnel, the NASA-AMES wind tunnellocated in California. The main advantage was a complete control over all inflow conditions. Most important was the so-called *blind comparison* in which a variety of design and analysis tools were used to predict the power and forces based on 2D data. Figure 15 gives an overview of all simulation for the shaft torque, which is proportional to the power, see [56] for broad discussion. Several important conclusions can be drawn from the results of this large-scale experiment:

- 10 tested aeroelastic codes (see Section 5) showed extremely large disagreement even in the attached flow regime. The measured low-speed shaft torque being 800 Nm, is scattered in the predictions from 200 up to 1400 Nm (factor of 7). This situation is only a little bit better at 10 m/s (30%) but became worse at the same level as before in the deep-stalled case.
- So-called performance codes (only three, being descended from the famous PROP Code) showed strong variations in the deep-stalled condition with one remarkable exception which gives almost the same result as one CFD investigation.
- Two wake codes give also non-uniform results.
- Most impressive were the CFD results of Sørensen [59]. There, without any 2D-profile data, results with the same degree of accuracy of the best BEM

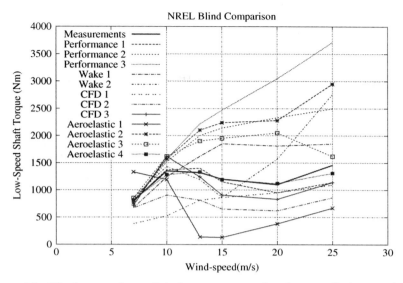

Figure 15: Blind comparison of shaft rotor torque of various predictions against measurement.

simulation (probably M. Hansen and T. Chaviaropoulus) were obtained. The effort was still rather expensive: 3M cells and about 50 h CPU on a 4-processor machine. Later Sørenesen [60] also introduced also transitional effects but with limited success.

4.4 MEXICO

The European answer to the NREL experiment was a model experiment in controlled conditions. The large-scale facility of DNW of $9.5 \times 9.5 \text{m}^2$ open section was used [57]. A complete analysis is to be worked out within the new established IEA Annex 29 *MexNext*, where scientists from several countries outside Europe will also participate. Most important is the additional measurement of the velocity field by Particle Image Velocimetry (PIV).

5 Aeroelastics

5.1 Generalities

Wind is transformed into forces by striking the solid structure of the blades. These are manufactured by regarding them as aerodynamic devices. Lift, drag and pitching moment result from the combined flow of wind and rotating flow. These forces now act ona flexible structure. The interaction of aerodynamics and structural mechanisms is called aeroelastics. As turbines become larger they necessarily have to become lighter, meaning that their lowest eigenfrequencies approach a certain limit. This limit is usually given by the excitation frequency of the rotor multiplied by the number of the blades. Therefore a model has to be constructed which includes both aerodynamics and structural dynamics.

Different stages of sophistication have to be identified

I Aerodynamic

1. Blade Element Momentum Code (BEM)
2. Wake Codes
3. Full 3D-CFD including turbulence modeling

II Structural Mechanics

4. Beam (1D) model
5. Shell (2D) model
6. Solid (3D) model

In principle $3 \cdot 3 = 9$ possibilities for coupling the various methods can occur. At the present time (1 with 4) coupling is one of the most commonly used in industry. Several industrial codes are available, some of which are:

- FLEX, from Stig Øye, DTU
- BLADED, by Garrad Hassan
- GAROS, by Arne Vollan, FEM
- Phatas, by ECN

and many more.

There are several difficulties in improving the accuracy of this approach. Beam-like parts of the turbine, like the tower and blades are easy to fit into the BEM Code. Not so easy to include are major parts of the the drive train, gear-boxes and the electromechanical parts such as generators. Also control systems which are used more and more to decrease loads have to be included. So in the forseeable future, coupling between more or less standardized and specialized software systems will be seen. Examples may be MATLAB/Simulink for electrical control, flexible coupling of multi-body systems like gear-boxes and shafts with standard FEM tools like ANSYS or MSC NASTRAN, to name only a few. From fluid mechanics it is not easy to see how BEM can be improved. CFD is rather time consuming and not a priori better when BEM is improved by empirical enhancement. At the present time, coupling of types (3 with 5) or (3 with 6) is the subject of ongoing research.

5.2 Tasks of aeroelasticity

By far the most important task of aeroelasticity is certification or type-approval. Most operators of wind-farms can only finance, operate and insure their turbine when appropriately certified. Usually during this process many aeroelastic load cases have to be simulated. They are documented within rules or guidelines.

Some of the tasks are to:

- ensure safety against aeroelastic instabilities such as divergence/flutter,
- investigate loads from extreme events expected only a few times within the estimated lifetime,
- accumulate the effects of loads from rapidly changing operating conditions during normal or electricity generation.

A turbine may be called optimized if it can resist equally against extreme winds and fatigue loads. Because a detailed description of recent procedures is rather exhausting to the beginner, he or she may start with a somewhat out-dated but classical text: *Wind turbine engineering design* by Eggleston and Stoddard [76]. There the reader will find a clear description of how the basic physics together with engineering requirements may be fit together in a computer code. A recent review more closely to pure aeroelastics was given by Hansen *et al.* [25]. On the other hand, it is worth noting how classical aeroelastics (usually coming from airplane design) now re-enters into recent wind turbine design for offshore applications [19].

5.3 Instructive example: the Baltic Thunder

Presenting a whole aeroelastic case study is far beyond the scope of this short introduction. Here it is tried to exemplify a description in a somewhat different way. In 2008 a Dutch organization called for a competition for a wind driven car for the first time. Six teams from four countries presented their design, one of them was the Baltic Thunder (see Fig. 16). Because one goal was to reduce parasitic drag as much as possible, the weight had to be reduced as far as possible. A light but flexible structure was the result. For other reasons a vertical axis rotor was chosen (see Section 7.1) This rotor is known to be particularly prone to aeroelastic instabilities. Therefore a Campbell diagram was produced by Vollan's Code GAROS. Due to the soft blade suspension a vertical bending mode of about 2.5 Hz was not to be avoided. This gives rise to a flutter type instability at 500 RPMs. Therefore the final design had carbon reinforced fiber (CRF) tubes as suspension giving much higher first blade eigenfrequencies. One of the various safety proofs included safety against a 18 m/s gust when operating in normal mode. Figure 18 shows the response of the main rotor tower. A static (constant) force of about 700 N superimposed on the dynamic response of the RPM excursion from about 200 to 300 RPM only. It has to be noted that a 18 to 12 m/s increase of wind speed is equivalent to an increase of power by factor of 3.4 if c_P remains constant. In this case maximum force excursion is only up to 1200 N so that at least the central column could be regarded as safe.

6 Impact on commercial systems

6.1 Small wind turbines

In general it is hard to see what effect the impact of scientific work will have on a specific commercial product. On the one hand, there is a huge number of engineering

Figure 16: Photograph of the Baltic Thunder wind driven vehicle prepared for the Racing Aeolus challenge at Den Helder, The Netherlands, August 2008.

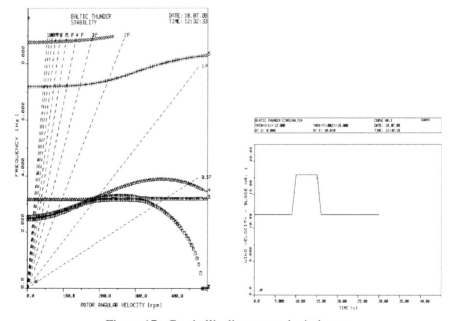

Figure 17: Cambell's diagram and wind gust.

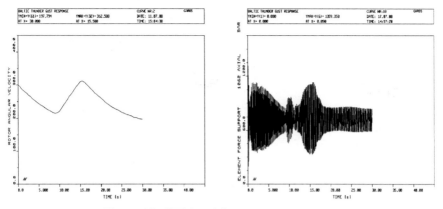

Figure 18: RPM and force response to gust.

experts working in large companies and on the other hand real testing is possible only for small wind turbines. An overview of small wind turbines is given first.

According to IEC, small WTs have a swept area smaller than 200 m^2. This gives a diameter of less than 17 m. Sometimes a subclass of WT which have swept area even less than 40 m^2 are also considered. These turbines have a diameter of 7 m or less. The main problems for these turbines are the manufacturing costs and a comparable aerodynamic performance to their larger relatives. Due to Reynolds number effects, the c_P^{max} of small wind turbines is approximately 0.3 instead of 0.5 for the best standard (2–3 MW) commercial turbines. This is mainly due to the much lower L2D ratio at Reynolds numbers below 10^5. Thus several attempts have been made to find thicker (> 10%) profiles giving a better performance. See Figs 19 and 20 for examples.

Further information can be found in [63].

6.2 Main-stream wind turbines

It is obvious that most efforts have been made to optimize so-called main-stream wind turbines with several MW rated power but below 3 MW. The author [50] was able to compare measurements of three different blade sets (see Fig. 21) on the same machine. Although the tip-speed ratio was 8.5 for one blade and only 7 for the other blades the c_P^{max} difference varies from 0.49 to 0.46 resulting in an improvement of 6%. As a conclusion this shows that *dedicated* or *tailored* profiles and blades can improve the performance significantly.

c_P^{max} currently seems to have settled down at around 0.52 for the best turbines as to be expected from the discussion in Section 2.

New developments therefore aim to reduce loads without losing power. This can be achieved by reducing c_L^{max} against c_L(L2D – max). In contrast a safe distance to c_L^{max} also prevents early stalling. During the 1990s a lot of new aerodynamic profiles were investigated in the US (Seri), in Delft (DU) and at Risø (Risø A,B

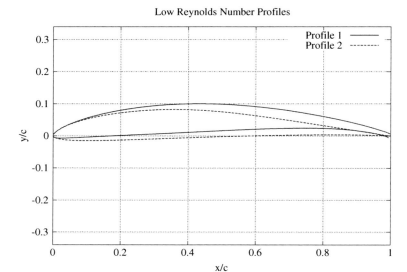

Figure 19: Outline of a low Reynolds number profile.

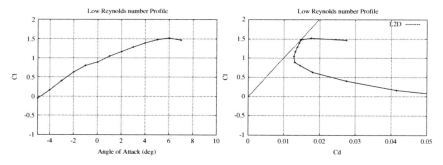

Figure 20: Polar of low Reynolds number profiles.

and P-series). As already and often mentioned, this has led to a c_P^{max} close to the theoretical maximum.

6.3 Multi MW turbines

Since around 2000 the off-shore use of wind energy has progressed, one of the first being the wind farm Middlegrunden close to Copenhagen, which started operation in 2000. It soon became clear that, due to the narrow time frame available to reach the wind farm off-shore, they have to be much larger than usual wind turbines. Especially in Germany the development of so-called Multi Megawatt turbines has started, these now reaching 6 MW rated power. In this connection several questions arise. Do the aerodynamic properties stay the same when upscaling the blades?

The Reynolds number of a 63-meter length (REpower 5M) blade may reach 6 million and more. A major project was started in 2002 at the UAS Kiel to measure the aerodynamic properties of a 30% thick slightly modified Delft profile [22,61,65]. As an outcome avariety of aerodynamic devices were measured as well as polars up to 10 million. This was achieved by cooling down the tunnel gas to only 100 K (= −173°C). Figs 22–24 give an impression of the results achieved so far. An attempt to model the results numerically are reported in [49].

7 Non-standard wind turbines

7.1 Vertical axis wind turbines

Two kinds of unusual turbines are dicussed briefly below as further examples of the application of aerodynamics to wind turbines. The first example is a wind

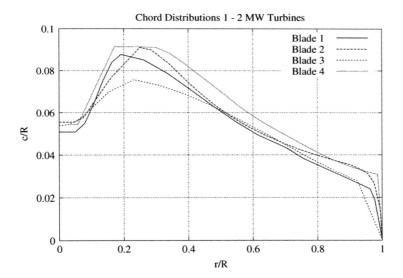

Figure 21: c/R from four blade sets for a 1.5 MW wind turbine.

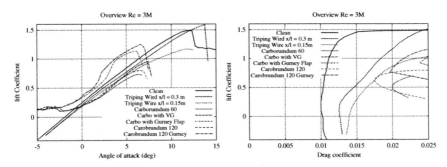

Figure 22: Polars of DU-W-300 mod at Reynolds number 3 million.

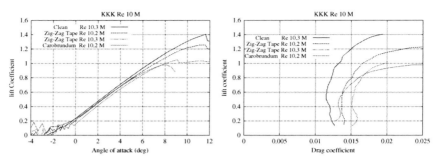

Figure 23: Polars of DU-W-300 mod at Reynolds number 10 million.

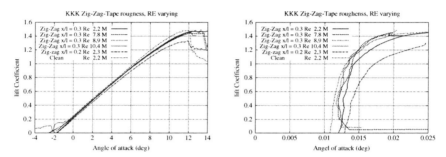

Figure 24: Polars of DU-W-300 mod with zig-zag tape, Reynolds number varying.

turbine with a vertical axis of rotation (VAWT) and the second are diffuser augmented wind turbines. A discussion of *counter-rotating* systems is ommitted here. The reader may find information in [48,32].

Two types of VAWT are possible (see Fig. 25): those driven by drag (Savonius) and those driven by lift (Darrieus). Aerodynamic modeling starts with the question whether an actuator **disk** modeling makes any sense. The total structure is an extended one, so at least two disks for up-wind and down-wind halves have to be used. This *double actuator disk* model was formulated by Loth and Mc Coy [70]. Only circumferential averaged values are discussed. The usual momentum theory for two *independent* disk was applied. As a result of these assumptions only one optimization parameter a – the (global) velocity interference factor – remains, giving $c_P^{max,dAD} = 0.64$. Further, the power extraction is seen to split into 80 and 20 % shares for the first and second disk respectively.

Figure 26 shows the findings in a graph. It is stricking that the total (normalized) power is >0.6 along the range $0.1 < a < 0.42$. Here a was defined as usual as $a := v_{disk}/v_\infty$. It should be noted that such high c_P s have not been observed so far. As an important design parameter the *solidity* $\sigma = Bc/R$ (B = number of blades; c = chord and R = rotor radius) is used. It can easily be shown that low σ corresponds to high TSR but unfortunately then there is **no** self-starting [67] behavior. Only when using high $\sigma \geq 0.4$ can a self-staring turbine be achieved.

Savonius-Rotor   Darrieus-Rotor   H-Darrieus-Rotor

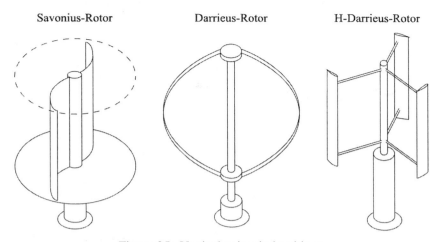

Figure 25: Vertical axis wind turbines.

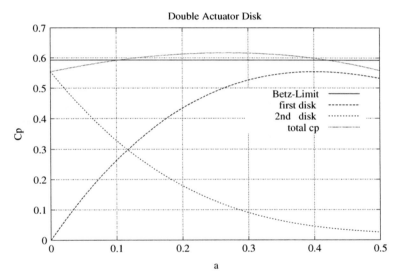

Figure 26: c_P (a) for a double actuator disk.

Figure 27 shows a comparison between various methods of calculating c_P. The first method [18] uses a simple procedure for the induced axial velocity which then is a function of circumferential angle φ. To simplify matters this is assumed to be $a(\varphi) = \lambda\sigma/2 \ sin(\varphi)$. This method results in a c_P^{max} which is only slightly smaller than c_P^{Betz}. Holme's [66] method uses a vortex approach to describe the flow field. Here the unsteadiness was avoided by distributing the bound circulation around a whole circle. This then gives a steady but non-symmetric model which also makes it possible to calculate transverse forces. Figure 27 shows, however, only minor changes to the first method, that of Wilson [18]. A simple blade element method was formu-

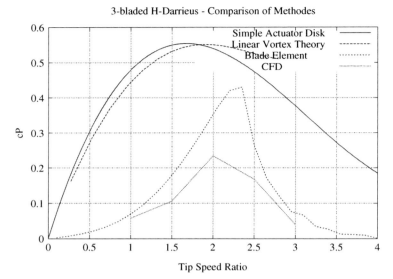

Figure 27: Comparison of different methods of describing a small VAWT.

lated by Stickland [74] in the 1970s. Here the the flow is divided into various stream tubes and an internal computer code interation ensures the actual balance of axial momentum. As a result of including drag the c_P (TSR) curve now becomes much narrower and the maximum also decreases to values common for HAWTs. In addition CFD methods can be applied to VAWT. To avoid time-consuming unsteady computations a special technique called *interface averaging* between two computational volumes with different frames of references were used. One volume in the vicinity of the rotor experiences inertial forces due to rotation. The delicate point now is how to pass over to the second volume where no inertial forces are applied. In summary a much smaller c_P^{max} may be predicted by this method. Meanwhile [71] 3D transient CFD simulation have been carried out.

Further research is obviously necessary to reduce the large differences obtained by the various methods. It should be noted that manufacturers of small wind turbines try to use the vertical axis principle again. Unfortunately no particular system seem tobe able to compete against HAWT. This is mainly because most systems do not possess a dynamically stable supporting structure.

7.2 Diffuser systems

Wind-concentrating systems such as shrouds, ring-wings and diffusers have been being investigated for a period of about 50 years. A summary has been given by van Bussel [20], together with a 1D momentum theory and a collection of experimental data. Special emphasis has been given to a discussion of the influence of the diffuser's exit pressure. It should be noted that most of the designs try to use a large input–output area ratio, whilst being relatively short. A permanent danger is the early separation of air flow through the diffuser as can be seen from Fig. 28 [41,47], where a relatively short diffuser was designed for an area ratio of approximately 3.

The author does not agree that a final theory has been formulated in which the specific problem has been solved. It may be necessary to extend the work of Loth and Mc Coy [70] who discussed an array of disks as a model of an actuator volume to a real continuum theory. This is an important task because a safe upper limit, such as Betz' law, is important for all investigations. A successful commercial design has also not been produced so far.

8 Summary and outlook

The history of wind turbine aerodynamics spans approximately 150 years. The most important influence has been given by modelling flow properties which then serve as tools for predicting loads on the structure. As a consequence safe and competitive wind turbines up to rated power of 6 MW have been constructed so far. First investigations of even larger blades feasible for rated power up to 10 MW are under consideration [29].

Further progress can be expected when CFD when taken in the strict sense of the expression as solving the Navier-Stokes equation without any further assumptions enters the usual design process in companies. Nevertheless full-scale experiments in wind tunnels and outdoors are equally as important as analytic theories.

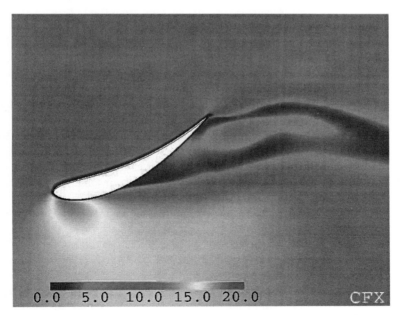

Figure 28: Sketch of the separation of flow through a short diffuser.

References

[1] Abbot I.H. & von Doenhoff A.E., *Theory of Wing Sections*, Dover Publications Inc.: New York, USA, 1959.

[2] Katz, J. & Plotkin A., *Low-Speed Aerodynamics*, Cambridge University Press, 2001.

[3] Thwaites, B. (ed), *Incompressible Aerodynamics*, Oxford University Press, 1960.

[4] White, F.M., *Viscous Fluid Flow,* 2nd ed., McGraw-Hill: New York, USA, 1991.

[5] Breslin, J.P. & Andersen, P., *Hydrodynamics of Ship Propellers*, Cambridge University Press: Cambridge, UK, 1986.

[6] Leishman, J.G., *Principles of Helicopter Aerodynamics*, Cambridge University Press: Cambridge, UK, 2000.

[7] Sparenberg, J.A. *Hydrodynamic Propulsion and Its Optimization*, Kluwer Academic Press: Dordrecht, The Netherlands, 1995.

[8] Theodorsen, Th., *Theory of Propellers*, McGraw-Hill Book Company Inc.: New York, USA, 1948.

[9] Betz, A., *Wind-Energie und ihre Ausnützung durch Windmühlen* Göttingen, Vandenhoeck & Ruprecht, 1926, in German.

[10] Brouckaert, J.-F. (Ed.), *Wind Turbine Aerodynamics: A State-of-the-Art*, von Karman Institute for Fluid Dynamics Lecture Series 2007-05, Rhode Saint Genèse, Belgium, 2007.

[11] Burton, T., Sharpe, D., Jenkins, N. & Bossanyi, E., *Wind Energy Handbook*, John Wiley & Sons, Ltd.: Chichester, 2001.

[12] Glauert, H., *The Elements of Airfoil and Airscrew Theory*, 2nd ed., Cambridge University Press: UK, 1993.

[13] Hansen, M.O.L., *Aerodynamics of Wind Turbines*, 2nd ed, Earthscan Ltd.: London, UK, 2007.

[14] Hau, E., *Wind Turbines: Fundamentals, Technologies, Applications, Economics*, J. Springer: Berlin, Germany, 2005.

[15] Paraschivoiu, I., *Wind Turbine Design – with Emphasis on Darrieus Concept*, Polytechnic International Press: Montreal, Canada, 2002.

[16] Spera, D.A. (ed.), *Wind Turbine Technology*, ASME Press, New York, 1998.

[17] de Vries, O., *Fluid Dynamics Aspects of Wind Energy Conversion*, AGARDograph, **242**, Neuilly-sur-Seine, France, 1979.

[18] Wilson, R.E., Lissaman, P.B.S., Walker, S.N., *Aerodynamic Performance of Wind Turbines*, Oregon State University Report, USA, 1976.

[19] Bir, G. & Jonkman, J., Aeroelastic instabilities of large offshore and onshore wind turbines, *Journal of Physics, Conf. Series*, **75**, 012069, 2007.

[20] van Bussel, G.J.W., The science of making more torque from wind: diffuser experiments and theory revisited, *Journal of Physics, Conf. Series*, **75**, 012010, 2007.

[21] Björck, A., Göran Ronsten and Björn Montgomerie, *Aerodynamic section characteristics of a rotating and non-rotating 2.375 wind turbine blade*, FFA TN 1995-03, Stockholm, Sweden, 1995.

[22] Freudenreich, K., Kaiser, K., Schaffarczyk, A.P., Winkler, H. & Stahl, B., Reynolds-number and roughness effects on thick airfoils for wind turbines, *Wind Engineering*, **28(5)**, pp. 529–546, 2004.

[23] Froude, R.E., On the part played in propulsion by differences in fluid pressure. *Trans. Inst. Nav. Arch.*, **30**, p. 390, 1889.

[24] Goldstein, S., On the vortex theory of screw propellers, *Proc. Royal Soc.*, **A123**, p. 440, 1929.

[25] Hansen, M.O.L., Sørensen, J.N., Voutsinas, S., Sørensen, N. & Madsen, H.Å., *State of the art in wind turbine aerodynamics and aeroelasticity*, *Prog. Aero. Sci.*, **42**, pp. 285–330, 2006.

[26] Hansen, C. & Butterfield, S., Aerodynamics of horizontal-axis wind turbines, *Ann. Rev. Fl. Mech.*, **25**, pp. 115–149, 1993.

[27] Hansen, M.O.L., Sørensen, N.N. & Flay, R.G.J., Effect of placing a diffuser around a wind turbine, *Proc. EWEC 1999*, Nice, France, 1999.

[28] Hillmer, B., Yun Sun Chol & Schaffarczyk, A.P., Modelling of the transition locations on a 30% thick airfoil with surface roughness, *Wind Energy, Proc. Euromech Coll.*, Bad Zwischenahn, Germany, 2005.

[29] Hillmer, B., Borstelmann, T., Schaffarczyk, A.P. & Dannenberg, L., Aerodynamic and structural design of multiMW wind turbines beyond 5 MW, *Journal of Physics, Conf. Series*, **75**, 012002, 2007.

[30] Himmelskamp, H., *Profiluntersuchungen an einem umlaufenden Propeller*, Mitteilungen aus dem Max-Planck-Inst. für Strömungsforschung, Nr. 2, 1950, in German.

[31] Hansen, M.O.L. & Johannsen, J., Tip studies using CFD and comparison with tip loss models, *Proc. EAWE Conference on The Science of making Torque from Wind*, Delft, The Netherlands, 2004.

[32] Herzog, R., Schaffarczyk, A.P., Wacinski, A. & O Zürchner, Performance and stability of a counter-rotating windmill using a planetary gearing: measurements and simulation, submitted to EWEC 2010, Warsaw, Poland, 2010.

[33] Hütter, U., Optimization of wind-energy conversion systems, *Ann. Rev. Fl. Mech.*, **9**, pp. 399–419, 1977.

[34] Jamieson, P., Generalized limits for energy extraction in a linear constant velocity fields, *Wind Energy*, **11**, p. 445, 2008.

[35] Johansen, J. & Sørensen, N.N., Aerodynamic investigations of winglets on wind turbines blades using CFD, *Proc. IEA Joint Action of Basic Information Exchange*, Kiel, Germany, 2006.

[36] Joukowsky, N., *Théorie tourbillonnaire de l'hélice propulsive*, Gauthier-Villars: Paris, France, 1929.

[37] Laursen, J., Enevoldsen, P. & Hjort, S., 3D CFD quantification of the performance of a multi-megawatt wind turbine, *Journal of Physics, Conf. Series*, **75**, 012007, 2007.

[38] Mikkelsen, R., *et al.*, Towards the optimal loaded actuator disk, *Proc. IEA Joint Action of Basic Information Exchange*, Kiel, Germany, 2006.

[39] Sørensen, J.N. & Okulov, L., Refined Betz limit for rotors with a finite number of blades, *Wind Energy*, **11**, pp. 415–426, 2008.

[40] Sørensen, J.N. & Okulov, V.L., Vortex theory of the ideal wind turbine, *Proc. EWEC 2009*, Marseille, France, 2009.

[41] Phillips, D.G., *An Investigation on Diffuser Augmented Wind Turbine Design*, PhD Thesis, The University of Auckland, Auckland, New Zealand, 2003.

[42] Prandtl, L., *Zusatz zu: A. Betz: Schraubenpropeller mit geringstem Energieverlust*, Nachrichten der Kgl. Ges. d. Wiss., Math.-phys. Klasse, Berlin, 1919, in German.

[43] Rankine, W.J., On the mechanical principles of the action of propellers, *Trans. Inst. Nav. Arch.*, **6**, p. 13, 1865.

[44] Göran Ronsten, Static pressure measurements on a rotating and a non-rotating 2.375 m wind turbine blade. Comparison with 2D calculations, *J. Wind. Eng. Ind. Aero*, **39**, pp. 105–118, 1992.

[45] van Rooij, R.P.J.O.M., *Open Air Experiments on Rotors*, in [10].

[46] Worobel, C., Wainauski, H. & Rohrbach, R., *Experimental and Analytical Research on the Aerodynamics of Wind Turbines*, ERDA COO-2615-76/2, Windsor Locks, Conn., USA, 1977.

[47] Schaffarczyk, A.P. & Phillips, D., Design principles for a diffusor augmented wind-turbine blade, *Proc. EWEC 2001*, Copenhagen, Denmark, 2001.

[48] Schaffarczyk, A.P., *Actuator Disk Modelling of Contra-Rotating Wind-Turbines*, CFD Lab Report, **32**, Kiel, Germany, 2003.

[49] Schaffarczyk, A.P. & Stahl, W., Experimentally and numerically deduced performed properties of a 30% thick roughened at high Reynolds numbers for use on multi megawatt blades, *Proc. 7th German Wind Energy Conference*, DEWEK 2004, Wilhlemshaven, Germany, 2004.

[50] Schaffarczyk, A.P. & Hillmer, B., *Leistungsvergleich der Rotorblätter LM34, NO134 und APX70 für die Windenergieanlage MD70*, CFD Lab Report, **44**, Kiel, Germany, 2005 (in German, confidential).

[51] Schaffarczyk, A.P., Investigations of 3d transition on blades of wind turbines, *IEA Annex XI and XX Expert Meeting on Aerodynamics for Wind Turbines*, Lyngby, Denmark, 2007.

[52] Schepers, J.G., *et al.*, *Final Report of the IEA Annex XIV: Filed Rotor Aerodynamics*, ECN-C-97-027, Petten, The Netherlands, 1997.

[53] Sharpe, D.J., A general momentum theory applied to an energy-extraction actuator disk, *Wind Energy*, **7**, pp. 177–188, 2004.

[54] Shen, W.Z., Hansen, M.O.L. & Sørensen, J.N., Determination of the angle of attack (AOA) on rotor blades, *Wind Energy*, **12**, pp. 91–98, 2009.

[55] Shen, W.Z., Mikkelsen, R. & Sørensen, J.N., Tip loss corrections for wind turbine computations, *Wind Energy*, **8**, pp. 457–475, 2005.

[56] Simms, D., Schreck, S., Hand, M. & Fingesh, L.J., *NREL Unsteady Aerodynamics Experiment in the NASA-Ames Wind Tunnel: A Comparison of Predictions to Measurements*, NREL/TP-500-29444, Golden, CO, USA, 2001.

[57] Snel, H., Schepers, J.G. & Montgomerie, B., The MEXICO project (model experiments in controlled conditions): the database and first results of data processing and interpretation, *Journal of Physics, Conf. Series*, **75**, 012014, 2007.

[58] Sørensen, J.N., Shen, W.Z. & Munduate, X., Analysis of flow states by full-field actuator disk model, *Wind Energy*, **1**, pp. 73–88, 1998.

[59] Sørensen, N.N., Michelsen, J.A. & Schreck, S., Detailed aerodynamic prediction of the NASA/AMES wind tunnel tests using CFD, *Proc. EWEC 2001*, Copenhagen, Denmark, 2001.

[60] Sørensen, N., CFD modelling of laminar-turbulent transition of blades and rotors, *Proc. EWEC 2008*, Brussels, Belgium, 2008.

[61] Timmer, W.A. & Schaffarczyk, A.P., The effect of roughness on the performance of a 30% thick wind turbine airfoil at high Reynolds Numbers, *Wind Energy*, **7**, pp. 295–307, 2004.

[62] Trede, R., *Entwicklung eines Netzgenerators*, Diploma thesis, Fachhhochschule Westküste, Heide, Germany, 2003, in German.

[63] Thor, S.-E. (ed.), Challenges of introducing reliable small wind turbines, *49th IEA Topical Expert Meeting*, Stockholm, Sweden, 2006.

[64] Vermeer, L.J., Sørensen, J.N. & Crespo, A., Wind turbine wake aerodynamics, *Prog. Aero. Sci.*, **39**, pp. 467-510, 2003.

[65] Winkler, H. & Schaffarczyk, A.P., *Numerische Simulation des Reynoldszahlverhaltens von dicken aerodynamischen Profilen für Off-Shore Anwendungen*, CFD-Lab internal Report, **33**, Kiel, Germany, 2003, in German.

[66] Holme, O., A contribution to the aerodynamic theory of the vertical-axis wind turbine, Paper C4, *Proc. (First) Int. Symp. on Wind Energy Systems*, Cambridge, England, 1976.

[67] Kirke, B.K., *Evaluation of self-starting vertical axis wind turbines for stand-alone applications*, PhD Thesis, Griffith University, Australia, 1998.

[68] Klimas, P.C. & Sheldahl, R.E., *Four aerodynamic prediction schemas for vertical axis wind-turbines: a compendium*, Sandia National Laboratories, Sand78-0014, Albuquerque, NM, USA, 1978.

[69] Lapin, E.E., Theoretical performance of vertical axis wind-turbine, *Trans. ASME Winter Meeting*, Houston, TX, USA, 1975.

[70] Loth, J.L. & McCoy, H., Optimization of Darrieus turbines with an upwind and downwind momentum Model, *J. Energy*, **7**, pp. 313–318, 1983.

[71] Mukinovic, M., Brenner, G., Rahimi, A., *Analysis of Vertical Axis Wind Turbines*, 16th DGLR Symposium of STAB, Aachen, Germany, 2008.

[72] I. Paraschivoiu, P. Desy, *Aerodynamics of Small-Scale Vertical-Axis Wind Turbines*, J. Prop. **2**, 282–288, 1986.

[73] Schaffarczyk, A.P. *New aerodynamical Modeling of a Vertical Axis Wind-Turbine with Application to Flow Conditions with Rapid Directional Changes*, Proc. Dewek 2006, Bremen, Germany, 2006.

[74] Stickland, J.H., *The Darrieus Turbine: A Performance Prediction Model Using Multiple Streamtube*, Sandia National Laboratories, Sand75-0431, Albuquerque, NM, USA, 1975.

[75] Templin, R.J., *Aerodynamic Performance Theory for the NRC Vertical Axis Wind Turbine*, LTR-LA-160, Ottawa, Canada, 1974.

[76] Eggleston, D.M. & Stoddard, F.S., *Wind Turbine Engineering Design*, Van Nordstrand Reinhold Comp., New York, USA, 1987.

[77] Lobitz, D.W. & Ashwill, T.D., *Aeroelastic Effects in the Structural Dynamic Analysis of Vertical Axis Wind-Turbine*, Sandia Report, SAND85-0957, Albuquerque, NM, USA, 1986.

[78] Vollan, A., *Aeroelastic Stability Analysis of a Vertical Axis Wind Energy Converter*, Bericht EMSB-44/77, Dornier System, Immenstaad, Germany, 1977.

[79] Vollan, A., *The aeroelastic behaviour of large Darrieus-type wind energy converters derived from the behaviour of a 5.5 m rotor*, Paper C5, *Proc. 2nd Int. Symp. on Wind Energy Systems*, Amsterdam, The Netherlands, 1978.

CHAPTER 4

Structural dynamics of wind turbines

Spyros G. Voutsinas

National Technical University of Athens, Greece.

Structural dynamics of wind turbines aims at assessing the integrity and reliability of the complete construction against varying external loading over the targeted life time. Since wind induced excitation is the most important, structural dynamics is closely connected to aeroelasticity. Most of structural analysis is based on modelling while tests are reserved for prototype assessment. To this end elastic modelling at component level and dynamic modelling of the complete system are required. The aim of the present chapter is to provide an overview on the theory and approximation tools of structural dynamic analysis of modern wind turbines on these two aspects. At component level beam models of varying complexity are discussed while at system level the multi-body approach is defined and its use in formulating the dynamic model of the complete wind turbine is explained. Next, after a short introduction to aeroelastic coupling, the issue of stability is outlined. Finally some indicative results show the kind of information structural analysis can provide. The focus of the presentation is on the procedures rather than the details of derivations for which appropriate pointers to the literature are given.

1 Wind turbines from a structural stand point

A typical wind turbine configuration includes as main components the rotor blades, the drive train and the tower, all considered as flexible structures. When considering the wind turbine as a system, all components are modelled as slender beams subjected to combined bending, torsion and tension with distributed material properties like stiffness, damping and density. At this level, the elastic behaviour of the different components is described by continuous displacement and rotation fields along the span of each component. In the discrete context the displacements and rotations will correspond to the *elastic degrees of freedom* (*DOF*). The beam like modelling associated to structural flexibility is discussed in Sections 3 and 7.

Clearly assessment of local stresses and strains especially when dealing with composites, requires more advanced modelling that account for the internal material structure.

Solely the elastic DOF cannot describe the complete dynamics of a wind turbine. There are also *rigid* DOF while the connections among the components further complicate the dynamics of the system. The rotor blades are connected to the hub (the one end of the drive train) allowing rotation with respect to the drive-train axis while the nacelle which houses the drive train, is connected to the tower top also allowing at least rotation in yaw. In addition to blade rotation and yaw there is also the blade pitch. Pitch can vary either collectively or cyclically or even independently for each blade. Finally in the case of two-bladed rotors, the rotor system as a whole is allowed to teeter. At the connection points, specific compatibility conditions are required which transfer the deformation state from one connected component to the other. This calls for a multi-component analysis which we detail in Section 4.

Most of the rigid body DOF are connected to the control system. At least in the more recent designs there is clear preference for variable pitch and variable speed control concepts. So when the wind turbine operates in closed loop, the variation of the pitch and the rotor speed will satisfy the control equations which therefore become part of the dynamic description of the system. In certain cases, there is need to also include concentrated properties like point masses, springs or dampers. For example, the yaw gear and the gear box in the drive train are modelled as combined point masses and springs while the hub is modelled as concentrated mass. Concentrated properties are associated to specific DOF for which we must formulate appropriate dynamic equations as in the case of the control variables.

Although reference has been made to the formulation of dynamic equations we did not specify the way this is done. The general framework is presented in Section 2. One way to proceed is to base the derivation on Newton's second law which requires the *dynamic balance* of forces and moments. It has the advantage of a direct interpretation in relatively simple systems but becomes cumbersome when dealing with complicated systems. Alternatively energy or variational principles, like the principle of virtual work or Hamilton's principle offer a general and systematic way for deriving the dynamic equations of even very complicated mechanical systems. It is true that depending on the complexity of the system, the derivations can become quite lengthy. However, the current availability of symbolic mathematics software has removed former reluctance in using variational principles.

Once the dynamic equations of the complete wind turbine are formulated, the next step is to include the aerodynamic loading on the blades. The main difficulty in this respect is that aerodynamic loads depend on the dynamics of the blades in a non-linear manner. A further complication is due to the onset of stall which because of the aeroelastic coupling becomes dynamic with clear lagging behaviour. A brief account on aeroelastic coupling is given in Section 5.

The usual way to solve the coupled final aeroelastic equations is to integrate them in time. Due to their non-linear character in principle we need an implicit iterative solver. Depending on the non-linearities considered and the level of detail we wish to include in the model, the computational effort can become high especially for industrial purposes. So, in order to reduce the computational cost, quite often linearization is introduced at various levels. There are both structural as well as aerodynamic non-linearities involved. Structural non-linearities are connected to large displacements and rotations, so a usual simplification is to consider the case of small deformations with respect to either the un-deformed state or a reference equilibrium deformed state. In this context significant computational cost reduction can be accomplished by introducing the structural modes of the system. A similar procedure can be followed for the non-linearities of aerodynamic origin. However, as already mentioned the onset of stall and the resulting load hysteresis complicates the whole procedure and linearization of the unsteady aerodynamic loading must be done carefully.

Linearized models are not only introduced for computational cost reduction. They are also used in carrying out aeroelastic stability analysis. Aeroelastic instabilities including flutter will appear whenever the system lacks sufficient damping. In systems with strong aeroelastic coupling the damping has two components; one linked to the material characteristics of the structure and the other contributed by the unsteady aerodynamics. Both damping contributions are subjected to changes. Because blades are made of composite materials, structural damping depends on the ambient temperature while it is known that ageing will degrade their damping. On the other hand, aerodynamic damping depends on the flow conditions and will decrease or even become negative as the flow approaches or enters stall. This means that close to nominal operating conditions, the effective angles of attack along the blade will approach their maximum and therefore the less favourable situation will occur. Wind speed fluctuations due to turbulence will force the blade to enter stall so if there is no sufficient structural damping the blades will flutter. In this connection linearization of the complete servo-aeroelastic system offers a direct and cost-effective way to assess stability. The option of using the non-linear form of the dynamic equations is also possible but at a much higher cost. Besides that the response recorded by a non-linear simulation depends on the excitation applied and in principle will excite all the modes of the system. In order to focus on a specific mode an appropriate excitation must be applied which is not always possible. Stability is discussed in Section 6.

Finally in Section 8 a brief presentation is given on the kind of information structural analysis can provide and what is their use in the design process. The presentation relies on simulations carried out in practice and whenever possible measured data are also included so as to have some insight into the quality of predictions current analysis can achieve.

2 Formulation of the dynamic equations

Structural dynamics is based on dynamic equilibrium (Newton's second law) stating that a solid volume V bounded by ∂V will accelerate as a result of volume and surface loading. The balance of forces and moments leads to the following equations[1]:

$$\int_V \rho \vec{g} \, dV + \int_{\partial V} \vec{t} \, dS = \int_V \rho \ddot{\vec{r}} \, dV, \quad \int_V \rho \vec{g} \times \vec{R} \, dV + \int_{\partial V} \vec{t} \times \vec{R} \, dS = \int_V \rho \ddot{\vec{r}} \times \vec{R} \, dV \quad (1)$$

In eqn (1), the current position $\vec{r}(\vec{r}_0, t)$ of a material point originally at $\vec{r}_0$, includes the elastic displacement and any rigid body motion while $\vec{R}$ denotes the distance from the point with respect to which moments are taken. (An overhead arrow will denote geometric vectors while matrices will be in bold.) Usually the volume loading term is due to gravity. Then as regards surface loading, when V is part of a solid body there will be two terms: the aerodynamic loading defined on the part of ∂V in contact with air, and the internal loading defined on the rest of ∂V in contact with the remaining of the solid body. Internal surface loading is directly connected to the stress tensor σ: $\vec{t} = \sigma^T \vec{n}$ with $\vec{n}$ denoting the outward unit normal, which together with Green's theorem allows transforming the surface integrals into volume integrals and thus deriving the dynamic equations in differential form. Elastic models are next introduced which relate stresses with strains so that at the end the equations are formulated with respect to the displacements and rotations contained in the definition of $\vec{r}$ (for further reading the reader can consult [14] out of the long list of modern textbooks on structural mechanics).

In structural analysis the differential form of the equations is seldom used. Instead the equations are reformulated in weak form based on the principle of virtual work. Weak formulations are the starting point of finite element discrete models of the Galerkin type ([2] and Section 3.2). An alternative and definitely more powerful way to formulate dynamic equations is to use Hamilton's principle [3]. Let T and U denote the kinetic and strain energies, respectively, defined by a set of displacements and rotations collectively denoted as $\mathbf{u}(t) = (u_1, u_2, \ldots)^T$. Assuming the presence of non-conservative loads F_i connected to u_i then

$$-\frac{d}{dt}\left(\frac{\partial(T-U)}{\partial \dot{u}_i}\right) + \frac{\partial(T-U)}{\partial u_i} + F_i = 0, \quad \forall i \quad (2)$$

This is also known as Lagrange equations. Each equation corresponds to balance of either forces or moments depending on whether the associated u_i is a displacement or a rotation. Among the various advantages Hamilton's principle has, of particular importance is that the final form of the dynamic equations is easily obtained using symbolic mathematical software [4, 5].

3 Beam theory and FEM approximations

3.1 Basic assumptions and equation derivation

Beam theory applies when one of the dimensions of the structure is much larger than the others [6, 7]. Using asymptotic analysis, it is possible to eliminate the dependence on the two shortest dimensions by appropriate averaging of the distributed elastic properties [8–11]. The problem is formulated with respect to the remaining third direction defining the axis of the beam also called elastic axis.

Beam theory considers combined bending, tension and torsion. There are several beam models of varying complexity. The simplest is the first order or Euler-Bernoulli model in which the elastic axis is considered rectilinear while cross sections originally normal to it remain so in the deformed state. As a consequence shear is eliminated. Shear will be included in Section 7.1 while in Section 7.2, second order theory will be briefly presented.

There are three steps to take: (a) define the deformation kinematics, (b) introduce the stress–strain relations, (c) apply dynamic equilibrium to a differential volume of the structure. To this end, consider a beam with its elastic axis along the y-axis of the co-ordinate system [O; xyz] which for this reason is called *beam* system. Bending takes place in the x (lead-lag) and z (flap) directions while tension and torsion both take place in the y direction. For any point of the structure, let $\vec{r}_0 = (x_0, y_0, z_0)^T$ and $\vec{r} = (x, y, z)^T$ denote its position in the un-deformed and deformed state, respectively. Assuming small rotations (Fig. 1 shows the situation in the xz plane):

$$\vec{U} = \vec{r} - \vec{r}_0 = \begin{pmatrix} U \\ V \\ W \end{pmatrix} = \begin{bmatrix} 1 & 0 & 0 \\ 0 & 1 & 0 \\ 0 & 0 & 1 \end{bmatrix} \begin{pmatrix} u \\ v \\ w \end{pmatrix} + \begin{bmatrix} 0 & z_0 & 0 \\ -z_0 & 0 & x_0 \\ 0 & -x_0 & 0 \end{bmatrix} \begin{pmatrix} \theta_x \\ \theta_y \\ \theta_z \end{pmatrix} \qquad (3)$$

where $\vec{u}(y,t) = (u,v,w)^T$ and $\vec{\theta}(y,t) = (\theta_x, \theta_y, \theta_z)^T$ are the displacement and rotation vectors of the section that define its deformation state. In the Euler-Bernoulli model, $\theta_x = \partial_y w, \theta_z = -\partial_y u$, so there are only four independent deformation variables $\mathbf{u}(y,t) = (u, v, w, \theta_y)^T$:

$$\vec{r} = \vec{r}_0 + \mathbf{S}^0(\vec{r}_0) \cdot \mathbf{u} + \mathbf{S}^1(\vec{r}_0) \cdot \mathbf{u}', \quad \mathbf{u}(y,t) = (u,v,w,\theta_y)^T, \quad (.)' = \partial(.)/\partial y$$

$$\mathbf{S}^0(\vec{r}_0) = \begin{bmatrix} 1 & 0 & 0 & z_0 \\ 0 & 1 & 0 & 0 \\ 0 & 0 & 1 & -x_0 \end{bmatrix}, \quad \mathbf{S}^1(\vec{r}_0) = \begin{bmatrix} 0 & 0 & 0 & 0 \\ -z_0 & 0 & -x_0 & 0 \\ 0 & 0 & 0 & 0 \end{bmatrix} \qquad (4)$$

Equation (4) defines the deformation kinematics with respect to the beam system.

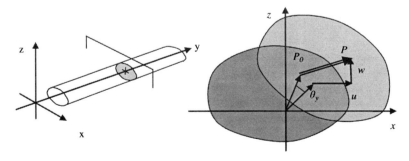

Figure 1: Deformation a beam cross section.

Next the stress–strain relations are introduced. Assuming linear material and linear Green strains, according to (4):

$$\sigma_y = E\varepsilon_y = E\frac{\partial V}{\partial y}, \quad \tau_{xy} = G\gamma_{xy} = G\left(\frac{\partial V}{\partial x}+\frac{\partial U}{\partial y}\right), \quad \tau_{zy} = G\gamma_{zy} = G\left(\frac{\partial V}{\partial z}+\frac{\partial W}{\partial y}\right)$$

By integrating the stresses over the cross section of the beam $A(y)$, the resultant sectional loads are obtained, for given averaged sectional properties of the beam structured here denoted by over bars $\overline{\Phi} = \int_A \Phi dA$ [9]:

$$F_y = \int_{A(y)} \sigma_y\, dA = \overline{E}v' - \overline{Ez}u'' - \overline{Ex}w'', \quad M_y = \int_{A(y)} (\tau_{xy}z_0 - \tau_{yz}x_0)\,dA = \overline{G(x^2+z^2)}\,\theta_y',$$

$$M_x = -\int_{A(y)} z_0\sigma_y\, dA = \overline{Ezx}u'' + \overline{Ex^2}w'', \quad M_z = \int_{A(y)} x_0\sigma_y\, dA = -\overline{Ez^2}u'' - \overline{Ezx}w'' \tag{5}$$

Then as regards F_x and F_z, in the Bernoulli model, their net contribution is derived by means of the balance of moments in the x and z directions:

$$dF_x = (M_z'' + (F_y u')')\,dy, \quad dF_z = (-M_x'' + (F_y w')')\,dy \tag{6}$$

The loads in eqns (5) and (6) represent the internal surface loads in (1) by taking V as a beam segment of width dy (Fig. 2). The remaining surface loads are due to the aerodynamic loading (mainly the pressure) which when integrated gives $\overline{dL}$ acting at the aerodynamic center $\vec{r}_a = (x_a, y, z_a)^{T}$ of the section considered. Note that the terms related to tension F_y are non-linear and second order. Tension is important because it contributes to bending which in the case of a rotating blade will lead to a reduction of the bending moments at blade root. So we need to retain this term in the otherwise first order model. Finally the following differential form of the dynamic equations is obtained:

$$\int_{A(y)} (\rho\, dA)(\mathbf{S}^0)^{T}\ddot{\vec{r}} - \left(\int_{A(y)} (\rho\, dA)(\mathbf{S}^1)^{T}\ddot{\vec{r}}\right)'$$

$$= [\mathbf{K}_{11}\,\mathbf{u}']' + [\mathbf{K}_{22}\,\mathbf{u}'']'' + [\mathbf{K}_{12}\,\mathbf{u}'']' + [\mathbf{K}_{21}\,\mathbf{u}']'' + \int_{A(y)} (\rho dA)(\mathbf{S}^0)^{T}\vec{g} + \mathbf{II}_a\,\overline{\delta L} \tag{7}$$

$$\mathbf{II}_a = \begin{bmatrix} 1 & 0 & 0 \\ 0 & 1 & 0 \\ 0 & 0 & 1 \\ z_a & 0 & -x_a \end{bmatrix}, \quad \mathbf{K}_{11} = \begin{bmatrix} F_y & 0 & 0 & 0 \\ 0 & \overline{E} & 0 & 0 \\ 0 & 0 & F_y & 0 \\ 0 & 0 & 0 & \overline{G(x^2+z^2)} \end{bmatrix}, \quad \mathbf{K}_{21} = \mathbf{K}_{12}{}^{T},$$

$$\mathbf{K}_{22} = \begin{bmatrix} -\overline{Ez^2} & 0 & -\overline{Exz} & 0 \\ 0 & 0 & 0 & 0 \\ -\overline{Exz} & 0 & -\overline{Ex^2} & 0 \\ 0 & 0 & 0 & 0 \end{bmatrix}, \quad \mathbf{K}_{12} = \begin{bmatrix} 0 & 0 & 0 & 0 \\ -\overline{Ez} & 0 & -\overline{Ex} & 0 \\ 0 & 0 & 0 & 0 \\ 0 & 0 & 0 & 0 \end{bmatrix}$$

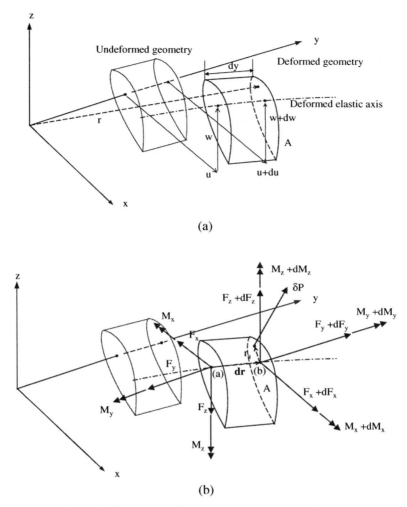

(a)

(b)

Figure 2: Loads and displacements of a beam element.

The above system is completed with appropriate boundary and initial conditions. Boundary conditions at the two ends of the beam will either specify the load (Neumann or static condition) or constrain the corresponding displacement or rotation (Dirichlet or kinematic condition). Unconstrained or free ends will have zero loading.

3.2 Principle of virtual work and FE approximations

Considering the dynamic equations in the form $F(\mathbf{u}) = 0$, then for any virtual displacement $\delta\mathbf{u}$ the work done by the loads $F(\mathbf{u})$ must be zero. Work is a projection operation defined by the inner product of integrable functions: $(f,g) \equiv \int f(x)g(x)\,dx$ with the integral defined on the domain of definition which in the present case is the length of the beam L. So,

$$\int_0^L \delta \mathbf{u}^{\mathrm{T}} F(\mathbf{u})\, dy = 0 \quad \forall \delta \mathbf{u}$$

The projected equations define the weak formulation of the problem. Note that demanding zero virtual work for any virtual displacements is equivalent to performing the projection with respect to a properly defined function basis. Proper in this sense means that boundary conditions must be taken into account. For all constrained DOF corresponding to kinematic conditions, we must set $\delta u_p = 0$. Note that static or load conditions will naturally appear in the weak formulation when integration by parts is carried out. Considering the bending term as an example, a double integration results in

$$\int_0^L \delta \mathbf{u}^{\mathrm{T}} (\mathbf{K}_{22}\mathbf{u}'')''\, dy = \left[\delta \mathbf{u}^{\mathrm{T}} (\mathbf{K}_{22}\mathbf{u}'')' - \delta \mathbf{u}'^{\mathrm{T}} (\mathbf{K}_{22}\mathbf{u}'')\right]_0^L + \int_0^L \delta \mathbf{u}''^{\mathrm{T}} (\mathbf{K}_{22}\mathbf{u}'')\, dy$$

The underlined terms correspond to the boundary terms and represent the virtual work done by the reacting force $(\mathbf{k}_{22}\mathbf{u}'')'$ and moment $(\mathbf{k}_{22}\mathbf{u}'')$ at the support points of the beam. If the displacement or rotation is specified then the term equals zero because δu or $\delta u'$ is zero. On the contrary, if the load is specified then either the force or the moment is set to its given value. All elastic terms in eqn (7) are integrated in the same way.

The most popular method for solving dynamic equations is the Finite Element Method [12–14]. It consists of projecting the equations using a basis of finite dimension. To this end first the beam is divided into elements. Then for each element the same local approximations are defined for $\mathbf{u}$ and $\delta \mathbf{u}$. In doing so, specific polynomial shape functions and discrete DOF $\hat{\mathbf{u}}$ are chosen (Fig. 3).

The choice of the shape functions depends on the order of the problem. The beam equations are second order for the tension and torsion so we can choose linear shape functions $(\gamma_n, n = 1,2)$, and fourth order for the bending so we use cubic functions $(\beta_n^a, n = 1,2, a = 0,1)$. The discrete DOF's usually correspond to the nodal values of $\mathbf{u}$ but can also include the values of its space derivative as in the case of bending. Taking as nodes the two ends of each element e:

$$\mathbf{u}_h^e(y) = \begin{pmatrix} u & v & w & \theta_y \end{pmatrix}^{\mathrm{T}} = \mathbf{N}^e(y)\hat{\mathbf{u}}_e$$

$$\hat{\mathbf{u}}_e = (\hat{u}_e, \hat{u}'_e, \hat{v}_e, \hat{w}_e, \hat{w}'_e, \hat{\theta}_e, \hat{u}_{e+1}, \hat{u}'_{e+1}, \hat{v}_{e+1}, \hat{w}_{e+1}, \hat{w}'_{e+1}, \hat{\theta}_{e+1})^{\mathrm{T}},$$

$$\mathbf{N}^e = \begin{bmatrix} \beta_1^0 & \beta_1^1 & 0 & 0 & 0 & 0 & \beta_2^0 & \beta_2^1 & 0 & 0 & 0 & 0 \\ 0 & 0 & \gamma_1 & 0 & 0 & 0 & 0 & 0 & \gamma_2 & 0 & 0 & 0 \\ 0 & 0 & 0 & \beta_1^0 & \beta_1^1 & 0 & 0 & 0 & 0 & \beta_2^0 & \beta_2^1 & 0 \\ 0 & 0 & 0 & 0 & 0 & \gamma_1 & 0 & 0 & 0 & 0 & 0 & \gamma_2 \end{bmatrix},$$

$$\xi = \frac{y - y_e}{L_e}, \qquad \beta_1^0 = 1 - 3\xi^2 + 2\xi^3, \qquad \beta_1^1 = L_e(\xi - 2\xi^2 + \xi^3),$$

$$L_e = y_{e+1} - y_e, \qquad \beta_2^0 = 3\xi^2 - 2\xi^3, \qquad \beta_2^1 = L_e(-\xi^2 + \xi^3),$$

$$\gamma_1 = 1 - \xi, \qquad\qquad \gamma_2 = \xi$$

(8)

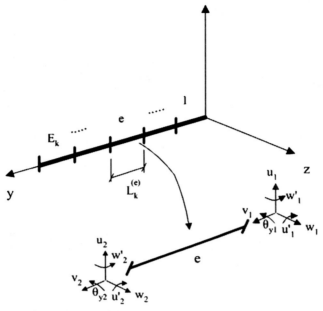

Figure 3: The finite element description of a beam.

Because the equations are in integral form, all calculations can be first carried out at element level and then proceed with the assembly. For example, the mass matrix will take the form:

$$\mathbf{M}_e = \int_{L_e} dy \left(\int_A dA\, \rho\, \mathbf{N}_e^T\, \mathbf{II}_s\, \mathbf{S}\mathbf{N}_e \right) - \int_{L_e} dy \left(\int_A dA\, \rho\, \mathbf{N}_e'^T\, \mathbf{II}_s\, \mathbf{S}\mathbf{N}_e \right)$$

The final result, after assembling the element matrices will have the usual form:

$$\mathbf{M}\ddot{\hat{u}} + \mathbf{K}\hat{u} = \mathbf{Q} \qquad (9)$$

The above system is integrated in time by a marching scheme. In most cases Newmark's second order scheme is sufficient. For stiff problems higher order schemes of the Runge-Kutta type can be also applied [15].

4 Multi-component systems

Next we consider a combination of several beams possibly in relative motion. By taking each component separately, we need first to include its motion and then add appropriate compatibility conditions that ensure specific connection constraints. This kind of splitting and reconnecting constitutes the so-called multi-body approach [16–18].

4.1 Reformulation of the dynamic equations

Motions are introduced by assuming that each component undergoes rigid motions defined by the position of its local system [O; *xyz*] with respect to a global (inertial) system $[O; xyz]_G$. Figure 4 shows the splitting of a wind turbine in its components and indicates the relation between local and global geometric descriptions.

Let $\vec{R}_k$ and $\mathbf{T}_k$ denote respectively the position and the rotation matrix of the local system $[O; xyz]$ of component k with respect to $[O; xyz]_G$. $\vec{R}_k$ and $\ddot{R}_k$ define the linear velocity and acceleration while the time derivatives of $\mathbf{T}_k$ introduce the angular velocity and the Coriolis and centrifugal accelerations. Then for a point P of component k, initially at the local position $\vec{r}_0$ and currently at $\vec{r} = \vec{r}_0 + \mathbf{S}(\vec{r}_0) \cdot \mathbf{u}(y,t)$, with respect to $[O; xyz]$:

$$\vec{r}_{G,k}(P;t) = \vec{R}_k(t) + \mathbf{T}_k(t)\vec{r}\,(P;t) \tag{10}$$

The dynamic equations with respect to $[O; xyz]_G$ will involve the global acceleration (for simplicity subscript k and dependencies have been removed):

$$\ddot{\vec{r}}_G = \ddot{\vec{R}} + \ddot{\mathbf{T}}\vec{r}_0 + \ddot{\mathbf{T}}\mathbf{S}\mathbf{u} + 2\dot{\mathbf{T}}\mathbf{S}\dot{\mathbf{u}} + \mathbf{T}\mathbf{S}\ddot{\mathbf{u}} \tag{11}$$

as well as the projection of the equations on the local directions. Projection on local directions is necessary because the elastic formulation is defined in the local (beam) system for each component. Local directions must be given with respect to $[O; xyz]_G$ which is done by the transpose of $\mathbf{T}_k$ So eqn (7) takes the form:

$$\int_{A(y)} (\rho\, dA)(\mathbf{S}^0)^\mathrm{T}\mathbf{T}^\mathrm{T}\ddot{\vec{r}}_G - \left(\int_{A(y)} (\rho\, dA)(\mathbf{S}^1)^\mathrm{T}\mathbf{T}^\mathrm{T}\ddot{\vec{r}}_G\right)'$$

$$= \int_{A(y)} (\rho dA)(\mathbf{S}^0)^\mathrm{T}\mathbf{T}^\mathrm{T}\vec{g} + [\mathbf{K}_{11}\,\mathbf{u}']' + [\mathbf{K}_{22}\,\mathbf{u}'']'' + [\mathbf{K}_{12}\,\mathbf{u}'']' + [\mathbf{K}_{21}\,\mathbf{u}']'' + \mathbf{II}_a\,\overline{\delta L} + \mathrm{BL} \tag{12}$$

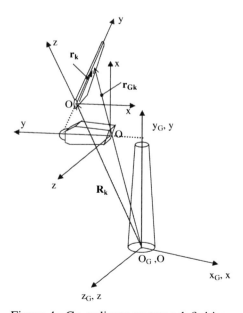

Figure 4: Co-ordinate systems definition.

"BL" in eqn (12) corresponds to the loads at the boundaries and is related to the connection conditions that must be satisfied when assembling the component equations into the final system. This point is explained in Section 4.3.

4.2 Connection conditions

Consider first two bodies connected at one point as in Fig. 5. If the connection is completely rigid, then the displacement and rotation at the connection point must be the same. Also the loading should be the same. The first constraint corresponds to a kinematic condition while the second to a static condition. Note that while it is allowed to either specify a kinematic or a static condition at an end point, we cannot do both. In fact each kinematic condition has a corresponding static condition associated to it. So if a displacement (or rotation) is specified then the associated reaction force (or moment) becomes part of the solution and therefore must remain free. Clearly the reaction will depend on the value specified for the displacement. If instead the force is specified, then the displacement will depend on the input load. So conditions appear in pairs and for each pair we can only specify one. At connection points the situation is somehow different because there are at least two bodies connected, each requiring its own boundary condition at the connection point. Also the entries to both conditions are unknown. Let q denote collectively the displacements and rotations of the connection point and Q the corresponding loads. By considering each body separately, it is possible to formulate separate solutions by setting q as kinematic condition to body 1 and Q as static condition to body 2. The solution for body 1 will provide Q as a function of q while body 2 will provide q as a function of Q. These two relations define the connection conditions needed. Note that any kind of combination is possible provided that the pairs of conditions are properly split.

If the connection is not rigid but allows free motion in certain directions, then the connection conditions do not apply in these specific directions. For example a

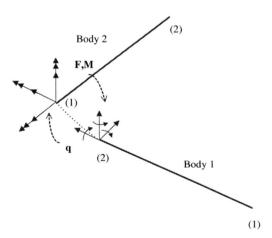

Figure 5: Realization of kinematic and dynamic coupling conditions.

two-bladed teetering rotor will transmit all loads to the drive train except for the teetering moment. Also the displacements and the two rotations at the end of the drive train will be communicated to the blades. Since the teetering moment is not transmitted it can only be zero and this corresponds to the dynamic equation that determines the teetering angle.

Certain connections, involve more than two bodies. In such cases note that while displacements and rotations should not be added, loads must. So if body 1 provides a displacement then it must receive the sum of the corresponding loads from all other connected bodies.

4.3 Implementation issues

In order to facilitate the code implementation for systems with several connections, the set of all kinematic DOF involved in the connections is introduced as an additional unknown denoted as q. Therefore extra equations must be defined. Note that q will contain not only elastic DOF but also all rigid body DOF like yaw, teeter or pitch. If a specific q_i is indeed an elastic degree of freedom u_p then we simply set $q_i \equiv u_p$ Otherwise an additional equation is needed. The condition of zero moment for the teeter angle is such an example. Another example is the pitch angle which is specified by the control system. In this case the extra equation would correspond to the controller equation (or equations).

The introduction of **q**, specifies the form of $\vec{R}_k$ and $\mathbf{T}_k$ Starting from the local system of component k, a series of system displacements and rotations will bring us to the global system:

$$\vec{r}_G = \vec{R}_{m,k} + \mathbf{T}_{m,k}(\cdots + \mathbf{T}_{1,k}(\vec{R}_{1,k} + \vec{r}_k)\cdots) = \vec{R}_k + \mathbf{T}_k\vec{r}_k = \vec{R}_k + \mathbf{T}_k\vec{r}_0 + \mathbf{T}_k\mathbf{Su} \quad (13)$$

In the above relation, each $\mathbf{T}_{m,k}$ may contain several consecutive rotations and therefore appear as a product of elementary rotation matrices of the type: $\mathbf{T}_*(\varphi)$ defined for a given direction $* = x, y, z$ and a given angle φ. As an example consider the case of the drive train.

$$\vec{r}_G = \left\{ \begin{matrix} q_1 \\ H_T + q_2 \\ q_3 \end{matrix} \right\} + \mathbf{T}_z(-q_4)\mathbf{T}_x(-q_5)\mathbf{T}_y(q_6 + \varphi_{\text{yaw}})\mathbf{T}_z(90)(\vec{r}_0 + \mathbf{Su})$$

where H_T denotes the tower height, q_1–q_6 denote the elastic displacements and rotations at the tower top and ϕ_{yaw} denotes the yaw angle. The 90° rotation is here added so that the axis of the drive train is in the y local direction.

Since $\vec{R}_k$ and $\mathbf{T}_k$ depend on **q**,

$$\dot{\vec{R}}_k = \partial_j\vec{R}_k\dot{q}_j, \quad \ddot{\vec{R}}_k = \partial^2_{ij}\vec{R}_k\dot{q}_i\dot{q}_j + \partial_j\vec{R}_k\ddot{q}_j$$

$$\dot{\mathbf{T}}_k = \partial_j\mathbf{T}_k\dot{q}_j, \quad \ddot{\mathbf{T}}_k = \partial^2_{ij}\mathbf{T}_k\dot{q}_i\dot{q}_j + \partial_j\mathbf{T}_k\ddot{q}_j \quad (14)$$

in which $\partial_j = \partial/\partial q_j$, $\partial^2_{ij} = \partial^2/\partial q_i\partial q_j$ and repeated indexes indicate summation. It is clear that all components of **q** are not always needed. Nevertheless eqn (14) reveals the complex non-linear character of multi-body systems and this is independent of the elastic modelling. Equation (14) is introduced in (10) and subsequently into (12). The resulting equations are discretized using FEM approximations. For each component of the system the dynamic equations will then take the following form:

$$\mathbf{M}_k \hat{\ddot{\mathbf{u}}}_k + \mathbf{C}_k(\mathbf{q}, \dot{\mathbf{q}})\hat{\dot{\mathbf{u}}}_k + (\mathbf{K}_k^G(\mathbf{q}, \dot{\mathbf{q}}, \ddot{\mathbf{q}}) + \mathbf{K}_k)\hat{\mathbf{u}}_k + \mathbf{R}_k(\mathbf{q}, \dot{\mathbf{q}}, \ddot{\mathbf{q}}) + BL = \mathbf{Q}_k \quad (15)$$

Note in (15) the appearance of a damping like term defined by $\mathbf{C}_k$, an extra stiffness term defined by $\mathbf{K}_k^G$ and a non-linear term $\mathbf{R}_k$ depending on $\mathbf{q}$ and its time derivatives.

The boundary loads BL are determined by the virtual work done by the reacting forces and moments $\vec{F}$ and $\vec{M}$ at the connection points:

$$(\delta\mathbf{u}^{\mathrm{T}}\mathbf{I}_F + \delta\mathbf{u}'^{\mathrm{T}}\mathbf{I}_M)\begin{Bmatrix}\vec{F}\\\vec{M}\end{Bmatrix} = \delta\mathbf{u}^{\mathrm{T}}[\mathbf{K}_{11}\mathbf{u}' + \mathbf{K}_{12}\mathbf{u}'' + (\mathbf{K}_{21}\mathbf{u}')'\mathbf{K}_{22}\mathbf{u}''] - \delta\mathbf{u}'^{\mathrm{T}}\mathbf{K}_{22}\mathbf{u}''$$

$$\mathbf{I}_F = \begin{bmatrix} 1 & 0 & 0 & 0 & 0 & 0 \\ 0 & 1 & 0 & 0 & 0 & 0 \\ 0 & 0 & 1 & 0 & 0 & 0 \\ 0 & 0 & 0 & 0 & 1 & 0 \end{bmatrix}, \quad \mathbf{I}_M = \begin{bmatrix} 0 & 0 & 0 & 0 & 0 & -1 \\ 0 & 0 & 0 & 0 & 0 & 0 \\ 0 & 0 & 0 & 1 & 0 & 0 \\ 0 & 0 & 0 & 0 & 0 & 0 \end{bmatrix} \quad (16)$$

after eliminating the virtual displacements and rotations. The boundary loads will introduce stiffness which will however depend on the DOF of the bodies connected to k. The final step is to assemble all component equations into one final system.

In view of obtaining a more manageable set of equations, linearization is usually carried out based on formal Taylor's expansions with respect to a reference state. The reference state can be either fixed as in the case of linear stability analysis, or represent the previous approximation within an iterative process towards the non-linear solution. By collecting all unknowns into one vector $\mathbf{x}$, the following form is obtained:

$$\mathbf{M}\ddot{\mathbf{x}} + \mathbf{C}\dot{\mathbf{x}} + \mathbf{K}\mathbf{x} = \mathbf{Q}, \quad \mathbf{x} = (\hat{\mathbf{u}}^{\mathrm{T}}, \mathbf{q}^{\mathrm{T}})^{\mathrm{T}} \quad (17)$$

In eqn (17) although there is no dependence indicated, $\mathbf{M}$, $\mathbf{C}$, $\mathbf{K}$ and $\mathbf{Q}$ depend on the reference state $\mathbf{x}_0$ and its time derivatives. The structure of the mass, damping and stiffness matrices is given in Fig. 6 in the case of a three-bladed wind turbine. The contribution of the local equations for each component are block diagonal. The kinematic conditions at the connection points appear in the out right column. The static connection conditions appear as isolated rows denoted as "dynamic coupling terms" and correspond to the terms appearing in eqn (16). Finally the extra equations for $\mathbf{q}$ appear last.

4.4 Eigenvalue analysis and linear stability

Eigenvalue analysis is a useful tool in structural analysis because it provides a concise dynamic characterization of the system considered. For a linear system without

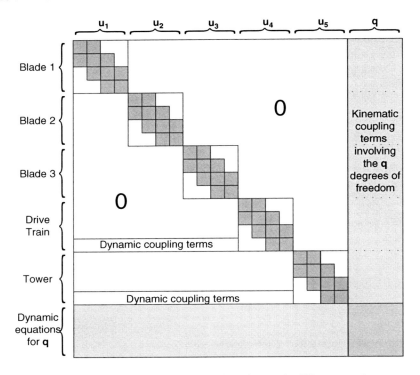

Figure 6: Structure of the mass, damping and stiffness matrices.

damping as described by eqn (9), the eigenvalues ω_i are defined as solutions of $\left| -\omega_i^2 \mathbf{M} + \mathbf{K} \right| = 0$ and for each one an eigenmode $\mathbf{e}_i$ is obtained by solving:

$$\mathbf{Ke}_j = \omega_j^2 \mathbf{Me}_j \qquad (18)$$

Eigenmodes are determined only within a multiple of themselves. Usually they are normalized to unity. The nice thing about eigenmodes is that they form an orthogonal basis for the space of possible responses in case the system is self adjoint as in the case of a conservative mechanical system. Although their number will be in the order of the system, in practice we only need to retain a small number corresponding to the lowest eigenvalues. Let $\mathbf{E}$ denote the matrix containing the retained eigenmodes as columns. Then by setting $\mathbf{u}(t) = \mathbf{Ea}(t)$ the dynamic equations can be formulated with respect to $\mathbf{a}(t)$ and so the size of the system is drastically reduced:

$$\mathbf{E}^T \mathbf{M} \mathbf{E} \ddot{\mathbf{a}} + \mathbf{E}^T \mathbf{K} \mathbf{E} \mathbf{a} = \mathbf{E}^T \mathbf{Q} \qquad (19)$$

Note that $\mathbf{u}(t) = \mathbf{Ea}(t)$ is equivalent to a projection of $\mathbf{u}$ on a reduced subset of the eigenmode basis. The error thus introduced will be small provided that all modes that can be resolve by the specific time step are included.

In case the system of equations has the form of eqn (17), the system is first transformed into first order with respect to $\mathbf{y} = (\dot{\mathbf{x}}, \mathbf{x})^T$:

$$\tilde{\mathbf{M}}\dot{\mathbf{y}} + \tilde{\mathbf{K}}\mathbf{y} = \tilde{\mathbf{Q}}: \quad \tilde{\mathbf{M}} = \begin{bmatrix} \mathbf{M} & 0 \\ 0 & \mathbf{I} \end{bmatrix}, \quad \tilde{\mathbf{K}} = \begin{bmatrix} \mathbf{C} & \mathbf{K} \\ -\mathbf{I} & 0 \end{bmatrix}, \quad \tilde{\mathbf{Q}} = \begin{Bmatrix} \mathbf{Q} \\ 0 \end{Bmatrix} \tag{20}$$

Because $\tilde{\mathbf{M}}$ will be always invertible,

$$\dot{\mathbf{y}} = \mathbf{A}\dot{\mathbf{y}} + \mathbf{b}, \quad \mathbf{A} = -\tilde{\mathbf{M}}^{-1}\tilde{\mathbf{K}}, \quad \mathbf{b} = \tilde{\mathbf{M}}^{-1}\tilde{\mathbf{Q}} \tag{21}$$

The dynamics in this case will be determined by the eigenvalues λ_i and eigenmodes $\mathbf{e}_i$ of matrix $\mathbf{A}$:

$$|\mathbf{A} - \lambda_i\mathbf{I}| = 0 \quad \text{and} \quad (\mathbf{A} - \lambda_i\mathbf{I})\mathbf{e}_i = 0 \tag{22}$$

The eigenvalues of $\mathbf{A}$ will be either real or complex in which case they appear in conjugate pairs. As before the set of $\mathbf{e}_i$ constitutes an orthogonal basis and therefore the solution of (21) admits a modal expansion $y(t) = \mathbf{E}a(t)$. Then the homogeneous part of the solution can be readily obtained using the initial conditions: $\mathbf{y}(0) = \mathbf{y}_0 = \mathbf{E}a_0$

$$\mathbf{y} = \mathbf{E}\exp(\Lambda t)\mathbf{a}_0 = \mathbf{E}\exp(\Lambda t)\mathbf{E}^{-1}\mathbf{y}_0 \tag{23}$$

in which Λ denotes the diagonal matrix of the eigenvalues. The presence of the exponential term indicates that the system will be stable if $Re(\lambda_i) < 0$ and unstable in the contrary. Thus $-Re(\lambda_i) / |\lambda_i| < 0$ defines the damping ratio while $\omega_i = Im(\lambda_i)$ is taken as the frequency of the mode and $\omega_{N,i} = |\lambda_i|$ as the corresponding natural frequency. Finally for the full solution,

$$\mathbf{y} = \mathbf{E}\exp(\Lambda t)\mathbf{E}^{-1}\mathbf{y}_0 + \int_0^t \mathbf{E}\exp(\Lambda(t-\tau))\mathbf{E}^{-1}\mathbf{b}\,d\tau \tag{24}$$

5 Aeroelastic coupling

The eigenvalue analysis so far made no reference to the aeroelastic coupling which plays a decisive role in the reliability and safety assessment of wind turbines. Aeroelastic coupling appears because the aerodynamic loads contributing to $\mathbf{Q}$ depend on the solution of the dynamic equations $\mathbf{x}$ and $\dot{\mathbf{x}}$. Even in the simplest blade element aerodynamic model, the sectional lift, drag and pitch moment along the blade all depend on the relative to the blade inflow direction or else the effective angle of attack α_{eff}.

Of course in defining the relative to the blade flow velocity we must include the blade velocity due to elastic deformation which constitutes the essence of aeroelastic coupling. In order to better understand this point, consider a blade section (Figure 7). The aerodynamic loads per unit spanwise length in the local system defined by the flap and lag directions, are given in terms of the blade elastic torsion Θ_y and the flow direction φ defined by the relative to the blade flow velocity components $U_{\text{eff},x}$, $U_{\text{eff},z}$.

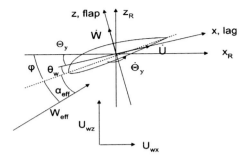

Figure 7: Basic notations on a blade section.

$$F_x = -L\sin(\varphi - \Theta_y) + D\cos(\varphi - \Theta_y)$$
$$F_z = L\cos(\varphi - \Theta_y) + D\sin(\varphi - \Theta_y)$$
$$M_y = M \tag{25}$$

In case the blade element momentum (BEM) aerodynamic model is used:

$$\tan\varphi = \frac{U_{\text{eff},z}}{U_{\text{eff},x}} = \frac{(1-a)\cdot(U_{wz} - \dot{W}\cos\Theta_y - \dot{U}\sin\Theta_y)}{(1+a')\cdot(U_{wx} + \dot{W}\sin\Theta_y - \dot{U}\cos\Theta_y)} \tag{26}$$

where a and a' denote the axial and circumferential induction factors, $\dot{U},\dot{W}$ denote the lag and flap deformation velocities and U_{xy}, U_{wz} the components of the relative wind inflow velocity which includes the wind inflow and the blade rotational speed (in the axial case for example the x-component will be the rotational speed of the blade while the z-component will be the wind inflow).

The lift L, drag D and pitching moment M of the section in eqn (25) will depend on the effective angle of attack $a_{\text{eff}} = \varphi - \theta_w - \Theta_y$ and the relative flow velocity $W_{\text{eff}}^2 = U_{\text{eff},x}^2 + U_{\text{eff},z}^2$. a_{eff} can substantially vary not only because the mean wind speed has a wide range, but also because of the dynamic response of the blades. According to (13), the blade velocity is subjected to its own deformation velocities and the dynamics of the configuration through $\bar{R}$ and $\mathbf{T}$ leading to the following expression for the local blade velocity:

$$\dot{\mathbf{U}} = (\dot{U},\dot{V},\dot{W})^{\mathrm{T}} = \mathbf{T}^{\mathrm{T}}\dot{\bar{r}}_G = \mathbf{T}^{\mathrm{T}}\left(\ddot{\bar{R}} + \dot{\mathbf{T}}(\bar{r}_0 + \mathbf{S}\mathbf{u})\right) + \mathbf{S}\dot{\mathbf{u}} \tag{27}$$

The coupling introduced by (25) and (27), is quite complicated and certainly non-linear. One complication is connected to its unsteady character which requires the use of *unsteady* aerodynamic modelling. Models of this type provide the sectional lift, drag and pitching moment coefficients as functions of the sectional steady polars and the dynamic inflow characteristics, namely the pitching and heaving motion of the section. So a link between the aerodynamic loads and the elastic response is established through explicit functional relations. Of wide use in wind turbine aeroelasticity are the Beddoes-Leishman [19] and the ONERA models [20]. They both are applicable over a wide range of angle of attack covering both attached and stalled conditions which explains why they are usually referred to as *dynamic stall* models.

Taking as example the ONERA model, loads are expressed by in total four *circulation* parameters: Γ_{1L}, Γ_{2L} for the lift which correspond to the attached and separated contributions respectively, Γ_{2D} for the drag and Γ_{2M} for the pitching moment, each satisfying a second order differential equation. This nice feature of the model allows including the blade aerodynamics into the system as extra dynamic equations. The spanwise piecewise constant distributions of the four circulation parameters become new DOF and are treated in the same way as any other. The combined set of the element DOF define the so-called aeroelastic element [21]. Consequently linearization with respect to a reference state can be extended to also include blade aerodynamics. The complete linearized system is the basis for linear stability analysis which in this case provides the coupled aeroelastic eigenmodes and eigenfrequencies. There is however a significant complication: the coefficients of the dynamic system are no longer constant but time varying.

6 Rotor stability analysis

Linear stability analysis of first order systems $\dot{\mathbf{y}} = \mathbf{A}\mathbf{y} + \mathbf{b}$ with varying coefficients is still possible using Floquet's theory provided that the system is periodic with period T: $\mathbf{A}\,(t - T) = \mathbf{A}\,(t)$, $\mathbf{b}\,(t - T) = \mathbf{b}\,(t)$ [22, 23]. The solution obtained takes a form similar to that given in eqn (26):

$$\mathbf{y} = \mathfrak{R}(t,t_0)\mathbf{y}_0 + \int_{t_0}^{t} \mathfrak{R}(t,\tau)\mathbf{b}\,d\tau \tag{28}$$

where $\mathfrak{R}(t,t_0)$ denotes the state transition matrix of the system with respect to initial conditions defined at t_0. The difficulty in applying eqn (28) is linked to the construction of the transition matrix. For a system involving a large number of DOF, for each one of them the equations must be integrated over one period. So depending on the size of the system this task can become exceedingly expensive.

Fortunately, for rotors equipped with *identical* blades and rotating at *constant* speed Ω stability analysis of the rotor system as a whole, is simplified significantly by applying the Coleman transformation [23, 24]. Let $\psi = \Omega t$ denote the azimuth position. Then for an M-bladed rotor, the expressions of any quantity defined on the blades $v^{(m)}\,(\psi)$ will be the same except for an azimuth shift $m\Delta\psi$, $\Delta\psi = 2\,\pi/M$. By introducing the following new variables:

$$\tilde{v}_0(\psi) = \frac{1}{M}\sum_{m=1}^{M} v^{(m)}(\psi) \quad \tilde{v}_{c,k}(\psi) = \frac{1}{M}\sum_{m=1}^{M} v^{(m)}(\psi)\cos(k\psi_m)$$

$$\tilde{v}_{M/2}(\psi) = \frac{1}{M}\sum_{m=1}^{M} v^{(m)}(\psi)(-1)^m \quad \tilde{v}_{s,k}(\psi) = \frac{1}{M}\sum_{m=1}^{M} v^{(m)}(\psi)\sin(k\psi_m) \tag{29}$$

it follows that:

$$v^{(m)}(\psi) = \tilde{v}_0 + \sum_{k=1}^{N}\left(\tilde{v}_{c,k}\cos(k\psi_m) + \tilde{v}_{s,k}\sin(k\psi_m)\right) + \underbrace{\tilde{v}_{M/2}(-1)^m}_{\text{if M even}} \tag{30}$$

where $\psi_m = \psi + m\Delta\psi$ and $N = (M - 1)/2$ if M is odd and $N = (M - 2)/2$ if M is even. Note that the coefficients in (30) depend on time through ψ. This time dependency is particular. For the usual case of a three-bladed rotor,

$$\frac{1}{M}\sum_{m=1}^{M}\cos(k\psi_m) = \begin{cases}\cos(k\psi), k = iM \\ 0\end{cases}, \quad \frac{1}{M}\sum_{m=1}^{M}\sin(k\psi_m) = \begin{cases}\sin(k\psi), & k = iM \\ 0\end{cases}$$

so the $\tilde{v}_*$ coefficients will only contain the harmonics that are multiples of the number of blades, i.e. $k = 3, 6, 9, \ldots$ Thus by taking as example the non-dimensional flap deflection at the blade tip, $\tilde{v}_0$ will be the cone angle of the tip-path plane while $\tilde{v}_{c,1}$ and $\tilde{v}_{s,1}$ will give the tilt and yaw angles of the rotor. The rotational transformation is not restricted to the DOF. The same transform is applied to the dynamic equations, an operation equivalent to considering each equation as a dependent variable to which eqn (30) applies.

In practical terms, the above analysis is carried out as follows. Consider the full set of non-linear aeroelastic equations. The first step is to construct a periodic solution. To this end, the non-linear equations are integrated in time until a periodic response (with respect to the rotor speed) is reached. If the conditions are close to instability the time domain calculations provide a response that contains significant components in all the basic frequencies of the system. In such a case, all frequencies besides the rotational frequency 1/rev and its basic multiple M/rev are filtered by means of Fourier transformation.

The next step is to linearize the problem. Based on the periodic solution obtained, the system is reformulated in perturbed form. To this system, the rotational transformation is applied on both the DOF and the equations. The end result of this procedure is a dynamic system with constant coefficients and therefore the standard eigenvalue analysis can provide directly its stability characterization.

The passage from the rotating to the non-rotating system affects the eigenfrequencies. For a simple system, the modes in the non-rotating system will be equal to those in the rotating system +/- the rotational speed. The modes produced in this way are called *progressive* and *regressive*. However in the case of a complicated system as that of a complete wind turbine, the regressive and progressive modes will be coupled with the non-rotating parts of the system and so their identification is more difficult [25].

Note that the perturbed equations are general and apply to both linear and non-linear contexts. In fact, non-linear responses can be obtained by iteratively solving the linearized set of equations until perturbations are eliminated. Therefore this kind of formulation can be also used for non-linear stability identification. Typically non-linear damping computations are based on either the assessment of the aerodynamic work [26] or signal processing of the transient responses [27]. The assessment of the aerodynamic work is applied at the level of the isolated blade, as a means of validating linear analysis results. In such a method the work done by the aerodynamic loads acting on the isolated blade is calculated for the blade undergoing a harmonic motion according to the shape and the frequency of the specific aero-elastic mode considered. It has been shown that the aerodynamic work is directly related to the damping of the mode considered [26]. So, it is possible to get a clear insight of the damping distribution along the blade span, as well as to identify the effects from the non-linearity of the aerodynamics.

7 More advanced modeling issues

The beam model presented in Section 3 is the simplest possible. Besides assuming small displacements and rotations it also suppresses shear (Bernoulli model) which can be important. Including shear leads us to the Timoshenko beam model while large displacements and rotations require upgrading of the model to second order. These two aspects of structural modelling are briefly discussed in the sequel.

7.1 Timoshenko beam model

In geometrical terms, introducing shear consists of assuming that the cross sections along the beam axis will no longer remain normal to the axis of the beam. This means that θ_x and θ_z in eqn (3) become independent resulting in extra shear strains γ_{zy}, γ_{xy}. Note that in the Euler-Bernoulli model shear is solely related to torsion. For example by suppressing torsion and bending in the x–y plane the following strains and stresses are defined:

$$\gamma_{zy} = -\theta_x + \frac{\partial w}{\partial y} \Rightarrow \tau_{zy} = G\gamma_{zy} = G\left(w' - \theta_x\right)$$

$$\varepsilon_y = \frac{\partial v}{\partial y} - z_0\frac{\partial \theta_x}{\partial y} \Rightarrow \sigma_y = E\varepsilon_y = E\left(v' - z_0\theta_x'\right) \tag{31}$$

which result in the following virtual work terms:

$$\int_0^L \left(\int_A \sigma_y \,\delta\varepsilon_y \,dA + \int_A \tau_{zy}\,\delta\gamma_{zy}\,dA\right) dy = \int_0^L \left(F_y\delta v' + F_z\left(-\delta\theta_x + \delta w'\right) - M_x\delta\theta_x'\right) dy$$

$$F_y = \int_A \sigma_y\,dA \quad M_x = \int_A \sigma_y z_0\,dA \quad F_z = \int_A \tau_{zy}\,dA \tag{32}$$

The virtual work in eqn (32) is equal to $\delta\mathbf{u}^T\mathbf{K}\mathbf{u}$, with $\mathbf{u} = (v, w, \theta_x)^T$. So by assuming that the same shape functions $\mathbf{N}$ are used for v, w and θ_x: $v = \mathbf{N}\hat{v}, w = \mathbf{N}\hat{w},\ \theta_x = \mathbf{N}\hat{\theta}$, the stiffness matrix of an element takes the form given in eqn (33). Note that $\mathbf{K}_e$ is no longer diagonal as in (7). In fact if the full (non-linear) Green strain is used, $\mathbf{K}_e$ will be fully completed. The above formulation can be similarly extended to also include shear in the x–y plane:

$$\mathbf{K}_e = \begin{bmatrix} \int_{L_e} \mathbf{N}'^T\overline{EA}\mathbf{N}'dy & 0 & -\int_{L_e} \mathbf{N}'^T\overline{EAz}\mathbf{N}'dy \\ 0 & \int_{L_e} \mathbf{N}'^T\overline{GA}\mathbf{N}'dy & -\int_{L_e} \mathbf{N}'^T\overline{GA}\mathbf{N}dy \\ -\int_{L_e} \mathbf{N}'^T\overline{EAz}\mathbf{N}'dy & -\int_{L_e} \mathbf{N}^T\overline{GA}\mathbf{N}'dy & \int_{L_e} \left(\mathbf{N}^T\overline{GA}\mathbf{N} + \mathbf{N}'^T\overline{EAz^2}\,\mathbf{N}'\right)dy \end{bmatrix} \tag{33}$$

In the case of a 3D beam structure of general shape, the same approach is followed but now the derivation involves the introduction of curvilinear co-ordinates which makes things a little more complicated. Such a need is a direct consequence of the fact that cross sections in the deformed state are no-longer normal to the beam axis and therefore the variables ξ, η, ζ we use in defining the Green strain are non-orthogonal (for additional reading see [8, 14]).

7.2 Second order beam models

Current design trends suggest that wind turbines will get bigger in future and this will eventually lead to more flexible blades. Therefore it could turn out that the assumption of small displacements and rotations will be no longer sufficient. One option is to upgrade the model into second order as already developed for helicopter rotors in the mid 70s. The derivation is too lengthy so we will only outline the main ideas (for further reading, see [28, 29]).

The formulation is carried out in the same three steps as in the first order model: first the displacement field is defined which is next used in order to determine the strains. Assuming linear stress–strain relations, the stress distributions are readily obtained. Finally the stresses are integrated over the cross sections of the beam material, and so the sectional internal loading is obtained. The main complication originates from the form the displacements take.

With respect to the [O; xyz] beam system, the elastic axis of the beam will lie along the y axis only in its un-deformed state. In order to describe the geometry of the beam in its deformed state, a local [O′ $\xi\eta\zeta$] system is defined that follows the pre-twist and elastic deflections of the beam (Fig. 8). The η axis of this co-ordinate system follows the deformed beam axis at any position whereas ξ and ζ define the local principle axes of each cross section. At the un-deformed state y and η will coincide while ξ–ζ will differ from x–z by the angle defining the principal axes of each section. The position vector of any point of the deformed beam with respect to the beam system [O; xyz] is given by

$$\vec{r} = \left(u, \eta + v, w\right)^{\mathrm{T}} + \mathbf{S}_2 \left(\xi, -f_w \theta', \zeta\right)^{\mathrm{T}} \tag{34}$$

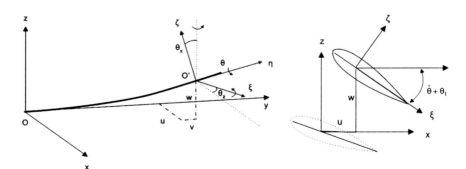

Figure 8: Beam co-ordinate system definition.

where f_w is the warping function of the cross section, u, v, w are the local displacements of the beam axis, θ is the torsion angle and S_2 is the transformation matrix between $[O; xyz]$ and $[O'; \xi\eta\zeta]$. By keeping the terms up to second order (with the exception of some higher order terms that are not always negligible), it follows that:

$$S_2 = \begin{bmatrix} \cos(\hat{\theta}+\theta_t - u'w')\left(1-\dfrac{u'^2}{2}\right) & u' & \sin(\hat{\theta}+\theta_t - u'w')\cdot\left(1-\dfrac{u'^2}{2}\right) \\[2mm] -u'\cos(\hat{\theta}+\theta_t)+w'\sin(\hat{\theta}+\theta_t) & 1-\dfrac{u'^2}{2}-\dfrac{w'^2}{2} & -(u'\sin(\hat{\theta}+\theta_t)+w'\cos(\hat{\theta}+\theta_t)) \\[2mm] -\sin(\hat{\theta}+\theta_t - u'w')\left(1-\dfrac{u'^2}{2}\right) & w' & \cos(\hat{\theta}+\theta_t)\cdot\left(1-\dfrac{w'^2}{2}\right) \end{bmatrix} \quad (35)$$

where θ_t is the local twist angle and

$$\hat{\theta} = \theta + \int_0^y u''w'\,dy \tag{36}$$

Even if warping is neglected the complication of the model is clear. There are several non-linear terms which will render the overall code implementation hard to follow especially in the multi-body context. For example, the expression for the tension force will become:

$$F_\eta = \overline{E}\cdot\left(v'+\frac{u'^2}{2}+\frac{w'^2}{2}\right)+\overline{E(\xi^2+\zeta^2)}\left(\theta'\theta_t+\frac{\theta'^2}{2}\right)$$
$$-\left(\overline{E\zeta}\cos(\theta+\theta_t)+\overline{E\zeta}\sin(\theta+\theta_t)\right)u'' - \left(\overline{E\zeta}\cos(\theta+\theta_t)-\overline{E\zeta}\sin(\theta+\theta_t)\right)w'' \tag{37}$$

Note that the loads will be defined in the $[O'; \xi\eta\zeta]$ system while the dynamic equations are defined with respect to $[O; xyz]$. Therefore a further transformation is needed, also using S_2, this time for the loads. The transformed loads are next introduced into the equilibrium equations and so the final system is obtained.

8 Structural analysis and engineering practice

In this last section, we discuss the kind of information structural and aeroelastic analysis can provide and how this is linked to the design of wind turbines. Due to the approximate character of all the tools we use, there is always need for validation which is carried out on the basis of measured data. Note that "absolutely" reliable measurements are not always easy to obtain. Full scale tests are subjected to a number of uncertainties mostly related to the uncontrolled inflow conditions. Besides that quite often there is lack of detailed structural input especially as regards the drive train. On the other hand laboratory testing on scaled models is subjected to similarity incompatibilities. Therefore comparisons must be carried out carefully, taking into account of all possible uncertainties.

8.1 Modes at stand still

Modes were introduced as the function basis with respect to which the behaviour of a linear system is fully described. There are three levels of modal analysis: the purely structural, the aeroelastic and the servo-aeroelastic. The case of a wind turbine at stand still is convenient for estimating the purely structural modes. At stand still, the rotor is blocked by the braking system so there is no rotation and the generator contributes only with its weight. Assuming almost zero inflow, aerodynamic loading can be neglected. One advantage of stand still conditions is that they can be easily reproduced in full scale. So measurements can be used as a basis for validating and fine tuning structural models. By setting the rotation speed equal to zero certain inertial terms are eliminated and therefore the stand still modes will not exactly represent the modal behaviour of wind turbines in operation. Of course wind turbines rotate slowly and therefore modes will not substantially change as in other rotor applications [23].

For a well balanced rotor the basic frequency is the rotation frequency p multiplied by the number of blades and its multiples. So for a three-bladed turbine the natural frequencies should be placed away from 3p and 6p in order to avoid resonance. The required margin depends on many factors but one should keep in mind that large margins are not possible. Besides the natural modes of the different components, coupled modes will appear as a result of their inter-connections. In Fig. 9 the natural frequencies of 3 three-bladed commercial machines taken from the late 1990s, are compared. The lowest two modes are the lateral and longitudinal bending modes of the tower which appear in between 1p and 2p. Such low values are due to the large mass placed on top of the tower. Next appears the drive train torsion mode which must be <3p and finally the lower blade modes. There are three flap bending modes: the symmetric and two asymmetric. The symmetric

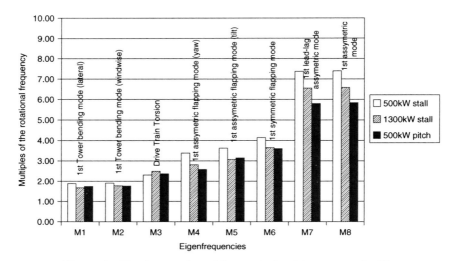

Figure 9: The first modes of three wind turbines at stand still.

mode corresponds to the natural frequency of the blades while the asymmetric are coupled modes; one is coupled with the tower torsion and concerns the yawing of the rotor while the other is coupled with the tower bending and concerns the tilting of the rotor. Because the tower modes are low, the coupled modes will be lower as compared to the symmetric one. In most designs the symmetric mode appears near 4p for three bladed rotors so that there is some margin to accommodate the coupled modes. In purely structural terms one would desire a stiffer blade, but this would increase the cost which is, in the case of wind turbines the most important design driver. Finally as regards the lead-lag (or edgewise) motion, due to higher stiffness the first stand still first mode should appear in the vicinity of 6p so that when in operation, the coupled modes are at 5 and 7p and thus 6p is avoided.

The quality of structural models based on beam theory can be quite good. In Table 1 predictions obtained for a commercial wind turbine are compared to measurements indicating a maximum error of 7%.

8.2 Dynamic simulations

Dynamic simulations refer to situations in which the excitation is time varying. Dynamic excitation on wind turbines is caused by the wind inflow and can be either periodic or non-periodic (the latter are usually referred to as stochastic). Typical periodic excitations are generated by the mean wind shear, the yaw misalignment, the blade–tower interaction and possibly the control. Non-periodic excitations are related to the turbulent character of the wind. Strictly speaking in practice the wind turbine is always stochastically excited. However it is possible to extract the periodic part of the response by averaging with respect to the azimuth angle. Azimuthal averaging can be performed with measurements and simulations. A result of this type is given in Fig. 10. The azimuthal variation of the predicted flapwise bending moment at blade root (left) and of the shaft tilting moment are compared

Table 1: Natural frequencies of the machine at standstill [30].

	Mode	Natural frequency (Hz)		Difference (%)
		Measured	Predictions	
1	First lateral tower bending	0.437	0.439	0.5
2	First longitudinal tower bending	0.444	0.448	0.9
3	First shaft torsion	0.668	0.674	0.9
4	First asymmetric flap/yaw	0.839	0.828	−1.3
5	First asymmetric flap/tilt	0.895	0.886	−1.0
6	First symmetric flap	0.955	1.024	7.2
7	First vertical edgewise	1.838	1.909	3.9
8	First horizontal edgewise	1.853	1.928	4.0
9	Second asymmetric flap/yaw	2.135	2.149	0.7
10	Second asymmetric flap/tilt	2.401	2.314	−3.6

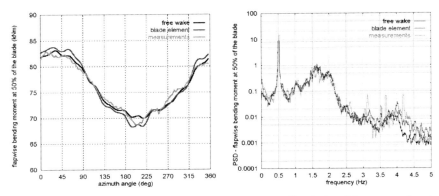

Figure 13: Fatigue loads spectrum for a 500 kW stall wind turbine operating in complex terrain at 13 m/s mean wind speed [32].

considered in Fig. 10. Note that the material fatigue data are given similarly and therefore it is possible to assess the fatigue strength of the design. Of course fatigue assessment requires the complete load spectrum so that the targeted life time of the machine is reproduced. The procedure is detailed in the IEC standard [35] and involves a long list of 10-min simulations covering the entire range of mean wind speeds.

8.3 Stability assessment

Stability appeared as a problem when wind turbines were first up-scaled to 600 kW. At that time most wind turbines were stall regulated and research identified dynamic stall as the main driver of aeroelastic instability [21, 36]. In order to analyze aeroelastic stability, specific tools were developed and validated against full-scale tests measurements; see for example [37–40]. The majority of these models combine the BEM aerodynamic theory with multi-body system dynamics using beam theory, similarly to the previously described analysis.

For stall regulated machines, the damping of the tower lateral bending mode as well as the blade lag modes is marginal at high wind speeds where stall is expected to have its major effect as shown in Fig. 14. Two sets of results are considered, both obtained for zero structural damping which in log scale is estimated to add 6–10%. The first set, denoted as "fixed RPM" corresponds to the modelling of the drive train as a spring. In this case, the damping of the tower lateral bending mode and the blade lag modes is negative at all wind speeds. Since there is no other source of damping except what unsteady aerodynamics contributes, it follows that this contribution is negative. Fixing the RPM is not realistic and the generator should be also included. A simple generator dynamics consists of assuming that the generator torque is proportional to the rotor speed which will add extra damping. The corresponding results are denoted as "generator". There is a substantial increase in the damping of the tower lateral bending mode and the symmetric lag bending mode of the blade. If structural damping is added the situation will become only marginally stable which explains why stall regulation has been substituted by pitch regulation in the design of multi MW machines.

Pitch regulation limits the onset of stall by keeping the mean angles of attack along the blade up to the level of maximum lift. However light stall will occur close to maximum lift flow conditions which correspond to wind speeds around rated power production. This is shown in Fig. 15. Even at fixed RPM, the range of wind speeds with negative damping is limited around the rated wind speed. When the generator is added the damping increases substantially and attains higher positive values as compared to the stall regulated situation.

So in conclusion, theory indicates that wind turbines are subjected to low damping in certain modes. In order to assess the validity of this result comparison with full scale measurements is needed. This is done in Figs 16–18 in terms of aeroelastic frequency and damping for the most critical modes of a ~3 MW pitch regulated commercial machine. The measured data were obtained through Operational Modal Analysis (OMA) [41] on a collection of normal operation time series of

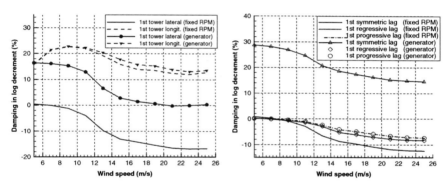

Figure 14: Damping variation for the tower bending modes (left) and the blade lag modes (right) of a commercial 500 kW stall regulated wind turbine [38].

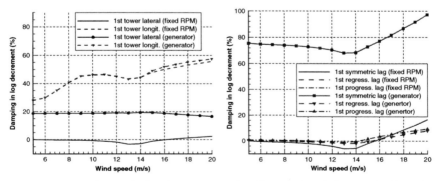

Figure 15: Damping variation for the tower bending modes (left) and the blade lag modes (right) of a commercial 3 MW pitch regulated wind turbine [38].

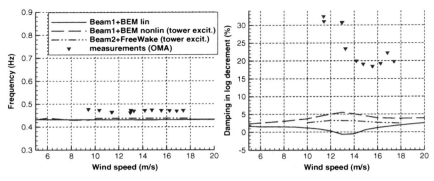

Figure 16: Frequency and damping variation of the tower lateral bending mode.

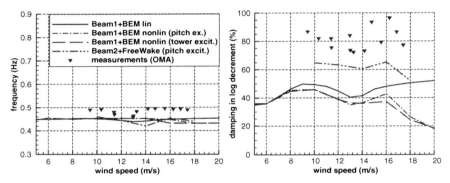

Figure 17: Frequency and damping variation of the tower longitudinal bending mode.

loads with a total duration of several hours. Three sets of simulation results are shown. The "Beam1+BEM" results were obtained using Euler-Bernoulli beam theory for the structure and BEM theory for the aerodynamics. Models of this kind represent the current state of art and can be used either in their linearized form (denoted as "lin") or in their original non-linear form (denoted as "nonlin"). The "Beam2 + FreeWake" results have been obtained using the second order beam theory outlined in Section 7.2 and the free wake aerodynamic model described in [34]. Free wake aerodynamic modelling is a 3D model with significantly higher computational cost and therefore is regarded as a research tool.

Figures 16 and17 show the results for the two tower bending modes. Both predictions and measurements clearly indicate that the lateral bending mode is the one less damped. However for this particular case predictions significantly underestimate the damping. In the specific simulations, this underestimation is due to the fact that the generator dynamics were not included which as already mentioned improves stability. Another point is that the scatter of the measurements was large and so of non-negligible uncertainty. Otherwise there is a noticeable mirror effect between the linear and non-linear results for the lateral mode near the rated wind speed (13 m/s). Contrary to linear models which give over this range minimum

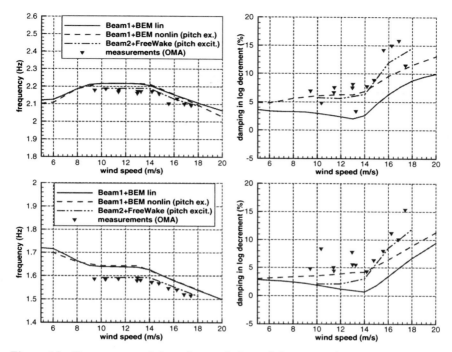

Figure 18: Frequency and damping variation of the regressive (upper line) and progressive (lower line) blade bending lag mode.

damping, non-linear models predict maximum damping. Note that in the longitudinal tower mode the non-linear results resemble to the trends of the test data. The best correlation was found for the "Beam2 + FreeWake" prediction which indicates that non-linear effects play an important role in terms of accuracy. This point is further supported by the results for the regressive and progressive lag modes of the blade shown in Fig. 18. Both non-linear models provide good damping predictions. In particular the "Beam2 + FreeWake" follows better the trend in the high wind speed range. So in conclusion the validation procedure indicates that stability modelling can reproduce the stability characteristics of wind turbines and that predictions are on the safe side which is very important in design.

References

[1] Meirovitch, L., *Methods of Analytical Dynamics*, McGraw Hill: New York, 1970.
[2] Zienkiewicz, O.C. & Taylor, R.L., *The Finite Element Method*, Elsevier Butterworth Heinemann, 2005.
[3] Goldstein, H., *Classical Mechanics,* Addison Wesley, 1980.
[4] Mathematica, http://wolfram.com.

[5] MathCAD, http://www.adeptscience.co.uk/products/mathsim/mathcad.

[6] Green, A.E. & Zerna, W., *Theoretical Elasticity*, Clarendon Press: Oxford, 1954.

[7] Timoshenko, S. & Goodier, J.N., *Theory of Elasticity*, McGraw Hill: New York, 1951.

[8] Wang, C.M., Reddy, J.N. & Lee, K.H., *Shear Deformable Beams and Plates*, Elsevier Science Ltd., 2000.

[9] Rivello, R. M., *Theory and Analysis of Flight Structures*, McGraw Hill, N.Y, 1969.

[10] Kunz D. L., Survey and comparison of engineering beam theories for helicopter rotor blades. *J. Aircraft*, **31**, pp. 473–497, 1994.

[11] Jung, S.N., Nagaraj, V.T. & Chopra, I., Assessment of composite rotor blade modelling techniques. *Journal of American Helicopter Society*, **44(3)**, pp. 188–205, 1999.

[12] Bathe, K.J., *Finite Element Procedures in Engineering Analysis*, Prentice Hall, 1981.

[13] Zienkiewicz, O. C., Taylor, R. L., *The Finite Element Method for solid and structural mechanics*, Elsevier Butterworth Heinemann, 2005.

[14] Crisfield, M.A., *Non-linear Finite Element Analysis of Solids and Structures*, John Wiley & Sons, 1991.

[15] Hall, G. & Watt, J.M., *Modern Numerical Methods for Ordinary Differential Equations*, Clarendon Press: Oxford, 1976.

[16] Pfeiffer, F. & Glocker, C., *Multi-body Dynamics with Unilateral Contacts*. John Wiley & Sons: New York, 1996.

[17] Bauchau, O. A., Computational schemes for flexible, non-linear multi-body systems. *Multibody System Dynamics*, **2**, pp. 169–225, 1998.

[18] Bauchau, O.A. & Hodges, D.H., Analysis of non-linear multi-body systems with elastic couplings. *Multibody System Dynamics*, **3**, pp. 166–188, 1999.

[19] Leishman, J.G. & Beddoes, T.S., A semi-empirical model for dynamic stall. *Journal of the American Helicopter Society*, **34(3)**, pp. 3–17, 1989.

[20] Petot, D., Differential equation modelling of dynamic stall. *Recherche Aerospatiale*, **5**, p. 59–7, 1989.

[21] Chaviaropoulos, P.K., Flap/lead–lag aeroelastic stability of wind turbine blades. *Journal of Wind Energy*, **4**, pp. 183–200, 2001.

[22] Peters, D.A., Fast Floquet theory and trim for multi-bladed rotorcraft. *Journal of American Helicopter Society*, **39(4)**, pp. 82–89, 1994.

[23] Johnson, W., *Helicopter Theory*, Dover Publications Inc.: New York, 1994.

[24] Coleman, R.P. & Feingold, A.M., *Theory of self-excite mechanical oscillations of helicopter rotors with hinged blades*, NASA-TN-3844, 1957.

[25] Hansen, M.H., Aeroelastic instability problems for wind turbines. *Wind Energy*, **10**, pp. 551–557, 2007

[26] Petersen, J.T., Madsen, H.A., Bjørk, A., Enevoldsen, P., Øye, S., Ganander, H. & Winkelaar, D., *Prediction of dynamic loads and induced vibrations in stall*, Risø-R-1045(EN), Risø National Laboratory, Roskilde, 1998.

[27] Simon, M. & Tomlinson, G.R., Use of the Hilbert transform in modal analysis of linear and non-linear structures. *Journal of Sound Vibration*, **96**, pp. 421–36, 1984.

[28] Hodges, D.H. & Dowell, E.H., Nonlinear equations of motion for elastic bending and torsion of twisted non-uniform blades. NASA TN D-7818, 1975.

[29] Hodges, D.H., A review of composite rotor blade modelling. *AIAA Journal*, **28(3)**, pp. 561–565, 1990.

[30] Fuglsang, P., *STABCON - Aeroelastic stability and control of large wind turbines*. EWEC 2003, Madrid, 16–19 June 2003.

[31] Riziotis, V.A. & Voutsinas, S.G., GAST: a general aerodynamic and structural prediction tool for wind turbines. *Proc. of the European Wind Energy Conf. & Exhibition*, Dublin, Ireland, 1997.

[32] Fragoulis, A., *COMTER.ID: Investigation of design aspects and design options for wind turbines operating in complex terrain environments*, Final Report for JOR3-CT95-0033 CEU project, 1998.

[33] Schepers, J.G., *VEWTDC: Verification of European wind turbine design codes*, Final Report for JOR3-CT98-0267 Joule III project, ECN, 2001.

[34] Voutsinas, S.G, Beleiss, M.A. & Rados, K.G., Investigation of the yawed operation of wind turbines by means of a vortex particle method. *AGARD Conf. Proc.*, vol. 552, pp. 11.1-11, 1995.

[35] IEC61400-1Ed.3, Wind turbines Part 1: design requirements, 2004.

[36] Petersen, J.T., Madsen, H.A., Bjørk, A., Enevoldsen, P., Øye, S., Ganander, H. & Winkelaar, D., Prediction of dynamic loads and induced vibrations in stall, Risø-R-1045(EN), Risø National Laboratory, Roskilde, 1998.

[37] Hansen, M.H., Improved modal dynamics of wind turbines to avoid stall-induced vibrations. *Journal of Wind Energy*, **6(2)**, pp. 179–195, 2003.

[38] Riziotis, V. A., Voutsinas, S. G., Politis, E. S. and Chaviaropoulos, P. K., Aeroelastic stability of wind turbines: the problem, the methods and the issues. *J. Wind Energy*, **7**, pp. 373–392, 2004.

[39] Riziotis V. A., Voutsinas S. G., Advanced aeroelastic modeling of complete wind turbine configurations in view of assessing stability characteristics. *Proc. EWEC'06*, Athens, Greece, 2006.

[40] Markou, H., Hansen, M.H., Buhl, T., Engelen, T. van, Politis, E.S., Riziotis,V.A., Poulsen, N.K., Larsen, A.J., Mogensen, T.S. & Holierhoek, J.G., Aeroelastic stability and control of large wind turbines – main results. *Proc. of the 2007 European Wind Energy Conf. & Exhibition*, Milan, 2007.

[41] Hansen, M.H., Thomsen, K., Fuglsang, P. & Knudsen, T., Two methods for estimating aeroelastic damping of operational wind turbine modes from experiments. *Journal of Wind Energy*, **9**, pp. 179–191, 2006.

CHAPTER 5

Wind turbine acoustics

Robert Z. Szasz[1] & Laszlo Fuchs[1,2]
[1]*Lund University, Lund, Sweden.*
[2]*Royal Institute of Technology, Stockholm, Sweden.*

This chapter presents some of the most important issues related to wind turbine acoustics. After a short background, an introduction to acoustics follows where we define the most important parameters that are met in the chapter. Next, the noise generation mechanisms are discussed from a physical point of view. Some aspects related to noise propagation and their perception are presented. A major part of the chapter consists of a review of the prediction models with different level of complexity together with experimental methods that are often used for wind turbine aeroacoustics. Finally, a short list of noise reduction strategies is also given.

1 What is noise?

Sound is the propagation of low-amplitude pressure waves traveling with the *speed of sound*. These pressure waves might be generated in several ways: for example the vibration of the vocal chords or the membrane of a loudspeaker, periodical vortex shedding in a cavity gives rise to a pure tone (i.e. having a single frequency). In contrast, turbulent flows may generate a broad spectrum sound, containing a wide range of modes with different amplitudes. Thus, the mechanisms for generating sound may be quite different resulting in sounds with widely different characters. Hence, a simple modal characterization of the sound might not be enough to evaluate its perception. When the generated sound is not wanted it is considered to be a *noise*. Since the perception of sound as pleasant or annoying is highly individual, there are no strict limits delimiting noise from acceptable sound.

2 Are wind turbines really noisy?

With the increasing demand of alternative energy sources the number of installed wind turbines increased exponentially in recent years. Due to the limited amount of space with favorable wind conditions, there are more and more plants installed close to residential areas. Together with the ever increasing size of the wind turbines, this reduced distance causes an increasing number of noise related complaints from residents. Even if most wind turbines respect the limits imposed for industrial noise emissions, the sound emitted is often perceived 'as a distant pile driving,' 'like a washing machine,' 'like a nearby motorway' or "like a B747 constantly taking off" [1]. Low-frequency sound (and in particular infra sound with frequency lower than 20 Hz) is known to have physiological effects (such as nausea and headache). Wind turbine noise is usually continuous and it contains also low frequencies. These low frequencies decay slowly and may reach to longer distances (depending on the terrain). In spite of its low level, wind turbine noise may have negative impact on humans and animals in their neighborhood [2]. The overall effects of wind turbines are not yet established through different independent studies. However, since there is documented evidence that individuals subject to wind turbine sound do report undesirable effects, it is definitely important to take into account noise considerations when new wind turbine farms are planned.

3 Definitions

Sound is strongly related to the compressibility of the fluids. When a pressure field is perturbed, the perturbation is propagating by molecular collisions to the surrounding fluid. On a macroscopic level this is perceived as a compression wave. Several physical parameters may be defined to quantify different characteristics of the sound perceived by an observer.

The part of the pressure fluctuations which is traveling as waves are called the *acoustic pressure* and is a function of time and space $p' = p'(t, x)$. The propagation speed of the sound waves is called the *sound speed*, which for an ideal gas can be written as [3]:

$$c^2 = \left(\frac{\partial p}{\partial \rho}\right)_S = \gamma RT \tag{1}$$

thus the speed of sound depends on the temperature. For air at atmospheric pressure and 25°C it is approximately 340 m/s.

For a pure tone, the acoustic pressure can be written as:

$$p'(t, x) = A \cos\left[2\pi\left(\frac{x}{\lambda} \pm \frac{t}{T}\right)\right] \tag{2}$$

where A is the amplitude in Pa, λ the wavelength in m and T the period in s. In eqn (2) the minus and plus signs denote waves propagating in the positive and negative directions, respectively.

Since the perceived loudness by the human ear may have a very large range of intensity one commonly uses a logarithmic scale. Thus, the relative acoustic pressure is obtained by normalizing the acoustical pressure with a threshold level that equals to the lowest hearable level (p_{ref}). The acoustic logarithmic scaling is given by

$$L_p = 10 \times \log\left(\frac{p'^2}{p_{ref}^2}\right) \tag{3}$$

where the reference pressure is $p_{-5} = 2 \times 10^5$ Pa. The sound pressure level can be expressed also in terms of the sound intensity, which is defined as the sound power transmitted through unit area and far from the source can be written as

$$I = \frac{p'^2}{\rho_0 c_0} \tag{4}$$

where ρ_0 and c_0 are the undisturbed air density and sound speed, respectively. Thus the sound pressure level can be written as

$$L_p = 10 \times \log\left(\frac{I}{I_{ref}}\right) \tag{5}$$

where the reference sound intensity is $I_{ref} = 10^{-12}$ W/m^2. To characterize the total power of a sound source, one has to integrate the sound intensity over a closed surface surrounding the source

$$P = \int_S I \, dS \tag{6}$$

Similarly to the sound pressure level, one can define the sound power level as

$$L_p = 10 \times \log\left(\frac{P}{P_{ref}}\right) \tag{7}$$

where the reference power level is $P_{ref} = 10^{-12}$ W. Thus for a point source with spherical wave spreading the sound intensity at a distance kR is I/k^2, I being the sound intensity at a distance R from the source (see Fig. 1). It is important to note, that the sound power level, L_w, is a characteristic of the sound source, it reflects

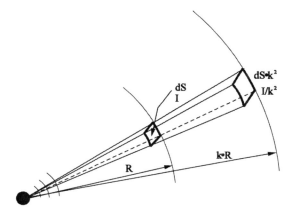

Figure 1: Illustration of sound intensity.

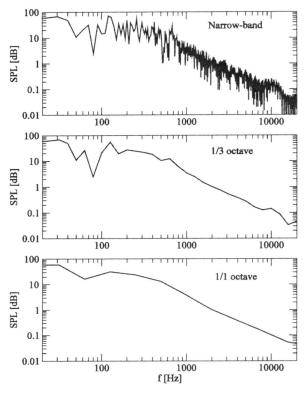

Figure 2: Narrowband (top), 1/3 octave band (middle) and 1/1 octave band (bottom) spectra of the same acoustic pressure signal.

the amount of acoustic energy generated by the source. In contrary, the sound pressure level, L_p depends on the observer position and quantifies the amount of sound energy reaching the observer.

Pure tones are rarely found in practice, most devices emit sound composed by a set of different frequencies:

$$p'(t,x) = \sum_i A\cos\left[2\pi\left(\frac{x}{\lambda_i} \pm \frac{t}{T_i}\right)\right] \tag{8}$$

Thus, the frequency spectrum of the acoustic pressure has to be determined as well to characterize the sound. There are three types of commonly used spectra: narrowband, 1/3 octave band and 1/1 octave band. In this context *band* refers to a given frequency interval over which the amplitudes are averaged. For an octave band the upper limiting frequency is exactly the double of the lower limiting frequency. For a 1/3 octave band the ratio of the upper and lower limiting frequencies is $2^{1/3}$. For a narrow-band frequency the width of the bands is constant and 'small' enough to capture pure tones and thus gives the most details about the spectrum. Figure 2

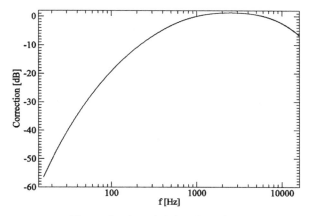

Figure 3: A-weighting function.

shows examples of narrowband, 1/3 octave band and 1/1 octave band spectra of the same acoustic pressure signal.

Human ear is sensitive to sound in a range of about 20–30 Hz and up to 20 kHz. This range varies among individuals and changes in time with reduction of the range with age. Additionally, the sensitivity is frequency-dependent. For sounds with frequencies of the order of 3–4 kHz the sensitivity is the highest, the threshold of hearing being around 0 dB, while at low frequencies it might be required a sound pressure level of 40 dB to be heard [4]. To account for this uneven sensitivity of the ear, weighted sound pressure/sound power levels have been introduced, the most commonly used being the A-weighted sound pressure level, computed as

$$L_{\mathrm{pA}}\left[\mathrm{dB}(A)\right] = L_{\mathrm{p}}\left[\mathrm{dB}\right] + L_{\mathrm{A}}\left[\mathrm{dB}\right] \tag{9}$$

where the weighting function varies as it is shown in Fig. 3. A-weighting is suitable for most applications. In cases where low-frequency noise is dominant, B or C weighting might be more appropriate. These weightings are similar to the A-weighting, just the shape of the weighting curve is different.

4 Wind turbine noise

Acoustic studies involve three main stages: sound *generation, propagation and reception* (see Fig. 4). In the case of wind turbines there are several different kinds of noise sources which are discussed in Section 4.1. The generated noise is propagating through the air, this propagation being influenced by the air properties, the landscape, vegetation, presence of different obstacles, etc. These issues are discussed in Section 4.2. The third stage is the sound immission, i.e. the perception of the sound at an observer position. Beside the sound pressure level there are other parameters as well which influence the perception of the

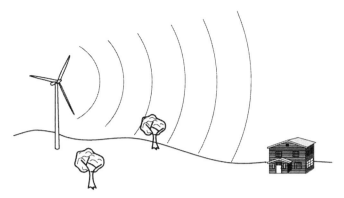

Figure 4: Illustration of sound propagation.

sound as acceptable or annoying. Details about these parameters are given in Section 4.3.

4.1 Generation

There are two main categories of wind turbine noise sources: mechanical and aerodynamic.

4.1.1 Mechanical noise

The mechanical noise is generated by the machinery in the nacelle. The main noise source is the gearbox, but important contributions are coming from the generator also [5]. The cooling fans and the other auxiliary devices contribute to a lower extent to the total noise.

Depending on the transmission path, the mechanically generated noise can be divided in two main categories (see Fig. 5). When the noise is directly radiated in the atmosphere it is called *air-borne*, like the noise emitted by the gearbox through the nacelle openings. Another part of the mechanical noise is due to vibrations propagated through the transmission elements and fittings to the nacelle casing and tower leg. This indirect noise is called *structure-borne*. The structure-borne gearbox noise constitutes the main contribution to the mechanical noise [5].

The noise and vibrations in the gearbox are due to the *transmission error* of the meshing gears. The transmission error is the difference between the desired and the actual position of the driven gear. Thus a more accurate production of the gearbox leads to lower noise levels. There are also differences among different gear types, helical teeth are less noisy than the spur ones. Also, epicyclic gearboxes are usually noisier, one of the gears being directly mounted on the casing. Due to the wear of the gears the amplitude of the generated vibrations increases with time. A doubling of the transmission error leads to an increase of approximately 6 dB in the noise level.

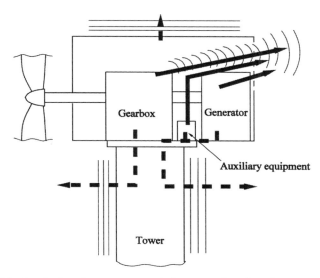

Figure 5: Mechanical noise sources: air-borne noise (continuous arrows) and structure-borne (dashed arrows).

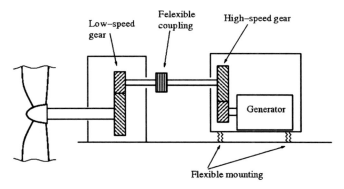

Figure 6: Mechanical noise reduction using elastic fittings (after [5]).

To reduce the gearbox noise, proper teeth profiles and high accuracy products are needed. To avoid amplification of the noise by the casing, the casing has to be designed to have eigenfrequencies far from the critical frequencies from the gears. To avoid the transmission of vibrations, elastic couplings have to be used between the mechanical devices. For large turbines, such couplings cannot transfer the torque and the elastic coupling is mounted between the low-speed and the high-speed stages of the gearbox. This approach is efficient, since most of the noise is coming from the high-speed gears, which can be elastically mounted using this solution (see Fig. 6) [5]. Further reductions of the emitted noise levels can be achieved by acoustic insulation of the nacelle.

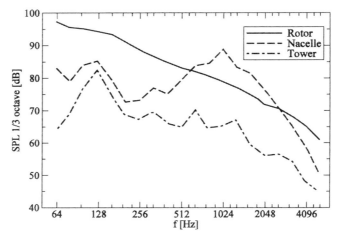

Figure 7: Typical frequency spectra from a wind turbine (after [5]).

Figure 7 shows typical 1/3 octave band spectra for a small 75 kW wind turbine measured by Ohlrich (cited in [5]). One can observe that the aerodynamic noise due to the rotor blades is dominating in almost all frequency bands, except around 1 kHz where the mechanical noise originating from the nacelle prevails. Also, the mechanical noise spectra from the nacelle and tower is more tonal, while the rotor noise is 'smoother,' it has a more broadband character.

As a conclusion, there are efficient ways to reduce the mechanical noise. Since mechanical noise is not increasing that fast with increasing turbine size like the aerodynamic noise [6], the current research is mainly focused on the reduction of the aerodynamic component of the wind turbine noise.

4.1.2 Aerodynamic noise

Early theoretical studies of wind turbine noise were based on analogies with acoustic studies of semi-infinite half-planes which were symbolizing elements of the rotor blade [7, 8]. Later the studies were based on results from non-rotating airfoil noise. Lowson [9] divided the wind turbine noise sources in the following categories:

- discrete frequency noise at the blade passing frequency and its harmonics
- self-induced noise sources
    - trailing edge noise
    - separation-stall noise
    - tip vortex formation noise
    - boundary layer vortex shedding noise
    - trailing edge bluntness vortex shedding noise
- noise due to turbulent inflow

The first group of noise sources contains low-frequency components due to the uneven loading of the blades due to wakes, large-scale structures or velocity gradients in the atmospheric boundary layer.

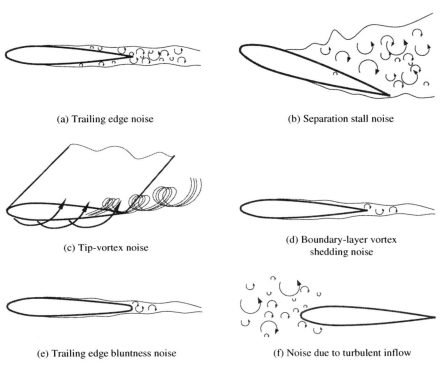

(a) Trailing edge noise

(b) Separation stall noise

(c) Tip-vortex noise

(d) Boundary-layer vortex shedding noise

(e) Trailing edge bluntness noise

(f) Noise due to turbulent inflow

Figure 8: Noise generation mechanisms for an airfoil.

The second group includes the noise sources generated by the airfoil itself, thus present even in a perfectly homogeneous inflow. The *trailing edge noise* (Fig. 8a) is due to the interaction of the turbulent eddies generated in the boundary layer and the trailing edge of the airfoil. The turbulent eddies themselves are relatively weak sources, but their effect is amplified when they are in the vicinity of a plate (in this case the trailing edge) [8]. *Separation-stall noise* (Fig. 8b) occurs when the angle of attack is too large and the flow separates. In such cases the separation-stall noise dominates the other noise sources [4]. At the tip of the blade a vortex is formed due to the pressure difference at the pressure and suction sides. The pressure fluctuations due to the presence of this vortex generate the *tip vortex formation noise* (Fig. 8c). The *boundary layer vortex shedding noise* (Fig. 8d) occurs when boundary layer instabilities develop along the airfoil. Such instabilities might lead to separation and the appearance of Tollmien Schlichting waves. While the trailing edge noise has a broadband character, the boundary layer vortex shedding noise is tonal. *Trailing edge bluntness vortex shedding noise* appears when the thickness of the trailing edge exceeds a critical limit, dependent on the Reynolds number and the shape of the airfoil (Fig. 8e). When this critical limit is exceeded a Karman-type vortex-street develops giving rise to a tonal noise.

The third group includes noise due to atmospheric turbulence. Due to the fluctuations in the incoming air stream, the blades suffer an unsteady loading, giving

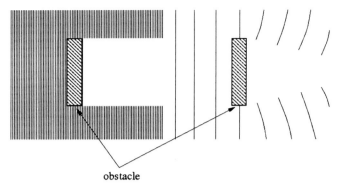

obstacle

Figure 9: Diffraction behind obstacles high-frequency (left) and low-frequency (right) waves.

rise to noise (Fig. 8f). One commonly divides the noise due to atmospheric turbulence into a low-frequency regime (when the length-scale of the fluctuations is much larger than the size of the body) and a high frequency regime (the length-scale of the fluctuations is much smaller then the airfoil). As it will be shown in Section 6.2, these two regimes can be treated separately in noise modeling [4]. Due to the chaotic behavior of turbulence, these models are not deterministic.

According to Lowson [9], the self-noise sources dominate at low wind speeds, near cut-in, while at the rated power the turbulence inlet noise source dominates. A detailed description of all these noise mechanisms can be found, e.g. in [4].

4.2 Propagation

In the previous section the noise generation mechanisms have been discussed. The knowledge of these noise *sources*, however, is not enough to predict the sound pressure level at a receiver. While the acoustic waves are traveling through the atmosphere, several factors influence the propagated sound pressure levels, the most important ones being the followings:

- The *distance* to the receiver. For increasing distance the acoustic energy is spread in a larger volume which decreases the sound pressure level.
- *Absorption* is due to the air viscosity and converts the acoustic energy into heat.
- *Reflections* due to the ground and surrounding objects.
- When a wave passes around a solid object *diffraction* occurs. For high frequencies (wavelength much smaller than the object size) a shadow zone occurs behind the object. The shadow zone decreases with decreasing frequency, completely disappearing for wavelengths much larger than the size of the object (see Fig. 9).
- *Refractions* are caused due to temperature gradients which cause different densities in different layers of the air, and as a consequence impose different propagation speeds of the sound waves.
- The *wind speed and direction* influences the directivity of the noise propagation.

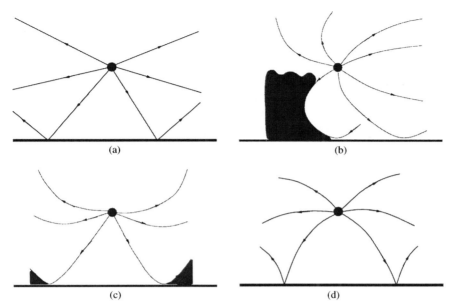

Figure 10: The influence of vertical temperature gradient and wind speed on noise
propagation: (a) no wind, no temperature gradient; (b) wind from left
to right, no temperature gradient; (c) no wind, negative temperature
gradient; (d) no wind, positive temperature gradient.

The influences of the last two parameters are visualized in Fig. 10. If there
would be no wind and no temperature gradients, the sound waves would propagate along straight lines (Fig. 10a). In windy conditions the noise propagation
paths are curved towards the wind direction (Fig. 10b). Negative temperature
gradients cause lower temperature regions at higher altitude, thus lower propagation speeds for the noise. As a result, the noise propagation paths will be
curved upwards (Fig. 10c) (the opposite effect happening for positive pressure
gradients, Fig. 10d). One can observe in Fig. 10 that in certain conditions shadow
zones appear where the noise will not propagate.

Compared to other industrial noise sources, wind turbines have two main special features [5]. First, the sources are located at an elevated level which leads to
reduced screening and ground attenuation effects. Secondly, in windy conditions
sound propagation is difficult to predict.

4.3 Immission

There are several factors influencing the level of annoyance from wind turbine
noise [6]. Both the odds of perceiving the wind turbine noise and the odds of being
annoyed by wind turbine noise increases with increasing sound pressure level [10].
Pedersen and Larsman [11] showed that the visual impact is important when wind

turbine noise is evaluated. When wind turbines are considered as ugly structures being in contrast with the surroundings, the probability of annoyance by the noise increases, regardless of the measured sound pressure levels. Even if the visual impact is not considered, the annoyance of the noise with the same equivalent sound pressure level can be rated differently. Waye and Öhrström [12] exposed several persons to noise registered from different wind turbine types, scaled to 40 dB $L_{eq}(A)$. The differences in annoyance response could not be explained by the psycho-acoustic parameters considered (sharpness, loudness, roughness, fluctuation strength and modulation). In regions with high background noise levels the wind turbine noise is considered less disturbing [6, 13]. Thus the acceptance level of wind turbines is higher in regions with large traffic, industrial areas or where a lot of noise is generated by vegetation or waves, while it is significantly lower in recreational and rural areas. For the same reason, wind turbine noise is considered more disturbing during the night than during the day. As it is emphasized by Grauthoff [14], not only the audible levels are important, high infrasound (below 16 Hz) levels might also be percepted as annoying.

4.4 Wind turbine noise regulations

A recent survey of noise regulations in several countries is presented by Pedersen [6]. Since it is considered unneeded to present all the details for the specific countries, here we summarize the major strategies adopted in the legislations. There are three different kinds of noise limitation strategies:

1. *Fixed values.* As an example, in Sweden the highest recommended sound pressure level originating from wind turbines is set to 40 dB, with a penalty of 5 dB for pure tones. Although this strategy is straightforward to apply it is the least flexible. In quiet areas even noise with sound pressure level corresponding to the 40 dB limit might lead to the annoyance of the people. Contrarily, in regions with high background noise levels even higher wind turbine noise levels would be accepted and a fixed value of the limit leads to suboptimal power production.
2. *Relative values.* Wind turbine noise limits in Great Britain recommend a maximum noise emission of 5 dB above the background noise levels. This approach is much more flexible, but difficult to implement practically.
3. *Variable values.* In the Netherlands, the maximum limits of the emitted noise vary as function of the wind speed. This method is more flexible than the first one and easier to implement than the second one. Nevertheless, there are further issues to be solved. Depending on the atmospheric stability conditions, the wind speed measured at a fixed height above the ground might not be an accurate indicator of the wind speed met by the rotor blades. Van den Berg [15] presents an example where the practically emitted sound pressure levels exceeded the ones predicted with 'standard' methods because of the higher velocities at rotor height than predicted.

5 Wind turbine noise measurements

There are two main categories of measurements related to wind turbine acoustics: on-site and wind-tunnel measurements. The first group gives more realistic data since it is applied for a whole wind turbine and gives directly the noise generated by it. The major drawbacks of the on-site measurements are the difficulty to install the instrumentation and the impossibility of prescribing desired flow conditions. This leads also to a limited repeatability of the measurements. Wind-tunnel measurements offer more control, but they have a limited size. As a consequence, the models have to be scaled down and sometimes only parts of the wind turbine can be studied. The scaling of the models is not straightforward, not all the non-dimensional parameters can be preserved. Also, the background noise from the wind tunnel is difficult to subtract from the measurements. A good review of the measurement methods applicable for wind turbine noise is presented in [4].

5.1 On-site measurements

5.1.1 Ground board

The ground board consists of a flat, acoustically hard plate with a microphone mounted horizontally, at the center of the plate [4]. The microphone has to be directed towards the turbine tower and is covered by a wind screen (see Fig. 11). The major advantage of using a ground board is that it diminishes the wind-induced noise and that the reflections from the ground are independent from the site. To remove the effect of reflections 6 dB is subtracted from the measured spectra at all frequencies.

5.1.2 Acoustic parabola

Acoustic parabolas (see Fig. 12) reflect and focus the acoustic waves in their focal point where a microphone registers the signal. The amplification depends on the

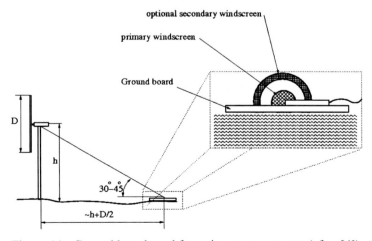

Figure 11: Ground board used for noise measurements (after [4]).

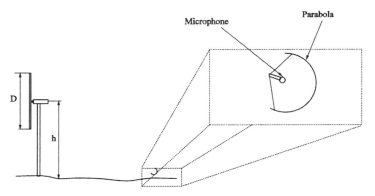

Figure 12: Acoustic parabola used for noise measurements.

Figure 13: Proximity microphone used for noise measurements.

angle of incidence, it amplifies more the noise coming from the direction where the parabola is aimed to.

In this way noise coming from a certain part of the wind turbine can be localized. If one is interested in the noise coming from a certain part of the rotating blades (e.g. tip region), the measured signal has to be windowed to post process only the data corresponding to the passage of the blade over the measurement spot. The amplification depends also on the wavelength of the acoustic waves, large wavelengths compared to the size of the parabola reduces the spatial accuracy.

5.1.3 Proximity microphone

Ground boards and acoustic parabolas are used far from the wind turbine. Proximity microphones (see Fig. 13) are installed close to the noise sources, either on the blades themselves or close to the blades on fixed devices. The disadvantage of mounting the proximity microphones directly on the blades is that it will be exposed to higher speeds of the air flow resulting in higher wind-induced noise.

If the microphone is not co-rotating with the blade a time-windowing technique has to be used to collect data only when the noise sources are in the vicinity of the

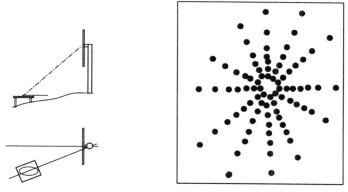

Figure 14: Sketch of the board position (left) and the arrangement of the micro-
           phones on the board (right) (after [16]).

microphone. The spatial accuracy of the proximity microphones depends on the
distance from the source, but, unlike acoustic parabolas, is independent of the
acoustic wavelength.

5.1.4 Acoustic antenna

Acoustic antennas are composed by a set of microphones in a row or in a matrix
which can be mounted on a ground plate or in the proximity of the wind turbine.
By computing correlations of the signals provided by different microphones, the
location of the acoustic sources can be determined.

Oerlemans *et al.* [16] used for example an array of 148 microphones mounted
on a board of 15 m × 18 m. The array had an elliptic shape to account for the circular motion of the blades and for the tilted angle of the board (see Fig. 14). Cross-
correlating the microphones data made possible to determine the distribution of
the noise sources in the rotor plane and along the rotor blades (individually for the
three blades). It was observed that most of the noise sources are located on the
right-hand side of the rotor plane, indicating that most of the noise is generated
during the downward motion of the blades. Furthermore, the contribution of the
hub was also clearly visible.

5.2 Wind-tunnel measurements

Wind tunnels have the advantage of providing better controllable flow conditions
and thus are preferred for systematic studies. Due to their size limitations, however, wind tunnels are limited to downscaled model studies. The large-scale facility at the National Renewable Energy Laboratory in the USA is equipped with
a wind tunnel with a measurement cross section of 24.4 × 36.6 m^2 which made
possible to carry out flow measurements on a full-scale (10 m rotor diameter) wind
turbine [17]. The wind tunnel, however, is not anechoic, thus is not well suited for
acoustic measurements.

Brooks *et al.* [18] measured the flow and the acoustic pressure around a set of NACA 0012 airfoils with different chord lengths in an anechoic chamber. Eight microphones have been used for the noise measurements. Based on the measurement data, semi-empirical models have been developed to predict airfoil self-noise (see Section 6.2).

In Europe there are two anechoic wind tunnels at the University of Oldenburg and at the National Aerospace Laboratory (NRL) in the Netherlands [4]. These facilities have been used also for the study of the airfoil self-noise, as well as for the study of the noise due to turbulent inflow.

The major drawback of wind-tunnel measurements is the self-noise of the wind tunnel itself. The errors due to the background noise can be reduced for rotating sources by using tracking methods. For stationary sources (e.g. airfoil cross sections) the error due to the background noise can be reduced by using multiple microphones and cross-correlating the signals at different observer positions.

Fujii *et al.* [19] studied the noise generated by the interaction of the rotor blades with the wakes of wind turbine towers. Towers with circular, elliptical and square cross sections have been considered. Their measurements showed that the tower with a slender elliptic cross section was the quietest, the loudest being the one with the square cross section.

6 Noise prediction

The solution of the compressible Navier-Stokes equations includes inherently also the generation and propagation of the acoustic pressure waves. This direct computational approach, however, involves several drawbacks, which make it applicable only for relatively simple cases and to small Reynolds numbers [20–22]. One issue is the small magnitude of the acoustic quantities (acoustic pressure and density fluctuations) compared to the hydrodynamic quantities of the flow. The low-amplitude acoustic fluctuations require discretization schemes with high accuracy which are computationally demanding. Also, the timesteps are limited by the sound speed and not the convection speed as it is the case of incompressible flow simulations. As a result, there was a need to develop aeroacoustic models.

Lighthill [23] was the first to derive a model for aerodynamically generated sound. By rearranging the continuity and momentum equations he obtained the following non-homogeneous wave equation for the acoustic density:

$$\frac{\partial^2 \rho'}{\partial t^2} - c_0^2 \frac{\partial^2 \rho'}{\partial x_i \partial x_i} = \frac{\partial^2 (\rho u_i u_j)}{\partial x_i \partial x_j} \tag{10}$$

where $\rho' = \rho - \rho_0$ is the acoustic density fluctuation defined as the departure from ambient conditions, and c_0 is the sound speed in the undisturbed ambient medium. In the derivation of eqn (10) it was assumed that the ambient conditions are constant and viscosity effects have been neglected. Equation (10) is called the Lighthill analogy and describes the propagation (left-hand side, LHS) and generation (right-hand side, RHS) of sound by the fluid flow. The term on the RHS is the

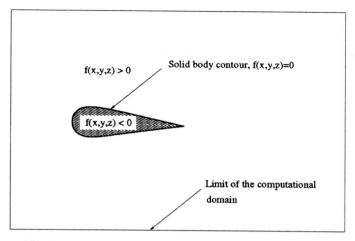

Figure 15: The value of the shape function in the solid and fluid regions.

second derivative of the Lighthill stress tensor, $T_{ij} = \rho u_i u_j$, which acts as a source of quadrupole type.

Lighthill's analogy does not take into account the noise generated by solid surfaces. To remedy this deficiency, Ffowcs Williams and Hawkings (cited in [24]) extended Lighthill's analogy by representing a solid object by a surface S defined as $f(t,x,y,z) = 0$ (see Fig. 15) and extending the fluid motion also into the object domain but restricting the velocity to the speed of the object itself. Using the generalized function theory the following equation has been derived:

$$\frac{\partial^2 p'}{\partial t^2} - c_0^2 \frac{\partial^2 p'}{\partial x_i \partial x_i} = \frac{\partial^2 (\rho u_i u_j)}{\partial x_i \partial x_j} - \frac{\partial}{\partial x_i}\left(p_{ij}\delta(f)\frac{\partial f}{\partial x_j} \right)$$
$$+ \frac{\partial}{\partial t}\left(\rho_0 u_i \delta(f)\frac{\partial f}{\partial x_i} \right) \tag{11}$$

The first term on the right-hand side is identical to the RHS of eqn (10) and is the Lighthill stress tensor. The second term on the RHS is proportional to the stress tensor p_{ij}, reflecting the force which the object exerts on the fluid and it has a dipole character. The third term on the RHS is a monopole term and is proportional to the acceleration of the fluid by the non-stationary solid surface.

Based on these theoretical models, researchers deduced semi-empirical models used for the prediction of the noise generated by non-rotating and rotating airfoils. The first models were developed to predict the noise generated by airplane wings and helicopter rotors. Later, these models were adapted for wind turbine applications. With the increase of the available computational power and evolution of the aeroacoustic theory more and more complex models were developed

Table 1: Category I models [4].

No.	Model
1	$L_{WA} = 10\log_{10} P_{WT} + 50$
2	$L_{WA} = 22\log_{10} D + 72$
3	$L_{pA} = C_1 \log_{10} V_{tip} + C_2 \log_{10} \left(n_b \dfrac{A_b}{A_r} \right) + C_3 \log_{10} C_T + C_4 \log_{10} \dfrac{D}{r} - C_5 \log_{10} D - C_6$

during the years. Lowson (cited in [4]) divided the prediction models in three categories:

- The Category I models predict the global emitted noise as a function of main wind turbine parameters, like rated power, rotor diameter, blade area, tip speed, etc.
- The Category II models are semi-empirical relationships which predict the total sound pressure level and spectral shape taking into account the various noise generation mechanisms.
- The Category III models are based on a detailed description of the rotor geometry and aerodynamics and give detailed information about the acoustic field.

6.1 Category I models

These models are the simplest ones. Based on simple algebraic relations they predict the emitted sound power level as functions of the wind turbine main parameters. Table 1 lists the category I models as it is summarized by Wagner *et al.* [4]. In Table 1 L_{WA} is the total A-weighted sound power level, P_{WT} the rated power of the turbine, D the rotor diameter, L_{pA} the A-weighted sound pressure level at a monitoring point located at a distance of r_0 from the tower base, V_{tip} the tip speed, n_b the number of the blades, A_b the blade area, A_r the rotor area, C_T the axial force coefficient, r the hub–observer distance. C_i are empirical parameters having different values in different references.

These Category I models are simple and fast, however they are by far not universal. As a consequence, these models were rapidly outranked by the computationally not much more expensive, but more accurate Category II models.

6.2 Category II models

The models belonging to this category are semi-empirical and consist of a set of models for the noise generation mechanisms listed in Section 4.1.2. Based on Lighthill's and Ffowcs Williams–Hawkings theory and experimental measurements algebraic relationships are deduced to model both the emitted sound power level and sound spectrum. There are a multitude of models in the literature for each noise generation mechanism, in the followings the most widespread models being presented.

6.2.1 Trailing edge noise

Grosvelds model (cited in [4]) is based on a frozen turbulent pattern convected downstream over the trailing edge and predicts the A-weighted sound pressure level as:

$$L_{pA} = 10\log_{10}\left(\frac{\delta\Delta_s U^5 \overline{D}}{r^2} n_b\right) + K(f) + C \tag{12}$$

The shape of the spectrum is given by

$$K(f) = 10\log_{10}\left\{\left(\frac{St'}{St_{max}}\right)^4\left[\left(\frac{St'}{St_{max}}\right)^{1.5} + 0.5\right]^{-4}\right\} \tag{13}$$

and the Strouhal numbers are defined as

$$St' = \frac{f\delta}{U}, \quad St_{max} = 0.1 \tag{14}$$

The empirical constant is $C = 5.44$ dB and the directivity factor is determined as:

$$\overline{D}(\theta,\psi) = \sin^2\psi\frac{\sin^2\theta/2}{(1+M\cos\theta)\left[1+(M-M_C)\cos\theta\right]^2} \tag{15}$$

Thus the sound pressure level depends on the fifth power of the velocity. Brooks *et al.* [18] developed a more complex model to predict the trailing edge noise, accounting for the length of the blade segment, angle of attack and separating the contributions from the suction end pressure sides:

$$L_p = 10\log_{10}(10^{L_{p,a}/10} + 10^{L_{p,s}/10} + 10^{L_{p,p}/10}) \tag{16}$$

where the individual contributions have the form:

$$L_{p,i} = 10\log_{10}\left(\frac{\delta_i^* M^5 \Delta s\overline{D}}{r^2}\right) + K_i\left(\frac{St_i}{St_{peak}}\right) + C_i \tag{17}$$

δ_i being the displacement thickness, M the Mach number, Δs the chord length, $\overline{D}$ the directivity function (given by a similar relationship to eqn (15)), K_i are shape functions for the spectra and C_i empirical constants. The peak Strouhal numbers St_{peak} have also empirical values.

Lowson [9] based his model on the work of Brooks *et al.* [18]:

$$L_p = 10\log_{10}\left(\frac{\delta^* s M^5}{r}\right) + K(f) + 128.5 \tag{18}$$

Thus the major difference is that the directivity effects are neglected. The spectral shape function is also simplified:

$$K(f) = 10\log_{10}\left[4\left(\frac{f}{f_{peak}}\right)^{2.5}\left(1 + \left(\frac{f}{f_{peak}}\right)^{2.5}\right)^{-2}\right] \tag{19}$$

6.2.2 Separation-stall noise

The literature lacks specific models for the separation-stall noise. Brooks *et al.* [18] used the same theory to predict separation-stall noise as the one used for the trailing edge noise presented in the previous section.

6.2.3 Tip vortex formation noise

For the tip vortex noise Brooks *et al.* proposed the following model:

$$L_p = 10\log_{10}\left(\frac{M^3 M_{TV}^2 l_{TV}^2 \overline{D}}{r^2}\right) + K_1\left(\frac{St}{St_{peak}}\right) + K_2\left(\frac{Re}{Re_0}\right) + K_3(a) \tag{20}$$

where the Strouhal number and M_{TV} are based on the tip vortex parameters,

$$St = \frac{f l_{TV}}{U_{TV}}, \quad M_{TV} = \frac{U_{TV}}{c_0}.$$

6.2.4 Boundary layer vortex shedding noise

This noise source is modeled by Brooks *et al.* [18] similarly to the trailing edge noise:

$$L_p = 10\log_{10}\left(\frac{\delta_p M^5 \Delta s \overline{D}}{r^2}\right) + K_1\left(\frac{St}{St_{peak}}\right) + K_2\left(\frac{Re}{Re_0}\right) + K_3(a) \tag{21}$$

where the three terms after the logarithm are empirical functions determining the spectrum shape K_1, and the influence of the Reynolds number K_2 and the angle of attack K_3.

6.2.5 Trailing edge bluntness vortex shedding noise

Grosveld developed two models for the bluntness shedding noise depending on the ratio of the trailing edge thickness, t^*, and the displacement thickness of the boundary layer, δ^*. If $t^*/\delta^* > 1.3$

$$L_{pA} = 10\log_{10}\left(\frac{U^6 t^* \Delta s}{r^2} \frac{\sin^2(\theta)\sin^2(\psi)}{(1+M\cos\theta)^6} n_b\right) + K(f) + C \qquad (22)$$

with the peak frequency:

$$f_{peak} = \frac{0.25U}{t^* + \delta/4} \qquad (23)$$

If $t^*/\delta^* < 1.3$

$$L_{pA} = 10\log_{10}\left(\frac{U^5 t^* \Delta s}{r^2} \frac{\sin^2(\theta/2)\sin^2(\psi)}{(1+M\cos\theta)^3 \left[1+(M-M_c)\cos\theta\right]^2} n_b\right) + K(f) + C \qquad (24)$$

and peak frequency:

$$f_{peak} = 0.1\frac{U}{t^*} \qquad (25)$$

The spectral shapes $K(f)$ are empirical functions.

Brooks et al. [18] used a single model for the trailing edge bluntness noise:

$$L_p = 10\log_{10}\left(\frac{t^* M^{5.5} \Delta s D}{r^2}\right) + K_1\left(\frac{t^*}{\delta_{avg}^*}, \psi\right) + K_2\left(\frac{t^*}{\delta_{avg}^*}, \psi, \frac{St}{St_{peak}}\right) \qquad (26)$$

where ψ is the angle of the trailing edge and $\delta_{avg}^* = 0.5(\delta_p^* + \delta_s^*)$.

6.2.6 Noise due to atmospheric turbulence

The atmospheric turbulence is highly dependent on weather conditions, the geometry of the landscape and ground roughness. As a consequence, the noise generated by the interaction of onflow turbulence and the blades is the most difficult to model. Indeed, the models used for the turbulence effects are very sensitive to the choice of appropriate turbulence scales. As an example, Glegg et al. [25] reports that using turbulent length-scales from the atmospheric boundary layer leads to an over-prediction of the noise levels, while assuming the turbulent length-scales to be of the order of the blade chord gave much better results.

Amiet [7] did a pioneering work deriving a theoretical expression for the far-field acoustic spectral density produced by an airfoil obtaining good predictions of experimental data.

One of the earliest models for the noise generated by inflow turbulence for wind turbines is developed by Grosveld (cited in [4]). This model is valid only for low frequencies because it is based on the assumption that the noise is generated by a point source located at the hub. Furthermore, neutral stability conditions and negative temperature gradient are assumed for the atmosphere. The sound pressure level is given by:

$$L_p = 10\log_{10}\left(\frac{\overline{w}^2 U^4 C R \, n_b \sin^2(\phi)\rho^2}{r^2} \frac{1}{c_0^2}\right) + K(f) + C \qquad (27)$$

where $K(f)$ (the shape of the spectrum) and C are determined empirically, $U = \Omega R_{ref}$ and the peak frequency is given by:

$$f_{peak} = \frac{16.6U}{h - 0.7R} \tag{28}$$

The reference chord location is at 30% of the blade length from the tip of the blade and the root mean square of the turbulence intensity is given by:

$$\overline{w}^2 = w_r^2 \left(\frac{hw_r}{V_w R(w_r - 0.014 w_r^2)} \right)^{-2/3} \tag{29}$$

where V_w is the wind speed, h the height above the ground, and the turbulence intensity is:

$$w_r^2 = 0.2(2.18 V_w h^{-0.353})^{1/1.185 - 0.193 \log_{10} h} \tag{30}$$

Lowson [9] derived a model based on Amiet's work [7]. Contrary to Amiet, Lowson has a single formula for the high- and low-frequency regimes, by introducing a correction term:

$$L_p = 10 \log_{10} \left(\frac{\rho^2 c_0^2 l_{IT} \Delta s M^3 \overline{w}^2 k^3 (1 + k^2)^{-7/3}}{r^2} \right) + 58.4 + 10 \log_{10} \left(\frac{K_{LF}}{1 + K_{LF}} \right) \tag{31}$$

where K_{LF} is the low-frequency correction factor, $k = (\pi f C)/U$, U includes contributions from the wind speed as well, $U = \sqrt{(\Omega R)^2 + (2/3 V_w)^2}$, and $M = U/c_0$.

6.2.7 Sample results

Moriarty and Migliore [26] used Category II models to predict the acoustic field generated by two-dimensional airfoils and from a full-scale wind turbine. Except the inflow turbulence noise the models from Brooks *et al.* [18] have been used. The inflow turbulence noise model was adapted from Lowson [9]. The tower–wake interaction and propagation effects are neglected. As it regards the directivity, it is assumed that all sources are propagated in the wind direction with a speed corresponding to 80% of the average wind speed. For the airfoil cross sections good agreements were obtained for high frequencies (around 3 kHz), while at lower frequencies the sound pressure levels were overpredicted by as much as 6 dB (see Fig. 16). For the full-scale turbine it was observed that the turbulence inlet noise had the largest contribution to the overall sound pressure level. The second most important noise source was found to be the blunt trailing edge noise or the laminar vortex shedding noise, depending on the wind speed.

6.3 Category III models

The models belonging to this category take into account the complex three-dimensional and time-dependent distribution of the acoustic sources. A recent

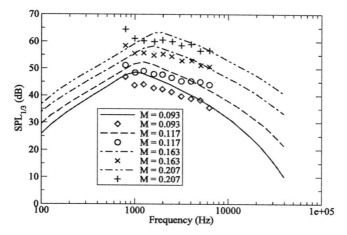

Figure 16: Comparison of two-dimensional NACA 0012 data to predictions for $a = 7.8°$ at several Mach numbers (after [26]).

review of the computational methods for aeroacoustic predictions is given in [27]. Direct solution of the acoustic field by solving the compressible Navier-Stokes equations being computationally too expensive for practical applications, the most promising approaches for aeroacoustic computations are the *hybrid* methods. The hybrid methods are based on the observation that for low Mach numbers the amplitude of the acoustic fluctuations is several orders of magnitude smaller than the hydrodynamic fluctuations. Based on this scale separation the governing equations can be divided into a set of flow equations and a set of acoustic equations having sources determined by the flow variables. Such hybrid approaches have several advantages as compared to direct computations. First, the acoustic field can be computed on a larger domain than the flow field, flow computations are needed only in regions where significant noise is generated. Second, the solution methods for incompressible/semi-compressible flows are faster than for compressible flows. Additional gain in computational effort is due to the possibility of having larger mesh spaces in the acoustic solver and the fact that for the acoustics commonly only one equation has to be solved.

Lighthill's analogy (presented earlier in this section) was the first method to separate the acoustics from the flow computations. Recently, a more systematic approach has been presented by Slimon *et al.* [28]. Their method is based on the expansion about incompressible flow. The compressible flow variables are expanded in a power series in $\varepsilon^2 = \gamma M^2$:

$$p = p_0 + \varepsilon^2 \overline{p_1} + \varepsilon^4 \overline{p_2} + \varepsilon^6 \overline{p_3} + \cdots \tag{32}$$

$$\rho = \rho_0 + \varepsilon^2 \overline{\rho_1} + \varepsilon^4 \overline{\rho_2} + \varepsilon^6 \overline{\rho_3} + \cdots \tag{33}$$

$$u_i = u_{i0} + \varepsilon^2 \overline{u_{i1}} + \varepsilon^4 \overline{u_{i2}} + \varepsilon^6 \overline{u_{i3}} + \cdots \tag{34}$$

$$T = T_0 + \varepsilon^2 \overline{T_1} + \varepsilon^4 \overline{T_2} + \varepsilon^6 \overline{T_3} + \cdots \tag{35}$$

Substituting the decomposed variables in the compressible Navier-Stokes equations and grouping terms with the same order in ε, equation sets with different approximations are derived. The zeroth order approximations ε^{-2} describe the thermodynamic field, the first order approximations ε^{-0} the hydrodynamic field and the second order approximations ε^2 describe compressibility effects. The results of this method can be written in the form of Lighthill's acoustic analogy. For further details the reader should consult [28].

Once the acoustic sources are determined, there are two main strategies for the computation of the acoustic field: analytical computation of the sound pressure level at the observer position or numerical computation of the noise propagation.

When the acoustic field is governed by a non-homogeneous wave equation having elementary acoustic sources (monopoles, dipoles or quadrupoles) of the form:

$$\frac{1}{c^2}\frac{\partial^2 p'(\vec{x},t)}{\partial t^2} - \frac{\partial^2 p'(\vec{x},t)}{\partial x_i \partial x_i} = Q(\vec{x},t)\delta(f) \tag{36}$$

and f is the surface over which the sources are distributed, using the free-space Green's function $\delta(g)/4\pi r$, with $g = \tau - t + r/c$ the solution can be written in an integral form:

$$4\pi p'(\vec{x},t) = \int\limits_{-\infty}^{t}\int\limits_{-\infty}^{\infty} \frac{Q(\vec{y},\tau)\delta(f)\delta(g)}{r}\,\mathrm{d}y\,\mathrm{d}\tau \tag{37}$$

In eqn (37) $\vec{x}$ is the observer position vector, $\vec{y}$ the source position vector, r the distance between the source and the observer and τ is the source time. This equation can be integrated after variable transformations. Further details can be found e.g. in [24] where the Ffowcs Williams–Hawkings equation is solved to determine the noise generated by helicopter rotors. Recently, Filios et al. [29] combined a low order panel method with prescribed wake-shape and the integration of the Ffowcs Williams–Hawkings equations to predict the noise generated by the NREL wind turbine. In their work only the monopole and dipole sources have been considered, the noise generated by velocity fluctuations (quadrupoles) being neglected.

This previously described integral approach is based on the assumption that the observer is far enough to consider the waves propagating from individual sources spherical. Also, propagation effects (reflection, absorption, refraction) are neglected.

An alternative method is to solve numerically the CAA equation(s) on an acoustic grid (which does not need to be identical to the flow grid). Although this approach is computationally more expensive, it allows the consideration of propagation effects and gives a more detailed three-dimensional picture of the radiated acoustic field. Moroianu and Fuchs used Large Eddy Simulations to compute the flow field around a single wind turbine [30]. The acoustic sources provided by the LES computations have been used in a separate acoustic solver to compute the corresponding noise generated by the wind turbine. Figure 17 shows an instantaneous snapshot of the radiated acoustic field and the sound pressure levels around the

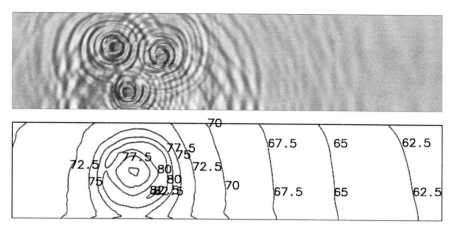

Figure 17: Instantaneous snapshot of the acoustic pressure fluctuation field (top) and the radiated sound pressure levels (bottom).

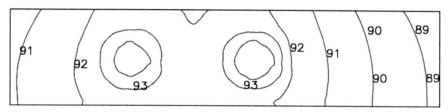

Figure 18: Sound pressure levels emitted by two wind turbines.

wind turbine. A similar approach has been used in [31] where the acoustic sources have been imposed in two instances to model the noise emitted by two wind turbines. The isocontours of the sound pressure level are shown in Fig. 18. The flow computations in [30, 31] are based on a finite volume approach with unstructured sliding grids leading to very large computational efforts. Szasz and Fuchs [32] used a flow solver based on the finite difference approach with Cartesian structured grids to compute the acoustic sources. The solid boundaries have been accounted for using the immersed boundary method, the acoustic solver was the same as in [30, 31]. This approach to compute the flow field turned out to be very efficient and it has been recently developed further to compute the flow around several wind turbines. In this way the effect of the wind turbine wakes on the noise generation of downstream wind turbine can be accounted for. Figure 19 shows the vortical structures around four wind turbines visualized by the λ_2 method.

6.4 Noise propagation models

As it was mentioned above, the solution of the compressible Navier-Stokes equations includes both noise generation and propagation. For large problems,

Figure 19: Large-scale vortices around four wind turbines visualized by the λ_2 method.

however, this approach is computationally costly, leading to the development of several noise propagation models during the years. A review of the noise propagation models are presented in [4].

6.4.1 Spherical spreading model

The simplest propagation model is to assume spherical spreading. This implies that the sources are considered point sources and all the parameters enumerated in Section 4.2 are neglected, except the distance to the observer. According to the spherical propagation model the sound pressure level decreases with 6 dB for a doubling of the source–observer distance.

6.4.2 VDI 2714

One of the most simple and widely used models is the VDI 2714 method, where the sound pressure level at a receiver is given by the following empirical formula [4]:

$$L_p = L_w + D + K - L_d - L_a - L_g - L_v - L_b - L_s \qquad (38)$$

where D accounts for the source directivity, K is a correction for the reflection of sound due to the presence of vertical walls ($K = 0$ dB for sound sources located in the free field, $K = 3$ dB above the ground, and $K = 6$ dB or $K = 9$ dB when two or three perpendicular surfaces are present, respectively). $L_d = 10 \log_{10}(4\pi r^2)$ accounts for the influence of the source–observer distance. The air absorption effect is represented by $L_a = ar$. The ground effects are modeled as $L_g = [4.8 - (2h_m / r)(17 + (300 / r))]$ where h_m is the average of the source and observer heights. The last three terms in eqn (38) account for the influence of the vegetation, buildings and screens, respectively.

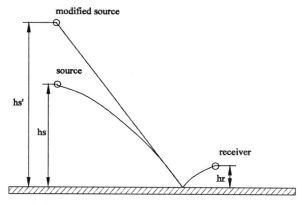

Figure 20: Modified source height due to curved sound path (after [4]).

6.4.3 Ray-tracing models

In order to account for complex landscapes and the meteorological effects, more advanced methods were needed. The ray-tracing models describe the propagation of the acoustic waves by the path along which the waves propagate (rays). For spherical spreading (no solid objects, no temperature gradients and no wind) the rays will be straight lines emerging from the source. When solid objects are present the rays will be reflected, while the presence of wind or temperature gradients will curve the rays.

The DELTA model (cited in [4]) has two steps. First, a modified source height is calculated based on the arrival angle of the ray, in this way accounting for ray curvature (see Fig. 20):

$$h'_s = h_s + d \cos \theta \tag{39}$$

The receiver height is not corrected, because it is much smaller than the source height and the error which is introduced by neglecting the correction is minimal. In the second step the noise propagation is computed as if no atmospheric effects would be present.

Recently, Prospathopoulos and Voutsinas [33] used a more advanced ray-tracing model to compute the sound propagation even from wind turbine parks. In their model the noise source is assumed to be located at the rotor center. The propagation of the noise is computed in three steps:

1. finding the eigenrays
2. calculating the energy losses along the eigenrays
3. reconstructing the sound pressure level by summing up the contribution of the eigenrays

When computing the energy losses, both the atmospheric and ground absorption, wave refraction and diffraction and atmospheric turbulence effects have been taken into account. Good agreements have been obtained with experimental measurements for several testcases.

7 Noise reduction strategies

In the previous sections it has been shown that there are a multitude of parameters which influence the generation and propagation of wind turbine noise. In the followings a short list of noise reduction strategies is given:

- *Blade–tower interaction.* For downwind turbines, the rotor blades passing the wake of the tower, the noise levels are significantly larger than for upwind turbines (see e.g. [19, 34]). For this reason, new wind turbines have upwind configurations, even if technically there are additional issues associated with them (like the bending of the blades because of the wind load towards the tower and not away like in the downwind case).
- As it was shown in Section 6.2., the emitted sound power levels are proportional to the fifth power of the blade speed. Thus, an efficient way of reducing the emitted noise levels is to decrease the angular speed of the rotor. This approach is currently employed, but it has the drawback, that it involves also a reduction of the generated electric power, thus the efficiency of the wind turbine decreases.
- The experiments of Oerlemans *et al.* [16] showed that *blade surface smoothness* affects the emitted noise levels, clean blades emitting considerably less noise than the rough ones.
- The shape of the *blade cross section* and the angle of attack is crucial on the emitted sound pressure levels. Large angles of attack and thick airfoils lead to separation and vortex shedding with increased noise levels. Marsden *et al.* [35] reported an 89% noise reduction from an airfoil by optimizing its cross section to eliminate the low-frequency vortex-shedding.
- *Trailing edge shape.* Blake (cited in [4]) observed that an airfoil with flat pressure side and beveled suction side produces the least noise. Serrations on the trailing edge have also the potential to reduce the emitted noise levels by breaking up the large-scale structures and reducing by this the low-frequency oscillations. Nevertheless further research is needed to find optimal configurations, since discrepancies are observed between the theoretically predicted and measured values [4].
- *Trailing edge material.* As it was discussed in Section 6.2, trailing edge noise is mainly attributed to the amplification of the turbulent vortex caused pressure fluctuations by the presence of the airfoil surface. By using porous or flexible materials there is a potential to reduce these reflections and by this also the emitted noise levels (see e.g. Hayden, cited in [4]).
- Several experimental measurements have been carried out to evaluate the influence of *blade tip shape* on the emitted noise levels but no profiles could be found which consistently reduced the noise levels for each measurement condition.
- *Masking.* Bolin [13] has shown that masking the wind turbine noise by adding 'positive' noise from natural sources (trees, waves) can be an efficient way for the increase of the acceptance of wind turbines, especially in quiet areas, where

even low sound pressure levels (below the admissible limits) can be experienced as annoying. Björkman [34] carried out longtime measurements at several wind turbine sites in southern Sweden. At the site located next to Lund low correlation was observed by the wind speed and the registered noise levels which were explained by the fact that the wind turbine noise was masked by other noise sources.

8 Future perspective

Recently, van den Berg [15] showed that even if the noise limitation standards were taken into account already in the design phase, the inhabitants living in the surrounding of a new wind park were still annoyed by the emitted noise levels. The two main reasons for the annoyance were found to be the inaccurate prediction of the equivalent noise levels and the impulsive character of the sound. Thus there is a need for improved models. First, there is need of improving the Category III models to achieve detailed knowledge about the generated sound field already in the design phase. The influence of atmospheric stability and turbulence level, landscape shape, type of vegetation or the presence of sea waves have to be evaluated. Also, the impacts of the interaction of wind turbine wakes in wind turbine farms are not well understood. Another important issue related to wind turbines operating in cold areas is the ice accretion on the wind turbine blades. Since the deposition of the ice affects the blade cross section it affects also the character of the emitted noise. Thus acoustic measurements might be used to detect ice deposition. Recently, Fuchs and Szasz used a coupled Lagrangian–Eulerian approach to model ice accretion [36]. From a legislative point of view there is a need for more accurate and flexible, but at the same time not too complicated models. Such models will lead to noise regulations which will increase the power production of the wind turbines without impairing the well-being of the nearby residents.

References

[1] van den Berg, G.P., The beat is getting stronger: the effect of atmospheric stability on low frequency modulated sound of wind turbines. *Journal of Low Frequency Noise, Vibration and Active Control*, **24(1)**, pp. 1–24, 2005.

[2] Pedersen, E. & Persson Waye, K., Wind turbines – low level noise sources interfering with restoration? *Environmental Research Letters*, **3**, pp. 1–5, 2008.

[3] Andersson, J.D., *Modern Compressible Flow with Historical Perspective*, McGraw-Hill, 1990.

[4] Wagner, S., Bareiss, R. & Guidati, G., *Wind Turbine Noise*, Springer, 1996.

[5] Pinder, J.N., Mechanical noise from wind turbines. *Wind Engineering*, **16(3)**, pp. 158–167, 1992.

[6] Pedersen, E., Noise annoyance from wind turbines – a review. *Naturvårdsverket*, Report 5308, 2003.

[7]   Amiet, R.K., Acoustic radiation from an airfoil in a turbulent stream. *J. Sound and Vibration*, **41(4)**, pp. 407–420, 1975.

[8]   Ffowcs Williams, J.E. & Hall, L.H., Aerodynamic sound generation by turbulent flow in the vicinity of a scattering half plane. *J. Fluid Mech.*, **40(4)**, pp. 657–670, 1970.

[9]   Lowson, M.V., A new prediction model for wind turbine noise. *Renewable Energy*, pp. 177–182, 1993.

[10] Pedersen, E. & Persson Waye, K., Wind turbine noise, annoyance and self-reported environments health and well-being in different living environments. *Occup. Environ. Med.*, **64**, pp. 480–486, 2007.

[11] Pedersen, E. & Larsman, P., The impact of visual factors on noise annoyance among people living in the vicinity of wind turbines. *Journal of Environmental Psychology*, **28(4)**, pp. 379–389, 2008.

[12] Persson Waye, K. & Öhrström, E., Psycho-acoustic characters of relevance for annoyance of wind turbine noise. *J. Sound and Vibration*, **250(1)**, pp. 65–73, 2002.

[13] Bolin, K., Wind turbine noise and natural sounds – masking, propagation and modeling, *PhD thesis*, Royal Institute of Technology, Stockholm, Sweden, 2009.

[14] Grauthoff, M., Utilization of wind energy in urban areas – chance or utopian dream? *Energy and Buildings*, **15-16**, pp. 517–523, 1990/1991.

[15] van den Berg, G.P., Effects of the wind profile at night on wind turbine sound. *J. Sound and Vibration*, **277**, pp. 955–970, 2004.

[16] Oerlemans, S., Sijtsma, A.P. & Mendez Lopez, B., Location and quantification of noise sources on a wind turbine. *J. Sound and Vibration*, **299**, pp. 869–883, 2007.

[17] Hand, M.M., Simms, D.A., Fingersh, L.J., Jager, D.W., Cotrell, J.R., Schreck, S. & Larwood, S.M., Unsteady aerodynamics experiment phase VI: Wind tunnel test configurations and available data campaigns. *Technical Report,* NREL/TP-500-29955, National Renewable Energy Laboratory, 2001.

[18] Brooks, T.F., Pope, D.S. & Marcolini, M.A., Airfoil self-noise and prediction. *Technical Report,* RP-1218, NASA, 1989.

[19] Fujii, S., Takeda, K. & Nishiwaki, H., A note on tower wake/blade interaction noise of a wind turbine. *J. Sound and Vibration*, **97(2)**, pp. 333–336, 1984.

[20] Colonius, T., Lele, S.K. & Moin, P., Sound generation in a mixing layer. *J. Fluid Mech.*, **330**, pp. 375–409, 1997.

[21] Freund, J.B., Noise sources in a low-Reynolds number turbulent jet at mach 0.9. *J. Fluid Mech.*, **438**, pp. 277–305, 2001.

[22] Bogey, C. & Bailly, C., Downstream subsonic jet noise: link with vortical structures intruding into jet core. *C.R. Mecanique*, **330**, pp. 527–533, 2002.

[23] Lighthill, J., The bakerian lecture. Sound generated aerodynamically. *Proc. of the Royal Society of London*, **267**, pp. 147–182, 1962.

[24] Brentner, K.S. & Farassat, F., Modeling aerodynamically generated sound of helicopter rotors. *Prog. Aerospace Sciences*, **39**, pp. 83–120, 2003.

[25] Glegg, S.A.L., Baxter, S.M. & Glendinning, A.G., The prediction of broadband noise from wind turbines. *J. Sound and Vibration*, **118(2)**, pp. 217–239, 1987.

[26] Moriarty, P. & Migliore, P., Semi-empirical aeroacoustic noise prediction code for wind turbines. *Technical Report*, NREL/TP-500-34478, National Renewable Energy Laboratory, 2003.

[27] Wang, M., Freund, J.B. & Lele, S.K., Computational prediction of flow-generated sound. *Annu. Rev. Fluid Mech.*, **38**, pp. 483–512, 2006.

[28] Slimon, S.A., Soteriou, M.C. & Davis, D.W., Development of computational aeroacoustics equations for subsonic flows using a mach number expansion approach. *J. Comput. Phys.*, **159**, pp. 377–406, 2000.

[29] Filios, A.E., Tachos, N.S., Fragias, A.P. & Margaris, D.P., Broadband noise radiation analysis for an HAWT rotor. *Renewable Energy*, **32**, pp. 1497–1510, 2007.

[30] Moroianu, D. & Fuchs, L., Numerical simulation of wind turbine noise generation and propagation. *Proc. of LES/DNS Workshop-6*, eds. E. Lamballais, R. Friedrich, B.J. Geurts & O. Métais, pp. 545–554, Springer Verlag, 2006.

[31] Moroianu, D. & Fuchs, L., Prediction of wind turbine noise generation and propagation based on an acoustic analogy. *Proc. of Euromech Colloquium*, eds. J. Peinke, P. Schaumann & S. Barth, **464b**, pp. 231–234, Springer Verlag, 2006.

[32] Szasz, R.Z. & Fuchs, L., Computational study of noise generation and propagation from a wind turbine. *Proc. of 3rd International Conference From Scientific Computing to Computational Engineering*, 2008.

[33] Prospathopoulos, J.M. & Voutsinas, S.G., Application of a ray theory model to the prediction of noise emissions from isolated wind turbines and wind parks. *Wind Energy*, **10**, pp. 103–119, 2007.

[34] Björkman, M., Long time measurements of noise from wind turbines. *J. Sound and Vibration*, **277**, pp. 567–572, 2004.

[35] Marsden, A.L., Wang, M., Dennis, J.E. & Moin, P., Trailing-edge noise reduction using derivative-free optimization and large eddy simulation. *J. Fluid Mech.*, **572**, pp. 13–36, 2007.

[36] Fuchs, L. & Szasz, R. Ice accretion on wind turbines. *Proc. of 13th International Workshop on Atmospheric Icing of Structures*, Andernatt, Switzerland, 2009.

PART II

DESIGN OF MODERN WIND TURBINES

CHAPTER 6

Design and development of megawatt wind turbines

Lawrence D. Willey

General Electric Energy Wind, USA.

Electric power generation is the single most important factor in the prosperity of modern man. Yet, increasing concerns over carbon emissions from burning fossil fuels have brought large-scale renewable energy technologies to the forefront of technological development. Framework for the successful design and development of large wind turbines (WTs) addresses the world's power generation needs. The motivation for this work, the broad framework for success, the best approach for product design, various techniques and special considerations, and development aspects are presented. Horizontal axis WTs operating inland or in near-shore applications are specifically addressed.

1 Introduction

Supplies of fossil fuels are inadequate to meet the growing need for more power generation, which is driven by increasing demand for electricity [1, 2]. The horizontal axis wind turbine (HAWT) is one of the most economical forms of modern power generation. HAWTs have numerous benefits: they conserve dwindling fossil fuel resources, reduce harmful emissions, and support a sustainable electric energy infrastructure for posterity. Large wind turbines (WTs) are an engineering challenge; they must endure some of the largest numbers of fatigue cycles for structures while meeting size and cost constraints [49]. New WTs must maximize reliability, availability, maintenance, and serviceability (RAMS) while having the lowest cost of energy (CoE) and highest net present value (NPV) for a customer and an original equipment manufacturer (OEM) [50]. One of the biggest motivations for accelerating technological development of WTs is in support for the U.S. 20% wind energy by 2030 initiative [3].

1.1 All new turbine design

What design decisions will make a new multi-megawatt (MMW) WT successful in today's power generation market? How big should it be? What are the governing parameters that drive its design and ensure that it operates without failure and for the budgeted cost? These are just a few of the many questions at the beginning of any new turbine design [54]. Providing the lowest CoE in the power generation market is a challenge to the WT industry, but ever-improving designs are meeting the challenge. MMW turbines often operate for over 20 years and are among some of the largest manmade structures, especially from the perspective of controllable moving parts. An all-new MW WT begins with conceptual design and follows a rigorous technology building block and toll gate approach for component and systems development, and then undergoes a validation process involving prototypes and pre-series projects prior to entering full production. Value analysis or value engineering drives this process [4–6], where cost and financial return are continually assessed to ensure success for the customer and the OEM.

1.1.1 Technology readiness levels (TRLs)

MW WT technology is advancing quickly, and concepts need to be characterized in a way that enables everyone to immediately understand the maturity of a particular technology. The TRL system (used by NASA and U.S. Government agencies) provides a common basis for assessing new concepts [7].

- TRL-1: Basic principles observed and reported
- TRL-2: Technology concept and/or application formulated
- TRL-3: Analytical and experimental critical function and/or characteristic proof of concept
- TRL-4: Component and/or breadboard validation in laboratory environment
- TRL-5: Component and/or breadboard validation in relevant environment
- TRL-6: System/sub-system model or prototype demonstration in a relevant environment (ground or space)
- TRL-7: System prototype demonstration in a space environment
- TRL-8: Actual system completed and "flight qualified" through test and demonstration (ground or space)
- TRL-9: Actual system "flight proven" through successful mission operations (space)

1.1.2 Technology proofs required (TRL-1, 2, 3, 4 and 5) – early phase

New turbine development cannot be undertaken effectively using unproven component technologies. The best development strategy for new technologies involves devoted internal research and development (IRD) programs. To mitigate risk, these programs should be completed before embarking on a new product introduction. New products released before proper validation and refinement may result in the OEM having to mitigate problems later in the field, which can cost the OEM a factor of 10 or more relative to doing it correctly up front. A roadmap of technologies and products relative to the marketplace and other manufacturers helps keep upfront costs down and minimizes overall development cycles. Organizations need not necessarily have their own IRD departments. There are examples of broad-based collaborative research programs among industry, university and government participants. Value analysis is critically important in understanding research priorities and where to apply funding. Regular re-evaluation is required to ensure market changes are reflected in re-direction or termination of programs.

1.1.3 Technologies in-hand (TRL-6, 7, 8 and 9) – late phase

Having a broad range of component technologies from wind or other sectors (that are well understood and scalable) is an important aspect of new WT design. Even with this, system integration readiness should not be underestimated. During the product development cycle, early assumptions about the suitability of a component technology may come under question, and new validation steps or even the invention of new component technology may become necessary. Again, continuous and ever-improving value analysis is the key activity for assessing technologies in-hand and identifying new development opportunities. Carrying the value analysis cycles throughout the development phases will help identify where additional technology proofs are needed and will ultimately yield the very best product.

1.2 Incremental improvements to existing turbine designs

Once a product is introduced, later market entrants force competition and cause a need for the original product to improve. Incremental product improvements can be more tractable for some risk adverse organizations from the perspectives of quick return on investment (ROI) funding allocation and minimizing investment. There comes a point in every WT design when up-scaling component technology, improving performance or further cost reduction are not possible, and a new breakthrough design is required to begin the cycle anew [46]. Continual value analysis provides the foresight needed to know which route to take and when, and it helps build a comprehensive multi-generation technology plan (MGTP) that is regularly reviewed and updated.

1.3 The state of technology and the industry

Thirty-five years into the modern WT era, less than half of this experience is in today's large utility scale machines. More than 27 GW of new wind capacity was commissioned worldwide in 2008, a 36% increase over 2007. By the end of 2008, global wind capacity grew nearly 29% reaching 121–123.5 GW with turbine and wind power plant (WPP) investments worth about US$47.5B (€36.5B) [8, 9]. The state of WT design today is similar to the state of automobile design in the 1920s and 1930s. Even the most advanced WTs are relatively immature when considering future advances for component integration, dry nanotechnology, quantum wires and advanced digital controls [1]. There are many choices to be made today,

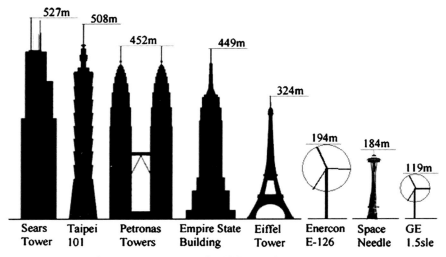

Figure 1: Megawatt-size WTs are large structures.

and with tighter R&D budgets and a shortage of qualified wind engineers and specialists in the industry, there is great opportunity for those who recognize this view of the future today. The need for more electrical energy is driving installation of more wind energy. Rapidly increasing numbers of government, university and industry collaborations are becoming involved with wind energy. All of these groups are quickly recognizing the importance of value analysis to set R&D priorities and guide innovation [55].

2 Motivation for developing megawatt-size WTs

Modern WTs have become very large structures that push the limits for structural engineering and require the use of lightweight, low-cost materials (Fig. 1) [52].

From first principles, the net power rating and size of a turbine grows with swept area; i.e. rotor diameter raised to the second power. At the same time, the amount of material (i.e. mass or cost) increases as the cube of the diameter. This is known as the square-cube law [10]. Aside from improving energy capture by accessing stronger winds at higher hub heights (HHs), the original motivation for going larger in power rating and rotor size was to lower the CoE through economies of scale. Today's most common turbine size is within the 2–3 MW rating, and perhaps as high as 4–5 MW for some of the newest designs. Larger machines beyond 5 MW should become economically viable with further advances in material and design technology, but a sudden change is unlikely over the next 5-year period.

To better understand where the large WT market is today in terms of size, an "industry study set" of key design parameters for more than 150 utility scale turbines is used to characterize trends and provide a basis for setting targets for new WT designs [11]. This data is plotted in terms of rotor diameter and rated net

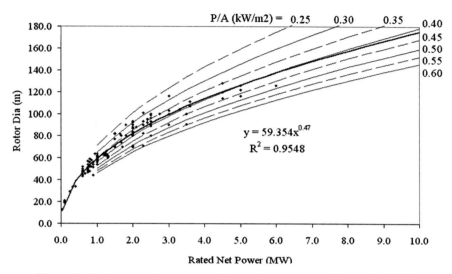

Figure 2: Industry study set w.r.t. lines of constant specific rating (P/A).

power in Fig. 2. Specific rating is defined as the rated net power of the turbine divided by the rotor swept area (P/A) and is typically used to characterize the overall ability of a design to economically extract electrical energy from the wind. The majority of today's large WTs have a specific rating in the range of 300–500 W/m^2. The general trend passes from specific ratings of less than 400 W/m^2 for machines below 4–5 MW sizes to greater than 400 W/m^2 for WTs above 4–5 MW sizes.

This result is due to plotting all the turbines within the study set without regard for their particular International Electrotechnical Committee (IEC) type class (TC). See IEC 61400-1 for an explanation of WT design TC [12, 13]. In general, turbines installed in the EU are typically designed to IEC TC3 and turbines for the U.S. to TC2. Offshore turbines are typically designed to meet the requirements for IEC TC1. Today's largest machines are generally for offshore or IEC TC I applications. These TC1 data points pull the average trend line towards higher specific ratings with increasing rated net power.

To better illustrate this point, Fig. 3 shows the same data plotted by IEC TC, together with the study group trend line. The TC data shows that today's utility scale turbines have specific ratings independent of machine size; TC3 turbines (7.5 m/s) have an average specific rating of 327 W/m^2, TC2 (8.5 m/s) of about 362 W/m^2 and TC1 (10 m/s) at 425 W/m^2. The standard deviation for these results is approximately 7%, 10% and 13% for TC3, TC2 and TC1, respectively.

The industry study set dimensions for 10 turbines along the mean characteristic for the 150+ turbine data set are plotted in Fig. 4. While it cannot be asserted that turbines in the 5–10 MW range will end up having precisely these dimensions, this does help to characterize a reasonable starting point. The tower height, rotor diameter and rating are consistent with the natural progression of marketable designs that have so far balanced higher electrical output per kg of material, cost

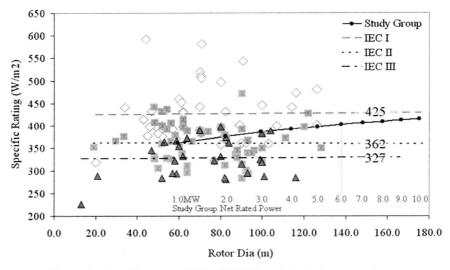

Figure 3: Specific rating (P/A), IEC TC and the industry study set.

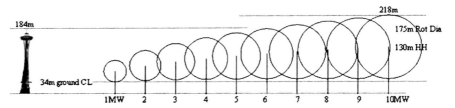

Figure 4: Turbine size geometry for the industry study set.

of transport to the project site, installation and operations and maintenance (O&M) costs. More importantly, this generalized set of turbines (herein after referred to as the 10-turbine analysis group) provides a very useful datum for comparing value analysis results with actual or derived mass and cost data.

2.1 Value analysis for wind

Based on extensive value analysis experience for a wide range of real project economics, up-scaling today's most popular turbine architectures and component technologies beyond 4–5 MW results in unfavourable customer economics. The very best value analysis procedures today are computerized in a format that allow individual component owners to keep their models updated. These procedures ensure that the modules maintain their functionality relative to each other and in the overall roll-up to the WPP and electrical grid views.

2.1.1 The cost for value analysis

A rigorous value analysis program is an investment that an OEM must make to ensure that every dollar spent on new WT technologies or products will result in satisfied customers and maximum shareholder value. Not provisioning for a rigorous ongoing value analysis program is foolish. Experience shows that a value analysis program with an annual budget of at least 1–1.5% of an organization's overall engineering budget will establish and maintain the system of continuous innovation and value assessment needed for successful technology development and new product designs.

2.1.2 Concepts and definitions

- AEP – Annual energy production; kWh or MWh
- Turbine price – Cost of the WT equipment to the project developer
- BOP price – Cost of the balance of plant to the project developer; everything else needed to build a WPP beyond the WT equipment
- Other capital costs – Financing terms and requirements
- Operational costs – O&M required to produce and deliver electricity
- CoE – The cost to produce 1 kWh (or 1 MWh) of electricity for a given time; e.g. for a given year
- LCoE (levelized CoE) – Average CoE over the WPP life; e.g. 20 years
- Energy price – Power purchase agreement (PPA) – wholesale price paid to the WPP owner for the electricity produced
- Taxes – Paid to local and federal government (in some instances may be an income to the WPP owner in the early years of the WPP life)
- PTC – Production tax credit; any of a number of government policy incentives that provide income to the WPP owner. Typically a function of power production and over a specified time schedule
- OEM margin – The amount of profit required by the OEM to satisfy shareholders and provision for long-term sustainability of the business
- Customer value –NPV for the customer
- OEM value – NPV for the OEM
- Total value (TV) – NPV for the customer and OEM combined

A simplified illustration of the main contributors to value for both the customer and OEM are shown in Fig. 5. While hundreds of parameters are considered in a

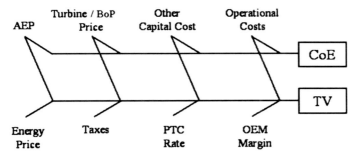

Figure 5: Main contributions determining CoE and TV.

complete value analysis for a WT design and its use in an overall WPP, this simplified diagram shows the main framework. As an example, lease payments to a landowner for the land the WTs are built upon may be accounted for in the operational costs, and so on.

2.1.3 Value analysis outline

The most important aspect of new WT and WPP design is the calculation procedure and the inclusion of all applicable business functions. This is a challenging task that can only be realized if a considerable investment is made to establish a digital environment to capture, maintain and continually update these inputs. An example of a methodology that has been successful in new WT design is shown below:

1. Input high level turbine parameters and design driving wind conditions
2. Select the component technologies that make up the new WT design
3. Select the WPP size and description governing installation including BOP and site construction characteristics
4. Select the territory that the WPP is to be analysed and check the default settings for the latest permitting requirements, financial terms and policy incentives.
5. Repeat steps (1)–(4) if a range of design parameters are to be explored
6. Select any of a number of pre-defined WTs from a library of existing products that you wish to compare the current results
7. Submit the overall inputs for analysis to compute a wide range of output that include value metrics such as CoE and TV relative to the pre-defined turbines

The analysis method computes loads for the turbine, relates the loads to the mass required for components to perform for their intended life, and then assigns the cost required to procure the components and assemble the system. This is all done in an environment that ensures that component engineers continually update their physics-based transfer functions for loads and mass, sourcing and advance manufacturing update their cost data, and finance and marketing keep the project funding and regional settings up to date.

2.1.4 Simplified equations for CoE and TV [14]

$$CoE = (Ops_{lev} + Int_{lev} + prin_{lev} + Dwnpymt)/AEP \qquad (1)$$

where $Ops_{lev} = f$(Term, DR, VAT Recovery Interest, Property Tax, Landowner Royalty, Utilities Cost, Insurance Cost, Interconnect Service Charge, Other Royalty Performance Bond, Warranty, LTCSA, Machinery Breakdown Insurance, Management Fee, Independent Engineer, Revenue Tax, O&M); $Int_{lev} = f$(Term, DR, Debt Type, Interest Rate, Takeout Cost, Debt %, Debt Term); $Prin_{lev} = f$(Term, DR, Interest, PTC Income, Takeout Cost, Debt Term); Dwnpymt = f(Takeout Cost, Debt %, O&M Cost, O&M Reserve Months, # of Turbines, Financing Fee %, Other Upfront Fees); AEP = Annual Energy Production, MWh at the meter (billable)

$$TV = An(Rev_{lev} - Ops_{lev} - Int_{lev} - prin_{lev} - Tax_{lev} + PTC_{lev} - Dwnpymt) + OEM \quad (2)$$

where **An** = NPV factor = $((1 + DR)Term - 1)/(DR*(1 + DR)Term)$; Rev_{lev} = f(Term, DR, Initial PPA Rate, PPA Escalation, Merchant Index Point, Initial Merchant Rate, Merchant Escalation, Initial Green Tariff Rate, Green Tariff Escalation, Green Tariff Term, Annual Energy Production, VAT %, Takeout Cost, VAT Method, Capital Subsidy Type, Capital Subsidy Cap, Capital Subsidy %); Tax_{lev} = f(Term, DR, Rev, Ops, Int, Depreciation Type, Depreciation Straight Line Term, Depreciation Declining Balance %, Carry Forward Loss Duration, Duration Corporate Tax Start, Duration Corporate Tax Duration, AMT Rate, Tax Rate); PTC_{lev} = f(Term, DR, Initial PTC Rate, PTC Inflation Factor, PTC Escalation, PTC Inflation Reference, PTC Rate Adjustment, Annual Energy Production); OEM = f(Turbine Cost, Margin %, # of Turbines) [48].

2.2 The systems view

Common parameters to study:

- WPP size range – 50 MW, 100 MW, 200 MW or more
- WT net power rating in relation to WPP size and markets
- Land-constrained versus MW-constrained markets
  - Land-constrained – Fixed amount of land, not limited by MW that can be accepted by the local grid
  - MW-constrained – "Unlimited" land, fixed MW connection
- IEC TC – Turbine designed to meet specific fuel (wind) attributes

A systems view is required to ensure that the final outcome is as expected. Focusing on individual components and managing interfaces is not enough. A MW WT system is complex, and it takes viewing it from different perspectives in an iterative way in order to arrive at an answer that is truly a systems solution. The successful turbine designer will need to circle back many times, zooming in and out from system level to component level and back again.

2.3 Renewables, competitors and traditional fossil-based energy production

A WT designer must understand the competition and continually focus on TV and CoE. Assumptions change quickly, and re-assessment must be provisioned in the schedule and crosschecked often. In addition to modern wind power, what are the alternatives for producing electricity? It is important to keep the excitement for investing in new WT designs alive and builds commitment from all team members across an organization. This is not easy, as there is always fear of the unknown, and for many organizations there is a deep entrenchment in the old way of doing things.

As with any new product development, the revenue potential will be the strongest motivation for an OEM. Companies will invest in technology that will bring the best returns to their stockholders. Large WTs have become very competitive

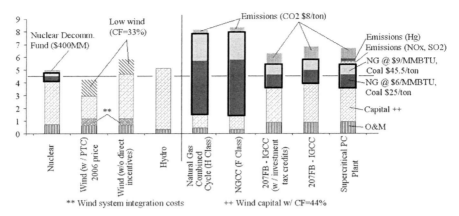

Figure 6: Generating alternatives − North America 20 Yr LCoE, $c/kWh.

with traditional forms of electricity production. Figure 6 is an example showing the current financial merits for a range of mainstream electric power generation technologies [15].

There will not be pressure to produce new and better WT designs if the economics are unfavourable. When fossil fuel prices are low, most customers will not be motivated to purchase renewables unless there are incentives, usually resulting from government policy. Whether it is a WT or a traditional fossil fuel-based electricity option, using value analysis is the key to guiding an organization's decision to proceed.

2.4 Critical to quality (CTQ) attributes

The first step of a new turbine design is to identify the critical few characteristics that make all the others seem trivial. These provide a common set of goals and clarify the critical attributes that will help focus the design across a wide range of operational environments.

Entry into the 21st century coincided with the maturity of large-scale WT manufacturing capability. Prior to this, WT designers and manufacturers were relatively small start-ups by entrepreneurs and environmental enthusiasts. For the most part, these start-ups did not fully understand the importance of value analysis and risk mitigation. This has all changed with the entry of industry participants striving for new WT designs that have the lowest capital expense, the lowest O&M costs, and the highest availability.

3 The product design process

Truly great new products follow a disciplined design process. This is especially true for large rotating equipment such as a modern MW-size WT. A large collection of people and skills must be brought together in an environment where the people can exchange information and build on each other's ideas. The primary elements of a successful new WT design include strong support from stakeholders, financial backing to see the project through, and an adaptable framework for the teams to work together and get things done. This ensures world-class results that are fully optimized for the different system level and component views.

3.1 Establishing the need

In order to begin a design for a new turbine, studies must be performed to find the optimum concepts and characteristics for a proposed market. The process starts with the definition of customer needs, and to this end the following are defined:

- Market target segment; e.g. regions, customer types, WT design TC
- Market settings; e.g. prevailing finance terms, government policies, incentive and tax schemes, local supplier content requirements
- Company internal goals; e.g. amount of build-to-print versus spec, unit volume scenario's, sourcing strategies including supplier base, strategic partners
- A list of technologies ready (or will be in time) for product implementation
- Target internal design metrics for concept selection and optimization
- Range of turbine ratings, rotor diameters and HHs to scan for optimum configurations
- Target metrics relative to other products including the competition.

At a minimum, the last item should include the CoE and the specific TV (STV) of WPPs that assume different numbers of turbines. The CoE (c$/kWh) can be determined for each year of the turbine's life or expressed as levelized CoE (LCoE), and the specific TV ($/kW) is the customer NPV plus OEM margin normalized by plant rating. Market and customer needs outline an optimum solution for a turbine design.

3.2 The business case

What does a company want to be known and remembered for? Are better returns for stakeholder investments available? Are the risks worth the time and effort? These and many more questions form the foundation of business growth and vitality. There are a great number of opinions throughout an organization, and usually a large number of factors are continually changing, making consensus complicated. A new WT design may be judged too costly or an organization may answer these questions with alternative profit strategies, but increasingly the economics of MW WTs and WPPs provide strong profitability. More important than today's profit expectations are actions that promote sustainability. Achieving a balance among all of these factors should be the ultimate goal of the WT designer. Tax benefits/ government incentives?

3.3 Tollgates

Tollgates are the orderly process for developing new products and technologies. They are applicable for component development as well as the higher-level system. TG's include the all-important cross-functional and business level interaction that ensures a new product development effort remains focused and relevant. Effective tollgate reviews cut across department and business unit boundaries and integrate all skills and requirements throughout all phases of the WT development program.

The following example tollgate structure can be used for new product and component technology programs:

- TG1: Investigation and justification of a business opportunity
- TG2: Product options identified and cross-functional buy-in secured
- TG3: Conceptual design
- TG4: Preliminary design
- TG5: Detailed design
- TG6: Factory test or validation + pre-series production
- TG7: Validation or redesign + product introduction
- TG8: Customer feedback, field experience and resolution

Some notes on TGs:

1. TGs 1–3 require an overall technical platform leader for a new turbine product design that's not a part of the customer-facing engineering department; e.g. advanced technology or conceptual design department separate from product engineering. The hand-off to product engineering is upon completion of TG3.
2. This does not mean that product and component engineers are left out of the early TGs; on the contrary, their buy-in and contributions as part of the consensus-based decision process is the key, but they do not lead or overwhelm the early processes of value analysis and developing the business case for a potential new turbine design.
3. Be flexible when applying any TG process for specific turbine design programs. Do not be afraid to combine some TGs or opt for a best effort design through to component and even full prototypes with a strategy to learn and provision for a cycle of optimization prior to committing drawings to pre-series or serial production.
4. Limit the stakeholders or final decision makers to one key leader representing each of the major departments; e.g. marketing, product line management, engineering, sourcing, manufacturing and services.
5. A strong Chief Engineering Office is the conscience of the technical community and becomes increasingly important as the TGs progress through the design phases [53].
6. Do not be afraid to partner with suppliers early under non-disclosure agreements (NDA) to leverage the best and the brightest early in the process.
7. The spirit and intent of the foregoing elements are in complete alignment with concurrent engineering (a.k.a. simultaneous engineering), an approach in which all phases of the product development cycle operate at the same time. Product and process are coordinated to achieve optimal matching of cost, quality and delivery requirements. There are a large number of resources readily available on this topic such as Curran *et al.* [16] and CERA [17].

While some OEMs have demonstrated faster WT product development cycles than others, it generally takes 4–5 years to progress from TG1 through TG7 for new designs. Concurrent engineering techniques will compress time in the product development and business cycles in general. Every OEM has basic cycles that govern the way that documentation is processed, decisions are made and parts are manufactured. Another major factor that affects this timing is the level of investment a company makes in basic research and funding stability from year to year. At one extreme is a "Manhattan" type project where experts and resources are assembled for a specific purpose, while the other extreme is a corporate research laboratory that may not have much coordination in terms of a specific vision or product development program. What is recommended is a blend of both worlds and the ability for the OEM to adjust the approach as needed without having to recreate either realm from scratch.

3.4 Structuring the team

The best chance for success is an organization that includes all points of view from the beginning. It is important to choose a new platform leader who is well seasoned and has the respect of the organization. This is the single most important position on the conceptual design engineering team, particularly for executing TGs 1–3. Although the platform leader works closely across the entire organization, there is particular focus in the early TGs between the product line leadership and the conceptual design engineering team. A general outline for a new turbine organization is:

1. Marketing and product line management – commercial departments
2. Conceptual design engineering – leads for TGs 1–3
3. Platform leader as member of conceptual design engineering (TGs 1–3)
4. Advance manufacturing, sourcing, service engineering
5. Systems engineering and component engineering – system engineering leads for TGs 4–8
6. Platform leader as member of systems engineering (TGs 4–8)
7. Validation department, project management
8. Fleet engineering
9. Services, aftermarket opportunities

3.5 Product requirements and product specification

The product requirements document (PRD) specifies the commercial and product requirements that are to be achieved by the new product development program. The development of the PRD starts during the business opportunity investigation phase (TG1), continues during the identify product options phase (TG2), and gets refined and frozen during the concept design phase (TG3). Different from the product specification document (PSD) at TG3, the PRD specifies what the product should do, not how technical details are chosen to answer the requirements.

The PSD is created during the conceptual design (TG3) to collect all of the detailed technical specifications for components and their aggregation into an overall system that satisfies the PRD. The PSD establishes the roadmap for completing the design, ensures everything that can possibly be considered is included, and builds the framework for measuring progress towards completion. It is important to note that once the PSD is established, changes are accepted only if the entire team (including Conceptual design engineering) are brought back together for discussion and approval. The document includes outline models, component technology selections and value assessments. The PSD is one of the key hand-offs from conceptual design engineering as the systems engineering team embarks on the preliminary design phase of the new product development program (TG4).

3.6 Launching the product

Obtain and document buy-in from all the stakeholders. Initial commitment for funding and resources are needed for the best possible start. Long-term commitment for a sustained level of funding is the bedrock of the best technical organizations. Regardless, if specific development programs are carried through or replaced with emergent business opportunities, the best talent that is fully engaged in an environment of creativity cannot be maintained if too much energy is focused on continually developing funding opportunities.

3.6.1 Re-assessing on a regular basis

Most if not all of the conditions and assumptions that had initially supported a decision to proceed on a new MW WT design will change over time. The importance for having some level of ongoing technical competitive surveillance cannot be emphasized enough. Plan regular stand-downs across the functional organizations to absorb, vet, and act on emergent information that changes any of the previous conditions and assumptions. This is the key to know when to change course, or even terminate a particular MW WT program.

3.7 Design definition: conceptual → preliminary → detailed

Conceptual design (TG1–3) permits the rapid exchange of ideas and proposals for how to physically achieve the goals laid out for a business proposition. Conceptual design is physics-based, realistic and timely. Successful conceptual design is dependent on the creation and maintenance of a computer-based environment for MW WT and WPP value analysis.

Preliminary design (TG4) builds on the foundation provided by the conceptual design. Conceptual design transitions to preliminary design when all of the stakeholders agree that the current and estimated future market conditions and assumptions, together with the best conceptual design, continue to meet the program CTQs and the business goals. Preliminary design's primary thrust is to test all of the conceptual assertions with real world suppliers and advance manufacturing teams.

Detailed design (TG5) builds on the foundation provided by the preliminary design. Preliminary design transitions to detailed design when all of the stakeholders agree that the latest and estimated future market conditions and assumptions, together with the best preliminary design, continue to meet the program CTQs and the business goals. Detailed design carries through every aspect of the new wind plant design including analyses, proofs and modelling. Concurrently, sourcing, suppliers, and manufacturing are key players in finalizing component drawings. Detailed design is not complete until the first qualifications for all the parts and components are substantially finished and the first prototype turbine is successfully running. This approach permits learned improvements to be put into up-revised drawings and specifications prior to final release for serial production.

3.7.1 Design parameters for value and CTQ attributes

Market specifics that are provided by the OEM's Marketing Department include opportunity size, permitting limits and timing. The conceptual design team:

1. Develops turbine design parameters flowing down from CTQs and vets them across the team.
2. Evaluates WPP alternatives for a range of turbine sizes and ratings.
3. Facilitates iterative improvement cycles to refine the turbine specifications.
4. Evaluates a range of design boundaries for best NPV and LCOE.
5. Documents all results, including internal design records for advance studies or a PSD for a new product, as applicable.

Results are also compared with competitors to see where machines may be improved. Develop your own strategy of future improvements that could be introduced for your design at a later time. Setting design targets is part science and part market savvy. Regardless of the initial values set at the beginning of a design campaign, be ready to modify specific ones with an eye on maintaining the overall optimum design – this includes re-assessing the market profitability profile. The following sections provide a starting point for new MW WT design targets based on the industry study set trends.

3.7.2 Technology curves

The industry study set introduced earlier in this chapter includes all of the various technologies of today such as single-bearing, double-bearing, geared, direct drive (DD), and so on [51]. Larger turbines in the 3–6 MW ranges generally have architecture that departs from the traditional architecture of early 1–3 MW turbines. Architecture changes with size results in a natural adjustment to the industry trend

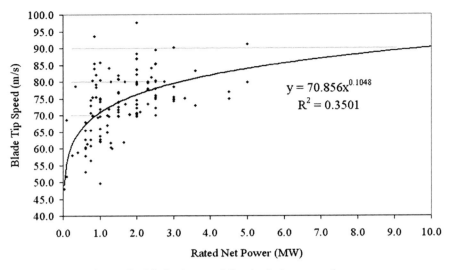

Figure 7: Blade tip speed for the industry study set.

when projecting forward to the 7–10 MW ranges. This projection is probably on the conservative side of what can be expected, assuming similar technology progress over the last decade projected to larger machines of the next decade. It is also reasonable and highly desirable to beat this trending with an acceleration of breakthrough technology. This scenario is becoming more likely due to the massive increase in public and venture capital funding that is beginning to be injected into the wind industry, not to mention the recent exponential response by academia to shift research to the renewables sector.

Noise emission is one of the first considerations when setting out to design a new turbine. Blade tip speed is considered one of the primary drivers of noise with sensitivity proportional to the tip speed raised to the 5th power [10]. Figure 7 shows the range of blade tip speeds for today's industry study set as a function of turbine power rating. The curve fit shown is not very good, in part because of the wide range of blade lengths for machines with the same generator rating, but the correlation is better than other types of fits. Its character is also consistent with other related parameters such as the main shaft rotational speed. Tip speeds of 70–85 m/s are being used, with the higher end of the range tending to be for larger machines. The trend for this data suggests that tip speeds much beyond 90 m/s are unlikely for machines in the 7–10 MW sizes. Other noise reducing considerations would need to be developed if advantages for even higher tip speeds proved desirable.

As shown in Fig. 8, there is a much better correlation for values of rated rotor speed with turbine size. Today's machines have rated rotor speeds in the low 20s to the high teens, but speeds are not projected to fall below about 10 RPM for 7–10 MW sizes. This is also consistent with the practical limit of torque that increases with lower speeds.

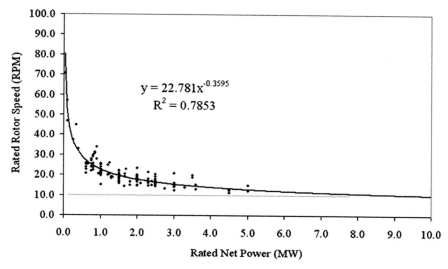

Figure 8: Rated rotor speed for the industry study set.

To best illustrate the range of MW WT design parameters and how they influence technology choices for turbines and WPPs using value analysis methodology, a 10-turbine analysis group was prepared by aggregating the trends (e.g. *P/A*, rotor diameter, rated rotor speed, etc.) from the industry study set. Table 1 contains the summary listing for this value analysis group. Many of the results shown in the remaining sections of this chapter are based on this uniform group of basic MW WT design parameters.

3.7.3 Influence coefficients and functional relationships

The design wind conditions and market settings are first entered into the value analysis tool and sweeps of turbine design cases yield trends for the output performance and financial merit parameters. These trends, and their corresponding sensitivities to variation, are next used to assess the relative merits for component technology decisions. This process is repeated for combinations of component technologies to arrive at conceptual design proposals that are further vetted by all the various groups within the organization. Once a number of these studies have been performed, a pattern of influence coefficients (ICs) and functional relationships result. These provide a convenient guideline for the WT design TC and market conditions of prime interest, and can be quickly applied to assess emergent component technologies without the need to go through the rigorous value analysis calculation. However, use these parameters with caution, always keeping in mind the assumptions that were used to generate them. If the main assumptions are different enough, revert back to full sweeps using the value analysis tool. Over time, this process of developing "rule of thumb" ICs will emerge again and again, with better appreciation across the organization as more people gain experience using them.

Table 1: 10-Turbine analysis group inputs for the value analysis program.

P/A = net rating* 1e6/((rotor dia^2)*0.7854)	Given	Rotor dia = 59.354*(net rating^0.47)	RPM = 22.781* (net rating^-0.3595)	Tip Spd = rotor dia *PI()* RPM/60	HH = rotor dia /(0.225* (rotor dia^0.3464))	Twr base dia = HH/18.60 **
P/A (W/m^2)	Net rating (MW)	Rotor dia (m)	RPM	Tip Spd (m/s)	HH (m)	Twr Base OD (m)
361.4	1.00	59.35	22.78	70.80	64.11	3.45
376.8	2.00	82.21	17.76	76.43	79.33	4.26
386.0	3.00	99.47	15.35	79.94	89.85	4.83
392.8	4.00	113.87	13.84	82.52	98.15	5.28
398.1	5.00	126.46	12.77	84.58	105.11	5.65
402.4	6.00	137.78	11.96	86.30	111.17	5.98
406.2	7.00	148.13	11.32	87.78	116.56	6.27
409.4	8.00	157.73	10.79	89.09	121.44	6.53
412.3	9.00	166.70	10.34	90.25	125.92	6.77
415.0	10.00	175.17	9.96	91.31	130.06	6.99

**18.60 is taken as the standard tower height aspect ratio = GE –1.5sle 80m HH/4.3m base dia.

To illustrate a nominal set of ICs for today's mainstream MW turbines, the forgoing procedure was followed with the following assumptions:

1. Market settings for the U.S. w/ average construction conditions
2. Average results of value analysis using 15, 50, 100, 200 and 300 MW blocks of power (land is not a constraint)
3. WTG ratings in the 1–5 MW size range (note that the 4–5 MW rated machines assume logistics cost is no worse than 2–3 MW machines), i.e. no significant increase in logistics capability or cost
4. Assumes 20-year project financials that are levelized
5. 6–10% discount rate
6. 8.5 m/s AMWS w/ 12% TI
7. 1.55 $c/kWh levelized PTC
8. 5.50 $c/kWh levelized PPA

Examples of the ICs that result from this generalized value analysis are shown in Table 2 [18]. The ICs are very useful for making first assessments of component technology effects on the overall MW WT design. For example, a 1 pt increase in AEP for a particular technology enhancement results in a 0.08 or about 1/10th of a penny reduction in LCOE and an increase in TV of $20.3K/MW of machine rating. Thus, if a 2.5 MW turbine was to be used in a given WPP specification; 2.5 × 20.3 = $50.75K/turbine of TV results.

Similarly, if there were a 1 pt increase in turbine cost, this would result in a 0.05-cent increase in LCOE and an $8.6K/MW decrease in TV. Using the 2.5 MW turbine example, this would result in a $21.5K/turbine reduction in TV. Taken together with the AEP example, if one was able to introduce a component to a new WT design that increased AEP by 1 pt with a cost increase of 1 pt, the net result would still be a benefit of $50.75K − $21.5K = $29.25K of TV per turbine. Clearly any number of scenarios can result for evaluating new technologies, and careful appraisal using ICs for well understood value analysis assumptions are very useful for determining the merits of potential technology.

3.8 Continual cycles of re-focus; systems–components–systems

One of the most important aspects of a successful MW WT program is the organization's ability to continually re-focus across all aspects of the design from systems to

Table 2: ICs for assessing new component technologies.

Variable	LCOE $c/k/Wh /1 pt increase	Total value k$/MW/1 pt increase
AEP	-0.8, +/- 0.005	20.3, +/- 0.13
Turbine cost	0.05, +/- 0.004	-8.6, +/- 0.71
BOP cost	0.02, +/-0.002	-4.7, +/-0.55
O&M cost	0.005, +/-0.001	-0.9, +/- 0.37
Transportation	0.003, +/- 0.000	-0.7, +/-0.07
	+/- 95% confidence	+/- 95% confidence

components and back again. This process is repeated throughout the design phases to ensure the best of all requirements are considered and optimized, while remaining agile to ever changing external factors; e.g. markets, emergent technologies, etc.

Table 3 provides a simplified listing of considerations as a starting point for the turbine designer and illustrates just how challenging this process can be. There are additional levels of systems and sub-systems and all of these and their inter-relationships should be thoroughly understood. Many of the items in the system level have direct counterparts in the component level, while others must be considered in both systems and components. Without exception, truly optimized MW WT products have component level characteristics that are performance enablers for meeting overall system level requirements.

4 MW WT design techniques

This section pulls together all of the pieces that go into a well-organized new turbine design. Requirements are emphasized and optional approaches presented. The 10-turbine analysis group is used together with the value analysis methodology to estimate turbine loads and their effect on component mass and cost. Component and turbine results are plotted for the 10-turbine analysis group against the backdrop of the entire industry.

4.1 Requirements

The customer's highest priority is investing in equipment and power plant infrastructure that is affordable to finance and continues to operate cost effectively for its entire 20-year lifetime. Thus, it is imperative that new turbine designs consider reliability, availability and maintenance from the very beginning. The ease of operating and deriving profits from the WPP is as important as the initial cost. The need for specially trained personnel or specialists is another important consideration early in the equipment selection process. The successful turbine designer accounts for all of these and the individual significant requirements of prospective customers.

4.1.1 Turbine
While there are numerous possibilities for WT system configurations, the majority of modern equipment consists of a horizontally configured drivetrain with two or three blades attached to an upstream facing rotor. The rotor is supported at the tower top using at least one rotor bearing and a main shaft that drives a generator system for power conversion. The drivetrain and power conversion system are located on top of the tower within a nacelle enclosure to protect the power conversion equipment from the environment. The nacelle is equipped with a yaw system that is used under normal operating conditions to point and maintain the rotor facing upstream into the wind.

It is beyond the scope of this article to discuss the details for why this configuration has become the de-facto standard. Suffice to say that many other configurations have been investigated, and in almost all cases, the reasons boil down to longevity (through relentless unsteady operational conditions and extremes), reliability and

Table 3: Example elements for successful MW WT design.

System Level	Component Level	System Level	Component Level
Customer Needs/Requirements/Economics		OEM/Company Design Practices/IEC 61400-1 Edition 3	
OEM/Company Goals/Economics		Grid/Building codes/Environmental Compliance, CE/UL/others	
Market/Region/Policies/Incentives		Technologies Ready for Implementation	
	Target Metrics -- Value Analyses		Blades
	Wind Class & Turbulence		Pitch
	Rating and Rotor Diameter		Hub
Energy Production	Component Efficiencies	Integration Boundaries	Shaft and Bearings
Loads	Stress/Deflections/Vibrations/Fatigue	Turbine Control Interfaces	Machine Structure
	Noise Emissions Requirements/Technical & Economic Implications		Yaw
	Reliability, Availability, Maintainability & Serviceability (RAMS)		Tower and Foundation
	Operation & Maintenance/Life Cycle Costs		Gearbox/Transmission
	Environmental Health & Safety (EHS)	WPP Controls/Grid Interface	Generator and Converter
	Intellectual Property (IP)	Turbine Controls Hdwr & sftwr, SCADA	Sensors
	Analyses/Simulations	Accident Scenario Review/Safety System Validation	
	Thermal Management	Component & System Prototype Validation/Documentation	
	Risks/Mitigation	Configuration Matrix/Quoting Limits/Customer Documentation	
System & Component Definition -- Conceptual, Preliminary & Detailed		As-running Configuration Mgnt & Service Records	

costs. This again comes back to the central theme of value analysis, without which others will continue to follow paths of whimsical or even elegant WT concepts, only to eventually return to the same place as the rest of the industry – three-bladed upwind WTs for reliable and economical WPP installations. This is not to say there will not be a future breakthrough that shatters this paradigm, but a strong value analysis must accompany any such breakthrough.

4.1.2 Reliability
Reliability is the ability of the WT to perform at expected cost levels while producing the expected annual energy production (AEP) based on the actual wind conditions. Poor reliability directly affects both the project's revenue stream through increased O&M costs and reduced availability to generate power due to turbine downtime [19].

4.1.3 Availability
If a turbine were always ready to produce electricity when the wind speed is within the range of the power curve, it would then have 100% availability. This has been an area of dramatic improvement in the past few years with 98% availability typical for the most popular machines today versus 85% just 6–8 years ago. Proven component technologies must be integrated into an overall system that is value-effective and robust for high availability.

4.1.4 Maintainability
According to wind-energy-the-facts.org [20], "Typical routine maintenance time for a modern WT is 40 hours per year." What are the value-effective ways to reduce O&M time and costs? Maintainability and serviceability are closely related design goals that provide the backstop for ensuring the highest possible availability. All machines require some amount of maintenance over their design life. However the best machine finds the value-effective minimum maintenance strategy and stream-lines the maintenance that must be done. Should a part or component fail for any reason, cost-effective design provisions that minimize service and logistics costs should be integrated into the design.

4.2 Systems

The overall product requirements and design features for achieving business objectives include different system level views – continually revisit these levels throughout the product development process to ensure optimal results.

4.2.1 New WT design
A wind turbine generator (WTG) converts the kinetic energy of the wind into mechanical shaft power to drive a generator that in turn produces electrical energy. A WT is composed of five main elements:

- Rotor made up of rotor blades that use aerodynamic lift to convert wind energy into mechanical energy.
- A rotor bearing fixed on a structure that causes a defined rotation of the rotor and leads to conversion of the aerodynamic wind energy into a rotational shaft torque.

A yaw system maintains the horizontal rotor axis pointing upstream into the wind.
- A power conversion system that converts the low-speed rotational energy into suitable shaft power to drive an electrical generator.
- A tower and foundation structure to support the rotor and generator system at a height that harvests the most amount of energy for an acceptable capital cost.
- An electrical power distribution system that supplies the energy to the consumer in compliance with local grid code and system requirements.

4.2.2 Mechanical–electrical power conversion architecture

Figure 9 shows the four main types of WT architecture that have been employed according to Wojszczyk *et al.* [21]. Types A and B are older and less prevalent configurations, while Types C and D comprises the mainstay of modern MW WT offerings.

Type A is the oldest and utilizes a squirrel cage induction generator (SCIG) directly coupled to the electrical grid. It may incorporate a soft starter to limit in-rush current during start-up conditions. The rotor is connected to the generator through a gearbox and can often run at two different (but constant) speeds. This is achieved by changing the number of poles of the stator winding. The generator always consumes reactive power, and therefore this has to be compensated by capacitor banks to optimize the power factor and maintain the voltage level.

Type B introduces a variable resistor in the rotor circuit of a wound rotor induction generator (WRIG). This can be done using slip rings or the resistors and electronics can be mounted within the rotor. The variable resistors control the rotor currents to maintain constant power output and allow the WT to have dynamic response during grid disturbances. Some degree of self-protective torque control and energy capture range is provided by ±10% speed variation.

Type C is known as the double fed induction generator (DFIG). It improves on the Type B design by adding variable excitation (instead of resistance) to the rotor circuit. The generator stator winding is directly connected to the grid and the rotor

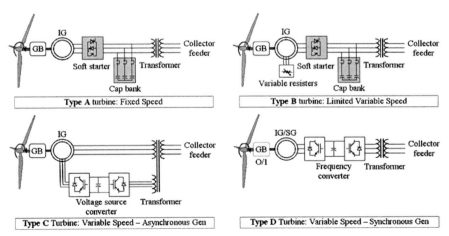

Figure 9: The four main types of WT mechanical–electrical power conversion.

winding is fed from the grid using voltage source converters. A rotor-side converter is connected back-to-back with a grid side converter and requires only about 1/3 of the overall turbine power rating to accomplish a wide range of output power control. The system permits relatively fast response to significant grid disturbances and advantages for reactive power supply to the grid, power quality and improved harmonics. Type C is a more expensive architecture than Type A or B designs, but it is popular for its advantages.

Type D uses a synchronous generator that can have either a wound rotor or can be excited using permanent magnets. This is also known as "full-power conversion" and provides design and operational flexibility including the option of eliminating the gearbox. Variable speed operation is achieved by connecting the synchronous generator to the grid through the frequency converter. Type D turbines are better than Type C turbines despite the higher cost, as they are decoupled from the grid via power electronics, resulting in excellent capability for power quality and reactive power. Use of permanent magnets in the generator rotor offers the advantage of greater efficiency and high power density, resulting in more compact generators.

4.2.3 Governing dimensional considerations

One of the most important design drivers of a WT is the closest approach of the blade tip to the tower. The baseline axial distance from the blade tip to the tower wall for the fully unloaded blade in the 6 o'clock position is known as the "static clearance." The amount of clearance remaining for the actual turbine operational conditions is a function of the loads imparted on the rotor blade and tower, the blade geometry, stiffness, mass, and how the blades are positioned relative to the rotor hub and the turbine [22]. Typical minimum operational blade-tower clearance is 1/3 of the static clearance or a distance of one tower radius relative to the tower diameter at the same elevation of the passing tip [59].

Blade-tower operational clearance considerations are shown in Fig. 10 relative to the wind flowing from the left to the right. Starting from the left, sketch /A/ is

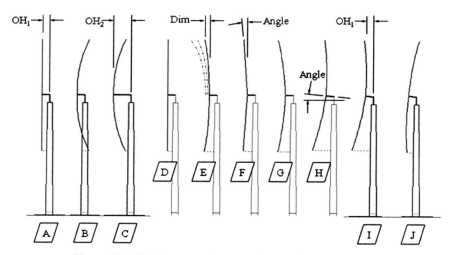

Figure 10: Blade-tower clearance for upwind facing WT.

for the baseline condition of zero wind and rotor speed. There is a set amount of rotor "overhang" (OH_1) and the blades are positioned within a vertical rotor plane. Sketch /B/ depicts an extreme load case for the swept surface of the rotor plane under operational wind conditions. Clearly this scenario is not acceptable, as the blades will interfere with the tower. In an attempt to gain clearance by moving the rotor forward, sketch /C/ shows that even with significant increase of the overhang dimension OH_2, the minimum clearance is still not acceptable. Increasing overhang is very costly as it requires significantly more mass in the machine head (MH) structure to support the rotor mass and operational loads due to the longer lever arm.

Sketch /D/ shows the same starting assumption as /A/ with the rotor in a vertical plane. Various types and amounts of "zero wind" geometry are depicted for sketches /E/ through /H/:

- /E/ – Blade prebend, an amount of forward reaching curvature characterized by the axial dimension measured from the blade tip vertical plane to the rotor centre vertical plane. Typical values are 1–3 m (larger values are anticipated as rotor diameters continue to increase).
- /F/ – Blade cone, an amount of forward reaching cone characterized by the cone angle measured between the normal vector at the blade root attachment and the rotor centre vertical plane. Typical values are 2–6°.
- /G/ – As typically the case use a combination of blade prebend and cone.
- /H/ – Tilt the rotor and the drivetrain by a small angle relative to the horizontal plane. Typical values are 2–6°.

Using rotor /H/, sketch /I/ depicts a return of the rotor overhang to the more economical OH_1 value. Extreme loadings for this arrangement result in sketch /J/. Since the blade-tower minimum clearance exceeds 1/3 the static clearance, this combination of design parameters provides a solution to the OH_1 constraint. All new turbine concepts must go through this type of optimization process, and be revisited throughout the structural design phases to ensure loads and deflections are accommodated.

4.2.4 WT system modelling for determining design loads

There are a number of WT structural calculation programs currently being used by WT designers. These include readily available models using commercial finite element analysis (FEA) programs such as ADAMS, ANSYS and NASTRAN. Some of the most widely used programs specifically developed within the WT design community are BLADED and FLEX5. FLEX5 has the advantage of wide industry support and open source code, making it the preferred platform for many OEMs.

One drawback for FLEX5 from the conceptual design perspective (where rapid solution times are needed to explore a large design space) is the relatively long run times. Usually 1–2 h are required to run a turbine configuration through a reduced suite of load cases using a dual-processor workstation. This can be improved with ever-faster hardware, but the preferred method for performing value analyses sets up a large number of cases using the method of design of experiments (DOEs). These results are used to populate the design space with solutions that can later be used to interpolate conceptual designs in seconds instead of hours. This "DOE

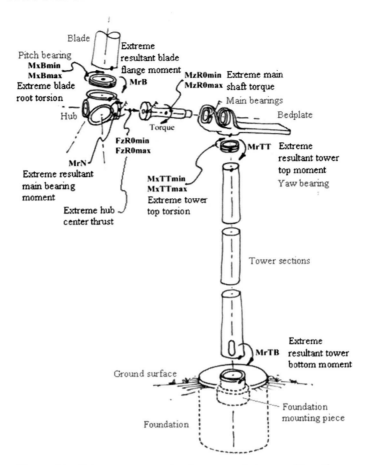

Figure 11: Primary extreme loads nomenclature and locations.

loads engine" makes it possible for the turbine designer to explore more options in a shorter period of time and quickly adapt the overall value analysis program for emergent component technology assessment.

Today, FLEX5 has as many as 34 degrees of freedom modelled for the upwind three-bladed WT. There is a great deal of data available from the normal solution results, but there is a smaller subset of this data that has the greatest impact on the initial sizing of turbine components. Using FLEX5 naming conventions, Fig. 11 shows these primary forces and moments and the components they are related to and act upon.

4.2.5 Load calculations

To gain an appreciation for the level of extreme loads developed in a WT, and how they change with turbine size, a 10-turbine analysis group of today's industry average technology was undertaken. These turbine design cases are geometrically scaled along the industry trend projected up to 10-MW size. The basic WT parameters were presented in Table 1 earlier in this chapter.

One of the most important features for a value analysis program is the selection of component technologies that are used to describe the turbine configurations. Some components, such as bearings and gearboxes, exhibit some degree of discrete changes with increased size or loadings such as the number of main shaft bearings or the number of stages within the gearbox. These details are important to address in subsequent design iterations, but for illustrative purposes the value analysis is performed using a blend of turbine component technologies to avoid discontinuities and demonstrate smooth nominal values for loads, mass and value metrics.

4.2.6 Load cases

There are a large number of different types of load cases that are examined throughout the various design phases. Load cases account for every conceivable loading situation, and all of the possible configuration variations, operational conditions and extremes that could be encountered over the 20-year design life for the turbine and its installation environment. It has been found that a specific subset of these cases actually affect the conceptual design of a WT, and working with this subset greatly simplifies this early phase in the design process. Expected extreme loads will have an impact on the primary or first-order sizing of components. Expected dynamic loads affect the fatigue life of components and consequently, the final sizing of components. The full range of design load cases should ultimately be examined in order to gain a complete understanding for the behaviour of the WT, particularly for the detailed design phase. A set of specific load cases will eventually be required by a certification agency (if applicable) such as Germanischer Lloyd (GL), "Guideline for the Certification of Wind Turbines" [23].

4.2.7 Load levels

To gain a better appreciation for the level of forces and moments for these large machines, it is useful to make an analogy with another machine. An M1A2 Abrams main battlefield tank has a mass of 63 tonnes fully loaded [24]. This can be used as a convenient reference for the level of loads calculated throughout the WT; e.g. 63 tonnes acting at the end of a 1-m moment arm is "one tank-meter."

4.2.8 Impact to system design

Load calculations lead to the provisioning of strength requirements for the WT, and flow down into component requirements. At the system level, the turbine must be able to resist being pushed over by the thrust imparted on the rotor in the downwind direction, while at the same time carrying the low-speed shaft torque from the rotor through to the generator. Rotor loads must be carried into the MH structure and passed into the tower top. These loads and the top head mass (THM) are carried through to the tower bottom and foundation, with the soil bearing pressure being the last interface to distribute these loads into the earth.

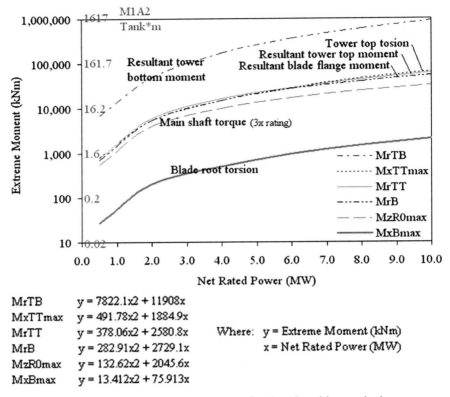

MrTB	$y = 7822.1x2 + 11908x$	
MxTTmax	$y = 491.78x2 + 1884.9x$	
MrTT	$y = 378.06x2 + 2580.8x$	Where: y = Extreme Moment (kNm)
MrB	$y = 282.91x2 + 2729.1x$	x = Net Rated Power (MW)
MzR0max	$y = 132.62x2 + 2045.6x$	
MxBmax	$y = 13.412x2 + 75.913x$	

Figure 12: Calculated extreme moments for the 10-turbine analysis group.

Figure 12 shows the primary extreme moment results for the 10-turbine analysis group. The tower overturning moment is the largest value due to the thrust force acting over the long lever arm of the tower height.

The main shaft torque dominates the loads carried by the MH structure. While many of the load cases impart significant forces and moments to this structure, the main shaft torque drives the overall level. A 3-MW WT must resist about 7330 kNm of moment about the shaft in the MH. This is equivalent to about 12 tank-meters. These loads increase nearly fivefold for a turbine rated at the 10 MW level, or about 55 tank-meters. One quickly appreciates the level of loads that this represents by imagining 55 M1A2 Abrams tanks at the end of a 1-m moment arm trying to twist-off the top of the turbine.

The amount of thrust absorbed by the rotor and turbine support structure is shown in Fig. 13. A 3 MW rated WT will have about 816 kN of axial force applied at the main shaft location, or the equivalent weight force of about 1.3 M1A2 Abrams tanks.

The previous examples are for extreme loads, but more important to consider are the time-varying or fatigue loads, which mechanically work the structure,

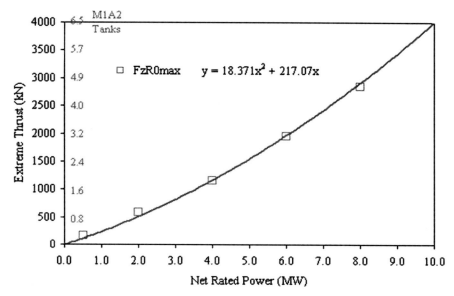

Figure 13: Calculated extreme thrust for the 10-turbine analysis group.

particularly at the different attachment and component transitions throughout the turbine. Goodman diagrams are used to estimate fatigue life for various stress ratios experienced throughout the critical locations [47]. These fluctuating loads often dictate the final sizing of structural members (beyond what is required for extreme loads). The utility scale WT sees more fatigue load cycles than any other manmade structure or machine – many millions of cycles over its 20-year design life – more than automobiles, ships, aircraft or rockets.

4.3 Components

Figure 14 illustrates the top-level mass results from the value analysis program. The extreme and time-varying loads for the 10-turbine analysis group (i.e. Table 1) are converted into the amount of material or mass that the various components need in order to meet the design requirements for the turbines. The results are for a selection of component technologies that are most often used today and include:

1. Nominal glass fibre reinforced plastic (GFRP) blades optimized for structural properties and lowest weight.
2. Nominal characteristics of a main shaft incorporating a single forward main bearing or a double-bearing design.
3. Nominal gearbox with the number of stages required carrying the torque for a speed-up ratio in the range of 1:100 or more.
4. Electrical generator spanning the range of wire wound doubly fed induction to permanent magnet full-power conversion topology.

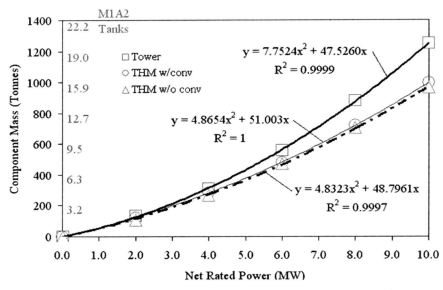

Figure 14: Calculated tower and THM results for the 10-turbine analysis group.

5. Typical cast or large fabricated bedplate used in today's workhorse turbines.
6. Tubular steel tower, with the tower bottom diameter scaled for the same aspect ratio versus height, with the height taken along the industry study set trend as a function of turbine net power rating.
7. Simple scaling rules for shipping logistics assumed to apply across all the components, even for the high end of the rated power range. This last assumption is the most difficult to accept as the larger onshore machines in the 7–10 MW size range could not actually ship 80 m long blades or tower bottom diameters approaching 7 m. However, the turbine solutions are valid in terms of physical parameters in situ, and provide a convenient benchmark for the turbine designer striving to overcome the logistics problem with new technology and innovation.

The effect of including the converter up-tower is hardly noticeable but the tower itself outweighs the THM by 25% in the higher 7–10 MW size range. Tower and THM in the 500–1000 tonne range (weight of 9–16 Abrams M1A2 tanks) are not economically viable using today's readily available technologies. This view provides a starting point for imagining what materials and construction technologies are needed to make these large machines possible.

Figure 15 further breaks down the THM into the major sub-systems. The partial nacelle is the heaviest and includes the massive bedplate for transferring the rotor bending moments into the tower structure, as well as the covering (i.e. nacelle) that keeps the weather from the drivetrain components. The lightest components are the generator, bearing and converter. As mentioned before, the converter is positioned up-tower in some designs, although the most prevalent location is down-tower.

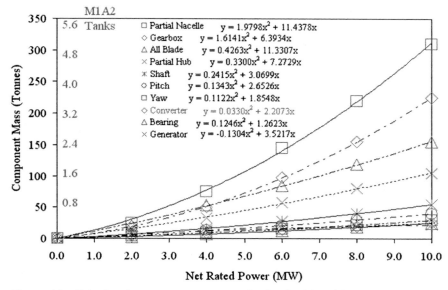

Figure 15: Calculated component mass results for the 10-turbine analysis group.

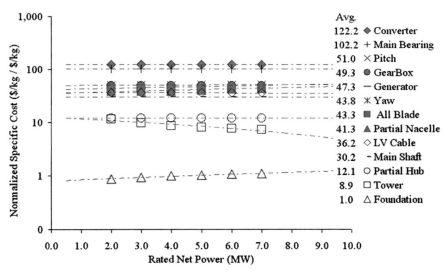

Figure 16: Normalized specific cost results for the 10-turbine analysis group.

4.3.1 Impact to component cost

The next step in the value analysis method is to convert the component mass requirements into their respective cost. This is more difficult than estimating the loads and mass of components due to the endless change in costs for factors such as sourcing options, materials, labour, transportation and currency exchange. Figure 16 shows the levels and relatively flat trends for component specific cost as a function of the WT rated net power. On a cost per kg basis, the converter, main bearing and pitch systems are the most expensive. At the other end of the spectrum, the partial hub, tower, and foundation are the least expensive. The foundation is considerably less expensive than any other component on a $/kg basis. However, the foundation is by far the heaviest component – typically two to three times the mass of the WT itself.

Accurate cost data for the actual component technologies under consideration are crucial for making proper assessments in the value analysis methodology. The best way to ensure the accuracy of this information is for the OEM's Sourcing Department to provide these figures (normalized to eliminate the largest variations) and for routines to be established to store the data for the value analysis program to automatically use the latest information.

4.3.2 Power performance calculations

The method used to calculate the performance of a WT is to use a power curve together with the annualized wind distribution for a given project location. The power curve plots the power produced by the WT as a function of wind speed. The power production quoted by an OEM is typically given at either the low voltage (LV) or high voltage (HV) terminals of the medium voltage (MV) transformer depending on the supply scope for the customer. Losses from cables and converter are more commonly included in the turbine losses along with other conversion losses within the turbine OEM scope of supply. Sometimes an OEM may quote output at the generator terminals, and additional losses for the LV cables, converter, MV transformer, MV cables and wind plant collection and substation must be properly accounted. The end goal is to ensure the accurate determination of electricity production (i.e. "at the meter") used to compute revenue.

4.3.3 Acoustics and vibrations

WT acoustic and vibration design falls into three major categories:

* WT rotor system acoustics
* Machinery structure-borne noise
* Machinery and airborne noise

The rotor blade and operating controls impact the rotor system acoustics. The direct acoustic pressure radiation and the forces that drive vibrations in the machinery structure contribute to airborne noise emission. Noise is an important factor during the conceptual design of most new MW turbines, particularly for onshore machines targeted for land-constrained markets. Today's 1–3 MW turbines are designed to emit maximum noise in the range of 102–106 dBA at HH. This produces actual measurements of 35–45 dBA observed 300–400 m from the tower base at ground level, similar to the background sound experienced in a bedroom at night or quiet conversation around the dinner table.

4.3.4 Thermal management

Electrical components and sub-components housed in close quarters can quickly lead to over-temperature and unnecessary production loss. MW WTs are subjected to the full range harsh environmental conditions, and these need to be accounted to ensure reliable operation and maximum revenue.

4.3.5 Dynamic systems analysis

Turbines and components, like all physical objects, have natural frequencies. External load sources, operational scenarios and moving parts are potential stimulus and dynamic reinforcement sources. Relative deflections and stiffnesses for large structures is an extremely acute consideration in the design of robust MW WTs.

WT tower stiffness is a key consideration to the overall structural design of WTs. To avoid resonance, the tower's natural frequency (f_0) must not coincide with the frequency of the cyclical loading resulting from the rotor frequency (f_r) and the frequency for the blades passing in front of the tower (f_b). Towers and turbine systems are classified as soft–soft, soft–stiff, or stiff–stiff, depending if the tower's natural frequency is less than the rotor frequency ($f_0 < f_r$), greater than the rotor frequency but less than the blade passing frequency ($f_r < f_0 < f_b$), or greater than the blade passing frequency ($f_0 > f_b$), respectively [25]. This is best viewed relative to a Campbell diagram for the tower and stimulating frequencies, where to avoid resonance the turbine power producing speed region is defined using margins no less than ±5% [23]. The majority of large WTs are designed for the soft–stiff regime. Soft–soft or soft–stiff are preferred over stiff–stiff because much more material and cost are required for this later approach.

Sensors and controls are also used to avoid or mitigate resonance vibration, and to ensure safe operation. This is a rapidly evolving development area for large WTs, and holds promise for becoming a key technology for enabling larger "smart" structures. These designs should be capable of withstanding higher loads with superior performance while using considerably less material for a lower cost.

4.4 Mechanical

Materials for structures must be inexpensive, readily available, and easy to fabricate, require minimum maintenance over the turbine's 20-year or longer life, and in the best scenario, be recyclable. When a conceptual design shows promise and feasibility, the advanced mechanical design (AMD) function works out enough of the details to warrant advancing further in the process.

4.4.1 Blades

Blades are one of the most important components in MW WT design. They directly affect AEP and the loads imparted to the entire turbine structure. Blades endure a large number of cycles for wind speed and direction, extreme gusts, and, with every revolution, load reversals from their own weight. To be economically viable, the cost of the material and manufacturing needs to be a fraction of the cost of blades for aircraft or aerospace counterparts.

Aerofoil sections have max thickness to chord ratio (i.e. aspect ratio) of around 2–3 near the root and 5–7 near the tip. By comparison a wooden 2 × 4 has a finished

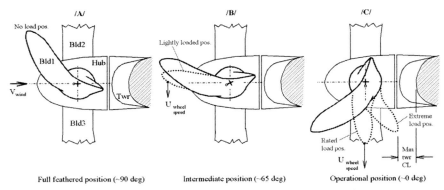

Figure 17: Blade design elements and pitch positioning.

aspect ratio of 3.5"/1.5" or 2.33 and a 2 × 8 is 7.25"/1.5" or 4.83. In either case it is easier to deflect the span in the flat or shorter direction as opposed to the longer dimension. For wind blades these load directions are referred to as flapwise and edgewise, respectively.

Blades are constructed to be as light as possible yet still providing the strength, stiffness and life required by the system. This is achieved with internal structures that incorporate either a box spar or shear webs. The box spar is a radial beam that the aerofoil skins are bonded around. The shear web approach uses the aerofoil skins as part of the spanwise structure with the internal shear webs transferring loads from one side of the aerofoil to the other to form a deflection resistant box structure. Box spar construction has an advantage for longitudinal strength and uniformity, but it is heavier and structurally inefficient relative to the shear web approach. Both techniques are successfully used in today's GFRP blades.

Figure 17 illustrates a number of typical blade design features as viewed by an observer looking up from the base of the tower with the wind approaching the turbine straight-on and travelling from left to right. Sketch /A/ shows the blade in the 85–90° or fully feathered pitch position. In this position the blade is primarily experiencing edgewise loading from the wind. Sketch /B/ shows the blade pitched to the intermediate blade angle of 65–68° (as will be explained in subsequent sections, this is the position used to start and reinforce rotor rotation for the case of increasing winds above cut-in wind speed). The blade in this position is experiencing a combination of edgewise and flapwise loads from the oncoming wind and wind gusts. Sketch /C/ shows the blade pitched to the full operational angle of around 0° – the position maintained throughout the variable speed region of the power curve. In this position the blade primarily experiences flapwise loads due to the wind.

The combination of blade prebend, cone, drivetrain tilt and overhang can also be seen in Fig. 17. These design features are not of significant consequence for the pitch positions of sketches /A/ and /B/, where the blade is lightly loaded and spanwise deflections are small. However, blade spanwise deflections are greatest for

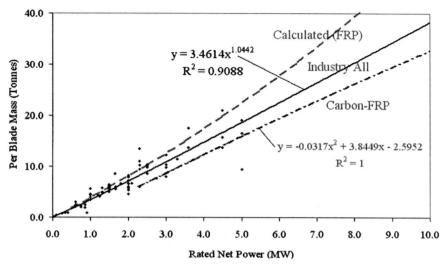

Figure 18: Blade mass – 10-turbine analysis compared to industry study set.

sketch /C/ and for winds above rated wind speed. The maximum deflection outline shows another view of the importance for the minimum tower clearance discussed earlier in this chapter.

Figure 18 shows the 10-turbine analysis group (i.e. calculated), a curve fit for the industry study set using GFRP, and the industry study set for carbon spar and GFRP hybrids. Clearly if one just scaled today's average technology to 10 MW, the individual blade mass would be in excess of 40 tonnes. Since the larger machines in the industry study set tend to incorporate a carbon spar or utilize some form of an advanced GFRP construction, the industry trend projects the 10 MW blade to be less than 40 tonnes. A carbon–GFRP hybrid blade should be able to achieve 32–34 tonnes per blade. Based on the past industry progress going from 1 to 5 MW, and with new technology yet to be discovered, it may be possible for a 10 MW blade to be designed in the 25 tonne range.

The majority of today's blades are made from GFRP incorporating either a box spare or shear web construction. Figure 19 shows the typical mass and cost breakdown for an average sample of blades incorporating shear web construction [26]. The glass fibre and epoxy or vinyl resin comprises the vast majority of the mass. Manufacturing these blades requires a large amount of man-hours such that labour accounts for nearly 1/3 of the total cost for a blade.

Today's mainstream blade construction technology requires significant investment in mould tooling to form, cure and assemble large WT blades. A steel subframe and backing structure is used build-up the basic upper and lower mould shell tools. Curing heaters (electrical or temperature controlled fluid channels) are arranged throughout the surfaces prior to establishing the final mould surfaces. These are typically completed using a prototype blade (i.e. plug) to provide a form for the final tool surface made from high temperature epoxy resin within the upper

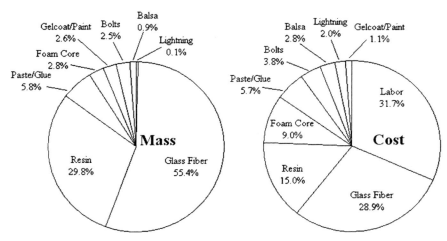

Figure 19: Mass and cost breakdown for nominal GFRP blades.

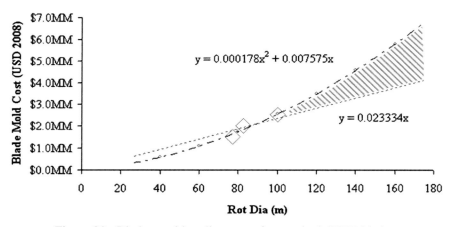

Figure 20: Blade mould tooling costs for nominal GFRP blades.

and lower blade half moulds. The mould halves are hinged together to facilitate the final closure and bonding operation for normal blade production.

As shown in Fig. 20, the cost of a blade mould for one of today's typical 1–3 MW WTs is around $2MM or more. A single mould can manufacture 600–1000 blades before it will require refurbishment at a cost of $120K or more. Moulds can usually be refurbished at least two to three times before new replacements are needed [27]. These trends imply that larger turbines in the 7–10 MW size would require moulds that cost $4–6MM unless alternative technologies can be found. In reality moving beyond rotor diameters of 120–130 m (i.e. 60–65 m long blades ignoring the hub diameter) using GFRP blade construction technology is generally viewed as too heavy and impractical so that new large blade technology will be needed to improve this trajectory [58].

Future large turbine blade technology may incorporate elements of dry nano-technology, hybrid construction (i.e. inboard structure of one material joined through one or more joints to outboard sections made of alternate materials) or some form of repeating and panellized spaceframe structures that may include self-erecting and self-healing features.

4.4.2 Pitch bearing and drive system

The main functions of the blade pitch system are to keep the WT operating within a designed speed range and to unload the rotor bringing it to a shut-down condition. This is accomplished by rotating (i.e. "pitching") the blades about their longitudinal axis relative to the hub. The pitch bearing is the mov-able or "slewing" interface that permits this rotation; while at the same time safely transmits the rotor loads into the hub, main shaft and support structure. This angular positioning of all the blades for a rotor is more or less coor-dinated simultaneously throughout the operational range; however there are some design concepts that deviate from this and deliberately operate each blade slightly different (largely per revolution) to optimize energy capture and minimize loads.

Should the blade pitch actuation and control system be hydraulic or electric? Figure 21 shows the concept of an electrical pitch system where the pitch drive bull gear is driven by an electric motor through a gearbox ratio sufficient to ensure enough drive torque for the proper range of operation meeting the requirements for blade aero torque and rotor loads transfer across the pitch bearings. Typical electrical pitch drive systems have pitching rates as fast as 7.5–8°/s.

A little more than half of all MW WTs running today use hydraulic pitch sys-tems instead of electrical. OEMs appear to make this choice and stick with it for reasons that are not clearly established. Although value analysis shows that there may be a slight cost advantage for hydraulic pitch systems, the potential control-lability advantages for electric systems and lack of environmental concern for hydraulic fluid leakage offsets this view.

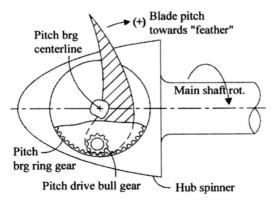

Figure 21: Hub and blade pitch gear – blade shown in "running position".

turbine operating conditions, the main shaft must transfer the rotor plane bending moments into the tower supporting structure and ultimately into the foundation (see Fig. 11). In addition, it must sustain transient and highly dynamic loads caused by grid failure, over speed events, breaking, and emergency stops, as well as loads due to extreme wind, gusting and environmental conditions.

The 10-turbine analysis group mass results for the main shaft and nominal bearing arrangement are shown in Fig. 24. The main bearing trend is an average for a single main bearing and a double main bearing type design. A single-bearing design (e.g. GE1.5) uses one large bearing at the front of the shaft with the gearbox and torque arms used to support the aft end of the shaft. A double-bearing design (e.g. GE2.5) has a second independent bearing located at the aft end of the main shaft and the main shaft supports the "floating" gearbox.

Figure 24 suggests a 10 MW main shaft to have a mass of nearly 55 tonnes and a main shaft bearing or bearings of about half that amount or approximately 25 tonnes. Using similar reasoning for technology advancement on the way to a 10 MW size machine, it appears possible to achieve 37 and 18 tonnes for the main shaft and bearings, respectively.

Based on the industry study set, the amount of steady-state torque being carried by the main shaft to a generator or a combined gearbox–generator is shown in Fig. 25. The effect of overall mechanical and electrical losses of 9 and 13% is included to illustrate the effect on the steady-state main shaft torque. Assuming that the losses for advancing technology while designing for a 10 MW machine are in the 8–9% range, about 10,000 kNm of torque capability is needed at the rating point. This is equivalent to about 16.2 Abrams M1A2 tank-meters of torque. Due to grid and

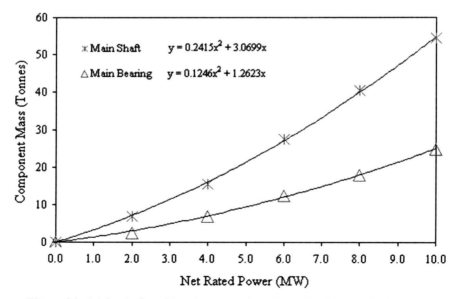

Figure 24: Main shaft and bearing mass from the 10-turbine analysis group.

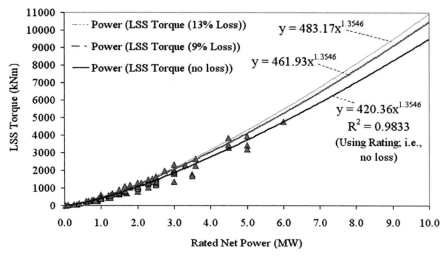

Figure 25: LSS rated torque for the industry study set.

transient operational considerations, max design torque of 2.5–3.0 times the steady-state level is provisioned, so that the shaft and reaction configuration would actually be designed to accommodate 30,000 kNm or about 48.6 tank-meters.

Future large turbine main shaft and bearing technology will likely need to be more integrated and consider cast or alternative approaches to today's typical main shaft forging. Taken together with the requirements for larger and heavier rotor hubs, the fixed axle arrangement should also be evaluated.

4.4.5 Mainframe or bedplate

The rotor and drivetrain support structure is a direct compliment and enabler to the main shaft and bearings. The design goal is to achieve as light a structure as possible, with allowable deflections within the overall materials and system specification, yet strong and capable of resisting fatigue damage with a maintenance free life. The amount of rotor overhang relative to the tower top centreline (see OH_1 and OH_2 in Fig. 10) directly affects the size and weight of the mainframe required for a given turbine design.

Figure 26 shows (1) today's industry study set and (2) moving from today's fabricated or cast bedplate towards a canard or spaceframe structure may save significant mass [28]. Further system benefits should be possible for integrating the drivetrain within these types of spaceframe structures. The heavy trend line shown in Fig. 26 represents the conventional fabricated or cast bedplate from the industry study set. This represents today's technology, and is more than 20% lighter than the double T frame used in the earliest MW WT designs. Even so, the bedframe alone would be more than 140 tonnes at the 10-MW size if today's technology were to remain unchanged.

Moving towards spaceframe type designs may reduce bedplate mass on the order of half, but this will also depend on the amount of stiffness reduction that can

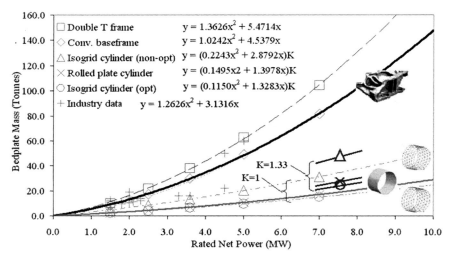

Figure 26: Bedplate mass – options compared to the industry study set.

be accommodated in the actual spaceframe design. The factor K illustrates the idea of normalizing entitlement of the isogrid and tubular cylinder for adding mass back into the idealized tubes to account for the main shaft bearing support, drive-train torque and yaw deck features relative to the conventional and double T frame. This makes it more of an apples-to-apples comparison. The 1.33 factor is purely an estimate to make the point, and may need to be higher to yield an acceptable design. The "optimized" isogrid cylinder would have mass strategically added to address local high stress regions for a particular spaceframe design making it comparable to the rolled plate cylinder.

To minimize mass to the greatest extent possible, and take full advantage of an enveloping spaceframe type structure, large 7–10-MW WT designs will likely need to incorporate some form of spaceframe-integrated drivetrain technology. There are some OEMs starting to move in this direction.

4.4.6 Machine head mass

Today's mainframe and drivetrain components and their protective housing (i.e. nacelle) are collectively referred to as a MH. The MH mass (MHM) is an impor-tant consideration for larger WTs because of shipping logistics, field assembly and installation crane requirements among other things.

The solid line in Fig. 27 is an estimate of the industry study set trend for increas-ing MHM with larger WTs. The dashed line is the trend for the 10-turbine analysis group. Future large WTs in the 7–10-MW size range will need the overall MHM targeted for the solid line or below to ensure favourable WPP economics. The considerable divergence for the calculated trend at the larger MW ratings illus-trates why straight scaling of existing drivetrain, bedplate and MH technologies will not result in a cost-effective WPP.

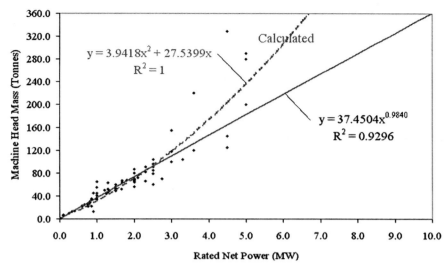

Figure 27: MHM – 10-turbine analysis compared to industry study set.

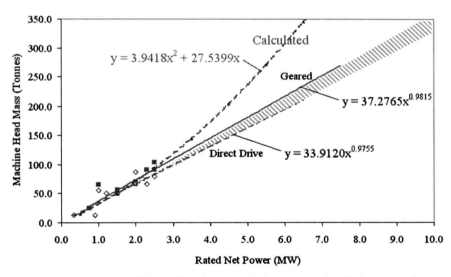

Figure 28: Geared and DD – 10-turbine analysis compared to industry study set.

Figure 28 plots a subset of the industry data presented in Fig. 27 for MHs that employ either a gearbox or DD. The mass of today's DD generator technology is nearly as much as the mass of a gearbox and high-speed generator combined. While it appears the DD technology has a slight mass advantage over the geared turbines, it is not significant for large machines. It should be possible for a 10 MW MH (that would otherwise be 330–370 tonnes) to be in the 230–270 tonne range or better for new low-speed generator breakthroughs. It may also require some level of field assembly to meet all of the logistics and crane requirements. Achieving the lowest possible MHM while still delivering all of the functional and operational needs is a fundamental design goal, and it will require new technologies and innovative integration approaches to develop an economical large WT MH.

4.4.7 Gearbox

A gearbox is needed to transmit torque and increase blade rotor rotational speed to match the requirements of an electrical generator. Today's high-speed generator technology requires speeder gear ratio of 1:100 or more. Torque transmission must be done with limited vibration and noise effects. Mechanical efficiencies have to be as good as economically feasible in order to minimize power losses in the system.

An emerging form of gearbox is a compact geared drivetrain where planetary gearing and medium speed (speeder ratios of 1:30 to 1:40) generator technologies are combined into one mechanical–electrical system. The integrated geared generator is lighter and more compact than conventional three-stage gearbox–generator systems and, with fewer gears and bearings, is both more reliable and more efficient [29].

Future large turbine gearbox technology will likely incorporate elements of lower overall gear ratio, reduced number of stages, or elimination of the gearbox altogether.

4.4.8 Drivetrain dynamics

Today, WT drivetrains undergo significant loading during WT operation due to changes in environmental conditions, such as wind gusts. Future large turbine drivetrain dynamics technology will likely incorporate elements of adaptive response and control. As speed control technology improves, loads on the drivetrain may be reduced, improving reliability [56].

4.4.9 Rotor lock – low-speed and high-speed shafts

A rotor lock is a device that prevents shaft rotation. In the case of a main shaft or low-speed shaft, the rotor lock is used during construction and during maintenance of the WT. The high-speed rotor lock is used to prevent rotation of the gearbox output shaft.

Future large turbine rotor lock technology will likely be replaced with some form of distributed system, and DDs obviate the need for a high-speed rotor lock.

4.4.10 Disk brake system and hydraulics

Future large turbine disk brake system and hydraulics technology will likely incorporate elements of integrated generator brake design or elimination of the disk brake system altogether due to advances in more reliable blade operation and advanced rotor controls.

4.4.11 Flexible torque coupling

Shaft couplings are used to transmit torque from one shaft to another, as is the case for today's designs that incorporate a gearbox and separate generator. Flexible couplings are used to balance radial, axial, and angular displacements. In addition, couplings may also provide for electrical current isolation, damping of torsional vibrations, and absorb peak torques. Compact geared drivetrains, where the gearbox and generator are combined into one mechanical–electrical power conversion unit [29], as well as DD designs do not require flexible couplings.

Future large turbine designs probably will not use flexible torque coupling technology.

4.4.12 Signal slip ring
A slip ring is at electro-mechanical device that allows transmission of power and electrical signals from a stationary component to a rotating component. In the case of a WT, the slip ring electrically connects the rotating hub to stationary equipment in the nacelle.

Future large turbine signal slip ring technology will likely be replaced with some form of contactless system or provide for a method of producing power onboard the rotating frame.

4.4.13 Yaw bearing and drive system
The yaw bearing performs the function of supporting the entire THM of the WT and permitting 360° rotation relative to the turbine tower. THM includes the MH, hub and blades. This angular yaw positioning is required to ensure the turbine rotor is always facing squarely into the wind. The primary considerations for the turbine designer include:

1. Support THM relative all possible load inputs (i.e. forces and moments)
2. Transmit rotor dynamic bending moments and loads
3. Permit full 360° rotation times some number of turns (i.e. 2.5× before requiring the generator electrical cable to undergo an "untwist" operation)
4. Minimize the amount of motor torque required to yaw while balancing the need for bearing joint stiffness for loads transfer, particularly wind gusts

Future large turbine yaw bearing and drive system technology may incorporate elements of bearing segments and rotor assisted yaw (i.e. "flying" the rotor into the wind).

The solid line in Fig. 29 is an estimate of the industry study set trend for increasing THM with larger WTs. The dashed line is the trend for the 10-turbine analysis group. Future large WTs in the 7–10-MW size range will need THM to be targeted for the solid line or lower to ensure favourable WPP economics. As can be seen, some of today's 4–5 MW size machines are considerably above the solid line, which is indicative of excessive THM and potentially poor economic performance. Provisioning a yaw bearing and drive system for excessively large THM only adds to the problem.

Achieving the lowest possible THM is a key design driver for the yaw bearing and drive system, as well as for the tower support structure (among other considerations).

4.4.14 Nacelle and nose cone
Today's nacelle and hub fairing (i.e. nose cone) are typically manufactured from GFRP. These are generally non-structural and protect the drivetrain components

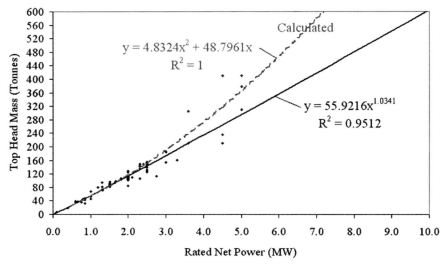

Figure 29: THM – 10-turbine analysis compared to industry study set.

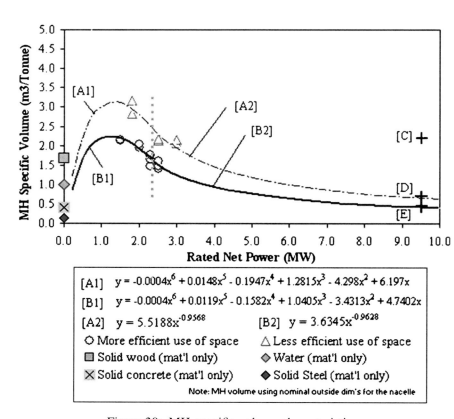

Figure 30: MH-specific volume characteristics.

from the weather. Hub fairings are largely aesthetic, although they do enhance the flow field for certain turbine designs that have large nacelles (e.g. DD) and provide anchorage for safety rails that are used to enter the hub.

Figure 30 plots the specific volume (m^3/tonne) for a number of WT nacelles. The combination of a drivetrain, bedplate and nacelle is typically referred to as the MH and the amount of volume required to house these components are indicative of material efficiency. There appear to be two characteristic trends for the data where the heavier line represents designs that better utilize nacelle volume. Curve [B1] illustrates the transition from sub-MW machines to 2–3 MW where personnel access and serviceability considerations peak in the 1–2 MW ranges. As turbines get larger, it is projected that specific volumes will gradually decline. This is due primarily to a smaller proportion of space required for personnel and service access and the desire to minimize frontal area for better performance and shipping logistics.

To illustrate this further, point /C/ of Fig. 30 is for a 9.5 MW turbine assuming the same specific volume of around 2.2 m^3/tonne for today's best 1–2 MW machines. Points /D/ and /E/ are for specific volumes of 0.7 and 0.5, respectively. Using a MHM of 305 tonnes (representing an improved technology DD at 9.5 MW size), points /C/, /D/ and /E/ would imply MH volumes of 671, 214 and 153 m^3, and represent cubes with side dimensions of 8.75, 5.98 and 5.34 m, respectively.

Future large turbine nacelle and nose cone technology are likely to trend towards their elimination. Designs using integrated drivetrain and structure will obviate the need for a separate nacelle covering and is consistent with lower MH-specific volumes for larger turbines. Elimination of a separate hub fairing should be possible for larger turbines. Most of today's turbine designs will not have a measurable performance impact for eliminating the nose fairing, so removal is further justified.

4.4.15 Tower

Towers are presently constructed using steel or concrete materials. The structure is typically tubular or lattice. Lattice towers require less material for a given strength than tubular towers, but for labour-intensive fastener and aesthetic reasons (among a number of others), tubular steel towers are the most prevalent. There are also many forms of hybrid towers, which combine varying amounts of these materials and construction types. The use of GFRP or other cost-effective materials yet to be identified may play a role in future large WT support structures.

As the industry trends towards larger power ratings and rotor diameters, towers must also increase in height and strength. Because the tower typically comprises over half the total mass of the WT itself (excluding the foundation), translating to about one-fifth of the cost, value analysis and searching for breakthrough concepts for new tower technologies represents a significant opportunity for improving large WT economics.

The solid line in Fig. 31 is an estimate of the industry study set trend for increasing tower mass with larger WTs. The dashed line is the trend for the 10-turbine analysis group. Future large WTs in the 7–10-MW size range will need the overall

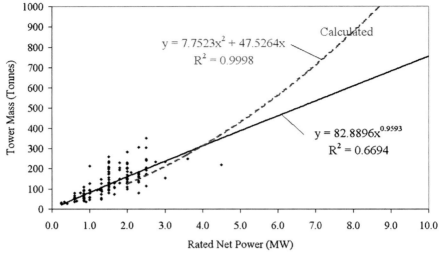

Figure 31: Tower mass – 10-turbine analysis compared to industry study set.

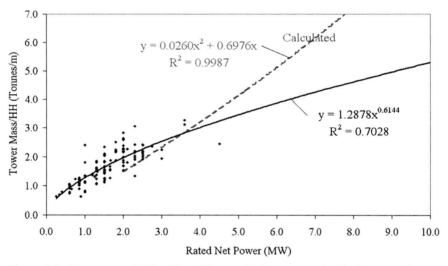

Figure 32: Tower mass/HH – 10-turbine analysis compared to industry study set.

tower mass targeted for the solid line or lower to ensure favourable WPP economics. The considerable divergence for the calculated trend at the larger MW ratings is particularly exasperated by the hefty tower base diameter (nearly 7 m for the 10 MW turbine) used for the larger machines in the 10-turbine study set. Achieving the lowest possible tower mass while still achieving all functional and operational requirements is an important design goal.

Figure 32 is derived from Fig. 31 in terms of tower mass divided by the WT HH. This alternative metric results in an average tower "mass per meter" of tower height. In practice the mass per meter of the lower elevations of the tower will be greater than the upper sections, but the average value yields an alternative view for the lowest possible tower mass design goal; i.e. target the average tower mass per meter for the solid line or lower.

The turbine designer must keep a large number of additional tower design goals in perspective when searching for the overall cost-effective solution. Some of these considerations include:

1. Dynamics, structural, machinery vibration damping and seismic
2. Internals, electrical cables, climb systems, platforms, packaged power modules (PPMs) or down-tower assemblies (DTAs)
3. Installation, erection methods, joints, fasteners and crane requirements

Future large turbine tower technology will likely incorporate lightweight space-frame construction incorporating multiple support legs that are spread apart. These configurations must remain simple; yet meet all of the design goals while complying with personnel health and safety requirements. Additionally, one should not underestimate the internals as these can significantly affect overall tower design; e.g. potential impact of a welded studs, Kt (stress concentration factor) on the tower plate thickness specification.

The use of hybrid materials and structural design configurations should become more prevalent for larger turbine sizes. Better tower and foundation integration should increase overall functional capabilities for lower total cost.

4.4.16 Structural bolted connections

Construction of today's WTs uses a large number of structural bolted assemblies including:

1. Blade attachment
2. Hub attachment
3. Drivetrain components
4. Bedplates
5. MH and yaw bearing attachment
6. Tower sections
7. Tower to foundation attachment

Flanged joints and machine elements are bolted together with pre-loads such that the flange does not separate. The flange friction takes all shear force and the bolts are only in tension. In practice, small amounts of bending moment due to flange machining, separation and misalignment can exist which needs to be accounted.

To finish a bolted connection, the torque process is the most commonly applied process. However, the torque process works against the frictional resistance on the bolt or nut face and in the thread resulting in inaccuracies of up to 100%. Another popular method is called "turn of the nut" or turn-angle. Bolts that are tensioned using the turn-angle method are one step better than the torque process. When the turn-angle process is used, the initial tension is provided by torque. The torque part of the process involves friction and provides about 20–30% of the maximum value.

The angular turn specified by the designer provides the finishing tension and results in finished inaccuracies of about 25–35%.

There has been huge progress in the past few years for pre-loading, locking and inspecting fasteners used in large WT design. Companies such as ITH-GmbH [30] offer a complete suite of bolt tensioning solutions using more accurate stretch type-tensioning tools. Bolt stretching processes are used for clamp-length ratio of 1:3 or more and for pre-tensioning large size bolts (e.g. M24 and greater). The clamp-length ratio is the ratio of the bolt diameter to the clamp length of the joint. Stretching procedures are best when (1) a high degree of accuracy (5–10%) is required or (2) when several bolts have to be pre-tensioned simultaneously.

Future large turbine structural bolted connection technology will probably not be too much different from today.

4.4.17 Fire detection system

Future large turbine fire detection technology will probably not change much from today, with the exception of better integration into the turbine control system such that predictive capability may be able to prevent a fire from happening to begin with.

4.5 Electrical

The electric power system can be classified into two main categories; (1) the turbine including the MV transformer and (2) the collection system and substation to the point of delivery to the grid. Future large turbine electrical system technology may incorporate elements of high voltage DC (HVDC) collection, centralized power conditioning and superconducting or quantum wire electricity transmission (for the right cost).

4.5.1 Turbine – generator and converter

The generator is where electricity production and the customer's revenue stream begin. As such, doing this efficiently for a justifiable cost is of prime importance and must be reliable and sustainable over the lifetime of the turbine.

Figure 33 presents the mass for the three main elements of today's typical WT power system that operates at relatively LV level (i.e. 575–690 VAC) between the up-tower generator and the down-tower power conditioning and MV step-up transformer. The mass for these components are in the same order of magnitude, with the LV cable mass roughly 2/3 of the mass of the generator and converter. The converter mass can vary widely depending on the particular converter topology, which can have a big impact on the amount of reactor mass required (reactor mass being copper and steel, and often dominating total mass).

Of all the 10-turbine analysis group component mass characteristics plotted as a function of net rated power, the generator is the only one to exhibit a negative squared term for the curve fit (i.e. diminishing mass increase for larger sizes). The 10-turbine analysis group varies gearbox ratio in order to hold the generator shaft speed constant. This is consistent with increasing generator torque rating and the corresponding electrical loading of the machine. This loading (electric current per

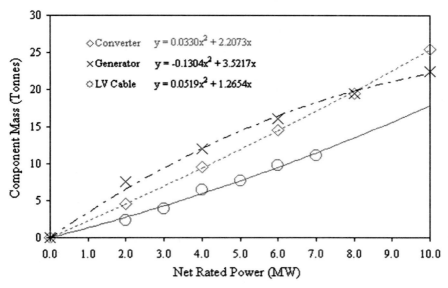

Figure 33: Generator, converter and LV cable – 10-turbine analysis results.

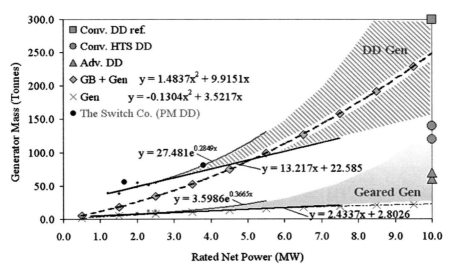

Figure 34: Generator mass – 10-turbine analysis compared to industry study set.

unit surface area of the rotor–stator air gap) also tends to increase resulting in increased torque density (Nm/kg) and hence the convex curve shape.

An alternative to this trend would be if the gear ratio were to be fixed instead of varied. Generator rotational speeds would then decrease with increasing power rating, which would increase the generator rated torque faster than the power rating. For this scenario, the curve shape would tend away from the convex shape [31].

The two shaded regions shown in Fig. 34 are for DD and geared (i.e. high speed) type generators. The lower bound represents the expected mass assuming historical technology progress with increasing MW rating, and the upper bound is for straight scaling of today's technologies. The shaded symbols at the 10 MW rating reflect recent public announcements and are unproven at this time. With respect to Fig. 34:

- DD generators using today's conventional wire-wound construction are more than five times heavier than the geared generator alone.
- The linear fits projected from industry mass data form the bottom edge of the shaded ranges. These mass trends should be achievable for continued technology advancements with increased MW rating (i.e. new technology is needed to avoid mass increase for straight scaling of today's technology originally developed at lower MW ratings).
- When the mass for the gearbox is included with the geared generator (a fairer comparison), the DD and geared configuration have the same order of magnitude mass.
- The geared configuration (i.e. GB + Gen) is lighter than DD below 5 MW and heavier above 5 MW.
- For a 10-MW size WT, an advanced DD using a projection of technology improvement from today's wire wound know-how, is at first a reasonable choice of about 150 tonnes. High temperature superconducting (HTS) DD generators (shaded circular symbols) are reported to be on the order of 20% lighter and advanced superconducting (shaded triangular symbols) are reported to be 50% lighter still, bringing them to almost the same mass as the best high-speed generator alone (i.e. without the GB accounted).

4.5.2 Turbine – electrical transformer

Should the electrical transformer be located up-tower or down-tower? The up-tower advantage is higher voltage and lower current for a given power level which translates into lower cost (i.e. smaller diameter) cable in the tower. The downside is that a higher voltage level results in more stringent design and environmental health and safety (EHS) requirements. Many turbines today employ relatively LV (e.g. 575–690 V) between the generator and a MV transformer. A down-tower transformer can be located inside the tower or installed on a concrete pad just outside the base of the tower. The vast majority of MW WTs installed in the U.S. utilize the later design.

MV/LV transformers are a key component of wind plant. Traditionally WT and MV/LV transformers have been protected using fused switches on and MV side and circuit breakers on the LV side. This solution has worked well with rated powers up to around 1.5 MW. WT ratings have grown well above 2 MW and WT distribution voltages up to 36 kV, normal for offshore, are now gaining interest for onshore application. Switch solutions have reached their limit for these higher voltage levels due to the nominal current ratings of the available fuse elements. There are advantages for using circuit breakers for the protection of MV/LV transformers. The use of modern digital protection relays associated with circuit breakers have advantages in terms of increased availability, more efficient maintenance and reduced downtime.

Transformers used in wind farm applications have been observed to fail at a higher rate (independent of manufacturer) compared to their use in other forms of power systems applications [32]. Some of the failure mechanisms include transient voltages on the LV side of the transformer causing overvoltages and abrupt loss of voltage quite regularly. Transients generate voltage surges in the MV winding leading to dielectric failures and thermal stress. Transformers subjected to continuous high power levels are often subjected to periods of overload due to wind gusts. This overloading can cause premature failure of the transformer. Some transformers are installed in the nacelle and therefore are subject to vibrations from the WT operation that may not be properly accounted in their design.

4.5.3 Turbine – grounding, overvoltage and lightning protection

Diverting lightning currents and conveying the energy safely to ground are accomplished by a lightning protection system [33]. The coupling effects of the high and extremely broadband frequency current in lightning are neutralized by means of screening. Surge voltages occurring on electric equipment are neutralized by means of lightning arrestors or surge arrestors.

Lightning receptors on a WT blade are intended to act as Franklin rods, but sometimes fail to intercept lightning strikes with subsequent damage and expensive repairs. When blade lightning receptors works as intended, but main shaft grounding brushes are inadequate, the current flow through main shaft bearings can cause significant damage [34].

4.5.4 Turbine – aviation/ship obstruction lights

It is necessary to account for these starting in the preliminary design to ensure no issues later on. Integrating these into the design is not trivial – poor planning can lead to costly nuisance issues in the field (e.g. cracking of light brackets or power supply mounts). These lights are required by most permitting authorities. However, in many instances, they are not required at every individual turbine for a given project.

4.5.5 Collection and delivery – WPP electrical balance of plant

Balance of plant (BOP) is defined to include the equipment and construction engineering beyond the WT itself (i.e. everything else beyond the typical OEM "scope of supply"). The wind park developer or the WT equipment customer normally supplies the BOP.

Companies are beginning to offer turnkey solutions. An example for the electrical portion is PACS Industries [35], who offer "Wind to Wire" electrical systems that are fully integrated electrical gear and enclosures with features and services such as:

• Tower switchgear through 38 kV
• Collector switchgear through 38 kV
• Arc resistant switchgear through 38 kV

- Outdoor electrical buildings
- Structural substations through 345 kV
- Installation engineering supervision
- Full equipment integration engineering services

4.6 Controls

WT control systems enable and facilitate smooth operation of the WT and the overall WPP. At the turbine level, the control system is responsible for reacting to changes in local wind speeds to generate the optimum level of power output with the minimum amount of loads. At the operational level, control systems collect data for automating decisions and for a number of downstream analyses. They also support communication and information flow throughout the turbine and power plant components. At the power plant level, control systems integrate all of the above with grid conditions. Control systems continue to rapidly evolve, resulting in steady performance enhancements from existing turbine hardware. These same advancements are providing new opportunities for future large turbine designs.

4.6.1 Turbine control

A WT control system is used to control and monitor the turbine sub-systems to ensure the life of components and parts, their reliability and meet the functional performance requirements. The fundamental design goals are safety, reliability, performance and cost. The turbine must produce the energy advertized for the wind conditions actually experienced – and do so for the life of the machine at the cost provisioned in the customer pro forma.

Variable-speed pitch controlled WTs are the most advanced control architecture and state of the art for today's MW WTs (see WT Type C, Fig. 9). To operate a WPP under optimum conditions, the individual turbine rotor-generator speeds are controlled in accordance with the local wind speeds using generator torque (electric current control) or blade pitch angle adjustment. Figure 35 illustrates the main considerations for a typical control scheme used in today's MW WTs.

- A turbine is in the standstill state with the high-speed rotor brake applied when the turbine is down for maintenance.
- The typical turbine condition with little or no wind is the idling mode with the blades feathered and the rotor near standstill or gently pin-wheeling.
- The spinning state can best be characterized as increasing winds starting from dead calm and approaching cut-in wind speed, while the blades are pitched to an intermediate blade angle (see sketch /B/ of Fig. 17) so that the rotor can proceed to accelerate during the run-up condition.

Figures 35 and 36 can be cross-referenced to better understand the continued sequence of control steps for the example case of increasing wind speeds (e.g. the approach and development of a storm front):

- Once the cut-in rotor speed has been achieved (condition (1) of Fig. 36), the blade pitch angle is advanced to the full operational position (see sketch /C/ of Fig. 17).

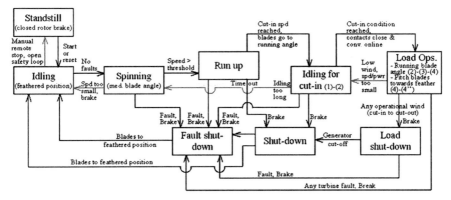

Figure 35: Generalized WT control diagram.

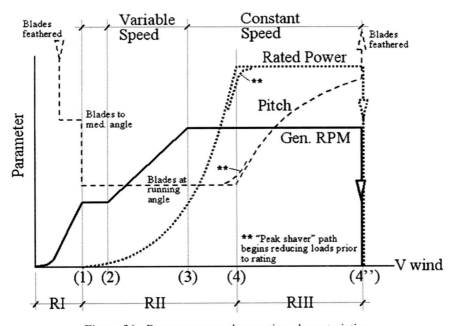

Figure 36: Power curve and operating characteristics.

- As a result of this transition, and with increasing rotor torque, condition (2) is achieved and the breakers close – bringing the converter online and power production begins.
- Now in region 2 (RII) of the power curve, increasing wind speed will advance the turbine towards conditions (3) and (4). The blades remain in the full operational position (fixed) throughout this period, as the increasing wind speed and increasing rotor speed (RPM) move together to achieve near optimal angle of attack at each radial position along the blades.

- As the turbine approaches rated power or condition (4), the blades are commanded to begin to pitch slightly back towards feather to lower the angle of attack and reduce peak loads. This prescribed schedule is also known as a "peak shaver."
- Implementation of a peak shaver comes with a reduction in generator output, so that the turbine designer must balance the benefits for reducing peak loads, material savings for the affected components, and the lower energy yield.
- From conditions (2)–(3), a torque command regulates the RPM for the optimal rotor tip speed ratio (TSR). The torque command is equal to a prescribed function of RPM.
- From conditions (3)–(4), a torque command regulates RPM based on the converter setting and the generator current.
- From condition (4)–(4"), where (4") is the turbine shutdown or cut-out wind speed, a current control maintains rated power output and the blades are pitched more and more towards feather to regulate the rotor RPM. This pitching is done to unload the blades with higher and higher wind speed to reduce WT loads by shedding the excess power that would have otherwise been captured (with massive loads to the turbine) beyond the generator rating.
- A typical cut-out wind speed for a modern MW WT is 25 m/s. When the machine reaches (4") and shuts down, the blades are pitched to the full-feathered position and the nacelle continues to be yawed into the wind. This keeps the turbine in a low drag configuration to ride out the storm.
- Referencing a WT designed to IEC TC1 criteria (as an example), wind speed increases beyond cut-out are provisioned throughout the turbine structure for survival wind speeds of 50 m/s (10-min average) and 70 m/s (3-s average). This is equivalent to hurricane intensities in the border region of category II–III and category IV–V in accordance with the Saffir-Simpson [36] scale, respectively.

Boes and Helbig [37] give an alternative description of this process, and is provided here for additional context – "There are two modes of operation: speed control at partial load operation (control of torque) and speed control at full load operation (pitch control). Torque control: To achieve the optimum power yield, the speed at partial load is adjusted to obtain an optimum ratio between the rotor speed at the circumference and the wind speed. The blades are set to the maximum pitch. The counter-torque at the generator controls the speed. Pitch control: After reaching the maximum counter torque at the generator (nominal power) at nominal wind speed, the speed cannot be maintained at the operating point by further increasing the generator torque. Thus changing the pitch from the optimum value reduces the aerodynamic efficiency of the blades. After reaching the maximum generator torque the blade pitch thus controls the speed."

Future control systems for large WT are likely to involve real-time measured signals in combination with physics-based control environments. Turbines would sense the actual imposed and reacted conditions and feed-forward these into smart proactive control actions. This type of approach should permit lowering design margins and enable designs to use less material throughout the structure and components.

4.6.2 SCADA hardware and software

The SCADA (Supervisory Control and Data Acquisition) system is the main system to operate a wind plant. Wind SCADA systems need several hardware components such as servers, modems, storage devices, etc. This rack-mounted hardware is typically located in the wind plant control room; however smaller wind projects may utilize standalone PCs. It provides the communication network and protocol for information flow between all components of the wind plant. At its simplest level, the SCADA network connects and controls the WT generators and enables collection of production and maintenance data.

- Receive data from individual turbines
- Send control signals to individual turbines
- Provide real-time data monitoring
- Alarm checking and recording
- Provide capability for historical data analysis
- Ability to model system (region-plant-cluster-unit) in hierarchy

It is used for configuration and commissioning of turbines, operation of the turbines, troubleshooting, and reporting. Commissioning technicians, service technicians, operators, owners, engineers and other experts use it [60].

SCADA knows everything that is going on in the wind farm: how much each turbine is producing, the temperature inside and outside of each turbine, wind direction, and if a turbine needs service or repair. It even records if lightning has struck a turbine. In the event the SCADA system detects a problem, it shuts down the machine or machines automatically and notifies the plant operator. Controllers inside the turbines also maintain the power quality of the electric current generated by the WT.

4.6.3 WPP control system

The WPP control system takes it to another level by integrating real-time grid conditions together with electricity production to ensure stable operation. Power quality is the stability of frequency and voltage and lack of electrical noise being supplied to the grid. The WPP control system monitor power production, and aggregates across the power plant and control regions ensuring power quality. It is not uncommon today to have very large parks in excess of 200 MW or more, and the WPP control system makes this possible. Examples of large projects include:

1. Horse Hollow Wind Energy Centre – The world's largest WPP with a capacity of 735.5 MW. It consists of 291 GE1.5 MW and 130 Siemens 2.3 MW WTs spread over nearly 190 km^2 (47,000 acres) of land in Taylor and Nolan County, Texas, owned and operated by NextEra Energy Resources (part of the Florida Power & Light (FPL) group) [38].
2. Titan wind project, a 5050 MW project announced for South Dakota consisting of 2020 Clipper 2.5 MW WTs [39].

4.6.4 Grid transient response

WT grid transient response relates to the operation control and protection of the transmission and distribution networks, interconnects, generators and loads. This is a very broad topic that has a number of specific points to consider from the WTG perspective. These include the fundamental frequency response and protection of the transmission and distribution system to faults that are also based on grid transients. The WT and the WPP must be able to absorb voltage spikes caused by lightning or switching events. This is addressed with proper grounding, overvoltage and lightning protection of the equipment.

4.7 Siting

There are many considerations when deciding where to site a WPP. Installation for flat terrain will often provide more favourable wind conditions and project economics, while irregular or difficult terrain has more difficult WT placement and increased installation cost. Turbine local geotechnical conditions are key when designing WT foundations. The amount of energy produced for a given project site is vital, but in many cases, more important are the regional electricity demand and pricing structures that will determine the overall WPP profitability. Land use, environmental regulations and permitting is the key when siting the individual WTs, and accounting for these requirements early in the new product development cycle will avoid costly miss-steps during production application.

4.7.1 Site-specific loads analysis

Turbines are often designed to one of the IEC TCs [13]. Specific sites may require more detailed analyses to ensure design adequacy. Some permitting agencies may require analyses certified by a licensed professional engineer for a specific turbine applied to the actual site conditions.

4.7.2 Foundations

Foundations are a crucial integral part of the overall MW WT design. They must account for the highly variable geotechnical conditions encountered in normal practice without adding unnecessary base cost. Foundations affect the natural frequency of the overall WT design. Many OEMs chose not to include the foundation in their scope of supply, but this does not mean that the foundation design can be ignored from the overall turbine design (Figs 37 and 38).

Figure 39 aggregates the major components of WTs into a single average technology trend derived from the 10-turbine analysis group. Against this backdrop the nominal foundation mass and combined WT and foundation mass trends are plotted with increasing WT size. The mass for some known foundations track reasonably well with the calculated "foundation only" trend. For a 10 MW machine these results project nearly 6000 tonnes for a monolithic foundation. However, a design goal of 4000 tonnes or below should be possible, especially considering numbers of smaller individual foundations supporting a multi-leg tower structure instead of a monolithic foundation. From Fig. 39, a good first estimate for the monolithic

Figure 37: Example of a circular raft type foundation under construction [40].

Figure 38: Tubular tower foundation after pour and backfill – tower ready [41].

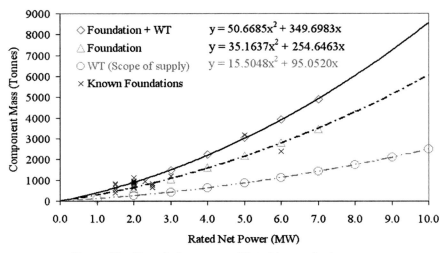

Figure 39: Foundation mass – 10-turbine analysis group.

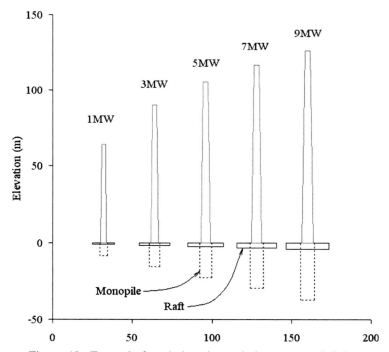

Figure 40: Example foundation sizes relative to tower heights.

foundation mass is about 2.48 times the total mass for the WT equipment above the tower base flange; i.e. the OEM's typical "scope of supply."

The term "monolithic" foundation (e.g. gravity foundation) refers to a single base for a WT. As shown in Fig. 40, this single base can be of a number of forms that include a monopile or raft configuration. Figure 40 relates the relative size of the required monolithic foundation dimensions relative to their respective towers for some representative turbines from the 10-turbine analysis group. Turbines in excess of 5–7 MW require excessive monopile depth or raft outer diameter, thus further supporting the thesis for multiple smaller foundations and multi-legged towers for larger turbines.

4.7.3 Offshore foundation

A significant amount of research is underway for offshore foundations, and many designers are attempting to leverage experience with offshore foundations in the oil and gas industry. Currently, two types of offshore foundations have been used with success in the wind industry – gravity and monopile. The gravity foundation was more common in the past, but it is more expensive and has a much larger footprint. More common today is the monopile foundation, which has a very small footprint and can be used for somewhat deeper water installations (up to about 30 m water depth).

As turbines become larger and near-shore locations become less available, the industry may need to look towards installations in deeper water. Many different types of foundations are being explored for these applications, such as different types of trusses, multiple legs and even floating foundations stabilized by ballasts and cable anchored to the seabed or lake bottom.

5 Special considerations in MW WT design

5.1 Continuously circling back to value engineering

The proportion of how total project cost is divided amongst the turbine, BOP, developer, and transportation costs vary as turbine rating is increased. Table 4 shows the breakdown in total project cost for a 100 MW WPP with flat terrain using a nominal technology MMW WT. Multiple trends in the proportions of project cost can be observed as the turbine rating is varied.

Table 4: Major cost breakdown for an onshore WPP.

| Rated Net Power | Wind Power Plant "all-in" cost fraction (total = 100%) | | | |
| | Turbine | BOP | Developer | Transportation |
[MW]	[%]	[%]	[%]	[%]
2.0	63.6%	25.0%	7.3%	4.1%
4.0	66.3%	22.5%	6.8%	4.4%
6.0	67.8%	20.9%	6.3%	5.0%
8.0	68.7%	20.4%	5.9%	4.9%
10.0	68.6%	20.5%	5.8%	5.2%

100MWWPP; Flat terrain w/easy access & good geotechnical; Nom technology MMWWT price; 10% BOP cont. inclusive. Number of loads determined by weight (no dim considerations) Larger components assumed capable of multiple loads.

A trade-off between the turbine and BOP cost is apparent as rating is increased. A site using a larger number of relatively inexpensive ($/kW) small turbines will have a lower turbine and higher BOP project cost proportion than a site using fewer expensive large turbines. Many BOP costs, such as roads and cabling, are more dependent on the total land area that the plant occupies than the rating of the turbines used. The larger turbine therefore has a BOP cost advantage on a MW-constrained site due to the reduced number of turbines required for a specified block of power. These factors reduce the specific cost ($/kW) of BOP as turbine ratings increase. Turbine cost behaviour further drives this proportionality trend as the specific cost of a turbine increases with rating (Fig. 41).

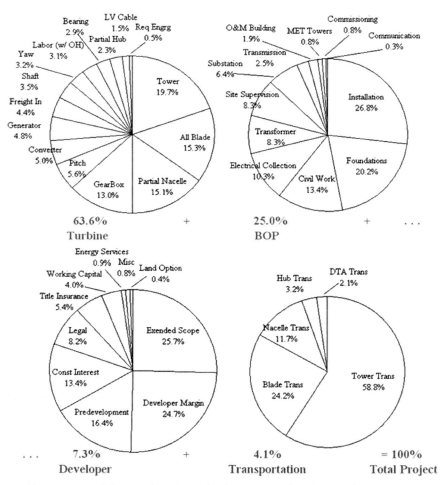

Figure 41: Additional detail – onshore WPP cost breakdown.

Notes: 100MW Wind Power Plant; Flat terrain w/ easy access and good geotechnical conditions; Nominal technology MMW Wind Turbine price; 10% BOP contingency inclusive

See Wiser *et al.* [42] for additional insight into U.S. wind power installation, cost, and performance trends.

The developer expenses proportion of total project costs drops as turbine rating is increased. Many developer costs, such as legal fees and title insurance, are not dependent on the machine selected. The total cost of these expenses are mainly dependent on the plant rating and not turbine rating. For a baseline or constant WPP rating, the total installed cost of a plant increases with turbine rating, so a reduced proportion of project costs for the developer are observed as turbine rating is increased.

The proportion of transportation cost increases slightly with rating, assuming technologies and designs for larger machines account for reasonable shipping limitations. The same savings found with BOP costs (because of reduced numbers of turbines for a fixed WPP size) are not realized in transportation. This is because transportation cost increases rapidly with rating due to the limited availability of 136 tonnes (150 ton) capacity 18-axle trailers in the United States. The 82-tonne (90 ton) capacity 13-axle trailers needed to ship the components of turbines with ratings less than approximately 2–3 MW are more readily available. Shipping a component 1000 miles on an 18-axle trailer is approximately five times the cost of using a 13-axle trailer. Additional costs are further observed for larger turbines because their components must be partially disassembled to allow transportation even on the large capacity 18-axle trailers [57].

5.2 Intellectual property (IP)

One of the most important activities for any technical organization is the ability to create and document IP as a natural part of their everyday engineering activities. For the wind industry today, obtaining patent protection for innovation must be a top priority. The industry is still young, and there is plenty of open IP landscape. Once new designs and methods are devised, these investments must be protected as they can provide additional income from licensing and sales.

Larger OEMs start the process of evaluating invention potential using an invention disclosure letter (IDL) process that contains the following elements:

1. A brief explanation of the invention.
2. Description of how the invention works (being very specific and including figures and images).
3. Description of the problem that is solved by the invention.
4. Descriptions of any prior attempts at solving the problem and how others may have tried to address the problem before.
5. Relating the technical and commercial advantages for the invention.
6. Explanation of how someone could design around the invention.

5.3 Permitting and perceptions

Know your markets – local ordinances typically specify max turbine rotor diameters, overall heights, and max noise emissions. These need to be accounted for new designs.

5.4 Codes and standards

Compliance with industry and equipment design regulations is one of the most overlooked, yet crucial details to ensuring widespread acceptance and trouble-free shipping and installation of new WTs. This is particularly true for personnel health and safety requirements that need to be considered from conceptual through detailed design [43, 44]. Det Norske Veritas (DNV) and Risø National Laboratory, Copenhagen, provide a good starting point in listing several relevant codes and standards for large WT design [45]. Since this is a rapidly developing and ever changing arena, it will likely pay for itself several times over to engage the services of a conformity assessment service prior to formally launching a new WT development program.

5.5 Third party certification

Unless adopting a strategy of self-certification, involving a third party conformity assessment vendor to provide certification guidance during new turbine preliminary and detailed design phases is critical to avoiding a large number of costly issues later. Some banks and insurance companies may require WT certification to specific standards (e.g. GL [23]) as a condition for customers securing their services. No matter the final path taken, third parties do provide their services under non-disclosure and offer a unique perspective for assessing design choices and leveraging a wide range of turbine configuration experience.

5.6 Markets, finance structures and policy

One should not underestimate the influence of financing options and government policy on MW turbine design. Some of the most potent influence parameters in value analysis of new WT systems are the target market settings. A machine and overall WPP that does well for some markets can be clearly disadvantaged in others. Strong guidance from the OEM's Marketing and Product Line Management (PLM) teams are crucial for engineering to account for the range of scenarios and sensitivities in the value analyses. An important corollary for this activity is that feedback for effective policy structures can be developed during the normal course of evaluations, and used by the OEM to proactively influence adoption of the best policies across the various market segments.

6 MW WT development techniques

Regardless of whether a turbine is large or small, the path to a successful new machine starts with conceptual design and value analysis. The value analysis includes a thorough understanding of where the market and competitor machines are positioned, and where any of the players can go in the future based on the technologies known or believed will become available. The path will not likely be a sweeping big step – rather, it is almost always the result of a number of iterations from an initial design. The trick is to develop the incremental technologies in the most economical way, and with as little exposure (i.e. lowest cost) to the customer as possible.

6.1 Validation background

Testing is a mandatory activity for any product development program. Validation is more than just collecting data for a new machine – it is about thorough collection and analytical reduction of specific data to make measurable changes to optimize a design.

6.1.1 Technology roadmaps

Maintaining technology roadmaps that identify pathways and timing are a key part of ensuring continuous improvement. Roadmaps should be maintained across the various systems as well as the individual components. This supports continuous improvement by showing how far the technology has come and providing a vision for the future of what technology needs to become.

6.1.2 Jugular experiments

Early on, new ideas are best demonstrated in their most basic setup – why spend more than is needed or mask understanding in too complex of a test apparatus? A "jugular experiment" is the first demonstration of the feasibility of a new technology that provides a proof of concept under the simplest conditions. While further testing may be necessary to determine whether an idea will be further developed or integrated into a new product, jugular experiments move the decision process along in the most cost-effective manner possible.

6.1.3 Technology demonstrators

When major component design or material changes are proposed – the best way to mitigate risk are limited trials of these components as incremental changes to existing machines. Demonstrate the technology and understand the design space before committing to serial production.

6.1.4 Prototypes

Product demonstration is the final step ensuring that the full system effects are accounted in the turbine design. A number of turbine prototype sites should be chosen to gather operating data for a range of environmental conditions. A small group of turbines (i.e. pre-series or limited production) undertaken after a period of successful prototype testing helps establish reliability and availability statistics, as well as power plant interaction effects.

6.2 Product validation techniques

Various techniques are used to validate products, and products must be validated at every level. Techniques that can be used include:

- Analytical experiments
- Jugular tests
- Sub-scale models
- Full-scale prototypes

Products should be validated at the system level and at the component level. The systems to components and back to systems approach should be used, as it supports continuous cycles of improvement.

6.2.1 System level validation

Systems level validation is the key activity that closes the loop between design and field performance and facilitates the process of continuous improvement for WT equipment. Validation also enables realistic noise and power curve specification for customer documents and project pro formas. In addition, turbine reliability improvements and optimized structural designs result from measuring load conditions and responses identified from validation activity.

6.2.2 Component validation

Unexpected interaction of components is revealed during tests. Certification testing under environmental extremes permits observations not easily possible in the field. Components are the building blocks for the larger system – get the components right, together with their interactions, and the system optimization will follow.

6.2.3 Rotor blade static and fatigue testing

Rotor blade static and fatigue testing are used to validate design assumptions and simulations used to predict the ultimate strength and 20-year life for the blade. There are a number of wind blade test facilities around the world that can be used to perform this type of testing, and plans for others have been announced to support the next generation of longer blades (i.e. >50 m in length). Of all the parts in today's modern turbines, the blades are perhaps the most fickle of all, requiring not only structural and aerodynamic design execution, but also the most critical of manufacturing and process control to ensure material and structural quality. Provisioning the cost and time for proper validation for blades is crucial before new blades are introduced into serial production.

7 Closure

Modern WTs are large complex structures that have achieved mainstream acceptance with rapid market growth and product standards development. The inexhaustible wind is a great fuel for electricity production, even with the challenges of turbulence, gusting, directional change and storm extremes. These impart the highest fatigue loadings to any manmade machine, and require turbine designers to carefully account for all effects to blades, hub, shaft, drivetrain, electricity generation system, support structure and the power plant system considerations that include interaction with the grid.

Value engineering or value analysis is the cornerstone process for identifying which innovations should be pursued and help the MW WT designer to focus on what matters most to their customers. It guides the OEM and ensures successful new products. The fundamental physics for the economic extraction of kinetic energy from wind results in rated net power densities on the order of 400 W/m^2 requires configuring rotor systems such that worse case blade tip deflections safely stay clear of support structures, and requires rotor thrust-induced overturning moments be accounted for every operational possibility and local geotechnical condition. Today's mainstream 1–3 MW WTs will give way to still larger turbines with the introduction of more and more advanced materials and technologies. Machines approaching 10 MW are within the realm of possibility. Successfully exploiting offshore wind resources in part depends on these larger machines becoming a reality, demands increased reliability, and the ability to install and maintain these machines at a price comparable to onshore.

Deriving power from the inexhaustible wind – it is truly a great time for the engineers that are taking up this challenge and for everyone striving to build a sustainable future for our heirs.

References

[1] Smalley, R.E., Our energy challenge, Walter Orr Roberts Public Lecture Series, Rice University, Aspen CO, July 8, 2003. http://www.archive.org/details/Agci-OurEnergyChallenge556

[2] Smalley, R.E., Future global energy prosperity: the terawatt challenge. *Material Matters*, MRS Bulletin, **30**, June 2005. http://www.mrs.org/s_mrs/bin.asp?DID=21838&CID=3682&SID=1&VID=2&RTID=0&DOC=FILE.PDF

[3] U.S. DOE, 20% Wind Energy by 2030 – Increasing Wind Energy's contribution to the US Electricity Supply; U.S. Department of Energy. http://www1.eere.energy.gov/windandhydro/pdfs/41869.pdf

[4] Miles, L.D., Techniques of value analysis and engineering. http://wendt.library.wisc.edu/miles/milesbook.html

[5] Miles, L.D., Dollar-sign engineering and value analysis. http://minds.wisconsin.edu/bitstream/handle/1793/3774/186.pdf?sequence=1

[6] SAVE, Devoted to the advancement and promotion of the value methodology, SAVE International. http://value-eng.org/

[7] Mankins, J.C., Technology readiness levels, Advanced Concepts Office, Office of Space Access and Technology, NASA, April 6, 1995. http://www.hq.nasa.gov/office/codeq/trl/trl.pdf

[8] Appleyard, D., Wind installations continue to break records across the globe, Renewable Energy World Magazine. http://www.renewableenergyworld.com/rea/news/article/2009/02/wind-installations-continue-to-break-records-across-the-globe-54658

[9] GWEC, Global Wind Energy Outlook 2008, Global Wind Energy Council. http://www.gwec.net/fileadmin/documents/Publications/GWEO_2008_final.pdf

[10] Burton, T., Sharpe, D., Jenkins, N. & Bossanyi, E., *Wind Energy Handbook*, John Wiley & Sons, pp. 329–330, 2001.

[11] Barr, A.L., Personal communication, October 2008, Wind Industry Data, GE Energy Wind, Greenville, SC.

[12] Gipe, P., IEC Wind Turbine Classes, Wind-Works.org. http://www.wind-works. org/articles/IECWindTurbineClasses.html

[13] IEC 61400-1, WT Design requirements, DNV. http://www.dnv.com/industry/ energy/segments/wind_wave_tidal/world_class_standards/

[14] Riley, P.S., Personal communication, 9 February 2009, Value Engineering, GE Global Research Energy Systems Laboratory, Niskayuna, NY.

[15] Lyons, J.P., CTO, Personal communication, LCoE for Electricity Generation Alternatives, Novus Energy Partners, Spring 2008.

[16] Curran, R., et al., Integrating aircraft cost modelling into conceptual design, Concurrent Engineering, Sage Publications, December 2005. http://cer.sagepub. com/cgi/reprint/13/4/321.pdf

[17] CERA, Concurrent Engineering: Research & Applications. http://www. ceraj.com/

[18] Powell, D.M., Personal communication, January 2009, Value Analysis, GE Energy Wind, Greenville, SC.

[19] Walford, C.A., Global Energy Concepts, LLC, Wind turbine reliability: understanding and minimizing wind turbine operation and maintenance costs, Sandia Report, SAND2006-1100, March 2006. http://www.prod.sandia.gov/ cgi-bin/techlib/access-control.pl/2006/061100.pdf

[20] Wind Energy the Facts. http://www.wind-energy-the-facts.org/en/part-i-technology/chapter-4-wind-farm-design/commissioning-operation-and-maintenance.html

[21] Wojszczyk, B., Herbst, D. & Bradt, M., Wind generation implementation and power protection, automation and control challenges, Power-Gen International, December 11–13, 2007. http://modernpowerengineering.com/Resources_files/ Wind%20Generation%20Implementation%20and%20Power%20Protec-tion%20Automationand%20Control%20Challenges-POWER-GEN%20 2007.pdf

[22] Pesetsky, D.S., Personal communication, January 2009, Blade Tip Closest Approach to the Tower, GE Energy Wind, Greenville, SC.

[23] Germanischer Lloyd – Guideline for the certification of wind turbines. https://www.gl-group.com/wind_guidelines/wind_guidelines.php?lang=en

[24] Army-technology.com, M1A1 / M1A2 Abrams Main Battle Tank, USA. http://www.army-technology.com/projects/abrams/specs.html

[25] Kuhn, M., Soft or stiff, a fundamental question for designers of offshore wind energy converters. Proc. of EWEC '97, Dublin, Ireland, October 6–9, 1997, http://www.lr.tudelft.nl/live/pagina.jsp?id=01b3b117-df62-4cfe-af8b-94327f86ef40&lang=en&binary=/doc/Soft%20or%20stiff_001.PDF

[26] Wang, J., Personal communication, April 2009, Mass & Cost Breakdown for Typical GFRP Blades, GE Energy Wind, Greenville, SC.

[27] Savage, J.R. & Johnson, S.B., Personal communication, May 2009, GFRP Blade Mold Cost & Life Factors, GE Energy Wind, Greenville, SC.

[28] Subramanian, P., Personal communication, March 2009, Bedplate & Space-frame Structural Analysis, GE Energy Wind, Salzbergen, Germany.

[29] GET, GE drivetrain technologies unveils new wind drive train concept at Husum, GE Transportation, 2008. http://www.getransportation.com/na/

en/docs/919534_1221057006_IntegraDrive%20Press%20Release%20
090308%20(Final).pdf

[30] ITH, Bolting tools & solutions, Industrie-Technische Konstruktionen
      Hohmann GmbH, 2009. http://www.ith.de/

[31] Jansen, P.L., Personal communication, February 2009, Generator & Converter
      Mass Trends with WT Size, GE Energy Wind, Schenectady, NY.

[32] Hazel, T., Vollet, C. & Fulchiron, D., Medium-voltage circuit-breakers
      improve transformer protection. *Proc. of the EWEC 2006 Conf.*, Athens,
      Greece, February 28, 2006. http://www.ewec2006proceedings.info/allfiles2/
      103_Ewec2006fullpaper.pdf

[33] IEC 61400-24, "Wind Turbine Generator Systems – Part 24: Lightning
      Protection," 2002. http://www.iec.ch

[34] Glushakow, B., Effective lightning protection for wind turbine generators.
      IEEE Trans. on Energy Conversion, **22(1)**, pp. 214–222, 2007.

[35] PACS, "Power and Control systems," PACS Industries, Inc. http://www.
      pacsindustries.com/Default.asp

[36] NOAA, Saffir-Simpson hurricane scale, http://www.prh.noaa.gov/cphc/
      pages/aboutsshs.php

[37] Boes, C. & Helbig A., Intelligent Hydraulic Pitch Control Valve for Wind
      Turbines. http://www.mec.upt.ro/~hme2008/lucrari/L_46.pdf

[38] NextEra Energy Resources, Horse hollow wind energy centre, currently the
      world's largest operating wind plant, 2009. http://www.nexteraenergyresources.
      com/content/where/portfolio/wind/construction.shtml

[39] Renewable Energy World, Planned titan wind project, 2008. http://www.
      renewableenergyworld.com/rea/news/article/2008/08/titan-wind-project-to-
      produce-5050-mw-53232

[40] Enercon Wind Turbines, Technology & Service Brochure, pp. 44–45, 2007.

[41] Khatri, D., *Global Qualifications for Wind Energy Development*, URS
      Corporation, June 2008.

[42] Wiser, R. & Bolinger, M., Annual Report on U.S. Wind Power Installation,
      Cost and Performance Trends: 2007, U.S. Department of Energy, DOE/
      GO-102008-2590, May 2008. http://www.nrel.gov/docs/fy08osti/43025.
      pdf

[43] EHS regulations – European Standard – "WEA - Schutzmaßnahmen -
      Anforderungen für Konstruktion, Betrieb und Wartung - EN 50308."

[44] EHS regulations – BEWEA "Guidelines for HEALTH & SAFETY in the
      Wind Energy Industry", April 2005.

[45] DNV/ Risø "Guidelines for Design of Wind Turbines." http://www.dnv.in/
      Binaries/GuidelinesforDesign_tcm55-29412.pdf

[46] Smith, K., WindPACT Turbine Design Scaling Studies Technical Area 2 –
      Turbine, Rotor and Blade Logistics. *National Renewable Energy Laboratory*,
      Golden, CO, March 2001. http://www.nrel.gov/docs/fy01osti/29439.pdf

[47] Young, W.C. & Richard, G., *Budynas Roark's Formulas for Stress and Strain*,
      7th Ed., McGraw-Hill, 2002.

[48] Manwell, J.F., McGowan, J.G. & Rogers, A.L., *Wind Energy Explained*, John
      Wiley & Sons Ltd., 2002.

[49] Wind turbine, Wikipedia article, April 2003–Present. http://en.wikipedia.org/wiki/Wind_turbine

[50] Dodge, D.M., Illustrated History of Wind Power Development, Littleton, Colorado, 1996–2005. http://www.telosnet.com/wind/index.html

[51] Bearings, Wikipedia article, May 2003–Present. http://en.wikipedia.org/wiki/Bearing_(mechanical)

[52] New Wind Turbine Catalogue, World Wide Wind Turbines, 2007. http://www.worldwidewindturbines.com/en/wind-turbines/select-wind-turbine-capacities/

[53] Merritt, F.S., *Standard Handbook for Civil Engineers*, 3rd Ed., McGraw-Hill, 1983.

[54] Wind turbine design, Wikipedia article, December 2006–Present. http://en.wikipedia.org/wiki/Wind_turbine_design

[55] Winds of change, Early US Wind Turbine Research, 1975–1985. http://www.windsofchange.dk/WOC-usastat.php

[56] Dinner, H., Trends in wind turbine drive trains, EES KISSsoft GmbH, Switzerland, March 2009. http://www.ees-kisssoft.ch/downloads/Hanspeter-Dinner-Enviroenergy-Format-EES.pdf

[57] Transportation of Unique Over-dimensional Cargo – Superloads, Diamond Heavy Haul, Inc. http://www.diamondheavyhaul.com/index.htm

[58] Weber, T., *When blades are growing*, New Energy, Magazine for renewable energy, pp. 64–67, January 2009. http://www.newenergy.info/index.php?id=1884

[59] Wind Turbine Towers, Danish Wind Industry Association, September 2003. http://www.windpower.org/EN/tour/wtrb/tower.htm

[60] Modi, V., Personal communication, August 2009, GE Energy Wind, Schenectady, NY.

CHAPTER 7

Design and development of small wind turbines

Lawrence Staudt

Center for Renewable Energy, Dundalk Institute of Technology, Ireland.

For the purposes of this chapter, "small" wind turbines will be defined as those with a power rating of 50 kW or less (approximately 15 m rotor diameter). Small electricity-generating wind turbines have been in existence since the early 1900s, having been particularly popular for providing power for dwellings not yet connected to national electricity grids. These turbines largely disappeared as rural electrification took place, and have primarily been used for remote power until recently. The oil crisis of the 1970s led to a resurgence in small wind technology, including the new concept of grid-connected small wind technology. There are few small wind turbine manufacturers with a track record spanning more than a decade. This can be attributed to difficult market conditions and nascent technology. However, the technology is becoming more mature, energy prices are rising and public awareness of renewable energy is increasing. There are now many small wind turbine companies around the world who are addressing the growing market for both grid-connected and remote power applications. The design features of small wind turbines, while similar to large wind turbines, often differ in significant ways.

1 Small wind technology

Technological approaches taken for the various components of a small wind turbine will be examined: the rotor, the drivetrain, the electrical systems and the tower. Of course wind turbines must be designed as a system, and so rotor design affects drivetrain design which affects control system design, etc. and so no component of a wind turbine can be considered in isolation. In general small wind turbines should be designed to IEC61400-2, Design Requirements for Small Wind Turbines [2].

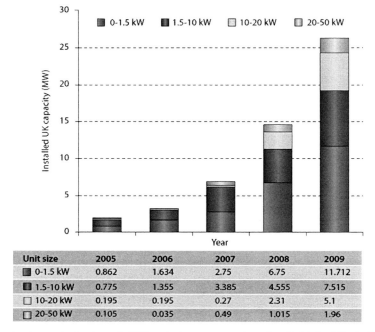

Unit size	2005	2006	2007	2008	2009
0-1.5 kW	0.862	1.634	2.75	6.75	11.712
1.5-10 kW	0.775	1.355	3.385	4.555	7.515
10-20 kW	0.195	0.195	0.27	2.31	5.1
20-50 kW	0.105	0.035	0.49	1.015	1.96

Figure 1: Annual deployed UK wind systems (credit: British Wind Energy Association).

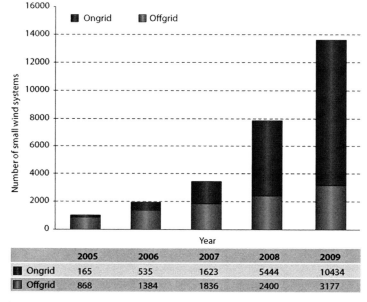

	2005	2006	2007	2008	2009
Ongrid	165	535	1623	5444	10434
Offgrid	868	1384	1836	2400	3177

Figure 2: On-grid vs. Off-grid UK wind systems (credit: British Wind Energy Association).

Figure 3: Whisper H40 (credit: AWEA, Southwest Windpower).

Figure 4: AIR Marine (credit: AWEA, Southwest Windpower).

1.1 Small wind system configurations

Just as has been the case with large wind technology, a number of attempts have been made to design vertical axis wind turbines – none of them commercially successful as of yet. Proponents of this technology for small wind point out important advantages: the ability to take cope with turbulent wind (as is found more often in small wind applications, due to lower towers, building mounting, etc.) and lower turbine noise. It remains to be seen as to whether a commercially successful vertical axis small wind turbine will emerge, and vertical axis machines will not be discussed further.

Unlike large wind turbines, which now exclusively use upwind designs (the blades upwind of the tower), there are successful upwind and successful downwind machines in the small wind turbine market. An early downwind design was the Enertech 1500 1.5 kW machine (which sold about 1200 units in the early 1980s, Fig. 5), the forerunner of the AOC 15/50 50 kW turbine (which sold between 500 and 1000 units in the 1980s and 1990s, Fig. 6), which was the basis for the current Entegrity EW50. The Scottish company Proven successfully use the downwind approach in its line of turbines (Fig. 7).

Virtually all small wind turbines use passive yaw control, i.e. the turbine requires no yaw motors and associated controls to orient the machine into the wind. In the case of upwind machines, a tail is used to keep the rotor upwind of the tower. The tail is often hinged to facilitate overspeed control (see Section 1.2.2). The tail becomes mechanically unwieldy as turbine size increases above

Figure 5: Enertech 1500 (credit: American Wind Energy Association).

Figure 6: AOC 15/50 (credit: AWEA, David Parsons).

Figure 7: Proven 600W (credit: AWEA, Leslie Moran).

about 8–10 m rotor diameter. Downwind passive yaw machines can be found up to about 15 m rotor diameter. Large wind turbines virtually are all of the active yaw, upwind design.

The upwind design appears more popular. The Jacobs design has been around for many decades (Fig. 8), and Bergey produce a well-known upwind turbine (Fig. 9).

Upwind turbines require a tail vane to orient the machine into the wind, whereas downwind turbines naturally track the wind without the need for a tail vane. The rotors on downwind machines are subject to "tower shadow" each time a blade passes behind the tower. The blade briefly sees reduced and more turbulent winds behind the tower, resulting in cyclical moments on the low-speed shaft and turbine mainframe which do not exist on an upwind machine. This increases fatigue cycles on the turbine. This must be traded off against the simplicity of the downwind design. In large wind turbines driven yaw is needed, and there is no reason not to have the rotor upwind of the tower.

1.2 Small wind turbine rotor design

In general the same issues in blade design exist for small turbines as for large wind turbines. These are discussed elsewhere in this book, and so only the unique aspects of small wind turbine blade design will be discussed in this section. These issues

Figure 8: Jacobs 10kW (credit: AWEA, Ed Kennel).

Figure 9: Bergey 10kW with tilt-up tower and furling tail (credit: AWEA, Don Marble).

can be put into three categories: rotor aerodynamics, rotor overspeed control, and rotor manufacturing considerations.

1.2.1 Rotor aerodynamics

In the early days of grid-connected wind turbines, rotors were usually "stall-controlled", i.e. maximum power was limited via aerodynamic stall. As wind turbines grew in size, pitch control has become the universal method to limit power output during high winds. Stall control is still commonly used on small wind turbines. Figure 10 shows two power curves illustrating the two types of power limitation.

The reason for the use of stall control is simplicity, and therefore low cost. The blades can be fixed to the hub without the need for pitch bearings and a pitch mechanism. Few small wind turbines feature pitch control.

Using blade pitch to effectively "dump" wind when turbine rated power is reached is a natural and obvious application of blade pitch. Stall control is simple and elegant, although somewhat less efficient. It is typically used on a constant speed turbine, e.g. one with an asynchronous generator. For example, suppose the blade tip speed is a constant 100 mph. In light winds the blade angle of attack would be very shallow, i.e. the wind is coming directly at the leading edge (the blade pitch angle is only a few degrees). In high winds the blade tip would see wind coming at it from a much steeper angle, and stall would occur, limiting power output with no moving parts. There are certain airfoils that exhibit a particularly useful stalling characteristic (e.g. the NACA44 series) which are commonly used on stall-controlled small wind turbines.

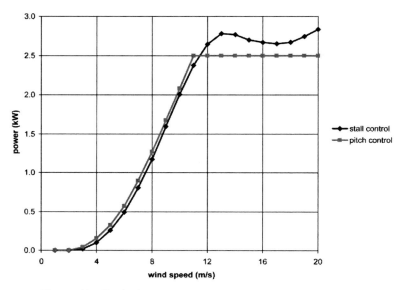

Figure 10: Typical power curves with stall and pitch control.

With the advent of grid-tie inverters, stall control is becoming less prevalent on the smaller wind turbines, power curves are beginning to resemble that of the pitch controlled turbine in Fig. 10, and power limitation accomplished in conjunction with the rotor overspeed control (see discussion below). In this case a stalling rotor design is no longer necessary.

It should be noted that it is difficult to get the same high aerodynamic efficiency on small turbines as on large turbines. This has to do with the formation of turbulent boundary layer on the surface of the blade. Towards the leading edge of the blade the boundary layer is laminar, but at some critical distance l across the blade surface the flow becomes turbulent. This distance is a function of flow velocity (V), Reynolds number (Re), viscosity (μ) and density (μ) as per the following equation:

$$l = (Re * \mu)/(V * \rho) \tag{1}$$

This turbulence reduces pressure drag (the primary effect) and increases friction drag (a secondary effect), with the net effect being a reduction in drag and an improvement in rotor efficiency. Small wind turbine blades have a small chord length and therefore there is a relatively smaller region across the blade surface where the flow is turbulent compared to large turbine blades, hence the higher efficiency from large turbine blades. C_p values on large turbines can be on the order of 0.5, whereas on smaller turbines it is on the order of 0.4.

1.2.2 Rotor overspeed control

Turbines (e.g. gas turbines, steam turbines, wind turbines, etc.), when unloaded, typically only have inertia to limit uncontrolled acceleration. Typically energy

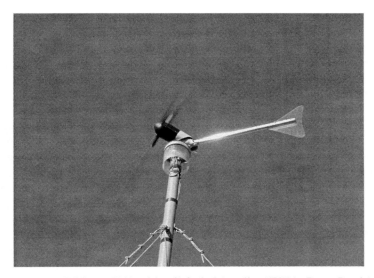

Figure 11: Whisper H40 with tail furled (credit: AWEA, Dean Davis).

supply (gas, steam) is then cut off via an independent overspeed control mechanism – a critical function since the overspeed condition can result in the destruction of the turbine. In the case of wind turbines, the energy supply (the wind) cannot be stopped, and so other means of overspeed control must be used. (It is interesting to note that, just as in the case of turbines in conventional power stations, the primary wind turbine speed control mechanism is the generator. Emergency overspeed control only comes into play, e.g. when the generator fails.)

The obvious way to prevent wind turbine rotor overspeed is to pitch the blades, and this is universally done on large wind turbines. It is possible to pitch the blades either way (toward "feather" or toward "stall"), and there is more than one small wind turbine using the pitch-to-stall approach for overspeed control (large wind turbines use pitch-to-feather, pitching the blades through about 90°). However only a few degrees of pitch variation in the other direction are required to achieve a stall condition, and this can be done e.g. through a hub hinge or through pitch weights mounted on a torsionally flexible blade. While both of these pitch-to-stall approaches are used in small wind turbines, neither approach is common.

The most common approach on small upwind turbines, as mentioned above, is the furling tail. Figure 11 shows a turbine with the furling tail actuated. The main features of a typical furling tail system are firstly the rotor has its centerline offset from the centerline of the tower, and secondly it has a hinged tail (capable of furling in one direction but not the other). At times of excessive rotor thrust (as occurs during overspeed), the thrust force causes the rotor to yaw "around the tower" and the tail to furl via the hinge. During normal operation, proper yaw orientation is maintained via the hinged tail. The hinge axis is typically slightly off of vertical, such that the tail must move "uphill" as it furls, i.e. gravity keeps it up against a stop (and directly behind the rotor) during normal operation. The above overspeed control methods eliminate the need for stalling airfoils when used in conjunction with a grid-tie inverter as discussed in the electrical system section below.

Another approach, not commonly used, is the so-called "tip brake" (see Figs 5 and 6). Centrifugally deployed flaps are mounted at the end of each blade, and an overspeed condition causes them to deploy and face the wind, resulting in significant drag at the blade tip, thus limiting rotational speed. Tip brakes typically do not automatically reset, as they should only deploy when other problems exist (brake failure, generator failure, etc.) which would require the attention of a service technician. It is interesting to note that tip brakes do not significantly degrade rotor efficiency during normal operation since, although there is increased drag at the blade tip, they tend to prevent blade tip losses.

Other braking systems whose main function is not rotor overspeed control will be discussed in Section 1.3.

1.2.3 Rotor manufacturing considerations

It is generally easier to build small wind turbine rotors than those for large wind turbines. The rotor weight plays a less role in the design, and there is more focus on using minimum cost manufacturing techniques (such as injection moulding for the smallest machines). While glass reinforced plastic is the most common material (as in large machines), it is also easily possible to use wood or recyclable thermoplastics.

Aerodynamic efficiency is sometimes sacrificed in favour of ease of manufacturing. For example, it is possible to extrude plastic blades, such that there will be twist but no taper. The effect on efficiency is illustrated on Fig. 12. The lower

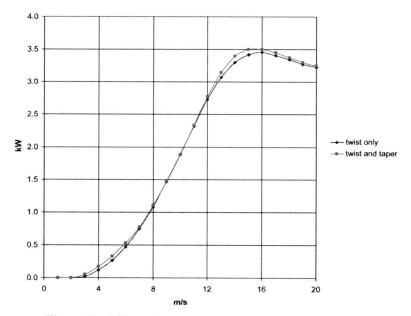

Figure 12: Effect of adding taper to a blade with twist only.

efficiency at lower wind speeds can be significant in terms of energy production, as a typical small turbine spends much of its life operating at low wind speeds.

On the smaller turbines, it is also possible to have more than three blades with relatively little impact on cost. Multiple blades allow higher starting torques, and lower operational speed (and therefore lower noise). This must be traded off against higher thrust loads and the slightly higher cost. On some of the smallest machines there is no rotor overspeed control at all, i.e. the machine is simply designed to survive the high rotational speed and thrust load of a "runaway" condition. In this case having multiple blades (e.g. the well-proven Rutland 913, which has six blades) limits the rotational speed to some extent.

The hub is part of the rotor, and small wind turbines typically have very simple hubs, as the blade pitch is typically fixed. On some rotors blade pitch is not adjustable, other rotors use shims to set the pitch, while others use a rotary adjustment at the blade root that is locked in place after final adjustment. Some small wind turbines have more complex hubs, consisting of springs and hinges (e.g. Proven wind turbines, which pitch the blades to stall for overspeed control). None of these features are typical of large wind turbines.

1.3 System design

1.3.1 DC systems
Traditionally small wind turbines use DC generators. The DC generator is now usually in fact a permanent magnet three-phase synchronous AC generator (alternator), with a diode rectifier either located up in the turbine (with two wires coming down the tower) or at the control panel (with three smaller wires coming down the tower). The rotor mounts directly onto the alternator shaft, and no gearbox is required. This remains the most common approach used by small wind turbine designers. With the advent of grid-tie inverters (see below), it is a solution that makes small wind turbines suitable for battery charging as well as grid-connected applications.

In the battery charging mode, DC systems operate at fairly constant speed. Figure 13 shows the simplified equivalent circuit of such a system. The voltage produced by the generator is proportional to rotational speed. If the sum of the circuit resistances (generator winding resistance, cable resistance and battery resistance) is relatively small, then $V_{gen} \cong V_{batt}$, i.e. the battery voltage effectively regulates the generator voltage, and therefore the generator speed, to be relatively constant.

In real applications the generator rotational speed increases noticeably with power output, as suggested by Fig. 13. When current is high, then voltage drop across the resistances is significant, and V_{gen} rises (and therefore generator rotational speed) with power output. This impacts aerodynamic performance and design. For example, if a stalling airfoil is being relied upon to limit power output, the power output at which stall occurs is a function of rotational speed as suggested by Fig. 14. When batteries reach a fully charged condition, the charge controller disconnects the wind turbine and the wind turbine freewheels, held back only by the overspeed control system.

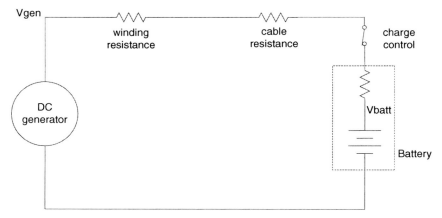

Figure 13: Simplified battery charging circuit.

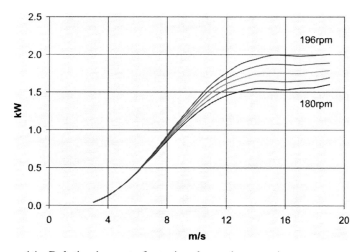

Figure 14: Relative impact of rotational speed on maximum rotor power.

Besides not needing a gearbox, DC systems can use the permanent magnet generator as a brake. The windings need only be shorted or connected across a small resistance, and this results in significant braking torque from the generator. It is typical for DC machines, during the commissioning process, to have their winding shorted until such time as the turbine is ready to rotate.

1.3.2 AC systems

1.3.2.1 Permanent magnet alternator with grid-tie inverter
This approach is commonly used for machines up to about 10 kW to produce grid-quality AC power. Above this level (and sometimes below), a gearbox

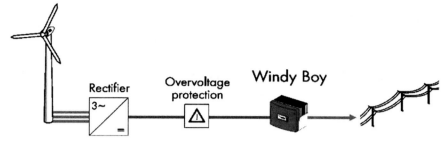

Figure 15: General arrangement of grid-tie inverter system (credit: SMA).

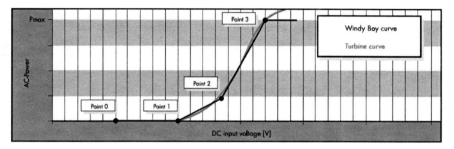

Figure 16: Grid-tie inverter programming (credit: SMA).

and an asynchronous generator are typically used (see below). Figure 15 shows the general arrangement for this approach. In this case three-phase power (variable voltage and variable frequency) comes down the tower. It is then rectified to DC, passes through an overvoltage protection relay and on to the inverter.

The grid-tie inverter loads the turbine (i.e. extracts power) on the basis of voltage (i.e. turbine rpm). In this way the turbine can be operated at or near the point of optimum system efficiency all along the power curve, as illustrated in Fig. 16. As the wind starts to rise, the DC voltage increases. As the voltage rises above point 1, the inverter begins to load the turbine according to the line between points 1 and 2. As the voltage rises above point 2, the inverter begins to load the turbine according to the line between points 2 and 3.

Above point 3 the rated power of the inverter is reached, and so regardless of voltage (i.e. turbine rotational speed), the turbine is only loaded to that power. This means that the turbine accelerates, since rotor power exceeds the power being withdrawn by the generator. In this case the overspeed control system comes into play, regulating the turbine rpm below a dangerous level. If the DC voltage rises to a value that exceeds the input rating of the inverter (if, e.g. the overspeed control mechanism fails), then the overvoltage circuit shown in Fig. 15 will disconnect the turbine from the inverter input. Inverter efficiency is poor at low power inputs, but then typically rises to a high level.

1.3.2.2 Induction (asynchronous) generator with gearbox

Induction generators are the most commonly used generators in large wind systems. They are simple, rugged, and relatively low cost. However they have high rotational speed compared to DC systems, and therefore necessitate the use of a gearbox. Small wind turbines rotate at higher speeds than large ones, and it is therefore easier to use the permanent magnet alternator design approach than on larger turbines. However it is also true that on small induction generator-based machines the gear ratio is lower than on large wind turbines, and so this can result in a somewhat simpler power transmission (fewer reduction stages) than on large wind turbines.

Small induction generator-based wind turbines operate at close to constant speed. This means that the wind system only operates at the peak of the $C_p-\lambda$ curve at one wind speed, therefore at other wind speeds the turbine operates at less than peak efficiency. Figure 17 illustrates the difference in rotor power between fixed and variable speed operation of the same rotor. When the efficiency of the inverter is taken into account, it can be expected that variable speed operation can result in 5–10% more energy capture.

1.3.3 Braking systems

Many small wind turbines have no braking systems at all (except for the rotor overspeed control discussed above). As mentioned above, permanent magnet generators are sometimes used as brakes, which can be accomplished by either simply shorting the windings or connecting the windings through a low electrical resistance.

It is also possible to use the generator as a brake on an induction machine (e.g. the AOC 15/50, see Fig. 6). This is sometimes called "electrodynamic braking".

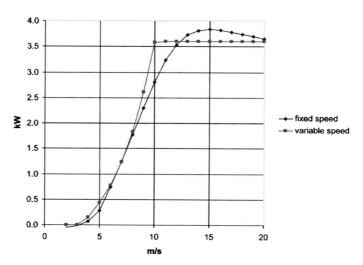

Figure 17: Fixed vs. variable speed power curves.

The generator is disconnected from the grid and then connected to a series-parallel combination of resistors and capacitors. The capacitors provide the excitation and the resistors dissipate the energy. It provides very smooth braking torque until excitation is lost at a low rpm, and a small disc brake is engaged.

A disc brake is commonly used on large wind turbines, and is sometimes used on small machines. The Enertech 1500 (Fig. 5) had a gearbox, induction generator, and a disc brake on the high speed shaft. The drawback of this arrangement was that the brake had to be sized for the maximum anticipated rotor torque, and so every time the rotor was halted the drivetrain (and particularly the gearbox) experienced maximum torque. Often a disc brake produces torque spikes as the discs come together, rebound and then finally settle. It is also possible to put the disc brake on the low-speed shaft, thereby eliminating strain on the gearbox, but a much bigger brake is required and so this is seldom used.

Other creative ideas have been tried, such as the so-called "hydraulic brake". In this case a hydraulic pump is coupled onto the high speed shaft rather than a disc brake. Hydraulic fluid is pumped through an open solenoid valve during normal operation, and when braking action is required the solenoid valve is closed and the flow diverts through a pressure relief valve. This produces very smooth braking torque, however there are pumping losses during normal turbine operation.

1.3.4 Power cabling

Typically on small wind turbines power cabling comes from the generator to slip rings, and then a separate power cable goes down the tower to a disconnect switch at the base of the tower. However the "twist cable" concept was introduced in the early 1980s, and is now typically used on large machines, and sometimes used on small machines.

The power cable is suspended from beneath the turbine, supported by a strain relief connection at the turbine. Depending on the cable length and flexibility, it is generally able to withstand many yaw revolutions in one direction before it is in mechanical stress. The cable is physically disconnected at the bottom of the tower and allowed to unwind during service visits to the turbine. In the case of large wind turbines, yaw revolutions are counted, and when a certain number have accumulated in one direction the turbine is stopped and the cable is "unwound" via the yaw mechanism.

1.3.5 Control system design

The control system depends very much on the application, and in general there are two applications: grid connected and battery charging (though there are other applications such as direct heating and direct pumping).

1.3.5.1 Grid-connected control systems

This is discussed to some extent in Section 1.3.2, and the controls depend on whether it is a grid-tie inverter system or an induction generator-based system. In the case of the former, the inverter effectively is the control system, loading the generator according to the DC voltage that it sees as discussed above.

The inverter would also include grid protection functions according to utility standards. In the European Union this would typically be EN50438 [3], whereas standards vary in the USA. Grid protection is primarily required to disconnect the wind system from the grid in the event that the power produced is outside of a frequency/voltage window (i.e. either the voltage or frequency going too high or too low). This would suggest that a utility power outage has occurred and that the inverter output is being fed into a "dead" line, thereby endangering utility linemen.

An induction generator-based system would also have grid protection functions according to EN50438. It would also have logic to connect and disconnect the generator from the grid, usually based on generator rotation speed (rpm).

Figure 18 shows the induction machine speed–torque curve. Torque is produced by the machine as it motors up to speed from a stopped condition. When driven beyond the synchronous speed (ω_s) it absorbs torque (and produces electrical power). Normal (full load) generating torque is indicated on the curve. Therefore the control action would allow the turbine to be driven up to synchronous speed by the wind, and at ω_s connect it to the grid. Similarly when the rotational speed drops below ω_s the machine is consuming power and so the control system disconnects it from the grid.

There are a number of possible variations on this approach. For example, it could be connected on the basis of rpm as above, but disconnected on the basis of power (i.e. when power is being consumed). Since precise connection at ω_s is needed (consider the torque spike that would result if connection occurred at ω_s + 10%), rate-of-change of ω could be integrated into the control algorithm. In some turbines (Enertech and AOC) the machine was actually motored up to speed (sometimes stalling airfoils have poor self-starting characteristics) when wind speed was deemed sufficient to generate power.

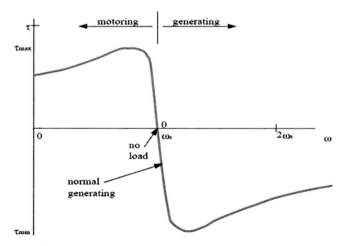

Figure 18: Induction machine speed–torque curve.

1.3.5.2 Battery charging control systems

In general, permanent magnet generators are used on small DC wind systems, and so the function of the control system is to intelligently connect or disconnect the wind turbine from the battery depending on the battery state of charge, which is a function of battery voltage (see Fig. 13). Control systems can also control charge rate, e.g. trickle charging the battery through resistors as the batteries approach full charge.

The control system may also connect the generator output to braking resistors to limit rotational speed in the case where the battery is fully charged. Commonly in this case, however, the turbine is simply electrically disconnected and the mechanical overspeed mechanism comes into play.

1.4 Tower design

Towers can be grouped into two categories: guyed towers and self-supporting towers. Large wind turbine towers are generally only the self-supporting type – typically tapered tubular steel towers. For small wind turbines, self-supporting towers generally come in three types: lattice towers (Figs 6 and 8), steel poles, and wood poles (phone poles). Self-supporting towers are more expensive than the guyed variety for a given height, and primarily for this reason the guyed towers are more commonly used in small wind turbine applications. However if there is inadequate room on the site to accommodate the guy wires, a self-supporting tower is used.

Guyed towers come in two types: guyed poles or guyed lattice towers. These are both mass-produced for use in the telecommunications industry. Guyed poles are illustrated in Figs 3, 7, and 11, and guyed lattice towers are shown in Figs 5 and 9. Figure 19 shows a tall guyed pole for a small wind turbine, with several levels of guys and an integral gin pole. Figs 9 and 20 show how the gin pole is used to erect the tower. This tilt-up technique makes it possible to erect and service a turbine without the need for a crane or climbing equipment.

It is often cost-effective to consider a rather high tower for a small wind turbine, as the incremental cost of increased height is low, and meanwhile the increase in wind speed (and reduced turbulence) increase production significantly. Consider the case of a Bergey XL1, with measured annual average winds of 5 m/s at a height of 10 m. Table 1 shows the energy production, indicative system cost, and payback time, assuming the energy is worth $0.20/kWh. On this basis it makes sense to purchase the tallest tower.

Recently roof-mounted systems have become somewhat popular (not unlike Fig. 4, but mounted on e.g. a gable end). Even though at first consideration it seems that this is an inexpensive way of getting a turbine up "into the wind", the concept suffers from several disadvantages; rooftops are not actually very high, winds at the rooftop tend to be turbulent, structural properties of roofs vary resulting in the possible need for (expensive) custom design work, and only the smallest turbines can physically mount on roofs which means energy capture will be limited.

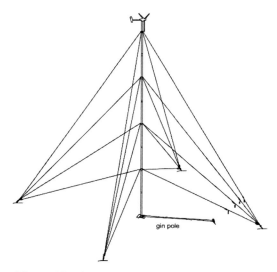

Figure 19: Guyed pole tower (credit: Bergey).

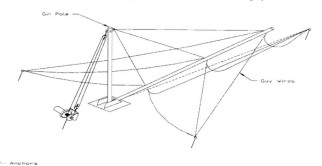

Figure 20: Tilt-up tower installation (credit: NRG).

Table 1: Effect of tower height on economics.

Tower height (m)	Annual kWh	System cost ($)	Annual energy value ($)	Payback (years)
18	2300	7800	460	17.0
24	2600	8250	520	15.9
30	2800	8600	560	15.4

The present author cannot recommend roof-mounted wind systems in general, and increasing numbers of people concur [4, 5].

2 Future developments

A number of areas are being explored by an ever-increasing number of small wind turbine companies, with a view toward increasing reliability and reducing cost.

Given the relatively limited resources of the industry at present, many design refinements can be expected in the future, in such areas as:

- Rotor design
    - Although most manufacturers have adopted the horizontal axis configuration, significant research is ongoing with vertical axis machines.
    - A number of overspeed control mechanisms have been considered, but the most prevalent (the furling tail) does not scale up very easily, and different concepts are being explored such as pitch-to-stall rotors, coning rotors, and deformable blades.
    - Most rotors are made of reinforced plastics, however there is active research into modified wood blades as well as other materials.
- Drivetrain design
    - Permanent magnet alternators are available up to about 10 kW, above which induction machines tend to be used. However, with advances in permanent magnet technology, it is possible that this technology will become available in larger size, along with larger grid-tie inverters.
    - Variable speed drives are commonly used in industry for variable speed motor applications using induction machines. It is possible though less common for "regenerative" drives to be used on e.g. elevator applications, where power is returned to the grid. This technology could be applied to induction generator-based wind turbines.
    - Though gearboxes have been traditionally used for power transmission in induction generator-based turbines, recent advances in timing belt materials mean that this technology can now be used, offering the advantages of low noise, high shock load capability and lack of need for lubrication.
- Tower design
    - Advances in laminated wood technology ("glulam") may result in the possibility for high-technology wood towers. This would result in low embodied energy/low CO_2 towers made from sustainable forests.
- Control systems
    - With mass production of small wind turbines even more advanced, microprocessor-based control and monitoring systems are expected. These will provide remote communications capability, detailed operator/owner displays and more advanced condition monitoring capability.

3 Conclusions

Although less significant in an energy sense than large wind turbines (which have the potential to supply a significant percentage of the world's electricity), small wind systems have a bright future. Demand is rising rapidly, and costs should correspondingly decline. Until now, R&D has been limited to the rather small budget of the few companies that have been able to survive in the small wind business. Therefore there will be a significant increase in R&D on small wind systems, which will result in improvements in reliability and value.

Small wind turbine applications are now becoming predominantly grid-connected, which opens up a mass market for this technology. Attempts at urban "building-mounted" systems so far have not proven to be a viable concept, and are unlikely to do so in the future.

As the cost of small wind turbines declines through mass production and technical advances, and as the cost of competing energy forms rises, installing small wind systems in windy rural locations will become increasingly viable.

References

[1] British Wind Energy Association. Small Wind Systems UK Market Report, 2008, www.bwea.com
[2] International Electrotechnical Commission. IEC61400-2, Design Requirements for Small Wind Turbines, 2006, www.iec.ch
[3] CENELEC. EN50438, Requirements for the Connection of Micro-Generation in Parallel with Public Low Voltage Distribution Networks, www.cenelec.eu
[4] Carbon Trust, Small Scale Wind Energy – Policy Insights and Practical Guidance, 2008, www.carbontrust.co.uk
[5] Gipe, P., *Wind Power*, Chelsea Green Publish Company: White River Junction, Vermont, 2004.

CHAPTER 8

Development and analysis of vertical-axis wind turbines

Paul Cooper

School of Mechanical, Materials and Mechatronic Engineering, University of Wollongong, NSW, Australia.

Vertical-axis wind turbines (VAWTs) have been demonstrated to be effective devices for extracting useful energy from the wind. VAWTs have been used to generate mechanical and electrical energy at a range of scales, from small-scale domestic applications through to large-scale electricity production for utilities. This chapter summarises the development of the main types of VAWT, including the Savonius, Darrieus and Giromill designs. A summary of the multiple-streamtube analysis of VAWTs is also provided to illustrate how the complex aerodynamics of these devices may be analysed using relatively straightforward techniques. Results from a double-multiple-streamtube analysis are used to illustrate the details of the performance of VAWTs in terms of turbine blade loads and rotor power output as a function of fundamental parameters such as tip speed ratio. The implications for VAWT design are discussed.

1 Introduction

Vertical-axis wind turbines (VAWTs) come in a wide and interesting variety of physical configurations and they involve a range of complex aerodynamic characteristics. Not only were VAWTs the first wind turbines to be developed but they have also been built and operated at a scale matching some of the biggest wind turbines ever made. VAWTs in principle can attain coefficients of performance, $C_{p,max}$, that are comparable to those for horizontal-axis wind turbines (HAWTs) and they have several potentially significant advantages over the HAWTs.

These advantages include the fact that VAWTs are cross-flow devices and therefore accept wind from any direction. Thus, in principle, they do not need a yaw mechanism to ensure that they are aligned to the wind as is the case with all horizontal axis machines. Another key advantage is that the mechanical load may be connected directly to the VAWT rotor shaft and located at ground level. This removes the need for a substantial tower to support the weight of equipment such as the gearbox, generator and yaw mechanism. There is also no need for slip rings or flexible cables to connect the generator to the load, which can be an important point for small-scale turbines.

From the 1970s to the 1990s a number of research groups and companies developed and built hundreds of VAWTs and a great deal has been learnt from that experience. But despite the inherent advantages of VAWTs they have fallen significantly behind HAWTs in recent years in terms of technical development and in the size and number of units manufactured. This has occurred for a number of reasons, not least because of some inherent disadvantages of VAWTs.

As the VAWT blades rotate about the main rotor shaft the velocity of the air relative to the blade is constantly changing in respect of both magnitude and direction. In addition, each blade will interact with the wakes of other blades, and possibly its own wake, when it passes through the downstream half of its path about the turbine axis. Both these effects result in fluctuating aerodynamic forces on the blades, which in turn lead to a potentially significant fatigue issue for the design of the blades and overall turbine structure. The fluctuating blade loads also lead to a varying torque transferred to the mechanical load.

Many designs of VAWTs produce very low torque when they are stationary and may produce negative torque at low tip speed ratios, so they must be powered up to a speed at which the aerodynamic torque is sufficient to accelerate the rotor to normal operational speeds. A further disadvantage is that parasitic drag losses may be high for a given VAWT design. This situation can arise when the VAWT blades need to be mounted on structures (spars, beams, cables, etc.) that rotate with the blades or are located upstream of the blades. The drag forces on these passive components can lead to significant parasitic losses in respect to rotor torque and power output. This has inhibited the successful development of a number of VAWT designs.

Nevertheless there continues to be widespread interest in VAWTs as a means of generating electrical and mechanical energy from the wind. Novel VAWT turbine designs appear relatively frequently at the time of writing and a number of small companies appear to be undertaking development of VAWTs for small-scale application, particularly in respect to domestic dwellings.

2 Historical development of VAWTs

2.1 Early VAWT designs

VAWTs appear to have been developed long before their horizontal axis cousins. One of the reasons for this is that the VAWT has a number of inherent advantages including the fact that a drive shaft may be connected directly from the rotor to a mechanical load at ground level, eliminating the need for a gearbox. The early pioneers involved in the development of wind turbines many centuries ago applied VAWTs to the milling of grain, an application where the vertical axis of the millstone could be easily connected to the VAWT rotor. Quite a number of excellent

Figure 1:   An example of VAWTs in the Sistan Basin in the border region of Iran and Afghanistan. Note in the right hand image how the upstream wall is used to expose only one half of the rotor to the wind (photographs taken in 1971 near Herat, Afghanistan, copyright: Alan Cookson).

review articles have been published in the past detailing the historical development of wind turbines of all types [1, 2]. Virtually all of these reviews suggest that the very earliest wind turbines were indeed VAWTs and it is thought that these were first used in Persia for milling grain more than 2000 years ago. These early wind turbines were essentially drag devices with a rotor comprising a number of bundles of reeds, or other simple blades, on a timber framework. The rotor was housed within a walled enclosure that channelled the flow of wind preferentially to one side of the rotor thereby generating the torque necessary to rotate the millstone. This type of device was still in use during the latter half of the 20th century and an example located in the border region of Afghanistan and Iran is shown in Fig. 1 [3].

The Persian and Sistan VAWTs had rigid vanes to generate torque whereas other designs have used sails that can effectively pitch with respect to their alignment on the rotor and thus can potentially increase efficiency. An example of a Chinese VAWT of the type used for many years for pumping applications, and which was described by King [4] for pumping brine for salt production, is illustrated in Fig. 2.

2.2 VAWT types

A wide variety of VAWTs have been proposed over the past few decades and a number of excellent bibliographies on VAWTs have been published that summarise research and development of these devices, including the survey by Abramovich [5]. Some of the more important types of rotor design are highlighted in the following sections.

Figure 2: A Chinese VAWT used for pumping brine (photo taken in early 20th century) from King [4].

2.2.1 Savonius turbines

The need to pump water in rural/remote locations has long been a driver for the development of wind turbines. In the early 20th century a number of innovations were developed by inventors such as Savonius who patented his device in 1929 [6]. This utilised a rotor made from two half-cylinders held by a disc at each end of the rotor shaft, as shown in Fig. 3. The Savonius turbine has been popular with both professional and amateur wind turbine developers over the years, not least because of its simple and robust construction.

Many variations of the Savonius rotor have been developed and tested. However, because of the inherently high solidity and hence high mass of the Savonius turbine it has not been used for large-scale electricity production. Nevertheless, it continues to find favour in a number of areas of application, including building-integrated wind energy systems which are now attracting attention as building designers seek to reduce the ecological footprint of building structures and their operations. Müller *et al.* [3], for example, explore the potential of VAWTs installed on buildings. Figure 4 shows an example of this type of application where Savonius turbines are mounted on the natural ventilation stacks of the landmark Council House 2 (CH2) Building in Melbourne, Australia. The low tip speed of rotors such as the Savonius has a number of attractions, not least that they are likely to produce less aerodynamic noise, which is an important issue for turbines included as part of inhabited structures. However, a number of considerable challenges remain to be overcome before building-integrated wind turbines can provide a cost-effective means of generating electricity. These include the fact that the urban environment is characterised by low wind speeds and high turbulence.

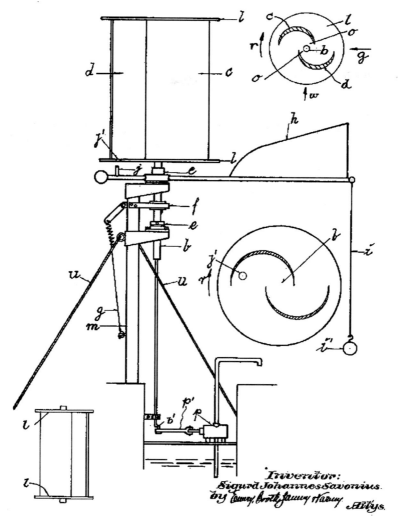

Figure 3: The principal embodiment of the Savonius VAWT patent [6].

The Savonius rotor is primarily a drag device with some inherent augmentation of the rotor performance available due to the air flow across each vane and mutual coupling of the two halves of the rotor. Like all drag machines it has a low operating tip speed ratio. This makes it less suitable for electricity generation than devices with higher tip speeds, since a high shaft speed is generally preferred to minimise the step-up ratio requirement of the gearbox coupling a rotor to a conventional electrical generator. Several new versions of the Savonius have been manufactured in recent years including devices with spiral blades of relatively short span mounted on a wide rotor hub.

Figure 4:   Savonius turbines used to assist ventilation and generate electricity on the Council House 2 (CH2) landmark building in Melbourne, Australia (photographs – copyright Pauline Anastasiou).

There have been many studies of the performance of the Savonius rotor, however, it would appear that the coefficient of performance, C_p, is modest. Modi and Fernando [7], for example, tested a wide range of Savonius rotor geometries in a wind tunnel with variations in the degree of overlap and separation of the blades and rotor aspect ratio. Modi and Fernando also carried out important tests to determine the blockage effect of the turbine in the wind tunnel. Thus, they were able to extrapolate their results to estimate the performance of Savonius rotors under unconfined conditions. Their conclusion was that the best coefficient of performance of a geometrically optimised Savonius rotor was likely to be $C_{p,max}$ ~ 0.3 at a tip speed ratio of λ ~ 0.7. Ushiyama and Nagai [8] carried out unconfined tests with various Savonius rotors located downstream of the exit of a wind tunnel. The maximum rotor coefficient of performance of $C_{p,max}$ ~ 0.23 at a tip speed ratio of λ ~ 1.0 was found to be less than that reported by Modi and Fernando although it was not made clear whether the effects of bearing frictional losses were accounted for in the experiments. More recent studies have also been conducted on a number of geometric variations including stacking rotors one above the other and Rahai has reported on optimisation of the Savonius design using CFD analysis [9].

2.2.2 Darrieus turbines

In 1931 the invention by Darrieus [10] of his rotor with a high tip speed ratio opened up new opportunities for VAWTs in regards to electricity generation. The fundamental step forward made by Darrieus was to provide a means of raising the velocity of the VAWT blades significantly above the freestream wind velocity so that lift forces could be used to significantly improve the coefficient of performance of VAWTs over previous designs based primarily on drag. Darrieus also foresaw a number of embodiments of his fundamental idea that would be trialled at large scale many decades later. These included use of both curved-blade (Fig. 5a) and straight-blade versions of his rotor. He also proposed options for active control of the pitch of the blades relative to the rotor as a whole, so as to optimise the angle of attack

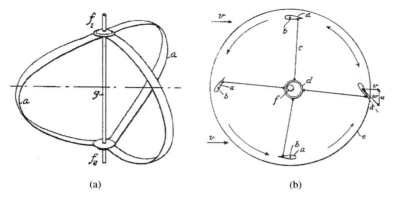

(a)   (b)

Figure 5: Images from the Darrieus VAWT patent [10]: (a) curved-blade rotor
           embodiment; (b) plan view of straight-blade rotor showing an optional
           active blade pitching mechanism.

of the wind on each blade throughout its travel around the rotor circumference (as
shown in Fig. 5b). However, it was not until the energy crisis of the early 1970s
that Darrieus' rotor was developed to the point whereby it became a commercially
viable wind turbine.

The Darrieus turbine can take a number of forms but is most well known in the
geometry sometimes called the "egg-beater" shown in Fig. 5a, where the two or
three blades are curved so as to minimise the bending moments due to centrifugal
forces acting on the blade. The shape of the curved blade is close to that taken by
a skipping rope in the absence of gravity and is known as the Troposkein ("spinning rope"). The Sandia National Laboratory wind research team was one of the
leading groups in development and analysis of the Darrieus curved-blade turbine.
One of the first devices tested was the 2-m diameter Sandia research turbine, which
was tested in a 4.6 m × 6.1 m wind tunnel and later field-tested at the Sandia
National Laboratories wind turbine site [11]. The match between the wind tunnel
and field tests was very close and this study indicated that a maximum power coefficient for this machine of $C_{p,max} \approx 0.32$ was achievable, which was promising for
a relatively small device.

A second 5-m diameter demonstration turbine was then developed, also with a
curved-blade rotor. This was superseded in 1975 by the 17-m diameter Sandia
Demonstration Darrieus turbine which ran successfully over several years and provided important experimental information on the aerodynamics and structural
dynamics of the Darrieus device. In 1988, a much larger 34-m diameter machine,
known at the 34-meter Test Bed, was commissioned at Bushland, Texas. This
device with aluminium blades had a rated power output of 0.5 MW and it provided
a wealth of data on the field performance of a large VAWT [12]. Indeed the Sandia
National Laboratories website [13] remains, at the time of writing, a rich source of
information on many aspects of Darrieus turbines, including analysis, design and
performance.

At least 600 commercial VAWTs were operating in California in the mid-1980s, the vast majority of these were Darrieus machines. The Flowind Corporation was probably the most successful manufacturer of VAWTs during this period and collaborated with Sandia Laboratories on the development of an enhanced Darrieus wind turbine [14] with superior aerodynamic and structural performance. By July 1995 Flowind were operating over 800 VAWTs in the Altamont and Tehachapi passes in California. However, the company's fortunes were to take a turn for the worse and they were bankrupt by 1997.

Canadian researchers also played a major part in the development of utility-scale Darrieus wind turbines. A recent book by Saulnier and Reid [15] details some of the pioneering work carried out by engineers of the Canadian Research Council (CNRC) and the Institut de Recherché d'Hydro-Québec (IREQ). The devices developed and tested included the 225 kW Darrieus turbine with a rotor 24 m in diameter and 36 m in height that was installed on the Magdelen Islands in the Gulf of St. Lawrence in 1977 and operated until 1983. A number of other research Darrieus turbines were also built and tested. A fully instrumented 50 kW Darrieus was constructed at IREQ in 1983 by Daf-Indal (Mississauga, Ontario) which was one of the series of commercial prototypes produced by the company and erected in many provinces of Canada under a program of the National Research Council of Canada [16]. However, the most significant Canadian VAWT project commenced in 1982 when IREQ, CNRC and other collaborators commenced work on the largest VAWT ever built. This was the curved-blade Darrieus Éole turbine rated at 4 MW with a two-bladed rotor, 96 m in height and 64 m in diameter (Fig. 6). The device operated successfully for over 30,000 hr during a 5-year

(a)   (b)

Figure 6:   The world's largest VAWT, the Éole 4MW Darrieus turbine located at Cap-Chat, Quebec: (a) view of the turbine as part of the Le Nordais/Cap Chat Wind Farm [20]; (b) view of the rotor (photograph – copyright Alain Forcione).

period from March 1988 and produced over 12 GWh of electricity in the Le Nordais/ Cap Chat wind farm (which also includes 73 NEG-Micon 750 kW HAWTs at the time of writing) [17].

One of the characteristics of the Darrieus family of turbines is that they have a limited self-starting capacity because there is often insufficient torque to overcome friction at start-up. This is largely because lift forces on the blades are small at low rotational speeds and for two-bladed machines in particular the torque generated is virtually the same for each of the stationary blades at start-up, irrespective of the rotor azimuth angle relative to the incident wind direction. Moreover, the blades of a Darrieus rotor are stalled for most azimuth angles at low tip speed ratios. As a result large commercial machines generally need to be run up to a sufficiently high tip speed for the rotor to accelerate in a given wind velocity. Nevertheless, two-bladed Darrieus machines do have the capacity to self-start, albeit on a somewhat unpredictable basis, and although this is advantageous in most circumstances it can cause problems. A case in point was in 1978 when the 225 kW Magdelen Islands Darrieus turbine was left for a few hours overnight without a brake engaged due to maintenance issues and the belief that such turbines would not self-start. The following morning researchers returned to find the turbine rotating at high speed with no load. As a result an energetic resonance developed in one of the rotor guy wires so that the guy came into catastrophic contact with the rotor which was entirely destroyed [17]. Another scenario where this can be a problem is when the turbine starts and turns initially at low tip speed ratio in a strong wind. If there is then a sudden decrease in the wind speed so the tip speed ratio increases the rotor power output may then be sufficient to cause rapid acceleration and damage may occur [18]. Self-starting capability may be enhanced through a number of strategies including: increasing solidity; using an odd number of blades; providing a form of blade pitch mechanism; and using blades that are skewed so that the blade azimuth angle is a function of axial distance along the rotor. A recent study of the self-starting characteristics of small Darrieus machines has been reported by Hill *et al.* [19].

In terms of noise generation there appears to have been very little experimental or theoretical analysis reported on Darrieus rotors or other types of VAWTs. Since Darrieus turbines have relatively low tip speeds compared to HAWTs one might expect the noise generated to be less problematic, as indicated by the analysis of Iida *et al.* [21].

2.2.3 Straight-blade VAWTs

The name Darrieus is usually associated with the curved-blade version of Darrieus' patent. However, a great deal of work over the past three decades has gone into the development and analysis of the straight-blade version of his original invention, which is sometimes known as the H-VAWT from the shape of the blades and supporting spars. One of the key researchers in the 1970s was Peter Musgrove who spent over 20 years working on wind turbines at the University of Reading.

Musgrove recognised that one of the key challenges facing VAWTs was the need to control the power output of the device at high wind speeds and that active pitch control of blades would result in an unnecessarily complex mechanical system for large devices. His research team developed a furling system whereby the straight blades could be hinged at their mid-point so that the angle of the blades relative to the axis of the rotor could be adjusted by mechanical actuators. A number of geometries were developed (including the V-VAWT machine and tested at small scales (e.g. diameter of order 3 m). However, it should be mentioned that the furling method described above can potentially lead to high transient vertical lift forces due to the effects of turbulence, which may in turn lead to high loads or failure of the supporting radial arms [18].

In the mid-1980s the UK Department of Energy supported the development of several VAWTs based on Musgrove's design. These were developed by a UK company VAWT Ltd. and several prototypes were built at the Carmarthen Bay test site of the Central Electricity Generating Board [22]. The first machine, the VAWT-450, was commissioned in 1986 and it had a 25-m diameter rotor with blades 18 m in length which provided a rated output of 130 kW at a wind speed of 11 m/s.

Subsequently a much larger version of this device was designed and built by VAWT Ltd. and also installed at Carmarthen Bay. The VAWT-850 had a 45-m diameter rotor and a rated output of 0.5 MW. The design did not include a furling system for the blades as previous experience with the VAWT-450 had demonstrated that this was unnecessary due to the inherent ability of the straight-bladed VAWT to avoid overspeed and excessive power generation through stall of the blades at high wind speeds. Although the VAWT-850 was a successful demonstration of the straight-bladed VAWT technology it suffered a catastrophic failure of one of the blades in 1991, apparently due to a manufacturing fault [22].

2.2.4 Giromills

One of the consequences of adopting a vertical axis for any wind turbine is that the apparent velocity of the wind at a particular location on a blade will change throughout each revolution of the rotor (Fig. 7). For example, when the blade is travelling upstream (i.e. when the azimuth angle $0° < \beta < 180°$) the resultant air velocity on the blade is greater than the tangential velocity of the blade relative to a stationary frame of reference, whereas, when the blade is travelling downstream ($180° < \beta < 360°$) the resultant wind speed is generally less than the tangential blade speed. This in turn means that the angle of attack on the blade is continually changing and is generally not optimal throughout its rotation about the axis of the turbine.

To improve this situation various means have been devised to optimise the blade pitch angle (i.e. the chord angle relative to a tangent to the path of the blade) as a function of azimuth angle, β. Systems have been devised to achieve this in a number of ways, including mechanical mechanisms with levers and/or pushrods connected between the blades and the main rotor shaft (as shown in Darrieus' original patent, Fig. 5b) or by means of aerodynamic mechanisms. Such turbines which seek to optimise the blade pitch angle are often known as Giromills [24], although some authors also refer to these systems as cycloturbines.

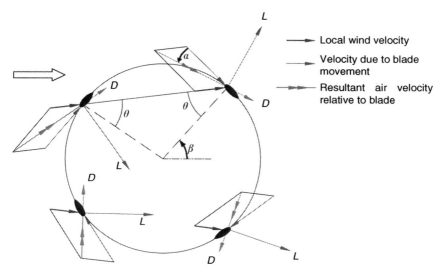

Figure 7: Plan view of velocity vectors and lift, L, and drag, D, vectors for a VAWT with freestream velocity, V_∞, after Sharpe [23].

Kirke [25, 26] conducted an in-depth study of a number of three-bladed giromills with aerodynamic/mechanical activation of the blade pitch mechanism. These devices were referred to as "self-acting variable pitch VAWTs" with straight blades. Each blade was mounted at its mid-span on the end of the rotor radial arm and counterweighted so the mass centre coincided with the pivot axis, located forward of the aerodynamic centre. The pitch mechanism was activated by the moment of the aerodynamic force about a pivot, opposed by centripetal force acting on a "stabiliser mass" attached to the radial arm, such that the aerodynamic force overcomes the stabiliser moment and permits pitching before stall occurs. Lazauskas [27] had previously carried out modelling of three different types of blade pitch actuation mechanism and had predicted significant improvement in turbine performance compared to fixed pitch VAWTs. Wind tunnel tests reported by Kirke and Lazauskas [26] were carried out on a prototype rotor of 2-m diameter and solidity $Nc/R = 0.6$ with three blades each 1.0 m long and with 0.2 m chord and NACA0018 profile. Comparison between the wind tunnel tests and numerical results for various blade pitch scenarios were quantitatively good. Following these laboratory tests a 6-m diameter demonstration turbine was built by Kirke with the three blades being 2.5 m in length using the NACA0018 aerofoil profile (Fig. 8). Unfortunately, this device performed relatively poorly compared to theoretical predictions. This was apparently due to: (i) high levels of freestream turbulence at the test site, so that blades were subjected to variable wind speeds and the relatively massive rotor was slow to respond to sudden gusts and lulls and therefore operated well away from the optimum tip speed ratio most of the time; (ii) low wind speeds and therefore relatively low blade Reynolds number; (iii) high parasitic drag losses.

Figure 8: Three blade variable pitch VAWT developed by Kirke [25] clearly
showing the counterweights incorporated in the blade pitch mechanism
(photograph – copyright Brian Kirke).

2.2.5 Other designs and innovations

The field of wind energy systems has attracted many inventors and researchers
over the years and in the field of VAWT innovation there has been a veritable
plethora of designs put forward and, in many cases, demonstrated. However, there
is little evidence in the academic literature that any of the relatively exotic designs
will eventually be developed to the point of being competitive for electricity
generation when compared with large, three-bladed HAWTs.

A class of VAWTs that have been investigated by a number of inventors include
drag-based machines, or panemones [1]. One interesting device utilises a mechanically pitched blade which pitches through 180° during the course of one revolution
of the rotor. The earliest technical paper related to this type of machine is thought
by the author to be the description of the "Kirsten-Boeing Propellor" by Sachse
[28], which was developed as a propeller for airships. Others have used the same
principle to devise a wind turbine, rather than a propeller, and this is illustrated in
Fig. 9 where it can be seen that the leading edge of a blade becomes the trailing
edge on successive revolutions of the rotor. It can also be seen that such a device
is strictly not a drag machine as lift may play a significant part in the development
of the torque over a substantial fraction of the blade travel around the rotor.

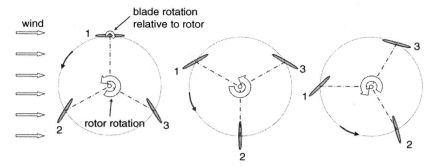

Figure 9: Principle of operation of a VAWT based on the variable pitch "Kirsten-Boeing Propellor" concept [28].

Theoretical modelling of this type of device indicates that the maximum coefficient of performance would be expected to be only $C_{p,max} \sim 0.2$ [29]. Like the Savonius rotor, this rotor does have the advantage of low tip speed ratio which will likely result in less noise and vibration issues, however, the high solidity ratio and hence material cost means that turbines of this type are highly unlikely to be a commercial success.

2.3 VAWTs in marine current applications

One of the hot topics in renewable energy at the time of writing is the development of marine current turbines (MCTs) to harvest the significant potential of tidal currents in various locations around the world. These devices are also known as hydrokinetic turbines, which include those operating on the same principles but in rivers and estuaries. Areas such as the English Channel and the north coast of Ireland have been identified as having great potential for this technology. The most common technology currently being applied in the field is that of horizontal axis MCTs such as that developed by Marine Current Turbines Ltd. [30] and Open-Hydro [31]. However, various research groups have investigated the feasibility of using vertical-axis MCTs which have obvious advantages in this application, particularly in that a yaw mechanism is not required to align the turbine with the wind [32–36].

3 Analysis of VAWT performance

As in the case of HAWTs, there are a number of levels of complexity with which one might analyse the performance of the VAWTs as outlined by authors including Touryan *et al.* [37], Strickland [38] and Wilson [24]. Allet *et al.* [39] classified the four main approaches to modelling of VAWTs as: (i) momentum models; (ii) vortex models; (iii) local circulation models; and (iv) viscous models, where the latter would include full viscous flow computational fluid dynamics (CFD) models. Other bibliographic sources of information on analysis methods include those of Abramovich [5] and Islam *et al.* [40]. An extremely useful and relatively simple momentum model uses the actuator disc/blade element momentum theory (or strip theory) in a similar manner to the momentum model used for analysis of HAWTs. This type of analysis is outlined below and can be implemented relatively easily by anyone with a basic knowledge of fluid mechanics.

3.1 Double-multiple-streamtube analysis

The performance of a VAWT may be estimated using a momentum analysis at one of a number of levels of complexity [24]. The simplest approach is where a single streamtube is considered in which the interaction between the air flow and rotor is treated as a single actuator disc located on the axis of the rotor, perpendicular to the incident air flow. However, it is preferable to use a multiple-streamtube analysis since the resultant air velocities and forces acting on the blades are strong functions of the blade azimuth angle, β (Fig. 7). Strickland was one of the pioneers of this approach [41]. However, to more accurately represent the flow through the rotor, a VAWT may be represented as an "actuator cylinder", whereby the cylinder surface is swept by the length of the rotating blades as shown schematically in Fig. 10. Air in the freestream interacts first with the upstream half of the cylinder and then the downstream half. A number of authors have described this double-multiple-streamtube methodology including Paraschivoiu [42–44] and Madsen [45]. Sharpe [23] in particular presented a very clear exposition of this method and his approach is summarised in the following section.

Using a methodology analogous to the actuator disc/momentum analysis for a HAWT one can define two velocity induction factors, a_u and a_d, which define the deceleration of the local air velocity relative to the freestream velocity, U_∞, on the upstream and downstream faces of the cylinder, respectively. A number of

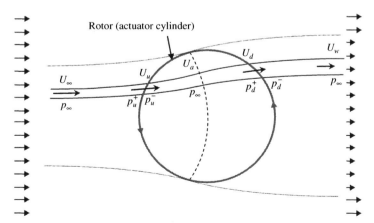

Figure 10: Plan view of a double-multiple-streamtube analysis of the flow through a VAWT rotor, after Sharpe [23].

streamtubes of rectangular cross-section are then followed from the upstream undisturbed flow to the upstream half of the cylinder where the flow in a particular streamtube is retarded to U_u such that:

$$U_u = U_\infty (1 - \alpha_u) \tag{1}$$

A local pressure drop from p_u^+ to p_u^- occurs across the upstream face of the actuator cylinder. Following the streamtube downstream the pressure recovers so that at a given point somewhere between the upstream and downstream faces of the cylinder the static pressure returns to the freestream value, p_∞ (Fig. 10). Here the local air velocity, U_a, is assumed to be

$$U_a = U_\infty (1 - 2a_u) \tag{2}$$

There is a further pressure drop at the downstream interaction with the actuator cylinder and the pressure then again recovers to the freestream datum some distance downstream in the wake which flows at velocity U_w:

$$U_d = U_a (1 - a_d) \quad \text{and} \quad U_w = U_a (1 - 2a_d) \tag{3}$$

The retardation of the flow through the domain of interest leads to an expansion in the cross-sectional areas of the streamtube, from A_u to A_d, through the rotor. The effect of this expansion is accounted for explicitly in the calculation of the magnitude and direction of the resultant wind direction at the turbine blades as illustrated schematically in Fig. 11.

In the case of VAWTs it is generally useful to resolve the lift and drag forces acting on the blades into components acting normally (radially) and tangentially (chordwise), F_n and F_t, respectively. In the case of a VAWT with blades of fixed pitch, the chord of the blades is generally held perpendicular to the radius from the rotor axis. The lift and drag coefficients, C_L and C_D, for a given aerofoil section may be manipulated to provide the non-dimensional normal

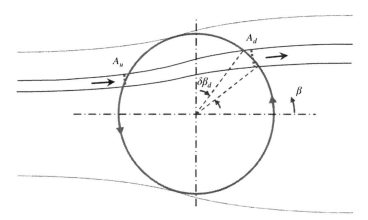

Figure 11: Illustration of streamtube expansion and nomenclature, after Sharpe [23].

and tangential coefficients, C_n and C_t, as a function of the angle of attack, α, as follows:

$$C_n = C_L \cos\alpha + C_D \sin\alpha, \quad C_t = C_L \sin\alpha - C_D \cos\alpha \tag{4}$$

The force acting on a blade in the direction of the local resultant air flow, W, is then:

$$F = \frac{\rho W^2}{2} c(C_n \cos\theta - C_t \sin\theta) \tag{5}$$

where ρ is the density of the air, W is the local resultant air velocity, c is the chord length and θ is the angle between the streamtube and the local radius to the rotor axis (see Fig. 7). Using the impulse-momentum principle the forces on the blade at the upstream and downstream portions of the cylinder may then be computed and related to the resulting deceleration of the flow to give expressions for the upstream and downstream induction factors:

$$a_u(1-a_u) = \frac{Nc}{8\pi R}\frac{W_u^2}{U_\infty^2}\sec\theta(C_{nu}\cos\theta + C_{tu}\sin\theta),$$

$$a_d(1-a_d) = \frac{Nc}{8\pi R}\frac{W_d^2}{U_\infty^2}\sec\theta(C_{nd}\cos\theta + C_{td}\sin\theta) \tag{6}$$

where N is the number of blades and R is the radius of the rotor. Equation (6) is solved iteratively together with the following auxiliary eqns (7) and (8) so as to find the unknown parameters in the problem. The angles of attack on the blade at the upstream and downstream locations are given by:

$$\tan\alpha_u = \frac{U_\infty(1-a_u)\cos\theta}{\Omega R + U_\infty(1-a_u)\sin\theta}, \quad \tan\alpha_d = \frac{U_a(1-a_d)\cos\theta}{\Omega R + U_a(1-a_d)\sin\theta} \tag{7}$$

where $\Omega = d\beta/dt$ is the angular velocity of the rotor. The local resultant relative velocities are then given by:

$$W_u = \sqrt{(\Omega R + U_\infty(1-a_u)\sin\theta)^2 + (U_\infty(1-a_u)\cos\theta)^2},$$

$$W_d = \sqrt{(\Omega R + U_a(1-a_d)\sin\theta)^2 + U_a(1-a_d)\cos\theta)^2} \tag{8}$$

The torque, Q, generated by the blade for each of the streamtubes may then be estimated from:

$$Q_u = \frac{\rho W_u^2}{2}\frac{Nc}{2\pi R}A_u\sec\theta\left(C_{tu}\cos\theta + \frac{C_{nu}}{4}\right),$$

$$Q_d = \frac{\rho W_d^2}{2}\frac{Nc}{2\pi R}A_d\sec\theta\left(C_{td}\cos\theta + \frac{C_{nd}}{4}\right) \tag{9}$$

and hence the total torque and shaft power from the rotor may be determined by integration of eqn (9) around the circumference of the rotor.

The remaining task is to relate the forces acting on the blade and the torque generated to the local rotor azimuth angle, β. However, independent variables, such as W, a_u and a_d, have to this point been calculated only as a function of the angle θ between the streamtube and the local radius arm from the rotor axis. The blade azimuth angle β is related to θ through the degree of expansion of all the streamtubes passing through the rotor. It is a relatively simple matter to determine β after the local streamwise velocities, U_u and U_d, have been found as functions of θ and the corresponding azimuth angles may then be computed as detailed by Sharpe [23]:

$$\beta_d(\theta) = \int_0^\theta \frac{2U_u}{U_u + U_d}\,d\theta, \quad \beta_u(\theta) = \beta_d\left(\frac{\pi}{2}\right) + \int_\theta^{\pi/2} \frac{2U_d}{U_u + U_d}\,d\theta \tag{10}$$

The double-multiple-streamtube model described above has been implemented by the present author and illustrative results are presented in Figs 12–14 for a straight-bladed VAWT with a rotor radius of 20 m and blades based on the NACA0012 profile with a rotor solidity of $\sigma = Nc/R = 0.15$. Lift and drag data have been taken at an average blade Reynolds number of $Re_m = 1.0 \times 10^6$ from the data provided by a key publication on the lift and drag characteristics of aerofoils for VAWTs [46].

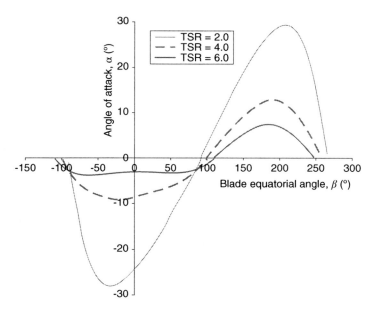

Figure 12: Angle of attack of resultant wind velocity, α, as a function of blade azimuth angle, β, and tip speed ratio. Predicted from a double-multiple-streamtube analysis of a straight-bladed VAWT (NACA0012H blade profile, $Re = 1.0 \times 10^6$, $R = 10$ m, $c = 0.5$ m, number of blades $N = 3$, $\Omega = 3.14$ rad/s).

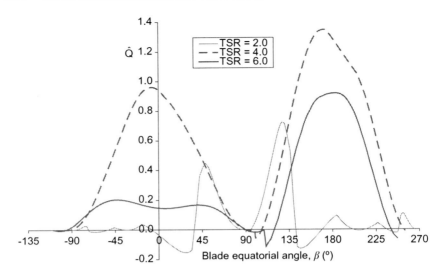

Figure 13: Non-dimensional torque coefficient, $\hat{Q}$, as a function of blade azimuth angle, β estimated from a double-multiple-streamtube blade element analysis of a VAWT (parameters as for Fig. 12).

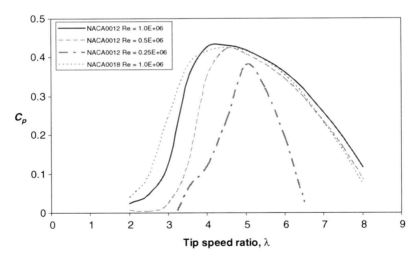

Figure 14: Double-multiple-streamtube results for the coefficient of performance, C_p, as a function of tip speed ratio, λ, for a straight-bladed VAWT turbine operating with different mean blade Reynolds numbers (solidity $cN/R = 0.15$, NACA0012 and 0018 profiles, with lift and drag data from [46]).

Figure 12 shows how the angle of attack on a blade varies throughout its travel about the rotor axis. The range of angle of attack is seen to decrease with increasing tip speed ratio. The non-dimensional torque per unit blade span generated at each azimuth angle, β, is shown in Fig. 13. This is a complex characteristic particularly at low tip speed ratios, λ. The complexity of the torque profile arises in part from the fact that NACA0012 blades will stall under steady-state flow conditions for any angle of attack, α, greater than approximately 14°. On the other hand, for the higher tip speed ratios of 4.0 and 6.0, the blade does not stall and there is a positive torque produced for the vast majority of azimuth angles. However, at a tip speed ratio of 2.0 the blades pass in or out of stall at four azimuth angles ($-75°$, 45°, 134° and 251°) and the blade is stalled for a very significant fraction of the total travel, which in turn results in limited overall torque generated and hence only a modest coefficient of performance, C_p. Although stalling of the blades in this way reduces C_p and causes significant fatigue loads, it does mean that an electrical generator connected to the rotor will benefit from passive overspeed protection at high wind speeds.

For the particular example chosen here with a mean blade Reynolds number of $Re_m = 1.0 \times 10^6$ the maximum coefficient of performance is $C_{p,max} \approx 0.43$ at an optimal tip speed ratio of $\lambda \approx 4$ as shown in Fig. 14. The performance of a VAWT rotor is strongly dependent on the blade Reynolds number as illustrated in Fig. 14 which serves to show that as the physical scale of a turbine is reduced so the maximum coefficient of performance decreases and the same is true of the range of tip speed ratios over which the turbine performs effectively. It could be said that VAWTs are particularly susceptible to reduction of C_p at low Reynolds numbers, since a lower Reynolds number limits the maximum lift coefficient that can be achieved with increasing angle of attack prior to stall. Thus, the effect of Reynolds number on the performance of small turbines may be more important for VAWTs as compared to HAWTs. It can also be seen that the thickness of the aerofoil has some influence on the C_p versus λ characteristic of the turbine (the NACA0012 aerofoil having a maximum thickness of 12% of the blade chord as opposed to 18% for the NACA0018).

The performance estimates from the double-multiple-streamtube methodology presented here do not account for a number of effects that may significantly reduce the output from a VAWT in practice. Parasitic drag loss is one of the key parameters that should be modelled by the designer of a VAWT. The loss of net power output due to the presence of components such as the radial arms on which the blades of a straight-bladed VAWT are mounted may be significant. Modelling such losses using a momentum model is relatively straightforward as the drag coefficients for beams and streamlined spars are well known [47]. These losses become increasingly important as the physical scale of the turbine is reduced. In addition, care should be taken in the interpretation of results from multiple-streamtube momentum models, particularly at high tip speed ratios, where large induction factors may be calculated which in turn lead to unrealistic wake velocity results.

3.1.1 Double-multiple-streamtube analysis of curved-blade VAWTs

The double-multiple-streamtube method described above considered a blade element located at a given radius from the rotor axis. This radius would be constant over the length of each blade for a straight-bladed VAWT. It is a relatively simple matter to adapt the equations above to other VAWTs such as curved-blade Darrieus machines where the blade radius is a function of elevation. Many curved-blade Darrieus machines have been constructed with a variant of the Troposkien blade shape where the ends of each blade comprise straight sections and the middle section has a constant radius. Whatever the actual shape of the blade may be the resultant velocity at a particular blade element is a function of the elevation from the mid-plane of the rotor (Fig. 15) and we can define $\zeta(z)$ as the angle of the blade element to the vertical.

Since only the wind velocity component normal (not spanwise) to the blade results in lift and drag forces, we require the magnitude of the local resultant velocity normal to the plane of the blade element, W, as illustrated in Fig. 15. W is then given by:

$$W = \sqrt{(\Omega r + U_\infty(1-a)\sin\theta)^2 + (U_\infty(1-a)\cos\theta\cos\zeta)^2} \qquad (11)$$

where $r(z)$ is the local radius of the blade element. The aerodynamic forces acting on the blade element in the horizontal plane can be determined by modifying eqn (5) so as to account for angle ζ as follows [23]:

$$F = \frac{\rho W^2}{2}c(C_n\cos\theta\cos\zeta - C_t\sin\theta) \qquad (12)$$

When analysing a curved-blade VAWT using the double-multiple-streamtube model, eqns (6) – (9) must also be modified so that C_n is replaced by $C_n\cos\zeta$, sec θ is replaced by sec θ sec ζ and the maximum rotor radius, R, is replaced by the local radius, r.

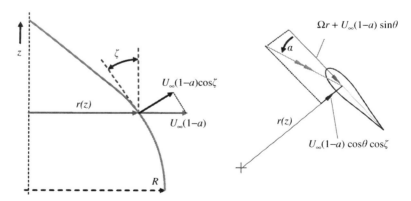

Figure 15: Schematic of kinematics of blade and wind velocity on a curved VAWT blade: (a) elevation of top half of blade showing component of wind acting normal to the plane of a blade element; (b) resultant velocities acting normal to the blade.

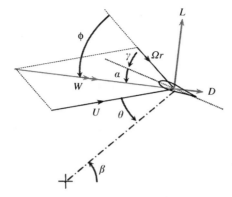

Figure 16:   Illustration of the kinematics of a variable pitch VAWT blade.

3.1.2 Variable pitch VAWTs

In the double-multiple-streamtube model described above the blade pitch is held constant with respect to azimuth angle, β, i.e. the chord is perpendicular to the radius arm of the rotor. However, it is a relatively simple matter to modify the double-multiple-streamtube model to incorporate passive or active pitch control [27, 29]. In addition, one can also model the effect that some investigators have reported whereby an improvement in performance for fixed pitch turbines can be achieved if there is a slight *toe out* of the blades, as this reduces stall on the upstream pass. However, the resolution of the lift and drag forces into the appropriate tangential and normal components can be algebraically tedious because of the need to introduce new parameters for the blade pitch angle, γ, and resultant wind velocity angle, ϕ, as illustrated in Fig. 16.

3.1.3 Flow curvature and dynamic stall

The double-multiple-streamtube momentum model described above is a quasi-steady-state model which relies on the lift and drag characteristics of the aerofoils determined generally from steady-state wind tunnel tests or from inviscid or viscous numerical simulations. It follows that this model does not inherently capture a number of flow phenomena that occur in VAWTs in practice, for example, flow curvature and dynamic stall.

The issue of "flow curvature" relates to the fact that the apparent air motion relative to a blade of a VAWT has a curvature due to the rotation of the blade about the rotor axis. This in effect changes the apparent angle of attack on the blade and can be treated from a quasi-steady standpoint. The rate of pitching of the blade relative to the undisturbed flow is equal to the rotational velocity of the rotor, Ω. Sharpe [23] proposes a correction to the normal force coefficient, δC_n, based on thin aerofoil theory to account for the pitching of the blade such that $\delta C_n = (dC_L/d\alpha)(c/R)(\Omega R/W)/4$. An indication of the magnitude of this effect is provided by Wilson [24] using previous work carried out at the Sandia National Laboratories, which showed that flow curvature may result in an offset in the apparent angle of attack on a VAWT blade of the order of $3°$ for $\lambda = 3.5$ and $c/R = 0.2$. Flow curvature has the effect of increasing the normal force on the blades on the upstream half of the actuator cylinder and decreasing this force on the downstream half [23].

The second important issue is what is commonly known as "dynamic stall". This complex transient phenomenon arises because of the rapidly changing angle of attack on a VAWT blade. At low tip speed ratios a hysteresis effect arises whereby stall occurs at higher angles of attack, α, than for steady-state flow (when α is increasing). Subsequent reattachment of the flow is also delayed for decreasing α. A number of empirical and theoretical models have been developed by authors such as Allet et al. [39], Oler et al. [48], Major and Paraschivoiu [49] and Liu et al. [50]. These models may be incorporated into a double-multiple-streamtube analysis to improve the modelling of the transient effects of stall and also into more complex vortex models such as the Sandia codes [51]. More recently Ferreira et al. [52] have reported on a detailed flow visualisation study of the dynamic stall phenomenon.

3.1.4 Application and limitations of the double-multiple-streamtube method

The double-multiple-streamtube analysis described above is relatively straight-forward and can provide quantitative results that are useful for optimisation of VAWT geometry in terms of fundamental parameters such as: operating tip speed ratio, blade profile, rotor solidity and aspect ratio. The model may also be used to estimate the forces on the blades which can then form the input to structural analysis and optimisation of the rotor. The accuracy of the double-multiple-streamtube model is comparable to that of more complex analysis methods. Sharpe [23], for example, showed that his prediction of $C_{p,max}$ for the Sandia 17-m diameter Darrieus turbine was within a few percent of the experimental results reported by Worstell [53]. Wilson [24] presents a comparison of the results from a number of double-multiple-streamtube analyses with the experimental data of Worstell [53] and very good agreement is seen for tip speed ratios, $\lambda > 3$.

It should be noted that care must be taken in the application of the momentum analysis methodology. In particular, it is possible for high induction factors to be predicted for VAWTs operating at high tip speed ratios which may lead to unrealistic wake flows. Corrections for some other flow phenomena not dealt with above, such as tip losses, can also be incorporated in the double-multiple-streamtube methodology.

3.2 Other methods of VAWT analysis

Inviscid flow models have been used by many of the key researchers in the field of VAWT analysis in years past and this approach has been summarised in the key overview article by Wilson [24]. While fixed wake models are relatively straight-forward to implement, free vortex simulations are extremely complex and costly in terms of computer processing time. Nevertheless the free vortex model methodology is accepted to be the most comprehensive and accurate method of modelling VAWTs [51]. This methodology has also been recently applied to the analysis of vertical-axis marine current turbine by Li and Calisal [54]. CFD codes have now developed to the point where this viscous flow analysis tool is available to most researchers in the academic and commercial sectors. However, application of this tool to VAWTs is not straightforward as full transient analysis and significant mesh refinement are necessary for meaningful results. CFD analysis of VAWTs does not appear to have been widely reported in the literature to date, although the research team at the École Polytechnique de Montréal have previously reported on their development of several viscous analysis codes for VAWTs [39] and more recently articles have appeared on CFD analysis of vertical-axis marine current turbines [33, 34].

4 Summary

This chapter has summarised the principles of operation and the historical development of the main types of VAWT. The Darrieus turbine remains the most promising of the vertical-axis rotor types for application to the utility-scale generation of electricity. The intense period of research, development and demonstration during the 1970s and 1980s did not lead to the development of a technology that is able to compete commercially with the three-bladed HAWTs that have come to dominate the market at large scale. Nevertheless new opportunities are opening up in the areas of marine current turbines and building-integrated wind turbines where the VAWTs may yet be competitive. In principle, the aerodynamic analysis of VAWTs is more complicated than that of HAWTs due to the significant variation of air velocity as a function of blade azimuth angle. The double-multiple-streamtube analysis summarised herein provides a tool that is relatively straightforward to use for those wishing to undertake an analysis of conventional VAWT designs.

References

[1]   Golding, E.W. & Harris, R.I., *The Generation of Electricity by Wind Power*, New York: John Wiley, 1976.

[2]   Shepherd, D.G., Historical development of the windmill. In: Spera D.A., ed. *Wind Turbine Technology: Fundamental Concepts of Wind Turbine Engineering*. New York: ASME, pp. 1–46, 1994.

[3]   Müller, G., Jentsch, M.F. & Stoddart, E., Vertical axis resistance type wind turbines for use in buildings. *Renewable Energy*, **34**, pp. 1407–1412, 2009.

[4]   King, F.H., *Farmers of Forty Centuries: Organic Farming in China, Korea, and Japan*, Courier Dover Publications, 2004.

[5]   Abramovich, H., Vertical axis wind turbines: a survey and bibliography. *Wind Engineering*, **11(6)**, pp. 334–343, 1987.

[6]   Savonius, S.J., Rotor adapted to be driven by wind or flowing water, US Patent no. 1697574, 1929.

[7]   Modi, V.J. & Fernando, M.S.U.K., On the performance of the Savonius wind turbine. *Journal of Solar Energy Engineering*, **111**, pp. 71–81, 1989.

[8]   Ushiyama, I. & Nagai, H., Optimum design con`uration and performance of Savonius rotors. *Wind Engineering*, **12(1)**, pp. 59–75, 1988.

[9]   Rahai, H.R., *Development of optimum design configuration and perfor-mance for vertical axis wind turbine: feasibility analysis and final report*, Long Beach: California State University, 2005.

[10] Darrieus, G.J.M. (inventor), Turbine having its rotating shaft transverse to the flow of the current. US Patent No. 1835018, 1931.

[11] Sheldahl, R.E., *Comparison of field and wind tunnel Darrieus wind turbine data*, Albuquerque, New Mexico: Sandia National Laboratories, Report No.: SAND80-2469, 1981.

[12] Ashwill, T.D., *Measured data for the Sandia 34-meter vertical axis wind turbine*, Albuquerque: Sandia National Laboratories, Report No.: SAND91-2228, 1992.

[13] Sandia, Sandia National Laboratories. http://www.sandia.gov/

[14] Sandia, *High energy rotor development: test and evaluation*, Albuquerque, New Mexico: Sandia National Laboratories, Report No.: SAND96-2205, 1996.

[15] Saulnier, B. & Reid, R., *L'Éolien: au coeur de l'incontournable révolution énergétique*: Multimondes, 2009.

[16] Saulnier, B., Personal communication, 2009.

[17] Forcione, A., Personal communication, 2009.

[18] Kirke, B.K., Personal communication, 2009.

[19] Hill, N., Dominy, R., Ingram, G. & Dominy, J., Darrieus turbines: the phys-ics of self-starting. *Proceedings of the Institute of Mechanical Engineering, Part A: Journal of Power and Energy*, **223(1)**, pp. 21–29, 2009.

[20] Accessed 14th August 2009; http://en.wikipedia.org/wiki/Darrieus_wind_turbine

[21] Iida, A., Mizuno, A. & Fukudome, K., Numerical simulation of aerodynamic noise radiated from vertical axis wind turbines. *Proceedings of the 4th ASME/JSME Joint Fluids Engineering Conference*, pp. 63–69, 2003.

[22] Price, T.J., UK large-scale wind power programme from 1970 to 1990: the Carmarthen Bay experiments and the Musgrove vertical-axis turbines. *Wind Engineering*, **30(3)**, pp. 225–242, 2006.

[23] Sharpe, D., Wind turbine aerodynamics. In: Freris L, ed. *Wind Energy Conver-sion Systems*. New York: Prentice Hall, pp. 54–117, 1990.

[24] Wilson, R.E., Aerodynamic behavior of wind turbines. In: Spera D.A., ed. *Wind Turbine Technology: Fundamental Concepts of Wind Turbine Engineering*. New York: American Society of Mechanical Engineers, pp. 215–282, 1994.

[25] Kirke, B.K., *Evaluation of self-starting vertical axis wind turbines for stand-alone applications* [PhD]. Gold Coast: Griffith University (Australia); 1998.

[26] Kirke, B.K. & Lazauskas, L., Experimental verification of a mathematical model for predicting the performance of a self-acting variable pitch vertical axis wind turbine. *Wind Engineering*, **17(2)**, pp. 58–66, 1993.

[27] Lazauskas, L., Three pitch control systems for vertical axis wind turbines compared. *Wind Engineering*, **16(5)**, pp. 269–282, 1992.

[28] Sachse, H., *Kirsten-Boeing Propeller*, Washington: Technical Memorandums National Advisory Committee for Aeronautics, Report No.: 351, 1926.

[29] Cooper, P., Kennedy, O.C. & Whitten, G., Aerodynamics of a novel active blade pitch vertical axis wind turbine. *Proc. IX World Renewable Energy Congress*, Florence, Italy: WREC, p. 6, 2006.

[30] MCT, Marine Current Turbines Ltd. Accessed 18th May 2009; http://www.marineturbines.com/

[31] OpenHydro, Accessed 22nd May 2009; http://www.openhydro.com/

[32] Camporeale, S.M. & Magi, B., Streamtube model for analysis of vertical axis variable pitch turbine for marine currents energy conversion. *Energy Conversion & Management*, **41**, pp. 1811–1827, 2000.

[33] Ishimatsu, K., Kage, K. & Okubayashi, T., Numerical trial for Darrieus-type alternating flow turbine. *Proc. 12th Int. Offshore and Polar Engineering Conf.*, Kitakyushu, Japan: ISOPE, 2002.

[34] Gretton, G.I. & Bruce, T., Aspects of mathematical modelling of a proto-type scale vertical-axis turbine. *Proc. 7th European Wave and Tidal Energy Conference*, Porto, Portugal, 2007.

[35] Gorlov, A., Helical turbine as undersea power source. *Sea Technology*, **38(12)**, pp. 39–43, 1997.

[36] Kirke, B.K. & Lazauskas, L., Variable pitch Darrieus water turbines. *Journal of Fluid Science and Technology*, **3(3)**, pp. 430–438, 2008.

[37] Touryan, K.J., Strickland, J.H. & Berg, D.E., Electric power from vertical-axis wind turbines. *J. Propulsion and Power*, **3(6)**, pp. 481–493, 1987.

[38] Strickland, J.H., A review of aerodynamic analysis methods for vertical-axis wind turbines, *Proc. 5th ASME Wind Energy Symposium*, New Orleans, USA, pp. 7–17, 1986.

[39] Allet, A., Brahimi, M.T. & Paraschivoiu, I., On the aerodynamic modelling of a VAWT. *Wind Engineering*, **21(6)**, pp. 351–365, 1997.

[40] Islam, M., Ting, D.S.-K. & Fartaj, A., Aerodynamic models for Darrieus-type straight-bladed vertical axis wind turbines. *Renewable and Sustainable Energy Reviews*, **12**, pp. 1087–1109, 2008.

[41] Strickland, J.H., *The Darrieus turbine: a performance prediction model using multiple streamtubes*, Albuquerque, New Mexico: Sandia National Laboratories, Report No.: SAND75-0431, 1975.

[42] Paraschivoiu, I., Desy, P., Masson, C. & Beguier, C., Some refinements to aerodynamic-performance prediction for vertical-axis wind turbines. *Proc. of the Intersociety Energy Conversion Engineering Conf.*, pp. 1230–1235, 1986.

[43] Paraschivoiu, I., Double-multiple streamtube model for studying vertical-axis wind turbines. *Journal of Propulsion and Power*, **4(4)**, pp. 370–377, 1988.

[44] Paraschivoiu, I. & Desy, P., Aerodynamics of small-scale vertical-axis wind turbines. *Proc. Intersociety Energy Conversion Eng Conf.*, pp. 647–655, 1985.

[45] Madsen, J.A., The actuator cylinder: a flow model for vertical axis wind turbine. *Proc. of the 7th British Wind Energy Association (BWEA) Conf.*, Oxford, UK, pp. 147–154, 1985.

[46] Sheldahl, R.E. & Klimas, P.C., *Aerodynamic characteristics of seven symmetrical airfoil sections through 180-degree angle of attack for use in aerodynamic analysis of vertical axis wind turbines*, Albuquergue, NM: Sandia National Laboratories, Report No.: SAND80-2114, 1980.

[47] Eppler, R., *Airfoil Design and Data*, Heidelberg: Springer-Verlag, 1990.

[48] Oler, J.W., Strickland, J.H., Im, B.J. & Graham, G.H., *Dynamic stall regulation of the Darrieus turbine*, Albuquerque, New Mexico: Sandia National Laboratories, Report No.: SAND83-7029, 1983.

[49] Major, S.R. & Paraschivoiu, I., Indicial method calculating dynamic stall on a vertical axis wind turbine. *Journal of Propulsion and Power*, **8(4)**, pp. 909–911, 1992.

[50] Liu, W.-Q., Paraschivoiu, I. & Martinuzzi, R., Calculation of dynamic stall on Sandia 34-m VAWT using an Indicial Model. *Wind Engineering*, **16(6)**, pp. 313–325, 1992.

[51] Berg, D.E., Recent improvements to the VDART3 VAWT code. *Proc. 1983 Wind and Solar Energy Conf.*, Columbia, MO, pp. 31–41, 1983.

[52] Ferreira, C.S., van Kuik, G., van Bussel, G. & Scarano, F., Visualization by PIV of dynamic stall on a vertical axis turbine. *Experiments in Fluids*, **46**, pp. 97–108, 2009.

[53] Worstell, M.H., *Aerodynamic performance of the 17-meter-diameter Darrieus wind turbine*, Albuquerque, New Mexico: Sandia National Laboratories, Report No.: SAND78-1737, 1979.

[54] Li, Y. & Calisal, S.M., Preliminary results of a vortex method for standalone vertical axis marine current turbine. *OMAE2007, Proc. 26th International Conference on Offshore Mechanics and Arctic Engineering*, San Diego, California, pp. 589–598, 2007.

CHAPTER 9

Direct drive superconducting wind generators

Clive Lewis
Converteam UK Ltd., Rugby, Warwickshire, UK.

There are plans for a large expansion of offshore wind energy, particularly in Northern Europe where there is limited space for onshore turbines. One means to reduce the cost of offshore wind energy is to build wind farms with fewer larger turbines, reducing the number of costly offshore foundations. The emerging next generation of HTS technology, which offers the prospect of low cost high volume HTS wire production, can be used to build compact and lightweight generators at high rating and torque. These new generators will become the enabler for very large, direct drive wind turbines in the 10 MW class. Direct drive turbines also offer an improvement in reliability *and* efficiency by removing the gearbox, which has been a troublesome component in many offshore wind farm projects, and replacing it with a much simpler mechanical system that is not sensitive to the misalignment or to fluctuations in the shaft torque. Reliability is particularly important in offshore turbines where access is difficult and expensive and often prevented by weather conditions. Converteam UK Ltd. are in the final stages of a project to design a direct drive HTS generator for this class of turbines, and to build and test a scaled prototype. Following on from this project will be the manufacture of a full size prototype and its demonstration on a 10 MW turbine. An economic analysis during earlier stages of the project calculated a reduction in the cost of energy of 17% from a 500 MW offshore wind farm by the use of this class of HTS direct drive turbines compared with the baseline case of 4 MW conventional DFIG turbines. This analysis did not include any additional cost reduction due to improved reliability and availability.

1 Introduction

The wind turbine market is large and rapidly growing; while at the same time there has been a trend towards larger and larger turbines. Larger turbines are attractive to the new generation of offshore wind farms currently under construction and planning. Recently there has been a development of direct drive generators by a number of turbine manufacturers in order to simplify the mechanical drive train and avoid reliability problems with gearboxes. Direct drive generators are required to operate at the very low speed of the turbine rotor, and hence very high torque. Since it is torque rather than power that predominantly determines the size of the generator, they are significantly larger than high speed generators. To date, most commercial direct drive generators have been very large conventional synchronous machines, most notably from the turbine manufacturer Enercon, but recently there has been the introduction of permanent magnet direct drive generators that are smaller and lighter for the same rating.

During the 1990s electrical machines began to be developed using high temperature superconducting (HTS) materials, the attraction being a significant reduction in the size and weight of the machines. Recently a 36.5 MW, 120 rpm HTS propulsion motor for the US Navy was tested, with a shaft torque similar to that of existing direct drive wind turbine generators. However, the cost of HTS wire has been too high for a cost sensitive market such as wind energy. A second generation of HTS wire (sometimes known as tape) is beginning to come into commercial production, offering an order of magnitude reduction in the cost of HTS wire when produced in volume. This new type of HTS wire opens up the possibility of using HTS generators in wind turbines to make the next significant step up in turbine power rating without the additional penalty of higher mass at the top of the tower.

2 Wind turbine technology

2.1 Wind turbine market

The wind energy market has been growing rapidly since the mid 1990s, with new installed capacity growing at an average rate of 28% in the years 1997–2004, and an average of 34% in the years 2005–2007 [1]. Total installed capacity stood at 93 GW by the end of 2007, and the wind industry expect this to increase to between 490 and 2400 GW by 2030 [2].

Along with this growth in installed capacity there has been a growth in the size and rating of wind turbines, with the largest turbines now being installed (2008) rated at 5 MW. Larger turbines are under development, with Clipper Windpower Plc developing the 7.5–10 MW Britannia turbine in the UK for the offshore wind market [3].

The UK, like many northern European countries is densely populated, so the number of acceptable sites for onshore wind farms is limited. However, the UK is surrounded by large areas of shallow sea with some of the best wind resource in the world. European countries are committed to increasing the share of renewables in energy consumption to 20% by 2020 [4]. In the UK and many other northern European countries, a large proportion of this energy is planned to come from offshore wind, and it is the UK government's intention to install up to 33 GW of offshore wind power to meet the 2020 target [5].

The requirements of the offshore market differ from those of the onshore market. Since the installation cost, and the cost of access for maintenance is significantly higher for offshore turbines, there is a preference for fewer higher power turbines in order to reduce the number of installations. The higher towers would also result in higher average hub height wind speed. Most of the offshore wind farms put into service around the UK to date have used 2, 3 or 3.6 MW turbines. The cost of turbine foundations is a significant proportion of the offshore wind farm, and since the mass of the nacelle has a significant impact on the cost of the most common monopole foundation, a low nacelle mass is important.

Due to the cost and limited availability of access to offshore turbines, reliability is most important. If a failure occurs in a turbine in the North Sea in the winter, the time of maximum energy production, it may be weeks or months before a suitable weather window provides access to enable major equipment repair.

2.2 Case for direct drive

Early wind turbines had low power ratings (100 kW or less) and typically used a fixed speed induction generator (a standard industrial induction motor), driven though a speed increasing gearbox. The turbine power was limited in high wind speeds by progressive aerodynamic stall of the blades. As turbine ratings increased to more than a few hundred kilowatts, the advantages of using a variable rotor speed with blade pitch control became apparent. The most popular solution was the doubly fed induction generator (DFIG), in which the stator is directly connected to the grid, and the rotor power (30–50% of the total power) is fed to and from the grid through sliprings and a variable frequency power converter [6]. This arrangement had the advantage of a smaller power converter at a time when the cost of power electronics was high. With turbine ratings of 2 MW or more, and with wind energy beginning to contribute a significant proportion of the grid generating capacity in some countries and the falling cost of power electronics, the fully fed generator became attractive. The DFIG began to be replaced with either a cage induction generator or a synchronous generator, and all of the power was transferred to the grid via a variable frequency converter. This system offered a number of advantages: the generator no longer had slip rings that required regular maintenance, and the fully fed converter made it easier to implement ride through grid fault capability and continue generation once the fault had cleared – essential for the security of power supply when wind contributes a significant proportion of generating capacity. Unlike the directly connected stator windings of the DFIG, the converter isolated the fully fed stator windings from the grid, so offered greater protection from grid faults for the generator, as well as the turbine mechanical components.

The speed increasing gearbox is a complex mechanical system requiring good mechanical alignment for reliable operation. It also has lubrication and cooling systems requiring maintenance. Gearboxes have been responsible for reliability problems in many wind turbines in the past. One investigation [7] found that gearbox development had not kept pace with the increasing size of wind turbines, results in more reliability problems with the newer larger turbines. The 2006 annual

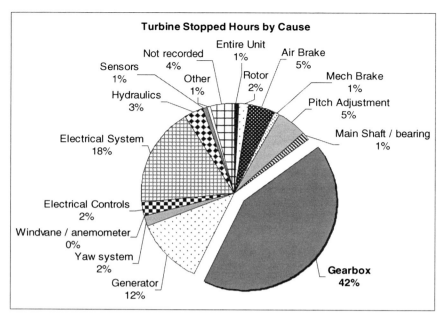

Figure 1: Down time for turbines in Germany.

report for the Kentish Flats offshore wind farm in the UK [8] reported that one-third of turbines were unavailable during that period due to gearbox problems.

There have been a number of studies on gearbox reliability Ribrant and Bertling [9] studied the cause of turbine failure in Sweden over the period from 1997 to 2005. They found that, while the gearbox was not the most common cause of failure, the long time to repair meant that it was responsible for the largest proportion of down time. There have been a number of studies looking at how to improve gearbox reliability. Musial *et al.* [10] have embarked on a long-term study to systematically investigate gearbox reliability problems. In the meantime problems continue. Figure 1 shows the proportion of turbine lost hours as a result of problems with specific components for turbines in Germany during the third quarter of 2008, data from [11].

A logical result of these problems is that a number of wind turbine manufactures and several independent studies have looked to direct drive where the generator rotates at the same speed as the turbine blades. By their nature, these generators also have fully fed converter system to connect to the grid. Polinder and van der Pijl [12] made a comparison study between wind turbines with direct drive synchronous, direct drive permanent magnet, single stage geared with both permanent magnet and DFIG, and three-stage geared systems with DFIG generators.

2.3 Direct drive generators

A direct drive generator for a wind turbine is characterized by having a very low rotational speed, and hence very high torque for a given power. Torque in an electrical machine is related to a magnetic shear stress in the airgap of the machine given by the following equation:

$$\sigma_g = \frac{\tau}{2\pi r_r^2 l_r} \tag{1}$$

where σ_g is the airgap shear stress, τ is the motor torque, r_r is the rotor radius, and l_r is the rotor core length [7]. The shear stress is effectively the mean value of the tangential component of the Maxwell stress tensor over the surface of the rotor, which is dependant on the square of flux density, as shown in eqn (2). This defines the Maxwell stress tensor in the cylindrical coordinate system for the magnetic field, assuming that the components due to electric fields can be ignored:

$$\sigma_{rt} = \frac{1}{\mu_0} B_r B_t - \frac{1}{2\mu_0} B^2 \delta_{rt} \tag{2}$$

where r is the radial direction, t is the tangential direction, σ_{rt} is the Maxwell stress tensor at a point, B is the magnetic flux density, μ_0 is the permeability of free space and δ is the Kronecker's delta. For this reason machines with a higher airgap flux density are capable of higher shear stress. A comparison of the shear stress obtainable in various types of electrical machine, including HTS, is given in [13].

If the torque exceeds the maximum overload shear stress capability of the generator then the machine will 'pull out' and cease to generate. The magnetic flux density in the airgap is limited to approximately 1 T in conventional machines by saturation in the iron magnetic circuit. Hence, for a given airgap flux density, the size of any given type electrical machine is largely determined by its torque rather than power. For this reason direct drive wind generators are large compared to their high speed geared equivalents. For a given wind speed and blade efficiency, the power obtainable from a wind turbine is proportional to the swept area of the rotor. Therefore, increasing the power of a wind turbine means increasing the diameter of the rotor, and since the blade tip speed is maintained within a certain limit for either mechanical or environmental (noise) reasons, this means a proportionally lower rotational speed and even higher torque as turbine power increases.

Wind turbines are commercially available with direct drive conventional synchronous generators, and turbines with direct drive permanent magnet generators (PMGs) are now beginning to appear, which have significantly greater torque density and hence smaller size and lower mass compared to conventional generators.

In 2008, Converteam UK Ltd. delivered a prototype direct drive PMG to Siemens Windpower in Denmark. This generator has been demonstrated on the Siemens 3.6 MW turbine at a test site in Denmark. Figure 2 shows this generator leaving the Converteam factory in Rugby, UK.

Converteam UK Ltd. are also in the process of producing a 5 MW direct drive PMG for the DarwinD offshore turbine.

Figure 2: The 3.7 MW, 14 rpm PMG.

3 Superconducting rotating machines

3.1 Superconductivity

Superconductivity is a phenomenon where electricity is conducted with zero resistance and zero loss, hence current in a loop of superconducting wire would continue forever. Superconductivity was discovered by H.K. Onnes in 1911 when he cooled mercury below 4.2 K (the boiling point of liquid helium) [14]. Temperatures in the fields of cryogenics and superconductivity are normally quoted using the Kelvin absolute temperature scale, where absolute zero is 0 K = −273.16°C. As temperature decreases the resistance of metals generally decreases. In the case of non-superconducting metals such as copper, a residual resistance value is reached and further temperature reduction does not result in and more reduction in resistance. Superconducting materials, on the other hand, have a critical temperature below which the resistance suddenly decreases to zero. When in the superconducting state the material has a critical current which increases as temperature decreases, above which superconductivity ceases, and also a critical magnetic field above which superconductivity ceases. Hence, the current carrying capacity of a superconductor is a function of temperature and magnetic field strength. They also exhibit a phenomenon, known as the Meissner effect, where all magnetic flux is excluded from within the superconductor when in the superconducting state [15].

The earliest superconducting materials (known as Type I Superconductors) were pure metals and had too low critical magnetic field to be of practical use. Later, another class of superconducting materials was discovered, consisting of metal alloys and known as Type II Superconductors, which were able to tolerate much higher magnetic fields. These materials allowed penetration of magnetic flux, which was then trapped within them by a mechanism known as "flux pinning".

The critical temperature of these materials can be up to around 25 K, but all need to be cooled to 4.2 K for practical use. The materials have been developed into practical wire products and are now commonly used in magnets for particle accelerators and in the commercial market for MRI scanners. A summary of the properties and manufacture of these low temperature superconductors (LTS) can be found in [16, 17].

Many studies were made into the use of these LTS materials for the field windings (magnets) of rotating machines [19], particularly in the 1960s and 1970s. The high magnetic field strength for superconducting magnets results in much smaller machines, and higher efficiency due to the zero loss in the field winding. However, the practicality and cost of cooling the field on the rotor of these machines using liquid helium meant that they never became a commercial proposition.

3.2 High temperature superconductors

Only small increases in the critical temperature of these LTS materials were achieved from their discovery in 1911 up until the 1980s. Then, in 1986 a material was discovered by Bednorz and Muller that became superconducting at a temperature of around 30 K [18], and very shortly afterwards (Fig. 3) many more materials were discovered with ever increasing critical temperature, although after 1990 this trend considerably slowed.

These discoveries brought the operating temperature of the superconductors into the range of liquid nitrogen, which is two orders of magnitude cheaper than the liquid helium used to cool LTS coils. All HTS materials are Type II Superconductors.

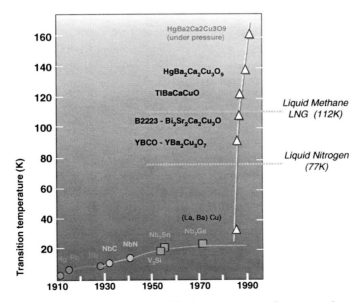

Figure 3: Development of the critical temperature of superconductors.

One common characteristic of these materials is they were all complex copper oxides and, more significantly, all ceramic materials. While it was relatively easy to manufacture such materials in bulk form, the technology to produce flexible wires that would be of use in electrical machine windings proved to be a considerable challenge. After considerable effort, the HTS material $Bi_2Sr_2Ca_2Cu_3O_x$ (more commonly referred to as BSCCO-2223) was successfully manufactured into practical wires during the 1990s.

3.3 HTS rotating machines

Rotating machines utilizing HTS materials have been under development for nearly 20 years, following the discovery of HTS materials in the late 1980s. HTS machines can use either HTS wires [19], or bulk HTS material [20], or even a combination of both [21]. Various types of HTS rotating machine topology have been proposed, including synchronous, homopolar [22] and induction [23]. Most large HTS machines projects to date have used a topology similar to conventional large synchronous machines, with a DC field winding on the rotor wound with HTS wire, and a copper AC stator winding at conventional temperature, as in [19].

The largest HTS machine that has been built and tested so far is a 36.5 MW, 120 rpm ship propulsion motor designed by American Superconductor (AMSC), and manufactured by AMSC, Northrop Grumman Corporation and Electric Machinery (now part of the Converteam group) for the US Navy [24]. This motor completed full load testing in January 2009. It has a rated torque of 2.9 million Nm, comparable to that of a 4 MW wind turbine. The 36.5 MW machine was the follow-on from a scaled prototype 5 MW, 230 rpm machine design and manufactured by AMSC and ALSTOM Power Conversion Ltd. (now Converteam Ltd.). The 5 MW machine was tested at full torque at ALSTOM, Rugby, UK [25], and at full load under simulated ship at sea conditions over a period of 1 year in the C.A.P.S. facility at Florida State University [26].

4 HTS technology in wind turbines

4.1 Benefits of HTS generator technology

HTS technology allows rotating machines to be constructed with significant increases in power density compared to conventional or permanent magnet machines. This advantage becomes greater as the size of the machine increases [27]. High power density is the result of the high current density that can be obtained in HTS coil, reducing the space required for the rotor field coils. The copper coils in a conventional machine typically operate with a current density between 3 and 5 A/mm^2, while the current density in the wire in a HTS coil can operate at 200 A/mm^2 or more. In HTS wire this is known as the 'engineering current density' which is the current density in the full cross section of the wire, but as the HTS material only forms a small proportion of the cross sectional area of the wire, the current density in the HTS material itself is much higher, up to 20,000 A/mm^2. Additionally, the ability to place many Amp-turns of field winding in a small volume, way beyond that which could be achieved using copper without unacceptable losses, can be used to increase the airgap flux density, allowing the airgap shear stress to increase. This offers the advantages of a direct drive generator at increasingly large turbine ratings without encountering practical difficulties due to the ever increasing size and mass of the generator. This reduction in mass of the largest turbine ratings is particularly important for the offshore wind market. A smaller, lower mass generator also enables the nacelle to be transported and lifted to the tower in one piece. The current generation of dedicated offshore wind turbine installation vessels have a lift capability of typically around 300 tonnes. The nacelle mass of some of the larger turbines currently available exceeds this. Lifting heavier components is possible, but becomes very expensive. The assembly of nacelle components at the top of the tower at an offshore location, particularly in a climate such as that in the North Sea, would be prohibitively expensive. An HTS direct drive generator of 6 MW or more would be approximately 20% of the mass of an equivalent conventional direct drive wound pole synchronous generator such as a rim generator design, or 50% of the mass of an optimized permanent magnet direct drive generator. Hence an HTS generator can make a direct drive feasible, with a similar nacelle mass to the traditional geared high speed generator, at very large turbine power ratings (>6 MW), where conventional or PMGs would become impractically large. HTS generator technology, therefore, can make very large turbines (8–10 MW or more) viable, resulting in a reduction in cost of offshore wind energy.

HTS generators also offer efficiency advantages at full load and particularly at part load when it is important to extract as much energy from the wind as possible. The value of efficiency in a wind turbine could be questioned, since the source of energy is free. However, a more efficient generator will generate more sales revenue from the same power at the turbine blades. There is an economic balance between the amount of energy generated by the turbine over its lifetime against the capital cost of the turbine. However if an efficiency gain, resulting in greater output for the same mechanical equipment, can be obtained without a corresponding increase in the capital cost of the turbine, it offers an advantage. A conventional machine has significant losses in the generator rotor, which, apart from a relatively small power requirement for the cooling system, the HTS machine does not have. The permanent magnet machine also has virtually no loss in the rotor and no power requirement for the cryogenic cooling system, but the airgap flux density is limited by the permanent magnet material and saturation in the iron magnetic circuit. In an HTS machine the increased flux density induces more e.m.f. per unit length in the stator copper coil, hence for a given copper section, and a given airgap diameter, the HTS generator output will be greater with same loss, hence higher efficiency.

A direct drive generator eliminates the gearbox, resulting in reduced maintenance requirements and increased reliability. Unlike a DFIG, but like the PMG, the HTS generator can have no sliprings requiring maintenance. A DFIG or conventional synchronous generator contains insulated windings on the rotor which are subject to elevated temperature and to thermal cycling whenever the load on the generator changes. This is a known source of failure on conventional generators [28]. In contrast, the rotor field winding of an HTS generator is maintained at a constant low temperature, except for periods of prolonged shutdown, and therefore does not see continuous thermal cycling [29]. Moreover, the operating temperature is such that the chemical processes that are responsible for the ageing of the electrical insulation have all but ceased.

The HTS windings need to be cooled by a closed loop cryogenic cooling system, with a cryocooler providing the cooling power. Cryocoolers with a suitable power rating are available as off-the-shelf commercial products. The present generation of cryocoolers do require periodic maintenance with intervals similar to those of many other turbine components; however in Europe there are projects working on the development of low maintenance or maintenance free cryocoolers. Converteam Ltd. is involved in one of these projects.

4.2 Commercial exploitation of HTS wind generators

In order for an HTS generator to be commercially competitive in the wind turbine market, a number of prerequisites must be met.

4.2.1 HTS wire

• *It must be possible to manufacture HTS wire in large volume at low cost.*

The volume of HTS wire required for a viable HTS wind market is many times greater than the present HTS production capacity. The production process of the previously commercially available BSCCO-2223 wire would not be scaleable to the required volume at the required cost. HTS wind generators will rely on the development of the 2nd generation (2G) of HTS wire described below.

The cost of HTS wire is normally stated as the cost of the wire needed to carry an amount of current over a certain distance, typically in $/kAm – the cost to carry 1000 A over 1 m. There is a further complication in that the current carrying capacity of the wire depends on its operating temperature and the operating magnetic flux density, therefore it is conventional to use the current carrying capacity at 77 K (boiling point of liquid nitrogen) with no applied magnetic field. In 2008 the cost of HTS wire was around 130 $/kAm. In order to be cost effective in HTS generators for wind turbines the cost needs be in the range 10–20 $/kAm.

4.2.2 Generator design

• *The generator design must be optimised for low cost volume production.*

The majority of cryogenic design experience has been with low volume specialist applications where cost is much less important. The exception to this has been the MRI scanner market using LTS magnets, which manufacture moderately high volumes. Design for manufacture techniques, and careful selection of materials and components, must be used to obtain a volume manufacture generator design.

4.2.3 Cooling system

• *The cryogenic cooling system must be reliable with extended maintenance intervals.*

Commercially available cryocoolers are designed for the laboratory or hospital environment, and would require some ruggedisation to be suitable for an offshore wind turbine environment. The coolers of the temperature range and power required for the HTS generator use either the Gifford-McMahon cycle, or Stirling cycle. Maintenance intervals are presently 9–18 months, which would need to be extended for offshore wind power applications. Zero maintenance has already been addressed for very small cryocoolers – there a large number small Stirling cycle coolers on spacecraft that have operated without maintenance or failure for up to 16 years. Newer cooler technology such as pulse tubes, which have no moving cold parts [30], and free piston Stirling cycle coolers [31], which have no wearing seals, are beginning to become viable in the larger power ratings required for HTS machines.

5 Developments in HTS wires

HTS generators cannot be a commercial success in the wind market without low cost volume production of HTS wire. All HTS wire produced to date has been more than an order of magnitude too costly to be considered. However, a new class of HTS wires have been developed and are currently at the early stage of commercialisation. These wires have the potential to meet the volume and cost requirements for the HTS wind generator. These new wires have become known as 2G HTS wire and the earlier wire as 1G.

5.1 1G HTS wire technology

Until very recently (2006) all commercially available HTS wire was based on BSCCO and manufactured using the 'Powder in Tube' method. The HTS precursor powder was placed inside a machined silver tube which was the drawn out until reduced to about 1 mm diameter. This was then cut into short lengths and a large number of these, typically 80–100, placed inside another silver tube, and drawn out again until about 1 mm diameter. This wire was then rolled flat to about 4 mm wide by 0.2 mm thick. The final process was a controlled heat treatment in a controlled atmosphere to produce the superconductor material inside the filaments. The resulting wire structure can be seen in Fig. 4.

This was an inherently costly process, requiring a large floor space for the drawing process. Early wire was priced at around 1000 $/kAm, with prices in 2008 around 130 $/kAm. The ultimate minimum price in volume for this wire is

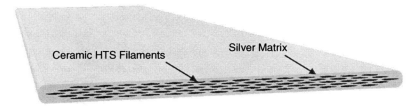

Figure 4: Composition of 1G BSCCO wire.

estimated to be more than 50 \$/kAm, which may be acceptable in niche rotating machine markets, but too expensive for the wind market.

5.2 2G HTS wire technology

The new 2G wires are based on the superconducting material $YBa_2Cu_3O_x$ normally referred to as YBCO-123 or simply YBCO. The structure and manufacturing method of 2G HTS wire is very different form that of 1G wire. YBCO was one of the first HTS materials to be discovered and is easily made in bulk form by growing a crystal in a similar manner to silicon. Development of YBCO-based wire began in the 1990s by attempting to deposit a crystal of YBCO onto a metal substrate tape. This technique has now been extensively developed by several manufacturers using a number of different processes. The wire structure consist of a substrate, typically a Nickel-Tungsten alloy, a very thin buffer layer onto which is deposited the YBCO superconductor to a thickness of 1–5 µm. Often an outer copper layer is added for stability. The overall wire thickness is between 0.1 and 0.2 mm thick depending on the manufacturer and product. The coatings are deposited on a wide strip of the substrate and then slit to the required tape width. This gives flexibility in the final width and current carrying capacity of the HTS tape, the most common being 4 mm for compatibility with 1G HTS materials, and 12 mm for higher current carrying capacity. A simplified wire structure is shown in Fig. 5, some processes introduce additional buffer layers. It is also possible to join two of these tapes back to back to produce a symmetrical duplex tape.

This type of HTS wire has the potential for volume production at low cost. It does not require the large floor space the 1G wire need for the drawing process, since the 2G process can be reel to reel, and once the correct process parameters are set up production remains almost entirely automated.

HTS materials are intrinsically anisotropic, and their sensitivity to magnetic field depends on the direction of the field relative to the surfaces of the HTS tape. It has been necessary to develop methods in the manufacturing process that minimise the effect of this anisotropic behaviour [32].

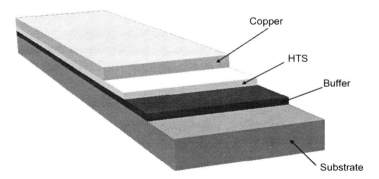

Figure 5: Simplified 2G wire structure – thickness scale exaggerated.

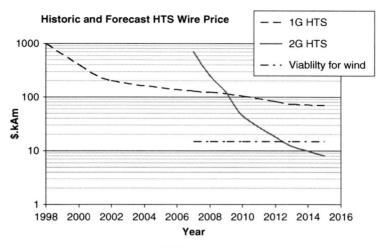

Figure 6: HTS wire price trend.

A number of different manufacturing processes are used to produce HTS wire of this composition. Superpower Inc., for example, use a vacuum deposition process [33], others use a mixture of chemical and vacuum processes. Zenergy Power Plc has been developing an all chemical deposition process, which they believe offers the potential for the lowest cost volume production [34]. Other manufacturers have also begun to look at all chemical processes.

5.3 HTS wire cost trends

HTS wire prices for 1G wire fell rapidly from the mid-1990s. Since 2004 the price of 1G wire has fallen more slowly as manufacturers have ceased production to concentrate on commercialisation of 2G wire.

2G wire first became commercially available in 2006–2007, although at a high cost, and with performance inferior to 1G wire. By the end of 2008 the performance of 2G wire was beginning to approach that of 1G, but with prices still higher. In 2009 the performance of the best 2G wire is expected to exceed that of 1G and the price to be comparable.

The historic and forecast prices are shown in Fig. 6, in which forecast prices were obtained from data supplied by a number of HTS wire manufacturers. The forecast shows that commercial viability for HTS technology in wind turbines is expected to occur after 2013.

6 Converteam HTS wind generator

In 2004 Converteam Ltd. (then ALSTOM Power Conversion Ltd.), undertook a feasibility study into a direct drive wind generator based on the use of low cost 2G HTS wires, expected to become available in commercial quantities and at an economic cost in the 2010–2015 time scale. This feasibility study resulted in a project to design and de-risk a full scale direct drive HTS wind turbine generator. The project is scheduled to complete in 2010, and will be followed by a program to prototype and industrialize a full size generator. It is partly supported by a grant from the U.K. Department of Trade and Industry (now Technology Strategy Board), and includes A.S. Scientific, a specialist cryogenic engineering company in Abingdon, U.K. and the University of Warwick, U.K., for their expertise in materials and volume manufacturing methods, as project partners.

6.1 Generator specification

The generator specification was based on the rating of the largest offshore turbines expected to be in production in 5–10 years time. The rating was chosen to be 8 MW at 12 rpm, which would be used on a turbine with a rotor diameter of around 160 m, and a blade tip speed optimised for far offshore application. This gives the generator a shaft torque of 6500 kNm, the largest torque of any HTS rotating machine project to date.

6.2 Project aims

The project was originally planned to extend over a period of 3 years although this was subsequently extended to 4 years to permit work on two other HTS projects concurrently. It was divided into three principle tasks:

1. The conceptual design of the full size generator during the first year of the project, followed by a gate review.
2. The detailed design, with cost and performance modelling, of the full size generator.
3. In parallel with the detailed design, a scaled model generator having a rated torque of up to 200 kNm, to be designed, manufactured and tested, employing the technology that will be used in the full scale design.

6.3 Conceptual design

The first stage of the project, involved the conceptual design of the full size generator, and was completed in November 2006. The resulting generator design was 5 m diameter with an overall length (excluding shaft extensions) of 2.2 m, and a mass of just over 100 tonnes. This stage of the project examined the technical, economic and market feasibility of the HTS generator, and aimed to provide a baseline design and one or more solutions to the technical challenges that could be used in the detailed design stages to follow. The completed concept design, with the rotor shown separately is shown in Fig. 7.

A preliminary study examined many of the synchronous HTS machine topologies, in order to determine the optimum design basis for the HTS wind generator. The design of HTS machines involves a broader range of skill than those that are

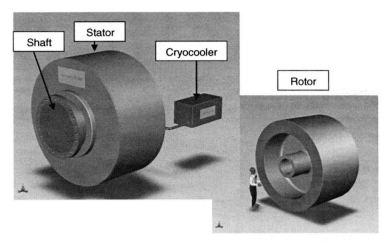

Figure 7: The conceptual 8 MW generator design.

necessary to design a conventional electrical machine. In addition to skills in the fields of electromagnetic and mechanical engineering needed for conventional machine design, skills are also required in the fields of cryogenics and vacuum technology. There are more options open to the designer of an HTS synchronous machine compared to a conventional machine. Since conventional electrical machine design tools are not applicable to some of these topologies, it is necessary to rely heavily on electromagnetic finite element analysis (FEA) for the design process. The limited magnetic circuit in most HTS machine designs give no defined path for the magnetic flux, which means that 3D electromagnetic FEA must be used. This is an order of magnitude more time consuming than 2D analysis.

The HTS synchronous machine can be classified into a number of different types with different characteristics, advantages and disadvantages:

1. Conventional stator with iron teeth and HTS rotor with magnetic pole bodies which can be either warm or at cryogenic temperature. The electromagnetic layout of this type is shown in Fig. 8 (components without an electromagnetic function are not shown). This type does not offer much improvement in size or mass compared to a conventional machine, but offers gains in efficiency due to almost zero rotor loss.
2. Conventional stator and HTS rotor with non-magnetic pole bodies (Fig. 9). The advantages are similar to type 1, but it requires more HTS wire to produce the necessary stator flux density. It avoids potential high cost cold magnetic materials or complex thermal isolation.
3. Airgap stator winding and HTS rotor with magnetic pole bodies (Fig. 10). This construction allows the flux density at the airgap significantly beyond what is possible with a conventional stator. The rotor iron can operate very highly saturated, since the flux is predominantly DC. This allows significant reductions in size and mass of the HTS machine. Since most ferromagnetic materials become

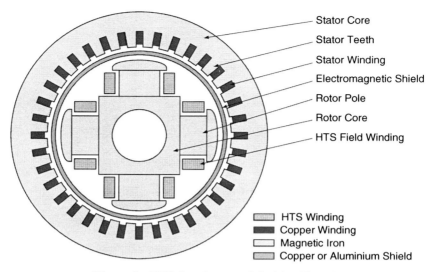

Figure 8: HTS Synchronous Machine Type 1.

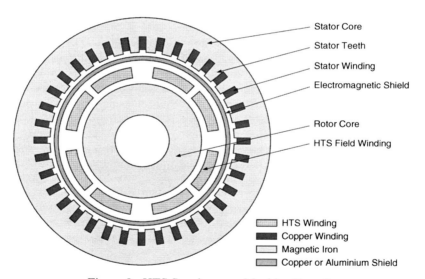

Figure 9: HTS Synchronous Machine Type 2.

very brittle at low temperature the choice of material is limited. Nickel-based alloys have satisfactory properties but are expensive.

4. Airgap stator winding and HTS rotor with non-magnetic pole bodies (Fig. 11). This construction also allows significant reduction in size and mass. It requires more HTS wire than type 4, but does not require expensive cold iron components since it is relatively easy for the rotor core to be warm and thermally isolated from the cold HTS field system.

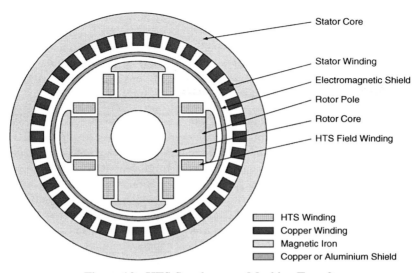

Figure 10: HTS Synchronous Machine Type 3.

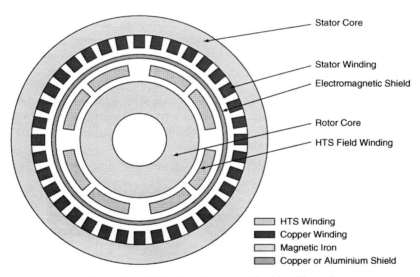

Figure 11: HTS Synchronous Machine Type 4.

The above options were studied for the direct drive wind generator, for which cost and low mass are important (size less so, apart from transport considerations). Types 3 and 4 offered the advantage of lowest mass. The cost balance between these two types to a large extent depended on the relative cost of HTS wire against other materials. Based on the predicted volume pricing for 2G HTS wire, type 4 was chosen for the Converteam HTS machine.

6.4 Design challenges

The direct drive wind generator presented a number of design challenges, which were identified as risks or potential stumbling blocks at the start of the project. However, the conceptual design stage identified solutions to all of them. A number of these challenges are described below.

6.4.1 Rotor torque transmission

The very high rated torque of this generator needs to be transmitted from the HTS coils at cryogenic temperature to the shaft at near to ambient temperature, without conducting an unmanageable quantity of heat from the warm parts to the cold parts. A typical cryocooler that can extract 100 W at 30 K would require in input power to the compressor at approximately 10 kW, although cryocooler efficiency is expected to improve over the next decade.

The conceptual design used a torsion rod system that could transmit rated and fault torque with only a little over 20 W of heat conduction to the cold parts.

6.4.2 Managing mechanical forces

The generator is a large machine, with a very high rated torque, operating at high magnetic flux density (>4 T in parts of the HTS coils). The large physical size means that stresses due to differential thermal contraction must be carefully modelled and managed to prevent excessive stress in rotor components, particularly in the HTS coils where excessive strain in the HTS wire could lead to a reduction in the critical current of the wire, which could lead to a quench, when the wire returns to its non-superconducting state.

The high operating current density in the HTS coils in combination with high magnetic flux density, leads to very high Lorenz ($\mathbf{J} \times \mathbf{B}$) forces acting on the HTS wire. Although the generator torque acts on the coil by this force, it represents only 10% of the total Lorenz force on the HTS wire, the remainder due only to applying rated field current. The force density on the HTS coil for this condition is shown in Fig. 12.

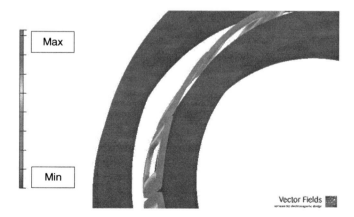

Figure 12: Force density on the surface of the HTS coils.

A coil geometry and support structure was chosen that met the criteria to control the mechanical forces, while minimising the flux density in the HTS wire, taking into account the anisotropic characteristics of the wire. Converteam has been supported by its superconducting partner Zenergy Power in develop the coil manufacturing process.

6.4.3 Wind turbulence

Wind does not blow at a constant speed, so a wind turbine is subjected to constantly changing wind speeds and load. The amount of turbulence is dependant on the site location and the wake effects from other turbines, with background turbulence much higher onshore than offshore. This results in a wind turbine generator being subjected to constantly changing speed and torque, which can induce eddy currents in the electrically conducting cold components, creating losses and hence unwanted heating of the cold parts. It can also result in fluctuating flux density at the HTS coils, causing AC loss in the superconductor.

A simulation was carried out on the conceptual design using a level of wind turbulence at the high end of what may be expected for an offshore location, as shown in Fig. 13.

A two-dimensional non-linear time stepping electromagnetic finite element (FE) simulation was carried out over a simulated time period of 10 min using Vector Fields Opera software. The model included an external circuit (outside of the FE mesh) which was continually varied to simulate the turbine control system. The solution was post processed to obtain eddy current loss in individual rotor parts and flux density variation in the HTS coil. The simulation included the effect continuous blade pitch control in the turbine to attempt to maintain the power supplied to the grid constant whenever possible, allowing the generator speed and torque to vary. A similar control method is described in [35]. The simulation also included blade pitch control compensating for the periodic torque variations due to wind

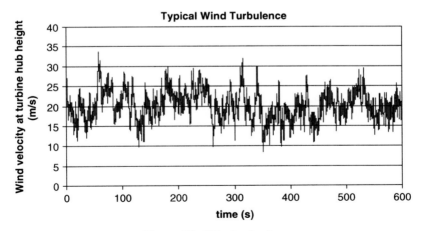

Figure 13: Wind velocity.

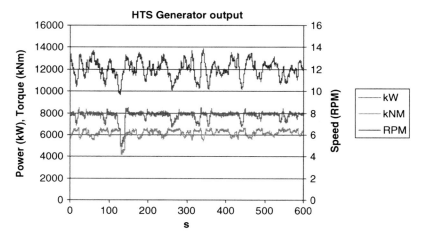

Figure 14: Generator output in simulated wind conditions.

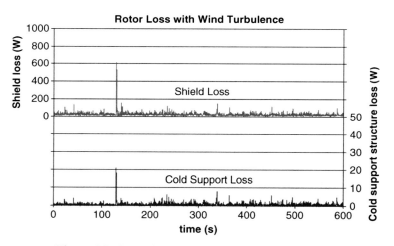

Figure 15: Rotor loss in simulated wind conditions.

shear giving different wind velocity between a blade at the top and bottom of its rotation and also of the effect of blades passing the tower. The resulting generator speed (top trace), output kW (middle trace) and torque (bottom trace) is shown in Fig. 14. The resulting eddy current losses in the rotor cold parts and the warm electromagnetic shield surrounding the rotor are shown in Fig. 15.

Although there is high instantaneous loss at the instant of sudden changes, the average loss is low, requiring negligible additional cooling power. The fluctuations in flux density in the HTS were also found to be small. AC components of current and magnetic flux density are known to induce losses in the HTS wire (known as AC loss), which was not included in the analysis. AC loss in superconductors is difficult to calculate and has been extensively researched. However, nearly all of this research was for AC current applications such as HTS power cables and transformers, where the HTS wire current and magnetic flux density is purely AC, and small relative to the DC critical current value, such as in [36]. In this situation there will also be hysteresis losses in the magnetic substrate of the 2G HTS wire. The 2 MW generator in [37] has a permanent magnet field and pure AC current in the HTS stator winding. In the Converteam HTS generator the HTS wire is operating with a DC current and in a very high DC magnetic flux density, not far from the wire critical current, with a very small (compared to the pure AC studies, and even slammer compared to the DC component) AC component superimposed. The 2G wire magnetic substrate would be fully saturated in this case, and would not experience hysteresis loss, but these will still be losses due to the changes in trapped flux within the superconductor (also known as hysteresis) and due to eddy currents in the wire substrate. Converteam Ltd. have commissioned the University of Cambridge, UK to carry out a theoretical analysis of AC loss for wire under these conditions backed up by tests using a variable temperature insert in a 5 T LTS magnet.

6.4.4 Cooling of HTS coils

It is essential that during operation the HTS coils are maintained at a temperature such that there is sufficient margin between the operating field current and the critical current of the wire. In order to minimise the power input to the cryocooler and make best use of its cooling capacity a temperature difference as small as possible between the cryocooler and the HTS coils is desirable.

Past HTS motor projects have used either closed circuit helium gas circulation [38], or phase change neon cooling systems. The neon-based systems, such as described in [39] condense the neon gas at the cryocooler at its boiling temperature of 27.2 K. Liquid neon is then supplied to the rotor and allowed to evaporate, removing heat from the rotor in the process, and returning to the cryocooler as a gas. This type of system has the advantage that it is a very effective cooling process and can operate as a thermosiphon, with no mechanical assistance to the circulation. One disadvantage to such a system is that the cryocooler cold head temperature varies with heat load, and it is necessary to introduce a heater to the system to prevent the temperature from dropping to 24.6 K and freezing the neon, wasting cryocooler power. A second disadvantage is that the coolant temperature is fixed at 27 K, and it is expected that with 2G HTS wire, that the operating temperature could be considerably higher, probably in the range 40–60 K. A third disadvantage is that cooling will be non-uniform when the rotor is stationary. This could cause undesirable stresses in coils and their support structure.

A helium gas circulation system was chosen for the HTS wind generator. While this had the disadvantage of requiring assisted circulation, it offered complete flexibility in the choice of operating temperature. Heat was transferred between the HTS coils and the cold helium in the rotor cooling circuit by conduction. In order to calculate the heat flow and to determine the coil operating temperature, it was necessary to use detailed computational fluid dynamics (CFD) and thermal FE models that also had to take into account the larger (order of magnitude) variation in material properties such as thermal conductivity and specific heat capacity with temperature.

6.4.5 Airgap stator design

In a conventional stator the radial and tangential forces act on the iron teeth, where the stator conductors only see a small force due to leakage flux, but in an airgap winding these forces act directly on the stator conductors. The stator coils not only have to withstand these forces, but the forces also have to be transferred from the coils using non-magnetic, non-conducting materials, since magnetic materials would saturate leading to high losses due to AC flux, and high eddy current losses would be induced in conducting materials.. These forces are also cyclic, so the stator teeth are subject to high cycle fatigue loads. Even when the total generator torque is steady, each individual coil side sees a force fluctuating at 2× the stator fundamental frequency with a pattern rotating around the machine with the rotor field. In fault conditions the patterns are continually changing in time as well as space, involving complex mechanical time stepping modelling techniques. An example of an electromagnetic and mechanical time stepping simulation of a short circuit fault is shown in Fig. 16, where the graph show the force on individual stator teeth against time, with mechanical FE output of the deflection. A number of composite materials have been investigated, and some glass-based materials have been found to offer acceptable properties.

The modelling produced a design with acceptable stress and deflection using composite material support structure. Further prototyping and fatigue testing is planned.

The high power density that is possible with HTS machines also means that careful design must be given to stator cooling. Due to the cost sensitive nature of

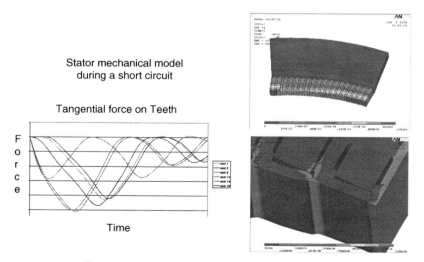

Figure 16: Force and deflection of stator coils.

the wind generators it was desirable to avoid complex liquid cooled systems. Extensive thermal and CFD modelling showed that the stator could be easily cooled by forced air ventilation.

6.4.6 Stator iron losses

The stator design contains a laminated iron core located radially behind then airgap winding. This serves three purposes:

1. It provides a means of mechanical support and rigidity close to the coil supports.
2. It shields external components from stray flux.
3. It provides an easy circumferential path for the flux passing behind the airgap winding, reducing the amount of HTS wire required.
4. It enhances the field in the active region of the stator winding.

A fully airgap design (type 4 above) machine has a significant component of magnetic flux in the axial direction near to and outside of the straight length of the machine. This can cause eddy currents to flow in the radial and tangential direction in the laminations, causing a high concentration of loss at the ends. The low frequency of the generator (<2 Hz) was expected to reduce these losses, but the phenomenon was still seen as a potential risk at the start of the project.

The conceptual design was modelled using Vector Fields Opera 3D electromagnetic FE software. The original design had a total end loss due to eddy currents of 55 kW, which may have been possible to remove by cooling, but would have had a detrimental effect on the efficiency on a machine of only 8 MW rating. Careful design of the stator core geometry and further modelling resulted in a design with this loss reduced to 6 kW.

6.5 The cost-benefit study

In order to justify the business case for the development of the HTS generator, Converteam Ltd. commissioned the independent wind turbine consultants BVG Associates Ltd. to analyse the cost/benefit of very large offshore turbines employing HTS direct drive generators and medium voltage power converters [40]. This study involved the design of a complete, notional, 8 MW, 12 rpm rotor wind turbine, with appropriate foundations by Sheffield Forgemasters, and turbine blade design, control and structural integration by Garrad Hassan Ltd.

Comparisons were made for a typical UK Crown Estate Round 2 offshore wind farm of 504 MW containing 4 MW geared conventional DFIG turbines (the baseline), 8 MW geared turbines, 8 MW direct drive PMG turbines, and 8 MW direct drive HTS turbines. To ensure consistent analysis, the study found first that on monopile foundations, conventional and PMG 8 MW turbines actually increased the cost of electricity from the wind farm compared to the baseline 4 MW turbines. Benefits of the Converteam design, low mass, 8 MW HTS generator resulted in an identical cost of energy compared to the baseline, even at pre-series volume costs.

Then, retaining the successful 8 MW HTS design, the foundation was replaced with an alternative more suited to such large turbines in relatively deep water. With this configuration, the 8 MW HTS turbines resulted in a 17% reduction in the cost of energy relative to the baseline, in the same pre-series volumes. The analysis included the efficiency benefits of HTS generators, but only at part load, since at wind speeds greater than the minimum required for full load output only part of the kinetic energy is extracted from the wind and the benefit of increased generator efficiency is limited to reducing the load on the mechanical components. Work is progressing with a wind turbine manufacturer towards the design and cost analysis of a new giant turbine, in serial volumes, prior to demonstrator deployment.

6.6 Model generator

Following the completion of the concept design, work started on the design of the scaled demonstration 'Model Generator'. Apart from the use of 1G HTS wire rather than 2G, due to the limited availability and performance of 2G at the time of the wire purchase, the model generator uses scaled components of the same type as in the full size generator. The rating of the machine was constrained by the project budget for HTS wire, and was chosen to be 500 kW at 30 rpm.

At the time of writing (December 2008) the design of the major components has been completed, and manufacturing was in progress. Manufacturing and testing of this generator is expected to be completed in 2010.

6.7 Material testing and component prototypes

An important part of the de-risking process for new technology is the manufacture and test of prototype components.

A test facility was established at the University of Warwick with a cryostat to enable mechanical and thermal testing of small components at room temperature and down to a temperature of 30 K. Other tests included material properties for the support teeth for the airgap stator winding. Figure 17 shows a carbon fibre cold-to-warm torque link for the model generator, prior to mechanical testing.

It was also considered essential to manufacture and test smaller versions of the HTS coils. During the concept design phase, three HTS coils were made to a design similar to the model generator and full size coils, which were tested in a

Figure 17: Torque link.

Figure 18: Large cryostat with a prototype HTS coil in a magnetic circuit at
            Converteam Ltd., Rugby.

simulated magnetic circuit in a small cryostat. These coils were successfully tested
up to an operating current of 230 A, compared to the 140 A operating current in the
model generator.

Following on from these tests Converteam opened a cryogenics laboratory in
their Rugby facility to support all HTS projects, with a large cryostat (8 m^3 vacuum
space) capable of accommodating the coils for the full size wind generator, shown
in Fig. 18 during the setting up of a test.

6.8 The full scale detailed design

Following discussions with a wind turbine manufacturer, the rating of the full size
generator was changed to 10 MW at 11.5 rpm, to suit their planned turbine development. The 10 MW design was scaled from the 8 MW concept design, resulting
in a slight increase in diameter to 5.4 m, an increase in length to 2.6 m, and a mass
of less than 140 tonnes.

The design is based on the expected performance of the best HTS wire in 2010,
the earliest time when wire will need to be ordered for the full size prototype generator. This is expected to require an operating temperature of 40 K. The helium gas
cooling system used in the design permits the operating temperature to be easily
changed when future improvements in HTS wire performance permit.

7 The way forward

The next logical step will be the manufacture and test of a full size prototype, followed by its demonstration on the 10 MW turbine. In parallel with this the supply
chain development and engineering of the manufacturing process for production
manufacturing will take place before the start of commercial volume production,
which could begin by 2015.

[20]   Hull, J.R. & Murakami, M., Applications of bulk high temperature super-conductors, *IEEE Proceedings*, **92**(10), pp. 1705–1718, 2004.

[21]   Masson, P.J. & Luongo, C.A., High power density superconducting motor for all-electric aircraft propulsion. *IEEE Transactions, Applied Superconductivity*, **15**(2), pp. 2226–2229, 2005.

[22]   Superczynski, M.J. & Waltman, D.J., Homopolar motor with high temperature superconductor field windings. *IEEE Transactions, Applied Superconductivity*, **7**(2), pp. 513–518, 1997.

[23]   Nakamura, T., Miyake, H., Ogama, Y., Morita, G., Muta, I. & Hoshino, T., Fabrication and characteristics of HTS induction motor by the use of Bl-2223/Ag squirrel-cage rotor. *IEEE Transactions, Applied Superconductivity*, **16**(2), pp. 1469–1472, 2006.

[24]   Buck, J., Hartman, B., Ricket, R., Gamble, B., MacDonald, T. & Snitchler, G., Factor testing of a 36.5 MW high temperature superconducting propulsion motor. *Proc. of the American Society of Naval Engineers (ASNE) Symposium*, 2007.

[25]   Snitchler, G., Gamble, B. & Kalsi, S., The performance of a 5 MW high temperature superconducting superconductor ship propulsion motor. *IEEE Transactions, Applied Superconductivity*, **15**(2), pp. 2206–2209, 2005.

[26]   Steurer, M., Woodruff, S., Boenig, H., Bogdan, F. & Sloderbeck, M., Hardware-in-the-loop experiments with a 5 MW HTS propulsion motor at Florida State University's power test facility. *Proc. of IEEE Power Engineering Society General Meeting*, pp. 1–4, 2007.

[27]   Malozemoff, A.P., Maguire, J., Gamble, B. & Kalsi, S., Power applications of high temperature superconductors: status and perspectives. *IEEE Transactions Applied Superconductivity*, **12**(1), pp. 778–781, 1992.

[28]   Stone, G.C., Advancements during the past quarter century in on-line monitoring of motor and generator winding insulation. *IEEE Transactions on Dielectrics and Electrical Insulation*, **9**(5), pp. 746–751, 2002.

[29]   Kalsi, S., Superconducting generators promise higher reliability and ease of operation. *Presented at American Society of Naval Engineers, Electrical Machines Technology Symposium*, Philadelphia, PA, 2004.

[30]   Dietrich, M., Yang, L.W. & Thummes, G., High power stirling-type pulse tube cryocooler: observation and reduction of regenerator temperature inhomogeneties. *Cryogenics*, **47**(5-6), pp. 306–314, 2007.

[31]   Vitale, N.G., Striling free piston cryocoolers, US Patent 5,022,229, 1991.

[32]   Malozemoff, A.P., Fleshler, S., Rupich, M., Thieme, C., Li, X., Zhang, W., Otto, A., Maguire, J., Folts, D., Yuan, J., Kraemer, H.-P., Schmidt, W., Wohlfart, M. & Neumueller, H.-W., Progress in HTS coated conductors and their applications. *Superconductor Science and Technology*, **21** 034005 (7 pp.), doi: 10.1088/0953-2048/21/3/034005, 2008.

[33]   Selvamanickam, V., Chen, Y., Xiong, X., Xie, Y.Y., Reeves, J.L., Zhang, X., Qiao, Y., Lenseth, K.P., Schmidt, R.M., Rar, A., Hazelton, D.W. & Tekletsadik, K., Recent progress in second-generation HTS conductor scale-up at superpower. *IEEE Transactions, Applied Superconductivity*, **17**(2), 2007.

[34] Lewis, C. & Muller, J., A direct drive wind turbine HTS generator. *Power Engineering Society General Meeting*, IEEE, 24–28 June 2007, http://ieeexplore.ieee.org/stamp/stamp.jsp?arnumber=4275835& isnumber=4275199

[35] Senju, T., Sakamoto, R., Funabashi, T., Fujita, H. & Sekine, H., Output power levelling of wind turbine generator for all operating regions by pitch angle control. *IEEE Transactions Energy Conversion*, **21**(2), 2006.

[36] Tsukamoto, O., Alamgir, A.K.M., Liu, M., Miyagi, D. & Ohmatsu, K., AC loss characteristics of stacked conductors with magnetic substrates. *IEEE Transactions Applied Superconductivity*, **17**(2), 2007.

[37] Fee, M., Staines, M.P., Buckley, R.G., Watterson, P.A. & Zhu, J.G., Calculation of AC loss in an HTS Wind Turbine Generator. *IEEE Transactions Applied Superconductivity*, **13**(2), 2003.

[38] Eckels, P.W. & Snitchler, G., 5 MW high temperature superconducting propulsion motor design and test results. *Naval Engineers Journal*, **117**(4), pp. 31–36, 2005.

[39] Frank, M., Frauenhofer, J., Gromoll, B., Hasselt, P., Nick W., Nerowski, G., Neumueller, H.-W., Haefner, H.-U. & Thummes, G., Thermosyphon cooling system for the Siemens 400 kW HTS synchronous machines. *AIP Conf. Proc.*, **710**(1), pp. 859–866, 2004.

[40] Valpy, B., Establishing the cost of energy benefits of the HTS generator technology, Report by BVG Associates for Converteam.

[41] Maki, N., Design study of high temperature superconducting generators for wind power systems. *Journal of Physics Conference Series* **97**, doi:10.1088/1742-6596/97/1/012155, 2008.

CHAPTER 10

Intelligent wind power unit with tandem wind rotors

Toshiaki Kanemoto[1] & Koichi Kubo[2]
[1]Faculty of Engineering, Kyushu Institute of Technology, Japan.
[2]Graduate School of Engineering, Kyushu Institute of Technology, Japan.

Wind is clean, renewable and home grown energy source of the electric power generation, and has been positively/effectively utilized to cope with the warming global environment. The propeller type wind turbines play main role in the wind power generation; however, it may have some weak points. To overcome the weak points of the traditional one, the authors have investigated the superior wind turbine unit which is composed of the tandem wind rotors and the double rotational armature type generator without the traditional stator. The large-sized front wind rotor and the small-sized rear wind rotor drive the inner and the outer rotational armatures respectively, in keeping the rotational torque counter-balanced. Such operating conditions may be able to make the output higher than the traditional wind turbines, and to keep the output constant in the rated operation mode without the brakes and/or pitch control mechanism. The first step toward practical use of this wind power unit, the characteristics of the double rotational armature type synchronous generator and the double rotational armature type doubly fed induction generator were investigated. Moreover, the desirable profiles of tandem wind rotors such as optimum diameter ratio (rear rotor diameter divided by front rotor diameter), blade numbers of both wind rotors and axial distance between both wind rotors were also investigated. Furthermore, the field tests of the wind turbine unit had been carried out on the pick-up type truck driven straight at constant speed.

1 Introduction

The propeller type wind turbines play main role in the electric power generation. The traditional propeller types may, however, have some weak points as follows:

1. The large-sized wind rotor generates the fruitful output but is not operated at the weak wind, while the small-sized wind rotor can be operated at the weak wind but the output is small.

2. It is necessary to equip the wind rotor with the brakes and/or the pitch control mechanisms, to suppress the abnormal rotation and the overload at the strong wind. To keep good quality of the electric power by increasing the rotational speed of the armature, it is necessary to prepare the large-sized generator with multi-poles or the accelerating gearbox.
3. There are few areas applicable to the traditional wind turbine with the rated mode at the wind velocity faster than 11 m/s. It is desired, for the next leap of the wind turbine technologies, to get the fruitful output at not only rich but also poor wind circumstances.

To overcome above weak points, the authors invented the superior wind power unit called "intelligent wind power unit". This unit is composed of the large-sized front wind rotor, the small-sized rear wind rotor and the peculiar generator with the inner and the outer rotational armatures without the traditional stator. The front and the rear wind rotors drive the inner and the outer armatures in the upwind type or drive the outer and the inner armatures in the downwind type, respectively. The rotational direction and speed of both wind rotors/armatures are free, and automatically and smartly adjusted pretty well in response to the wind circumstances. This unit is applicable to the area with not only rich wind circumstances such as Middlegrunden offshore [1], but also poor circumstances having weak and/or fluctuating wind.

2 Previous works on tandem wind rotors

The idea of the tandem wind rotors has been proposed [2], and Ushiyama *et al.* could also increase the output using the model tandem wind rotors, as follows [3]. The front wind rotor consists of six blades with the diameter of 0.8 m, and the rear wind rotor consists of three blades with the diameter of 1.2 m, as shown in Fig. 1. The front and the rear wind rotors are connected directly to the stator casing and the armature of the traditional generator, respectively. The outputs against the relative rotational speed are shown in Fig. 2. The maximum outputs are 2.6 W at the wind velocity $V = 6$ m/s, 19.8 W at $V = 8$ m/s, and 38.9 W at $V = 10$ m/s, in the experimental results. This paper discusses the output in the followings. The rated output of the generator is 36 W at the rotational speed $N_T = 950$ min^{-1} but the actual output is 38.9 W with $N_T = 768$ min^{-1} at 10 m/s. That is, the output is higher than the rated output even if the rotational speed is slower than the rated speed, and the relative rotational speed is increased by the counter-rotation. As a result, this paper concluded that the experimental results of the model demonstrated the technical possibility of the counter-rotating-type wind turbine generator. Jang and Heo brought recently the above ideas into his researches and has proposed the tandem type composed of the small-sized front wind rotor and the large-sized rear wind rotor [4].

In 2002, Appa Technology Initiative developed the counter-rotating type wind turbine (CRWT) and the prototype was supplied to the field tests in California. The prototype composed of tandem wind rotors, whose diameters are 4 m and blade numbers are 2 with the tip speed ratio $\lambda = 6$, and the isolated traditional generator

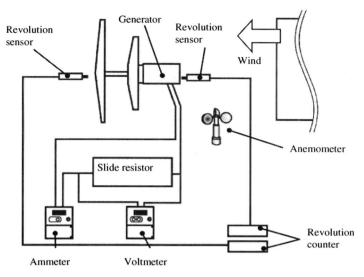

Figure 1: Model unit with tandem wind rotors [3].

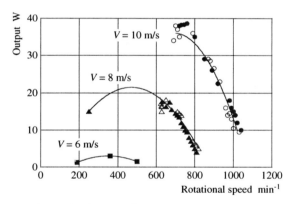

Figure 2: Output of tandem wind rotors [3].

of the rated output 3 kW (see Fig. 3, [5]). The each wind rotor rotates and the armature as usual, while the mechanical efficiency is 0.8. The numerical simulation showed that the prototype with the tandem wind rotors can get nearly 85% of the effective wind power, which is about two times higher than that of a practical single rotor system. The field tests suggested that the rear wind rotor can generate the additional output from 25 to 40% on the same stream tube and the efficiency is high at the comparatively lower rotational speed.

Continuously, Jung *et al.* has proposed the tandem wind rotors consisted of the front wind rotor with the diameter of 5.5 m, called the auxiliary rotor by them, and the rear wind rotor with the diameter of 11 m, called the main rotor by them [6].

The tandem wind rotors drive the armature of the traditional generator through the gearbox as shown in Fig. 4. The flows and the performances were predicted numerically, by the quasi-steady strip theory with the experimental wake model obtained in the wind tunnel, and were compared with the experiments in the full scale. The paper showed the optimum arrangement of the tandem wind rotors and the maximum output coefficient reaches as high as 0.5 in the full-scale test.

Furthermore, Shen *et al.* compared the energy production of the counter-rotating type with that of the shingle rotating type, in the numerical simulation at the actual velocity field in the island of Sprogo, Denmark [7]. The tandem wind rotors, whose diameters are 20.5 m, consist of three blades in the commercial Nordtank 500 kW rotor. The outputs were predicted by the commercial Navier-Stokes code Ellip-Sys3D, and Fig. 5 shows the predicted annual energy productions (AEP).

Figure 3: The 6 kW prototype wind turbine [5].

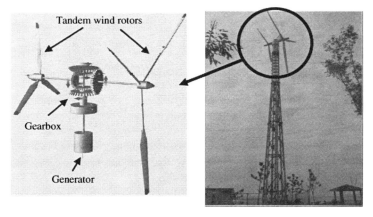

Figure 4: The 30 kW CRWT system [6].

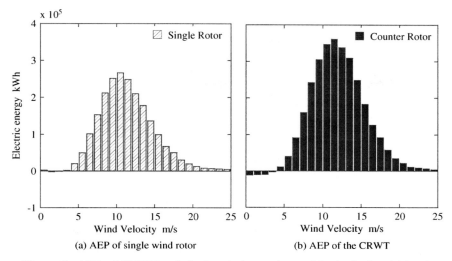

(a) AEP of single wind rotor   (b) AEP of the CRWT

Figure 5: AEP of CRWT and single wind rotor located in the isolated island.

The tandem wind rotors produce more energy than the single wind rotor, regardless of the wind velocity. The annual production at the wind velocity 5 m/s and over is 2966 MWh for the tandem wind rotors, while 2066 MWh for the single wind rotor. That is, the annual energy can be increased about 43.5%, as compared with the single wind rotor type.

It is very important to notice that the works of these previous tandem wind rotors quite differ from the works of the wind rotors which are boarded on intelligent wind power unit proposed by the authors. The proposed tandem wind rotors are in cooperation with the double rotational armature type generator.

3 Superior operation of intelligent wind power unit

Many researchers/engineers have proposed the tandem wind rotors to increase the output, as reviewed just before. Intelligent wind power unit proposed by the authors has the promising advantages that the output is not only increased but also controlled in response to the wind circumstances without accessories such as the gearbox, the pitch control mechanism, the braking system and so on. That is, the operation of the tandem wind rotors quite differ from the conventional wind rotors, and is in cooperation with the new type generator with double rotational armatures [8].

In the upwind type unit as shown in Fig. 6, the large-sized front wind rotor and the small-sized rear wind rotor drive the inner and the outer armatures in the peculiar generator, respectively, while the rotational torque is counter-balanced in both armatures. As for the tandem wind rotors, the rotational torque of the front wind rotor must be the same, in the opposite direction, as that of the rear wind rotor. The rotational direction and speed of both rotors/armatures, however, are free and adjusted automatically/smartly in response to the wind circumstances, as follows (see Fig. 7).

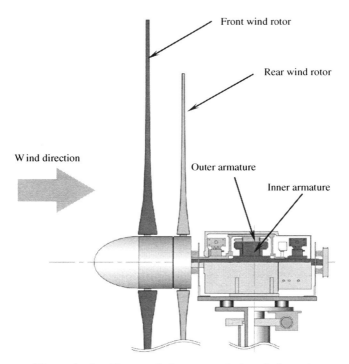

Figure 6: Intelligent wind power unit (upwind type).

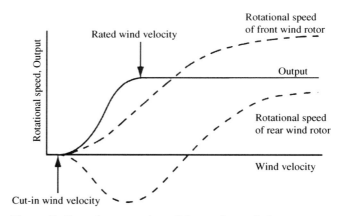

Figure 7: Superior operation of the tandem wind rotors.

1. Both wind rotors start to rotate in the counter directions at the low wind velocity, namely the cut-in wind velocity.
2. The increase of the wind velocity makes the rotational speed of both wind rotors increase, while the rotational speed of the rear wind rotor is faster than that of the front wind rotor because of its small size.

3. The rear wind rotor reaches the maximum rotational speed at the rated wind velocity. Under the rated wind velocity, both wind rotors work at the best efficiencies. Over the rated wind velocity, the rated operation mode starts, where the output is kept constant regardless of the wind velocity.
4. With more increment of the wind velocity, the rear wind rotor decreases its rotational speed gradually, stops, then it begins to rotate in the same direction of the front wind rotor, so as to its rotational torque coincides with the larger rotational torque of the front wind rotor. When the rear wind rotor rotates in the same direction to the front wind rotor, the flow in the blowing mode suppresses the abnormal rotation of the front wind rotor.

The above operating conditions enable successfully not only to increase the output but also to guarantee the quality of the electric power, namely power supply frequency at the rated operation without the traditional brake and/or the pitch control mechanisms. That is, the relative rotational speed between the front and the rear wind rotors affects directly the quality of the electric power. The power supply frequency can be kept constant when the wind rotors are designed so as to keep the relative rotational speed constant. The output can also be kept constant when the wind rotors are designed so as to keep the value, determined by the relative rotational speed and torque, constant. Moreover, the counter-rotation under the rated wind velocity makes the output higher in the poor wind circumstances, as reviewed before. That is a remarkable advantage for area which has no acceptable wind circumstance in the power generation.

4 Preparation of double rotational armature type generator

4.1 Double-fed induction generator with double rotational armatures

The doubly fed induction generator is installed in the large-scale wind power unit for the grid-connected electric power system [9]. Figure 8 shows the system diagram where the tandem wind rotors rotate the double rotational armature type generator. It is not necessary to control the wind rotor speeds, namely the rotational frequency f_w, mechanically because the revolving electromagnetic field is induced from the grid-connected electricity and its speed can be adjusted freely by the frequency f_2 modulation (the inverter) for the inner armature, namely the secondary feed. Besides, the system output P and voltage E can also be adjusted with the input P_2 and voltage E_2 for the secondary feed. That is, it is easy to keep the output, the frequency and the voltage constant as required in the grid system.

Figure 9 shows the trial model of the double rotational armature type doubly fed induction generator. The rated output is $P = 1.2$ kW at the synchronous rotational speed of 900 min^{-1}, while the output voltage is $E = 200$ V, the current is $I = 3.5$ A, the frequency is $f = 60$ Hz and the number of the pole is 8.

Figure 10 shows the bench tests for the trial model. The work of the tandem wind rotors was replaced by the motor. The motor shaft was connected to the shaft of the inner armature while the shaft of the outer armature was kept stationary

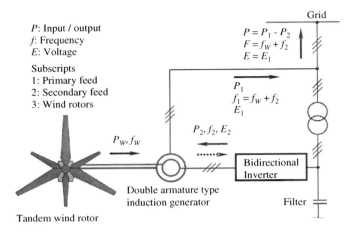

Figure 8: System diagram of double rotational armature type double-fed induction generator.

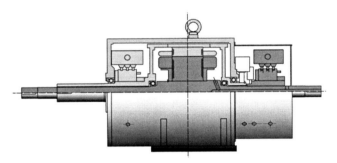

Figure 9: Trial model of the double rotational armature type double-fed induction generator.

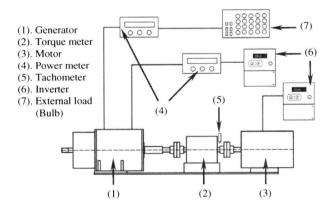

Figure 10: Bench test for the induction generator.

because the electric characteristics depend only on the relative rotational speed between both armatures. The rotational speed of the motor which represents the relative speed (in this case, rotational speed of inner armature) was controlled by the inverter. Another inverter and the transformer were used to control the input frequency f_2 and voltage E_2 of the secondary feed of the generator. The input to the generator, namely the output of the tandem wind rotors, was estimated with the rotational speed and torque. The output was measured by the power meter and consumed through electric bulbs (lamps).

In the bench test, the output frequency $f = f_1$ was kept constant at 60 Hz, and the output voltage $E = E_1$ was also kept constant at 200 V, regardless of the external bulb load P_{bulb}.

Figure 11 shows the characteristics in keeping the output voltage E_1 and frequency f_1 constant as described just above, where N_T is the relative rotational speed ($= N_F - N_R$, N_F and N_R are the outer and the inner rotational speed, and give the positive value in the N_F direction). The input frequency f_2 is in inverse proportion to the relative rotational speed N_T irrespective of the external load P_{bulb}, though the phase must be changed at the higher relative speed than the synchronous speed 900 min^{-1}. The input voltage E_2 depends not only on N_T but also on P_{bulb}, namely the output P_1. The input P_2 comes to be negative at the higher rotational speed than the synchronous speed 900 min^{-1}, as shown in Fig. 12. That is, P_2 changes fruitfully from the input to the output, and the net output from the generator is $P = P_1 - P_2$.

The output and/or the input are given in Fig. 13, for instance, when the tandem wind rotors with the relative rotational speed N_T and the rotational torque T against

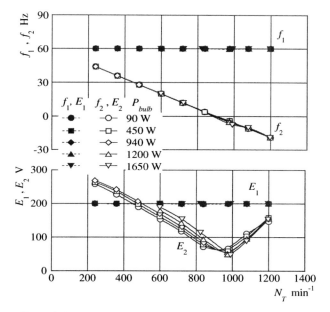

Figure 11: Effect of the external load and the relative rotational speed on the input frequency f_2 and voltage E_2.

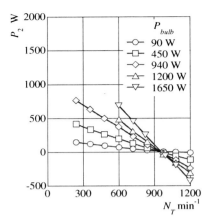

Figure 12: Effect of the external load and the relative rotational speed on the input P_2.

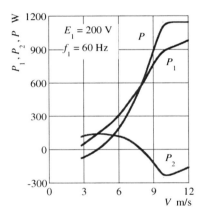

Figure 13: Output/input against the wind velocity.

the wind speed V as shown in Fig. 14 are prepared and the voltage E_2 and the frequency f_2 are controlled at $E_1 = 200$ V and $f_1 = 60$ Hz as shown in Fig. 15.

4.2 Synchronous generator with double rotational armatures

The model synchronous generator (3-phase, 4-pole, permanent magnet, AC generator with double rotational armatures) shown in Fig. 16, was prepared. The rated output $P = 1$ kW at $N_T = 1500$ min^{-1}, while the induced frequency $f = 50$ Hz, and the induced voltage $E = 100$ V.

Figure 17 shows the bench tests for the trial model. The armatures were driven by two isolated motors in place of the tandem wind rotors, to confirm experimentally the rotational torque of the inner and the outer armatures respectively. The shaft of the motor was directly connected to the shaft of the inner armature,

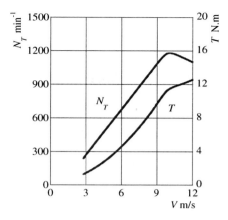

Figure 14: The relative rotational speed and torque of the tandem wind rotor against the wind velocity.

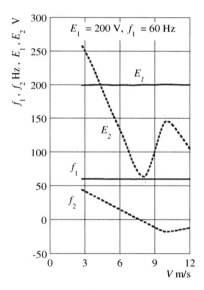

Figure 15: Control of the frequency f_2 and voltage E_2 against the wind velocity.

while the rotation of another separate motor is transmitted to the outer armature through the timing belt.

Figure 18 shows the relation between the outer armature rotational torque T_F^* and the inner armature rotational torque T_R^*, in keeping the rotational speed of the outer armature constant at $N_F = 750$ min^{-1}, while changing the rotational speed of the inner armature N_R at the various external loads. These rotational torques do not include the mechanical torques such as the bearings in the generator, and T_F^* is given by the absolute value though the direction of T_R^* is against T_F^*. It is obvious

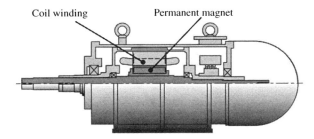

Figure 16: Trial model of the double rotational armature type synchronous generator.

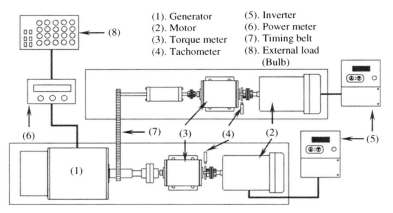

(1). Generator
(2). Motor
(3). Torque meter
(4). Tachometer
(5). Inverter
(6). Power meter
(7). Timing belt
(8). External load (Bulb)

Figure 17: Bench test for the synchronous generator.

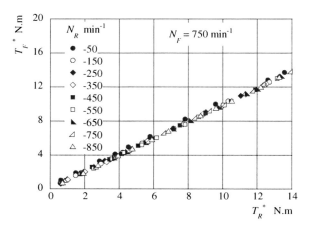

Figure 18: Relation between the torque of the inner and outer armatures.

that both torques show the same absolute value irrespective of the relative rotational speed and the output/load, because the rotational torque should be always dynamically counter-balanced between the inner and the outer armatures.

Figure 19 shows the rotational torques T_F and T_R at the various relative speeds N_T against the induced electric current I_G, where these torques include the mechanical torques in the generator and T_R is given by the absolute value. That is, these are the shaft torques in practical use. The rotational torques of both shafts T_F, T_R are directly proportional to the induced electric current I_G irrespective of the rotational speeds and/or the external load but the shaft torque T_F of the outer armature is slightly larger than T_R of the inner armature due to the mechanical torque of the larger bearings.

The output P and the induced voltage E against the induced electric current I_G are determined by the relative rotational speed N_T ($=N_F - N_R$), as shown in Fig. 20. The output increases with the increase of the induced voltage E at the same I_G, while E is proportional to the relative rotational speed N_T and I_G determines the rotational torque as presented in Fig. 19.

5 Demonstration of intelligent wind power unit

5.1 Preparation of the tentative tandem wind rotors

5.1.1 Front blade design
Figure 21 shows the drag and the lift coefficients C_D, C_L against the angle of attack a for MEL012 aerofoil [10]. The angle giving the maximum C_L is $a = 11°$, but the angle $a = 8°$ was chosen as the design dimension irrespective of the radial position, so as to avoid the flow stall on a large scale at the lower rotational speed. The chord of the blade was designed taking the traditional blade element theory [11] into

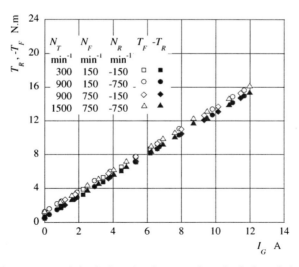

Figure 19: The output and the induced voltage against the induced electric current.

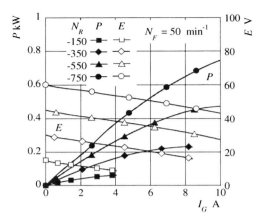

Figure 20: The output and the induced voltage against the induced electric current.

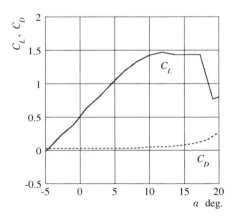

Figure 21: Drag and lift coefficient against the attack angle.

account. The flow through the wind rotor is contributed by the axial and the tangential induction factors (wind velocity deceleration ratio), and the relative flow angle at the inlet of the front wind rotor ϕ_F measured from the tangential direction and the chord C at the radius r are given as follows [12], where λ_r is the speed ratio at the radius r [=(rotational speed at r)/wind velocity].

$$\phi_F = K \tan^{-1}(1/\lambda_r), \quad K = \tfrac{2}{3} \tag{1}$$

$$C = \frac{8\pi r}{Z_F C_L}\left[1 - \cos(\phi_{Fin})\right] \tag{2}$$

Figure 22 shows the blade profile with MEL012 aerofoil elements [10] which is designed by the above procedures.

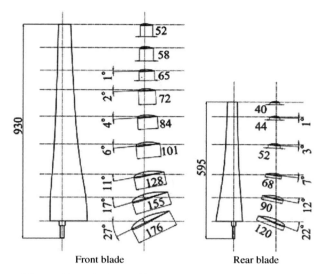

Figure 22: Blade profiles of the tandem wind rotors.

5.1.2 Rear blade design

The counter-rotational torque of the rear wind rotor must coincide with the rotational torque of the front wind rotor designed above, but the convenient design procedures for the rear blade interacting with the front wind rotor is not yet known. Accordingly, the flow around the front wind rotor was simulated by the commercial CFD code CFX-10 with k–ε turbulent model, to get the inlet flow condition of the rear wind rotor. In the simulation, the circularly cylindrical field with $20d_F$ radius (d_F: the diameter of the front wind rotor) was set, and was divided into two domains. One is the rotational domain surrounding the wind rotor with 80,668 tetra-type elements (mesh size is 0.5 mm in close to the blade surface) and 6294 prism-type elements in the radial direction with $2d_F$, while $1d_F$ upstream and $3d_F$ downstream of the rotor. The other is the stationary domain with 799,060 prism-type elements surrounding the rotational domain, and both domains were joined with the Frozen-Rotor interface keeping the relative blade position of the tandem wind rotors. The flow around the tandem wind rotors with two-dimensional blades was simulated in a trial by the above code and is shown in Fig. 23 accompanied with the plotted experimental results, where V_{Mtm} and $V_{\theta tm}$ are the axial and the swirling velocity components divided by the wind velocity V at the wind tunnel outlet, R is the dimensionless radius divided by $d_F/2$, Z_F and Z_R are the blade numbers of the front and the rear wind rotors, and the measurement sections M1–M3 and the blade profile are also given in the figure. It was confirmed that the above code can predict roughly the actual axial velocity, and the flow conditions are discussed in [13].

The flow around the front wind rotor equipped with three blades shown in Fig. 22 was simulated by the above code. Figure 24 shows the relative flow angles measured from the tangential direction, where ϕ_{Fin} and ϕ_{Fout} are the simulated relative flow angles averaged in the tangential directions at inlet and the outlet sections of the

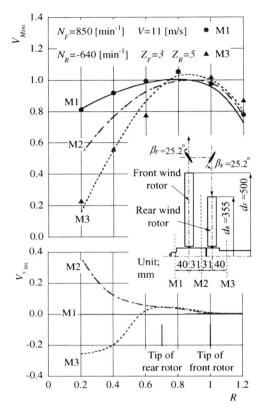

Figure 23: Flow around the tandem wind rotors composed of two-dimensional blades.

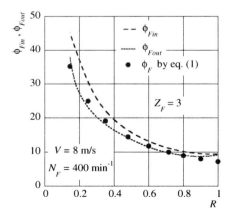

Figure 24: Relative flow angles measured from the tangential direction.

front wind rotor, ϕ_F is the outlet flow angle determined by eqn (1) at each radial position of the front blade. The angle ϕ_F is close to ϕ_{Fout}, that is, ϕ_F may give the outlet flow angle. Accordingly, the rotational torque of the rear wind rotor, namely the angular momentum change, was determined with the simulated outlet flow angle ϕ_{Fout} of the front wind rotor, which is corresponding to the inlet flow angle of the rear wind rotor, and the outlet flow angle of the rear wind rotor predicted by eqn (1).

Figure 22 also shows the designed rear blade with MEL012 aerofoil elements [10], where the chord is also derived from eqn (2) but the height was determined tentatively.

5.2 Preparation of the model unit and operations on the vehicle

The synchronous generator (Fig. 16) equipped with the tentative tandem wind rotors (Fig. 22) were mounted on the circularly cylindrical tower as the down-wind type, and boarded on the pick-up type truck as shown in Fig. 25. The wind velocity can be controlled by the driving speed of the truck, and also measured by the cup-type anemometer. The rotational speeds of the wind rotors, the output, the induced voltage and the induced electric current were automatically accumulated and stored in the personal computer every one second, in accompanying with the data of the wind velocity. The output was consumed by many bulbs, but these resistances depend mainly on the induced voltage, namely the rotational speed, and the induced electric current affecting the temperature of the filament. Therefore, the resistance was replaced by the specified power of the bulb at 100 V, P, called "bulb load" as the indication for the external load.

The driving site is at the seashore of Wakamatsu, Kitakyushu, Japan, where there are two straight roads of about 1.5 km in the north–south and the east–west directions. The wind is calm and its direction is nearly constant, that is, the site is suitable for making the experiments using the truck. The truck is used to be driven in three modes, which are the acceleration, steady state, then the deceleration. Figure 26 shows one of the rotational speeds of the front and the rear wind rotors N_F, N_R, and the wind velocity V

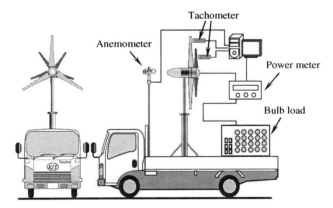

Figure 25: Field experiment on the pick-up type truck.

accumulated from start to stop of the truck. The data corresponding to the acceleration and deceleration of the truck were excluded for the evaluations of the experimental data, in order to get the results at the steady-state flow conditions.

5.3 Performances of the tandem wind rotors

The effects of the rear blade setting angle β_R on the performances are shown in Fig. 27, in keeping the front blade setting angle $\beta_F = 20°$ and the bulb load

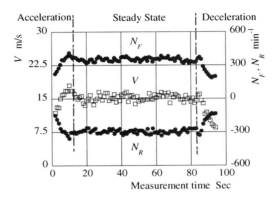

Figure 26: Accumulated rotational speeds while driving the truck.

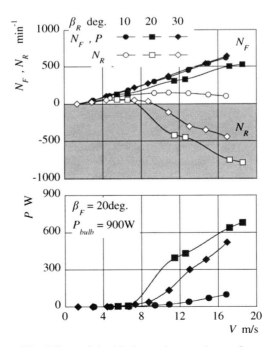

Figure 27: Effect of the blade setting on the performance.

P_{bulb} = 900 W, where the blade settings are defined as shown in Fig. 28 at the blade tips. The rear wind rotor never counter-rotates against the front wind rotor at the lower wind velocity V because not only the rear wind rotor cannot generate the sufficiently counter-rotational torque T_R corresponding to T_F of the front wind rotor but also the outer armature pulls the inner armature by the magnetic force. With the increase of the wind velocity, the rear wind rotor with β_R larger than 20° begins to counter-rotate successfully. On the contrary, the rear wind rotor with β_R = 10° never change the rotational direction and is in the blowing mode because the rear blade with the excessively larger angle of attack cannot generate the fruitful counter-rotational torque due to the flow stall on larger scale. Judging from the counter-rotation expected to this type unit, the rear wind rotor with β_R = 20° may be acceptable within the measured data.

Figure 29 shows the effect of the bulb load on the operating conditions at the blade setting angles $\beta_F = \beta_R = 20°$. The load P_{bulb} affects obviously the rotational speed of

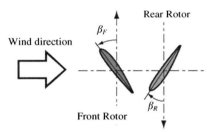

Figure 28: Blade setting angles.

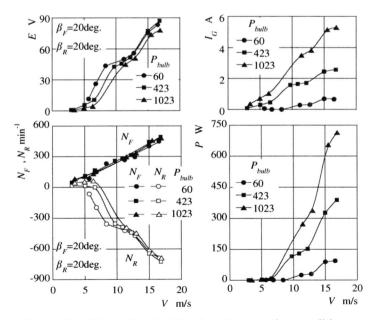

Figure 29: Effect of the bulb load on the operation conditions.

the rear wind rotor N_R in the lower wind velocity than 12 m/s, but scarcely affects the speed of the front wind rotor N_F regardless of the wind velocity V. Besides, the rear wind rotor starts to counter-rotate at the slower wind velocity with the decrease of the bulb load. The relative rotational speed $N_T = N_F - N_R$ contributes proportionally to the induced voltage E, the induced electric current I_G depends on the load, and the output P is obtained electrically from $\sqrt{3}EI$, while given mechanically by $T\omega_T$ ($T = T_F = -T_R$, ω_T: the relative angular speed). Within the measured data, the operating condition with the bulb load 1023 W gives the higher output P.

5.4 Trial of the reasonable operation

As recognized in the above discussions, the performances are affected not only by the wind velocity V but also by the blade setting angles β_F, β_R and the bulb load. Then, this wind power unit was operated impractically but reasonably, in trail, by changing the rear blade setting angle β_R and the bulb load, while keeping the front blade setting angle $\beta_F = 20°$. The output was kept constant at $P = 430$ W in the wind velocity higher than $V = 12.6$ m/s, and the unit was operated at the optimum rotational speed giving the maximum output in the slower wind velocity. Figure 30 shows the reasonable operation, where the adjusted rear blade setting angle and bulb load are denoted with (β_R (°), P_{bulb} (W)), and the experimental data are plotted and represented by the curve. The bulb load P_{bulb} is not related directly to the output P because the bulb resistances are affected by the filament temperature

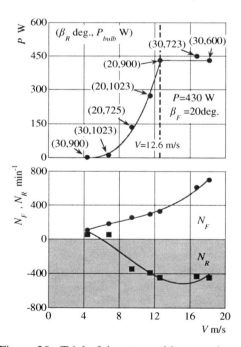

Figure 30: Trial of the reasonable operations.

as mentioned before. This type unit may be successfully operated by adjusting suitably the blade setting angle and the external load. Such operations, however, are not expected and not acceptable for the proposed wind power unit because the rotational speeds must be adjusted automatically in response to the wind velocity. The final target of the serial researches is to get smartly the above operation by the superior works of the tandem wind rotors in cooperation with the double armatures of the generator presented before. Consequently, it is necessary to improve more and more the blade profiles suitable for the front and the rear wind rotors.

6 Optimizing the profiles of tandem wind rotors

The tandem wind rotors have been investigated as reviewed before. In the proposed tandem wind rotorts, however, the rear wind rotors must counter-rotate against the front wind rotor, stop, and rotate in the same direction of the front wind rotor, with the increase of the wind velocity. This chapter is the status report for optimizing the wind rotor profiles. The authors pay attention, here, not only to the wind rotor profiles but also to the arrangements between both wind rotors.

6.1 Experiments in the wind tunnel

The diameter of the front wind rotor was kept constant at $d_F = 500$ mm, and the diameter of the rear wind rotor was changed from $d_R = 260$ to 560 mm, for the experiments in the wind tunnel. The axial distance between the front and the rear wind rotors (the twist centres of the blade) was also changed from $l = 40$ to 200 mm. The blade profiles of the front and the rear wind rotors are shown in Fig. 31. The blade E is supplied to the front and/or the rear wind rotors, and has not the camber and the twist. The front blade G is formed with MEL002 aerofoil elements [10] and has the twist to get the desirable angle of attack $a = 7°$, as the single wind

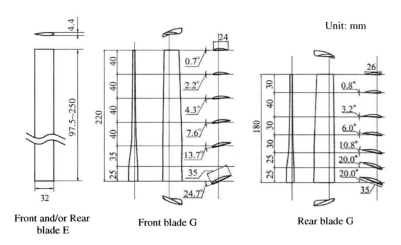

Figure 31: Blade profiles of model tandem wind rotors.

rotor with the blade tip speed ratio $\lambda = 6$ [=(blade tip speed)/(wind velocity)]. The rear blade G is also formed with MEL002 and is twisted taking account of the flow behind the front wind rotor. The front and the rear blade numbers are $Z_F = 3$, $Z_R = 5$ which are optimized in the previous paper [13]. Hereafter, the tandem wind rotors are called "Tandem wind Rotor EE" which is composed of the front blade E and the rear blade E, where the front and the rear blade setting angles are $\beta_F = \beta_R = 11°$ (see Fig. 28). "Tandem Wind Rotor GE" is composed of the front blade G and the rear blade E, while $\beta_F = 10°$, $\beta_R = 16°$. "Tandem Wind Rotor GG" is composed of the front blade G and the rear blade G, while $\beta_F = 5°$, $\beta_R = 15°$. "Single Wind Rotor G" consists of with the front blade G, while $\beta_F = 0°$. These blade setting angles were optimized at the preliminary experiments [14, 15].

The model wind rotors shown in Fig. 32 were set perpendicular to the wind direction, at 185 mm downstream of the wind tunnel outlet with the nozzle diameter of 800 mm. The front and the rear wind rotors were connected directly and respectively to the isolated motor with the inverter, in place of the peculiar generator. The rotational torques of the front and the rear wind rotors were counter-balanced by the rotational speed control. The performances of the model wind rotor in the following discussions are evaluated without the mechanical torques of the bearings and the pulley system. The flow conditions discharged from the wind tunnel outlet, namely the inlet flow conditions of the wind rotors, are almost uniform and axis-symmetry without the swirl.

The outputs of the model wind rotors may be smaller than those of the prototypes, as the Reynolds number estimated at the front blade tip is $(0.8–2.2) \times 10^5$ in the experiments.

6.2 Optimum diameter ratio of front and rear wind rotors

To know the optimum diameter ratio D_{RF} [$=d_R/d_F$], the performances of the tandem wind rotors were investigated by changing the diameter of the rear wind rotor, and are shown in Fig. 33. The output coefficient is defined by $C_P = P/(\rho A V^3/2)$ and

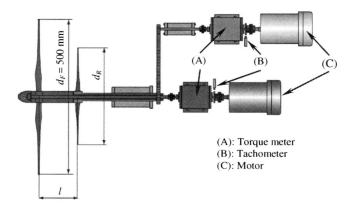

(A): Torque meter
(B): Tachometer
(C): Motor

Figure 32: Model tandem wind rotors for performance experiment.

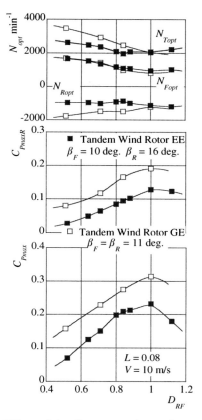

Figure 33: Effect of the diameter ratio on the performance.

the maximum output coefficient at each D_{RF} is denoted by $C_{P,\max}$, where P is the output, A is the rotational area of the front wind rotor, ρ is the ambient air density, V is the wind velocity, N is the rotational speed [subscript F, R, T: the front, the rear and the relative rotational speeds, $N_T = N_F - N_R$ (the rotational direction of N_F is positive), subscript 'opt' means the rotational speed giving $C_{P,\max}$ and L is the dimensionless axial distance between both wind rotors [$= l/d_F$]. Besides, the blade setting angles β_F and β_R of these wind rotors were optimized [14, 15]. With the increase of the rear rotor diameter, that is D_{RF} becomes larger, the rotational speed of the front wind rotor N_{Fopt} becomes slower, with almost the same speed, regardless of the blade profiles. The front blade profiles, however, affects markedly the rotational speed of the rear wind rotor N_{Ropt}, with the decrease of D_{RF}. The front blade G with the reasonable twist makes the rear wind rotor speed N_{Ropt} comparatively faster at the smaller D_{RF} (see Tandem Wind Rotor GE), in comparison with the front blade E without the twist. That may be caused from the difference of the flow condition at the rear wind rotor inlet, namely, the front wind rotor outlet. That is, the two-dimensional front blade E has unacceptable flow separation whose scale is large at the smaller radius due to the poor and negative angle of attack, and the separation disturbs the smoothly rotation of the rear wind rotor. Such rotational speeds resultantly contribute the relative rotational speed N_{Topt}.

The maximum output $C_{P,max}$ is larger with increment of D_{RF} and has the maximum value where D_{RF} is about 1.0, and $C_{P,max}$ of Tandem Wind Rotor GE composed of the twisted blades is higher than that of Tandem Wind Rotor EE. As recognized in Fig. 33, the maximum output coefficient $C_{P,max}$ is affected mainly by the output of the rear wind rotor $C_{P,maxR}$.

Figure 34 shows the rotational behaviours of the tandem wind rotors. The rotational directions are changed drastically at about $D_{RF} = 0.84$, in the smaller N_T. That is, the rear wind rotor rotates in the same direction as the front wind rotor, while D_{RF} is less than 0.84. On the contrary, the front wind rotor rotates unfortunately in the same direction as the rear wind rotor, while D_{RF} is more than 0.84. Such rotational behaviour of the front wind rotor is not acceptable as for the proposed intelligent wind power unit because the front wind rotor must keep the rotational direction as expected in Chapter 3 (see Fig. 7).

The maximum output coefficient and the rotational behaviours against D_{RF} are independent to the front blade profiles, as recognized in Figs 33 and 34. Therefore,

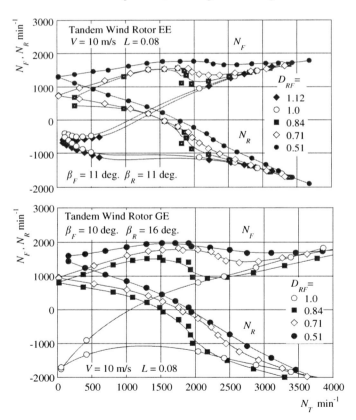

Figure 34: Effect of the diameter ratio on the rotational behaviours of the tandem wind rotors.

the desirable diameter ratio getting the sufficient output and playing the expected rotational behaviour is in close to $D_{RF} = 0.84$.

6.3 Optimum axial distance between front and rear wind rotors

Figure 35 shows the effect of the axial distance between the front and the rear wind rotor, measured at the twist centre of the blade, on the performances, where D_{RF} of Tandem Wind Rotor GE is 0.84 and D_{RF} of Tandem Wind Rotor EE is 0.71. The rotational speed of the front wind rotor N_{Fopt} becomes slower with the decrease of L because of the flow interaction between both wind rotors, regardless of the blade profile. The rotational speed of the front wind rotor with blade G is slightly slower than that of the front blade E, but the front blade G makes the rear wind rotor speed faster. Resultantly, the relative rotational speed N_{Topt} of Tandem Wind Rotor GE is slightly faster than that of Tandem Wind Rotor EE. Such characteristics are scarcely affected by the diameter ratio D_{RF}. The maximum output coefficient $C_{P,max}$ is higher at the smaller L because the rear wind rotor can get effectively the wind energy as recognized in Fig. 36.

Besides, $C_{P,max}$ of Tandem Wind Rotor GE is markedly higher than that of Tandem Wind Rotor EE because GE has not only the optimum but also the desirable blade profile.

These data suggest that the rear wind rotor should be set as close as possible to the front wind rotor, taking the bending moment and the vibration of the blade into account.

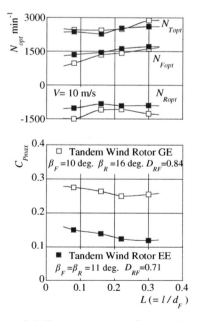

Figure 35: Effect of the axial distance between front and rear wind rotors on the performances.

6.4 Characteristics of the tandem wind rotors

Figure 37 shows the output coefficient C_P against the relative tip speed ratio λ_T [=(relative tip speed)/(wind velocity)], in comparison with C_P of Single Wind Rotor G. The relative tip speed ratio giving the maximum output of Tandem Wind Rotor GE is faster than that of Single Wind Rotor G, but the output coefficient of Tandem Wind Rotor GE is unacceptable. The latter is caused by the rear wind rotor without the twist. Then, the rear baled E was modified to the blade G (see Fig. 31) designed so as to get the desirable angle of attack ($a = 11°$) irrespective of the radial position taking account of the flow conditions at the rear wind rotor inlet. The maximum output coefficient can be improved successfully but the relative tip speed ratio λ_T may be unacceptable as shown in Fig. 38. That is, the blade profiles must be optimized more as the tandem wind rotors.

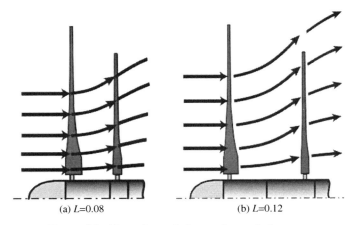

(a) $L=0.08$ (b) $L=0.12$

Figure 36: Flow through the tandem wind rotors.

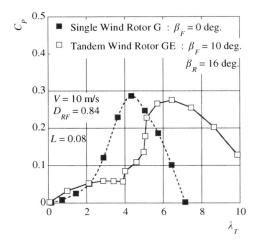

Figure 37: Output of tandem wind rotor GE.

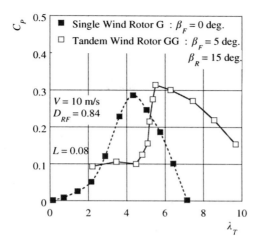

Figure 38: Output of tandem wind rotor GG.

Figure 39 shows the performances in the generating mode against the wind velocity, where the rated wind speed was set at $V = 10$ m/s and T is the rotational torque. The rotational speeds of the front and the rear wind rotors were adjusted at the best relative tip speed ratio giving the maximum C_P in the slower wind speed, and adjusted so as to keep the output constant in the rated operation. The output is controlled well with the both wind rotor speeds in place of the traditional mechanism. The abnormally high rotation of the front wind rotor N_F can also be suppressed well by the rotation of the rear wind rotor N_R in the blowing mode, at the faster wind velocity.

In the prototype connected to the grid system, the rotational speeds are adjusted easily by the double rotational armature type doubly fed generator with the secondary feed.

7 Conclusion

Intelligent wind power unit, which is composed of the tandem wind rotors and the double rotational armature type generator without the traditional stator, was proposed.

The performances of the preliminary model tandem wind rotors were investigated experimentally, as the first step to optimize the profiles of the unit. The desirable profiles as the tandem wind rotors are that the diameter ratio is D_{RF} = (rear wind rotor)/(frond wind rotor) = 0.84 and the axial distance between both wind rotors should be set as short as possible.

The authors are carrying on this research project to be put this wind power unit to practical use. The optimum profiles will be presented near future.

Some parts of this research project were co-sponsored by Fukuoka Industry, Science & Technology Foundation, JFE 21st Century Foundation, Research Project: Grant-in-aid for Scientific Research (c) (2) in Japan, Japan Science and Technology Agency, Iwatani Naoji Memorial Foundation and Research Project: Grant-in-aid for JSPS fellow.

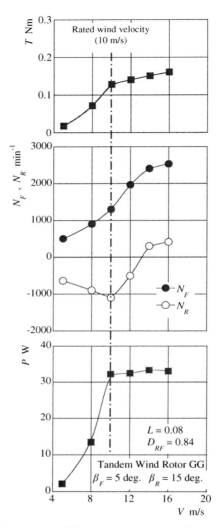

Figure 39: Performance of the tandem rotors in the generating mode.

References

[1] Soerensen, H.C., Larsen, J.H., Olsen, F.A., Svenson, J. & Hansen, S.R., Middelgrunden 40 MW offshore wind farm: a prestudy for the Danish offshore 750 MW wind program. *Proc. of 10th International Offshore and Polar Engineering Conference*, **1**, pp. 484–491, 2000.

[2] David A. Sper., *Wind Turbine Technology*, ASME Press: New York, p. 93, 1994.

[3] Ushiyama, I., Shimota, T. & Miura, Y., An experimental study of the two staged wind turbines. *Proc. of World Renewable Energy Conference*, pp. 909–912, 1996.

[4] Jang, T.J. & Heo, H.K., Study on the development of wind power system using mutually opposite rotation of dual rotors, *Proc. of the Renewable Energy 2008*, CD-ROM O-WE-015, 2008.

[5] Appa, K., Counter rotating wind turbine system. Energy Innovations Small Grant (EISG) Program Technical Report, 2002.

[6] Jung, S.N., No, T.S. & Ryu, K.W., Aerodynamic performance prediction of 30kW counter-rotating wind turbine system. *Renewable Energy*, **30**(5), pp. 631–644, 2005.

[7] Shen, W.Z., Zakkam, V.A.K., Sørensen, J.N. & Appa, K., Analysis of counter-rotating wind turbines. *Journal of Physics: Conference Series*, **75(012003)**, 9 pp., 2007.

[8] Kanemoto, T. & Galal, A.M., Development of intelligent wind turbine generator with tandem wind rotors and double rotational armatures (1st report, Superior Operation of Tandem Wind Rotors). *JSME International Journal*, Ser. B, **49**(2), pp. 450–457, 2006.

[9] Muller, S., Deicke, M. & De Doncker, R.W., Adjustable speed generator for wind turbines based on doubly-fed induction machines and 4-quadrant IGBT converter linked to the rotor. *Conf. Record of the 2000 IEEE*, **4**, pp. 2249–2254, 2000.

[10] Performance and geometry of airfoils supply system (PEGASUS), online, http://riodb.ibase.aist.jp/db060/index.html

[11] Kaya, Y., *New Energy Thesaurus* (in Japanese), Kogyo Chosakai Publishing: Tokyo, p. 434, 2002.

[12] Ushiyama, I., *Wind Turbine Technology* (in Japanese), Morikita Publishing: Tokyo, pp. 87–107, 2002.

[13] Kanemoto, T. & Galal, A.M., Development of intelligent wind turbine generator with tandem wind rotors and double rotational armatures. *JSME International Journal of fluid Science and Technology*, Ser. B, **49**(2), pp. 450–457, 2006.

[14] Galal, A.M., Kanemoto, T., Konno, Y., Ikeda, K. & Inada, Y., Intelligent wind turbine generator with tandem rotors applicable to offshore wind farm (flow conditions around tandem rotors equipped with two dimensional blades). *Proc. of the 16th International Offshore and Polar Engineering Conference and Exhibition*, **1**, 2006.

[15] Kubo, K. & Kanemoto, T., Development of intelligent wind turbine unite with tandem wind rotors and double rotational armatures (2nd report, characteristics of tandem wind rotors). *JSME International Journal of Fluid Science and Technology*, **3**(3), pp. 370–378, 2008.

CHAPTER 11

Offshore wind turbine design

Danian Zheng[1] & Sumit Bose[2]
[1]*Infrastructure Energy, General Electric Company, USA.*
[2]*Global Research Centre, General Electric Company, USA.*

Offshore wind plants have become a real option in the EU countries when onshore wind plants quickly occupy the prime land for wind energy generation. This chapter reviews the state of the art and some of the technical challenges anticipated during the development of U.S. offshore wind power plants in the U.S. It highlights challenges that are unique to the coastal U.S., e.g. hurricanes, in addition to the general challenges for the offshore wind industry, irrespective of geographical location. It will be shown that experience from the oil and gas industry is not completely applicable to the wind industry, although it certainly should be leveraged and modified to fit the unique needs of offshore wind. Finally, some of the research needed to further address these challenges are outlined.

1 Introduction

Offshore wind presents a tremendous opportunity for the United States and Europe. Onshore wind plants often face land use disputes, noise, and visual impact obstacles. Moving wind plants offshore not only mitigates these problems, but provides other important advantages including:

- Availability of large continuous areas suitable for major wind plants
- Higher wind speeds, which generally increase with distance from the shore
- Less turbulence, which allows turbines to harvest energy more effectively
- Reduced fatigue loads on the turbine
- Lower wind shear allowing shorter towers

The United States enjoys significant onshore and offshore resources. While onshore plants are more cost effective to develop than offshore ones, some of the onshore sites in the U.S. have the following disadvantages:

- They are away from the population centers in the U.S. More than half of the United States population lives at or near shore.
- Similarly (and perhaps as a consequence of 1 above), these onshore resources are away from the grid connections.
- They are mostly in areas which do not have significant electricity demand thus depressing electricity prices in those areas and upsetting the economical structure for wind developers.

On the other hand, some regions of New England and the Mid-Atlantic states have some of the highest electricity prices, thus making wind and other (generally more expensive) renewable energy sources more cost effective. East coast states also have high land usage preventing the development of onshore wind plants or even the development of fossil-based onshore power generation stations. These states also have readily available grid connectivity close to the shore where most of the population is located.

While onshore wind plants are well understood, offshore installations present unique challenges to the wind industry, particularly as coastal communities demand them to be placed further off shore. These challenges are discussed in the following sections.

2 Offshore resource potential

Significant offshore wind energy resources are available in the United States. Figure 1 shows the U.S. wind power resource with superb wind power availability around the coastal areas. Table 1 shows the offshore wind resources along U.S. coast. Looking at Fig. 1 and Table 1, one could argue the United States enjoys significant wind resources inland, and much of these resources are in mid-western states with limited land use or visibility issues. However, while development of these inland areas is more cost effective than offshore plants, they are unlikely to be fully developed for several reasons:

- Distance from population centers – more than half of the U.S. population lives at or near a coastline. Developing midwestern sites results in significant energy transportation costs.
- Distance from grid – similarly, available inland resources are not near grid connections.
- Low local demand – available midwestern sites primarily fall in areas which do not have significant electricity demand, thus depressing electricity prices in those areas.

On the other hand, it is clear that both New England and the mid-Atlantic regions have significantly more offshore wind resource than is available on land. In addition, New England and the mid-Atlantic states have the highest electricity prices, thus making wind and other renewable energy sources more cost effective. East coast states also have higher land usage, which tends to prevent development of inland

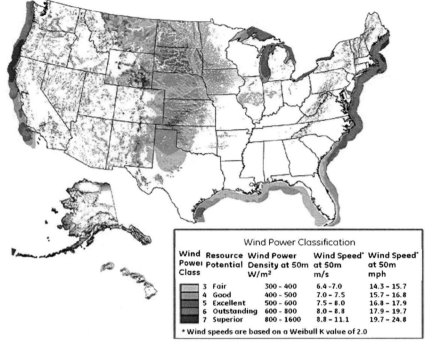

Wind Power Classification				
Wind Power Class	Resource Potential	Wind Power Density at 50m W/m²	Wind Speed* at 50m m/s	Wind Speed* at 50m mph
3	Fair	300 - 400	6.4 - 7.0	14.3 - 15.7
4	Good	400 - 500	7.0 - 7.5	15.7 - 16.8
5	Excellent	500 - 600	7.5 - 8.0	16.8 - 17.9
6	Outstanding	600 - 800	8.0 - 8.8	17.9 - 19.7
7	Superior	800 - 1600	8.8 - 11.1	19.7 - 24.8

* Wind speeds are based on a Weibull K value of 2.0

Figure 1: U.S. wind resource map [1].

Table 1: Offshore wind resource (MW) [2].

Region	5–20 Nautical Miles			20–50 Nautical Miles		
	Shallow < 30 m	Deep	% Exclusion	Shallow < 30 m	Deep	% Exclusion
New England	9,900	41,600	67%	2,700	166,300	33%
Mid Atlantic States	46,500	8,500	67%	35,500	170,000	33%
California	2,650	57,250	67%	0	238,300	33%
Pacific Northwest	725	34,075	67%	0	93,700	33%
Totals	59,775	141,425	67%	38,200	668,300	33%

wind power plants, or even development of other types of power generation stations. Finally, these states have readily available grid connectivity close to shore and the major population centers.

3 Current technology trends

Trends in the global wind industry are pushing toward larger, multi-megawatt turbines. Currently, Vestas has produced more than 740 MW of offshore turbines consists mostly of the V90 3 MW turbines. A new V120 4.5 MW turbine from the NEG MICON merger is in the works. GE holds a 3.6 MW turbine specifically for the offshore market and just acquired a 3.5 MW direct-drive product through merger with SCANWIND. REpower, Multibrid, and Enercon are known to be testing designs in the 5–7 MW and >100 m rotor diameter range. Nordex has a 2.5 MW N90 offshore product. On the more advanced pace, Siemens is demonstrating their 2.3 MW 93 m rotor turbine on a floating foundation suitable for deep water and has a 3.6 MW 120 m rotor turbine in the pocket. Gamesa recently introduced an advanced compact G10X 128 m rotor 4.5 MW turbine with a 120 m concrete tower that might serve as a low cost offshore wind turbine in the future.

The current industry trend to increase turbine power and blade sizes is based on two fundamental assumptions. First, the trend assumes that the cost of foundations and other balance-of-plant items do not increase linearly with the turbine's power. Second, the assumption is made that the cost of operating and maintaining a smaller number of bigger turbines is lower than operating and maintaining large number of smaller turbines.

In reality, offshore plants break both assumptions almost completely. With existing technology, foundation costs increases significantly as the plant is placed at deeper sites. Similarly, operations and maintenance costs also increase as plants are placed further offshore. Therefore, when considering offshore plants, it is no longer true that industry should blindly head toward larger turbines. With offshore plants, turbine size is not the only factor of importance unless the push toward increasing turbine sizes is coupled with significant reductions in foundation, operation and maintenance costs, as well as improvements in reliability and availability, to achieve the potential and viability of offshore wind power plants.

4 Offshore-specific design challenges

Significant wind turbine design innovations are required for offshore environment. Characterizing the potential development barriers and identifying the technologies to overcome them will be the important aspects of these innovations. Offshore wind plants present challenges that can be broadly grouped into three categories:

- Economic challenges
- 25-m barrier challenge
- Design envelope challenge

4.1 Economic challenges

The capital cost of offshore wind plant foundations increases as sea depth increases. While the current onshore wind plant foundations represent around 10% or less of the total overall capital cost, the foundation cost jumps to 20% simply by placing the turbine offshore. As shown in Fig. 2, the foundation contribution to the overall capital cost steadily increases as the sea depth increases. As foundations do not directly contribute to any increase in generating power, the increased cost of foundations directly increases the cost of electricity.

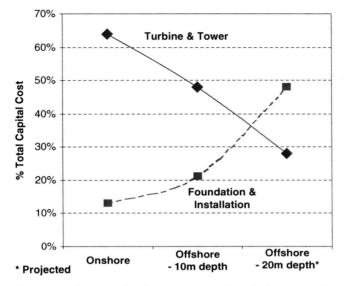

Figure 2: The foundation contribution to the overall capital cost steadily increases as the sea depth increases [3].

Table 2: Dutch North Sea offshore wind available area vs. distance to coast [4].

Distance to coast	Area (km^2)	
	Depth < 20 m	Depth < 40 m
>8 km	1,700	22,000
>12 miles	680	20,000

Due to the cost increase along with increasing depth, the often-encountered local public demands to place the offshore plants further from the coastline results in increased sea depth due to the reduction of available shallow seabed. For example, the area with depth less than 20 m off Dutch North Sea coast, which is well fit for offshore development, reduces about 2/3 from 8 km to 12 miles away from the coast (Table 2). Hence moving wind plant further into the sea will in general increase the cost of foundation and eventually the cost of electricity.

4.2 25-m barrier challenge

The U.S. wind energy industry currently faces what is termed the "25-meter barrier" challenge. Simply put, this means that wind plants placed deeper than 25 m are essentially outside of the existing technical capabilities for manufacturing and construction limitations as well as design limitations.

• Manufacturing and construction limitations:
Even in benign soil conditions, monopile foundation, which is the most popular offshore wind turbine foundation currently, needs a diameter that reaches 5.5 m

with a thickness extending to 90 mm at 25 m water depth from around 5 m and 50 mm accordingly at 10–12 m depth. The resulted weight could be 250 metric tons. This level of dimensions and weights stretch the maximum capability of the fabrication factories and installation vessels to their limits. In fact, only a few vessels in the world can even handle the hammering requirements of such a large pile.

• Design limits:

At 25 m depth, and for a hub-height of approximately 75 m above sea level, a monopile foundation's natural frequency becomes dangerously close to the turbine's natural frequency. The engineering solution is to increase either pile diameter or thickness, which is either at the manufacturing limit or at the construction limit already.

4.3 Overcoming the 25-m barrier

Due to these manufacturing and design limitations, the 25-m barrier is quite real, especially as coastal communities demand wind power plants to be placed even farther offshore. However, overcoming the 25-m depth is not without precedent and this gives strong hope for the wind industry to cross the barrier.

Before 1996, the gravity foundation was the industry's standard offshore foundation. However, in 1996, the industry reached the gravity foundation potential, approximately 7.5 m deep. Danish research performed in 1996–1997 recommended Danish wind industry to use monopile foundation for the deeper sites. Since then, the monopile has become the primary foundation choice for the offshore wind industry.

As an additional benefit from the 1997 monopile development, overcoming the gravity foundation depth limit gave Danish companies a significant lead in the offshore wind development race. While the offshore challenges are more significant for the United States as compared with European sites for reasons such as higher waves, susceptibility to hurricanes, and increased wave breaking energy, the 25-m barrier presents the United States with a significant opportunity for technology break through and leadership in the offshore wind industry.

The opportunity exists for the United States to spearhead a development effort similar to the European effort to break the 7.5-m barrier. Because of the coastal community demands for farther offshore deployment location, overcoming the 25-m barrier will be the enabler for the growth of the offshore wind power plants market in the United States.

The offshore wind challenges are not without parallel in other U.S. industries, particularly oil and gas. Given its offshore oil and gas experience, the United States is ideally suited to assume the lead on attacking the 25-m barrier challenge. However, it would be a mistake to assume that the solution amounts to the direct translation of the oil and gas experience to the wind industry.

The magnitudes of loads affecting oil and gas structures are much higher than those affecting wind energy structures. However, force magnitudes are neither the sole factor nor the most important one affecting foundation design. The ratio between over-turning moment and vertical load is known as "eccentricity" and is by far the most important factor governing foundation geometry and shape. This ratio is several orders of magnitude higher in wind energy structures than in the oil and gas structures.

Hence the foundation design to break the 25-m barrier needs significant research and development efforts pertinent to the wind industry-specific situations on top of leveraging the oil and gas experience.

4.4 Design envelope challenge

Wind turbines operate in uncertain environments. There is currently an insufficient understanding of the design envelope that accurately and probabilistically characterizes the extreme conditions off the United States northeast and mid-Atlantic coasts. Because of this, several challenges must be considered when developing new wind turbines.

4.4.1 Wind speed assumptions

The currently accepted wind turbine and foundation design envelope, such as the one used for the GE 3.6 platform, was developed by the IEC and adopted by Germanischer Lloyd AG in Hamburg, Germany. This design envelope specifies Class I wind turbine plants be designed for 50-year recurring, 10-min average wind speeds of 50 m/s and 5-s wind gusts of 70 m/s. However, it is not clear how closely these extreme design assumptions, developed based on European onshore experiences, will apply to U.S. regions that host offshore plants.

4.4.2 Turbulence ratio

A common wind industry practice calls for utilizing a 12% turbulence intensity for offshore sites [5]. This is based, again, on European experiences and it is not yet clear whether this is truly representative of potential U.S. sites.

4.4.3 Hurricanes

U.S. east coast is susceptible to hurricanes. The presence of hurricanes or typhoon is unique to the U.S. and Asian-Pacific areas because there is no similar scale storm hazard in Western Europe where the offshore wind technologies were pioneered. Although U.S. offshore construction and foundation design codes of practice have acquired significant expertise in understanding the impact of hurricanes in the Gulf of Mexico, this expertise may not be all transferable to the wind industry along the upper east coast.

In the northern hemisphere and because of their counter-clockwise rotation, hurricanes generally have lower wind speed at their western or left side compared with their eastern or right side as shown in Fig. 3. In the Gulf of Mexico where landmass largely extends east to west, oil and gas installation must be designed for the maximum wind speed of the hurricanes on its eastern side. For the wind industry, which focuses mainly on the east coast of the United States currently, landmass extends roughly south to north. It is less likely that the eastern side of a hurricane will affect a wind power plant unless the hurricane makes prior landfall and hence losing half or more of its energy (Fig. 4).

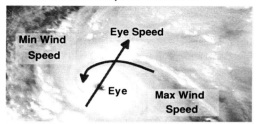

Northern Hemisphere Hurricanes

Counter-clockwise Wind Structure

Figure 3: Hurricanes wind speed distribution [3].

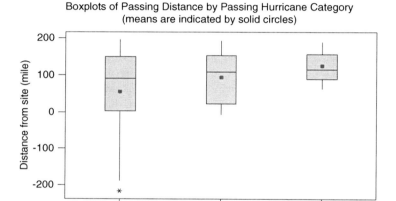

Figure 4: Probability of hurricanes as function of distance from Western Tip of
Long Island [3].

Nevertheless, the probability of encountering the sustaining maximum hurricane wind speeds on its eastern side increases as wind power plants move further offshore, i.e. further east. An internal study at GE showed that although the eastern tip of Long Island is unlikely to sustain hurricanes above level 2, moving 150 miles westward increases that probability significantly (Fig. 4). The effect of hurricanes on the low cycle fatigue and design wind gust speed is not known and represents one of the risks in developing offshore wind projects in the northeast and mid-Atlantic U.S. Technologies and methods need to be developed to accurately characterize the impact of hurricanes on the design envelope for both the turbine and the foundations in the target offshore locations.

4.4.4 Geotechnical conditions

While designing wind power plants, the geotechnical conditions account for most of the rest of uncertainties encountered other than Hurricane. Engineers normally demand to collect borehole soil samples at each turbine locations. However, this increased the plant development costs significantly. Moreover, it is not clear if a geotechnical program comprising boreholes at every turbine site, which is referred to as "full coverage", would significantly reduce the overall geotechnical risk. Studies performed by the oil and gas industry show the full coverage might be unnecessary as long as adequate information exists regarding the overall geological structure in the area, which can be synthesized utilizing non-intrusive geophysical measurement techniques. There exists a need in the wind industry to quantify the increased risk associated with less than full coverage of all turbine sites based on the synthesized geological and geophysical knowledge on the general plant area. Subsequently the marginal risk reduction from each additional borehole can be traded-off with the increasing investigation costs. Some boreholes will always be needed to get an idea about the real soil conditions across the site. However, there may not be a need for a borehole under each turbine.

4.4.5 Mud-line evaluation

At the end of foundation design life or immediately after severe storms, no techniques exist to evaluate foundation conditions below the mud-line. Effective above-the-mud-line evaluation techniques do exist to identify cracks using eddy circuits or visual observations. However, these techniques cannot be easily transferred to foundation sections below the mud-line. Moreover, there are no techniques to assess soil condition, which is an important element of the overall system condition. Technologies such as digitally extract soil density images with high-frequency ultrasound techniques before and after severe storms need to be developed to answer the critical questions over wind power plant residual life.

The above challenges are barriers to synthesize accurate design envelope that can be used to design the wind power plant system including the turbine and the foundation. These challenges stem from the uncertainty associated with the offshore wind environment. In developing that design envelope along with the uncertainty associated with the above challenges, it is the key to work within a probabilistic framework that provides feedback on the changing risks and costs. This will aid wind power plant developers, owners and design engineers to trade-off effectively based on cost and risk for decisions.

4.4.6 Sediment transportation

At the offshore wind power plant seabed level, the wave and current constantly transport the sediments around to morph the landscape. This effect could expose the embedded portion of the turbine foundation and produce problem later on during the plant life along with scoring. Anti-scoring typically involves expensive operations to lay stone layer around the foundation. New innovations utilizing plastic/rubber mat material has been tested to mitigate this problem. Its long-term effectiveness need to be checked. More global sediment transportation effects need to be evaluated additionally during the site survey period.

4.4.7 Unique U.S. challenges and oil and gas experiences

There are other challenges that are more significant in the U.S. compared with potential European sites:

1. Because of the longer fetch length, wave heights are significantly higher in the open Atlantic Ocean than in the most challenging European sites in North Sea. The most utilized extreme wave power spectral density function is modified Pierson-Moskowitz (PM) spectrum [6]. It assumes a fully developed sea, i.e. all the wind energy has been imparted to the sea over an infinite fetch length. However, in many coastal situations, that may not be the case and much smaller fetch lengths may exist. The Joint North Sea Wave Project (JONSWAP) proposed the JONSWAP spectrum [7, 8] that includes scale parameters for the fetch length. The marked difference between PM and the JONSWAP spectrums is that JONSWAP is significantly more peaked. The shorter the fetch length, the more peaked JONSWAP becomes. Over long fetch lengths, the sea receives more energy from wind shear and has more opportunity to develop random waves with more variable spectrum of wave heights and lengths. So the JONSWAP spectrum experience may need amendments when it is applied to the U.S. coastal wave evaluation.

    For offshore structures, Germanisher Lloyd has certification requirements for applying both wind and wave loads on offshore wind turbine without being too conservative. For fatigue analysis, load spectra are to be determined which include the influences to be considered for the wind turbine plus those from wave, currents, and sea-ice [9]. These load cases are also common with the oil and gas industry and are largely borrowed from that industry. However, it has to be carefully reconsidered and studied for the U.S. market. The first reason is that wave heights in the Atlantic are on the average higher than those in the North Sea since North Sea and Baltic Sea coasts have shorter fetch lengths and lower wind speeds compared with the U.S. North Atlantic coasts. Figure 5 shows the 95% confidence intervals for the long-term Weibull distribution of the probability of the extreme significant wave heights. Figure 5 shows that the

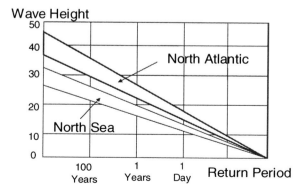

Figure 5: Wave heights in the U.S. east coast and Europe [3].

North Atlantic has 30% higher wave heights on the average compared with North Sea. It is worth noting that the 30% difference can be accounted for almost exactly using the JONSWAP spectrum [8]. The higher significant wave heights in U.S. pose many challenges for the structural modeling and design.

2. Back to the hurricane problem at the turbine design level, the current wind turbines are typically designed for up to IEC class 1 wind load, which can be roughly translated to a maximum wind speed between the Hurricane category 1 and category 2 conditions. However, many population centers along the U.S. east coast are hit by higher than category 2 hurricanes in a frequency higher than once per 50 years. So the safety of the offshore wind turbine erected near these regions could be a challenge to the system and structure design.

3. Wave breaking occurs when the ratio between the wave height and the sea depth exceeds a certain limit. Furthermore, in certain circumstances wave breaks as plunging wave instead of surging wave, which can cause significantly more fatigue load force on any structural element present at the breaking point. At a UK experimental offshore wind power plant in Blyth Harbour, one of the towers was erected at a wave breaking point and sustained significantly increased fatigue load. A study of wave breaking probability in U.S. sites vs. European sites shows that the probability of plunging wave breaking is about the same. However, because the U.S. sites have higher wave heights and are deeper in depth, the plunging wave could break with as much as four times the energy of its European counterparts.

There are many breaking wave theories available in the literature. The simplest and oldest model is:

$$H_b = 0.78h \tag{1}$$

where H_b is the wave breaking height and h is the sea depth. A little more sophisticated equation that considers the sloping seabed effect is termed the Goda expression:

$$H_b = 0.071\tanh(kh) \tag{2}$$

Many other theories exist in the literature with varying degree of complexity. Anastasiou and Bokaris [10] evaluated 19 of the hypothetical expressions available in the literature against their own data and found the two simplest equations above to be the most accurate ones. On the contrary, Kriebel [11] found that other equations to be better using different sets of data. So there is a high degree of disagreement that may never be resolved because of the complex and uncertain nature of the wave breaking phenomenon. However, it is possible to devise a methodology to generate a site-specific wave breaking prediction by fusing the site-specific short-term observations, the 3D bathymetry measurements, and the known wave breaking models. This task is not easy given the fact that the local current has impact to the wave breaking condition because increased current velocity increases the probability of wave breaking. The final prediction should be based on site-specific measurements to calibrate known physics and models. It should also be included as part of the wind turbine certification process for the site.

Table 3: Comparing oil & gas industry to wind industry requirements [3].

Category	Oil and gas Offshore Structures	Wind energy structures
Water depth	20–120 m and more	10–30 m
Vertical loading	5000–30,000 ton	100–300 ton
Horizontal loading	10–20% of vertical load	70–150% of vertical loads
Over-turning moment	Water depth × horizontal load	(Water depth + 75 m) × horizontal load
Number of installations	1	20–200
Structure width	O(10 m): >20 m	O(1 m): ~ 5 m
Power generated	O(100 MW)	Max of <5 MW per each installation

Above-mentioned challenges are certainly not new if we consider the significant oil and gas installations around the world. However, it is erroneous to make the assumption that overcoming these challenges amounts to direct translation of the oil and gas experience to the wind industry. Table 3 illustrates key differences between the two industries in terms of technical requirements.

Table 3 shows that magnitudes of loads affecting oil and gas structures are much higher than those affecting wind energy structures. However, the ratio between the over-turning moment and the vertical load, i.e. "eccentricity" which governs the foundation geometry and shape, is several orders of magnitude higher in wind energy structures than it is in the oil and gas structures. This problem is compounded by (1) the fact that non-dimensional ratio between eccentricity and the structure's width in the wind turbine structures will be even higher than in the oil and gas structures, and (2) the heavily dynamic nature of the loads. Therefore, a straightforward translation of foundation shapes and designs from offshore oil and gas structures to offshore wind structures is simply not feasible.

In addition to the design limitations, there are other factors that differentiate the oil and gas experience from the wind energy structures needs:

1. In contrary to one giant oil and gas offshore platform installation, a wind power plant normally consists of tens to hundreds of identical turbine installations. Therefore, experience with assembly-line techniques and the associated compartmentalization, modularization, and quality control with six-sigma methods become more relevant.

2. It is certainly economical to have $100 million worth of platform foundations to produce crude oil equivalent to hundreds of megawatts worth of electricity from one oil and gas platform. If the same foundations are translated to the wind industry, its cost must be in the neighborhood of one million dollar for the overall economics to work.

On the other hand, ruling out direct translation of foundation designs from the oil and gas industry to the wind industry does not mean that wind turbine foundation designs should or need to start from a clean sheet of paper. The new research and development for offshore wind industry need to be based on the oil and gas experience that can be leveraged. This requires collaboration between the wind turbine manufacturers, who command the necessary understanding of loads, mass production issues, and the overall system design and economics trade-off issues, and the world-class institutions with specific expertise in developing oil platforms foundation technologies.

4.5 Corrosion, installation and O&M challenges

Corrosion, installation and operation and maintenance challenges are also essential to the success of the offshore wind power plant.

Most offshore wind turbines components have been coated, cathode protected and/or sealed per the typical naval military standards (MIL) to protect them from the excessive corrosive environment offshore. However, severe corrosion problem has still been observed on a major EU OEM's electrical system and resulted in very expensive large-scale replacement of the electrical components in the field. Hence identifying the difference in the corrosion root cause between the naval vessel and the offshore wind turbine is something that needs further exploration.

On top of the corrosion issue, the installation of the offshore wind turbines, foundations and electrical infrastructures has been a major cost and schedule challenge due to

- The uncertainty in the weather, wave, current and soil conditions.
- The involvement of specialized large installation service equipments such as large hammers for pile driving and jack-up vessels for crane and general handling of the turbine.

To make large offshore wind project feasible, new installation concepts, technologies and equipments are needed to reduce the overall installation cost and extend the installation window period.

Furthermore, the operation and maintenance of the offshore wind plant requires special access capability to and survivability on the wind turbine in a variety of wind and wave combined conditions. Boat access is slow and the landing to the turbine can be dangerous even in the normal wave conditions. Helicopter-based access has been developed and preferred for its flexibility and speed. It requires special landing pad design for the turbine though. Many efforts have been invested in the condition-based monitoring (CBM) systems to minimize the need for human access.

4.6 Environmental footprint

There is a need to have independent and scientifically based studies on the true environmental footprint of offshore wind power plants. Expertises exist with regards to assessing environmental footprint, particularly on naval mammals and fish schools, in the north-eastern U.S. with institutions such as the joint marine biology program between MIT and Woods Hole Oceanographic Institution. These unique expertise need to be leveraged in order to gain better understanding of the environmental footprint of offshore wind power plants.

5 Subcomponent design

The two key technology drivers for any offshore wind project design are:

* Achieving maximum turbine reliability to minimize the total cost of ownership;
* Designing cost-effective foundations.

These technology drivers stem from the offshore wind power plant environment with offshore site location, extreme weather conditions and limited accessibility throughout the year. Given these drivers, the total cost of ownership per kWh of electricity produced is a top-level critical figure-of-merit. It is also constrained by the popular demand for the placement of plants farther offshore. Significant effort on design trade-off is needed to identify the most promising configurations of the following subsystems:

* Low cost foundations
* Advanced rotor design
* Advanced control, monitoring, diagnostic and repair system
* Reliable drive-train and power electric system

The critical challenges for each subsystem and component are discussed below.

5.1 Low cost foundation concepts

Controlling the cost of the offshore wind turbine foundations present unique challenges for the wind industry as coastal communities demand wind power plants placement further offshore.

The current state of the art in the wind industry utilizes two types of foundations depending on the site conditions and in particular the water depth:

1. *Gravity base structure* (GBS) type foundation is built onshore in a dry dock and requires the transportation of a concrete caisson to the plant location either by flotation or by a barge. The concrete caisson is then filled with concrete or other ballast materials and placed on the sea floor. The turbine tower is then erected on top and the nacelle and rotor are placed on top of the tower. This foundation type has two disadvantages: (1) it requires extensive seabed preparation and (2) the caisson top height must exceed the sea depth so that its top surface facilitates the cheaper dry installation of the tower. This limits its application to the wind industry rough definition of "shallow" as below 7.5 m water depth. Some of the largest wind plant in the world are near-shore plants in shallow depths and therefore utilized gravity base foundations.
2. *Monopile:* Under the supported of the Danish government a group of two Danish power companies and three foundation engineering firms studied the

most suitable foundation types for approximately 15 m sea depth in 1996–1997. This was viewed at that time as the next offshore foundation challenge. The study identified the steel monopile as the most suitable for this depth. The monopile installation requires hammering a steel pile into the soil. Then a so-called "transition piece" is grouted on top to correct for any lack of pile verticality. Tower and turbine top components are erected on top of the transition piece. Since its introduction in Europe, the monopile has become standard for sea depth ranging from 10 m up to approximately 20 m. Figure 6 shows the pile hammering operation at the Arklow Bay, Ireland offshore power plant that features seven GE 3.6 MW wind turbines.

As water depths increases, the capability of monopile is stretched to the limit primarily because of the manufacturability and constructability limitations. For sites deeper than 25 m, the monopile diameter reaches 5.5 m with thickness up to 90 mm, which could weigh upward of 250 metric tons. This level of dimensions and weights are near or exceeding the maximum capability of the fabrication factories, hammers and the installation vessels across the world.

To take advantage of the offshore wind energy in U.S., the wind industry must break the 25-m barrier in a profound way with technology development.

One recent trend practiced by some EU OEMs is to use tripod, quadpod and in general jacket type foundation structure to extend the service water depth to 50 m. These structures utilize O&G platform and shipbuilding welding technologies and tend to be very expensive in fabrication, transportation, and installation. We still need to wait and see if the new effort such as the REpower 5M jacket foundation in Beatrice as shown in Fig. 7 [12] and the new innovations such as the REpower 5M casted steel multiple-direction joints depicted in Fig. 8 will help to reduce the cost drastically and make this type of foundations economical.

Figure 6: Pile hammering at Arklow [3].

Figure 7: Repower 5M quadpod offshore foundation installed in 2006 [12].

Figure 8: Prototype of a REpower 5M offshore wind turbine in Bremerhaven,
Germany with casted joints [12, 13].

5.1.1 Level-controlled suction caisson foundations

Suction caisson foundations are currently utilized in the oil and gas industry for sea depths up to 50 m. Their installation involves transporting the suction caisson and placing it on the seabed. Water is then pumped out of the hollow dome in the caisson resulting in the caisson sinking into the soil and being anchored in place as shown in Fig. 9. Large hammers are not needed and the installation process takes hours in contrary to days with monopile. As such, this is very conducive to assembly-line operations.

There are risks to utilize the suction caisson foundation in wind power plants. They include:

- Performance under lateral dynamic wind loads – Two experimental wind turbines were constructed, in Denmark and England, using suction caisson foundations. Results indicate that performance under long-term lateral dynamic loads remains a key issue preventing large-scale utilization of suction caissons in deeper sites.
- Proper tower installation – Caisson verticality tolerance is not necessary in the oil and gas industry. However, wind turbine tower installation requires controlling the foundation top surface verticality within 1°. One degree off on verticality could result in several meters of eccentricity at the hub height. It not only results in increased over-turning moment, which is a key design criterion, but also affects turbine performance. One EU OEM suffered an installation failure for a near-shore test turbine because of the verticality problem from a hard spot under one side of the caisson bottom edge during the suction process.

5.1.2 Jetting and grouting in deep sea piling

With conventional monopile foundation, heavy vessels needed for pile driving account for 20% of the total foundation costs. Jetting and grouting eliminate the

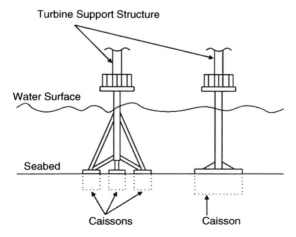

Figure 9: Level-controlled suction caisson foundations [14].

need of hammering and hence the heavy vessel. It is not only limited to monopiles but also can be easily extended to tethered foundations. Thus the costs could be reduced for both very deep sea wind power plants and the close to shore monopole-based wind power plant developments.

5.1.3 Tethered foundations

Tethered foundations are often used in oil and gas installations for water depth exceeding 100 m. Tethered foundations are quite scalable with water depth. However, their usage in wind power plants does present challenges particularly in the severe lateral loads, which is common in wind turbines and could affect overall system stability. Using ballasts and damping methods counters the lateral load with the risk of increasing the total cost to a level higher than the seabed-mounted foundations. This is particularly true at the lower water depths.

There are two types of tethered foundations for potential offshore wind applications.

1. *Tension leg platforms (TLP)*: With the TLPs, the hull and the tendon design is highly coupled since their natural frequencies are not significantly distinct. TLP stability comes from the buoyancy of the hull, which provides extra tension in the tendons and consequently provides adequate horizontal stiffness. TLPs as

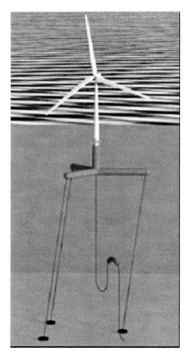

Figure 10: Tension leg platform [15].

shown in Fig. 10 have an edge over the other tethered or moored systems because of the minimum need for submersible hull depth. Therefore, they could be deployed in the shallower sites which could be just a bit too deep for the seabed-mounted foundation concepts.

As offshore wind foundation, the TLP could achieve extra hull buoyancy by increasing hull breadth instead off volume. It in turn enhances the lateral load carrying capacity for the overall system. There are two challenges for the TLP in offshore wind applications. Firstly, the overall hull-tendons-turbine system must be designed as one integrated unit because of the coupling of the dynamics. This makes the design process more complex. Secondly, the current wind turbine designs require maximum angular deformation at the sea level to be less than 1° to avoid imbalances in rotor loads. This deformation level may be difficult to achieve without using more advanced high-tension-high-buoyancy TLP, which could very well be the answer to the offshore wind developments in deeper sites. Oscillation damper technology and advanced high-strength tendon materials need to be evaluated to reduce the overall costs for this type of TLP.

2. *Spars*: As shown in Fig. 11, spars rely on ballast weight at the spar keel for stability. The ballast is much stiffer than the tendons. Therefore their designs are decoupled. Using spars in offshore wind may be problematic because their stability under lateral loads increases as the submersible length increases. There-

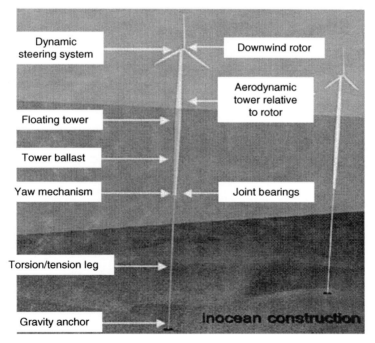

Figure 11: SWAY concept with 640 ft tall spar floating buoy [16].

fore, spars may not be suitable for sites less than 200 m, which could be way off limit for future wind power plants over the next 10 years. Spars also face other challenges similar to TLPs [16].

5.1.4 Hybrid tethered system

It is conceivable that the ultimate tethered concept may be a hybrid between the high-tension-high-buoyancy TLP and the spars in which lateral stiffness is provided by both ballast at the foundation keel and tension in widely separated tendons as seen in Fig. 12. The hybrid system will have the following features:

• Blasts at the keel ends, which is well assisted by the high-tension tendons, to provide lateral stability. The high tendon tension comes from the high buoyancy in the hull design.
• A hull with a liquid column oscillation damper, which consists of channels utilizing the differential inertia of water across different hull compartments to reduce lateral wave load. Liquid column dampers could either be actively tuneable or passively set at fixed frequency. Preference might be given to the passive system so as to reduce the reliance on the control system for the over-all stability.

It is still likely that the turbine's tolerance for lateral sea level angular rotation will have to be widened even after all the efforts. The appropriate implications to the control system design and the drive-train component design need to be addressed accordingly in turbine design.

Recently there have been efforts from several offshore engineering groups and wind turbine OEMs to put prototypes floating foundations with wind turbines on

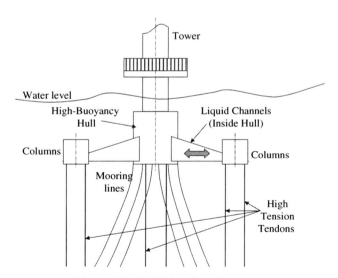

Figure 12: Hybrid tethered system.

top in the water. They include concepts such as SWAY, Blue-H and Hywind from StatoilHydro, etc. For the details please refer to [16].

5.2 Rotor design for offshore wind turbines

Rotor design for the offshore wind turbines requires new rotor structural concepts and aerodynamic models for blade geometry optimization.

5.2.1 Structural concept

Structural design for large offshore wind turbine blades calls for development of the hybrid carbon fiber/fiberglass blades at system ratings in the multi-megawatt to 5–7 MW range. Structural performance needs to be evaluated for various arrangements of the carbon blade spar. Critical performance aspects of the carbon material and blade structure need to be addressed. This type of rotor blade design will use carbon strategically. The goal is not just to reduce the weight and cost of such large blades, but to maximize the benefits of the introduction of aeroelastic tailoring, i.e. twist-bend coupling. These features combined will allow the blades to shed peak transient loads.

Earlier studies conducted by GE Wind Energy showed significant potential for relieving fatigue and extreme loads using aeroelastic tailoring. Research at TU Delft [17] in the Netherlands has further shown potential to use carbon to substantially reduce blade cost. The gains are particularly significant for large rotors with reductions of up to 10% in blade cost and 30% in weight when it is compared to the current practices.

As wind turbine technology evolving, the industry's optimal turbine size has been steadily increasing. For turbines to grow into the 5–7 MW range and beyond economically, rotor blades longer than 60 m will be needed. Longer blades will help make lower wind speed locations close to shore more economically attractive since rotor diameter is the single biggest design parameter affecting the amount of energy capture for a given wind speed. In such conditions, it may be possible to put an ultra-long blade on a conventional turbine without exceeding its design load capability offshore. Few fundamental barriers have been identified to cost-effectively scale the current commercial blade designs and manufacturing methods over the size range of 100–140 m diameter. Turbine designs with low specific rating need to be studied for the lower wind speed sites. As specific rating is decreased, i.e. blade lengths increase at a given rating, blade stiffness and the associated tip deflections becomes increasingly critical for cost-effective blade design. The WindPACT rotor study [18] predicted added costs of transportation and assembly adversely affect the cost of energy (COE) for machines rated above 1.5 MW. Constraints for transportation cost should be considered in all projects. An option for offshore may be to place the blade fabrication plant at a harbor, which is advantageous with cheaper blade shipment using barge only.

5.2.2 Aerodynamics and blade geometry optimization

A lower importance of noise emission, expected for offshore wind turbines, offers new opportunities in rotor aerodynamics and blade geometry optimization.

Specifically, blade geometry can be optimized for a higher maximum tip speed as compared with wind turbine rotors designed for land-based deployment. So the tip of the blades can be designed for maximum aerodynamic efficiency. A higher maximum tip speed allows for a higher design tip speed ratio for the blades, which will reduce rotor solidity and facilitating transportation as well as reduce drive-train size and cost. Using high-lift airfoils can be considered to further reduce rotor solidity while giving due attention to the impact of reduced solidity on the blade structure strengths and cost. Overall, blade geometry can be optimized to minimize COE using tools that simulate and evaluate detailed aerodynamic blade design with considerations to blade structures and cost of all main turbine components.

A blade design tip-speed ratio between 9 and 10 is expected to be optimum for minimizing the COE of the offshore wind turbine system and the optimization process can be used to survey the design space.

Traditionally, the tip of a wind turbine blade has been designed to provide a smooth unloading because of the noise considerations. In particular, the chord lengths over the tip region are tapered below the optimal point for maximum aerodynamic efficiency. This approach results in a loss in power output. Tailoring the tip geometry towards that goal could make a gain in aerodynamic efficiency. Tip losses due to the tip vortex amounts to roughly 3% in annual energy capture. The optimization of a tip for maximum energy capture is expected to cut this loss to 1.5–2%, which is a significant improvement given the value of 1% more energy at the current multi-megawatt size of utility-scale wind turbines.

5.3 Offshore control, monitoring, diagnostics and repair systems

Offshore wind turbine control system will consist of three highly integrated sub-systems:

• Diagnostic system
• Regulatory system
• Supervisory system

The offshore wind turbine system size will allow for a sophisticated control system at a smaller fraction of the total turbine cost than is possible for a smaller onshore turbine. The advanced diagnostic/CBM system will monitor sensor data from major components such as blades, drive-train bearings, gearbox and generator to detect faults and rapid degradations which could negatively impact the power capture or lead to premature critical component failure.

The advanced regulatory system can use feedback from deflection and load sensors as well as local wind conditions to increase power capture and reduce fatigue and extreme loads well beyond current industry practice. Including the improved regulatory control in the design phase will mean that these advantages could be conferred in the initial design envelope.

The more advanced supervisory system can reduce unnecessary hard stops, thus significantly increasing the life of drive-train subsystems. This can be achieved using redundant sensors and back-up systems, also to be incorporated in the design phase. This will ensure high availability and increased energy capture. It will offer component cost benefits through reduced fatigue load and the ability to mitigate or avoid some current design driving loads.

5.4 Drivetrain and electrical system

Improving overall system efficiency is a key goal of any offshore development effort. To obtain the lowest COE, losses must be reduced wherever possible. Large diameter, gearless generator technologies such as so-called "direct-drive" which uses permanent magnets generator are investigated by many companies as one solution for reducing losses and maintenance cost related to the traditional planetary gearbox-based system. Design trade-offs could be carried out for optimizing the cost and efficiency of drive train and the converter based on the generator side and grid-side converters location in the tower. Factors such as weight and cable loss should be considered.

The electrical system is the critical interface between the drive train and the utility network. It enables both the optimized drive-train design and the harvesting and transfer of power to the utility in accordance with grid regulations. An optimized electrical system design will result in higher system efficiency and lower system cost. Both of which affect the COE in a positive manner.

The following issues related to optimizing the design of a multi-megawatt to 5–7 MW offshore wind turbine need to be addressed:

- Design trade-offs for optimizing cost and efficiency of the drive train and converter: The power converter enables the electro-mechanical drive train to operate at the optimum power factor and deliver maximum output power through peak-power-point tracking. New circuit topologies and control, converter packaging and thermal management need to be explored to reduce cost and losses of the power converter as well as the generator. The power converter's physical location and associated electromagnetics that may include transformers in the tower will be considered as a design factor to minimize cost and losses in the cables as well as reduction of weight in the nacelle.
- Electrical interconnection of wind turbines in a wind power plant: Studies need to focus on optimizing the electrical interface between turbines in an offshore wind power plant. The feasibility of the medium voltage (MV) DC interconnect system, based on cost and dynamic stability, need to be studied.
- Transmission of power from offshore to onshore: Voltage level and frequency are factors which impact cost of cabling and converter interface at the sending and receiving ends of electricity. MV or high voltage DC transmission is an enabling technology for such bulk power transfers. Optimum voltage levels as well as converter topologies need to be achieved to minimize the overall systems cost.
- Protection of electrical system: Electrical system protection from grid-faults and component failures is critical to system design from availability and reliability

point of view, which directly affects the COE of an offshore plant. Advanced means of control for handling grid-faults and low/zero-voltage ride through as well as modular converter topologies that enable part-load/full-load operation and optimization of ratings of the protection equipments such as circuit-breakers along converter system need to be studied.

6 Other noteworthy innovations and improvements in technology

This section describes other key improvements that need to be achieved by the offshore wind industry.

6.1 Assembly-line procedures

Unlike today's construction method, i.e. building unique one-of structures, the future offshore projects will utilize assembly-line procedures to maximize cost control in a mass production manner.

6.2 System design of rotor with drivetrain

Current commercial wind turbines utilize a three-blade configuration. However, two-blade designs which incorporate alternative hub structures may see a rise in popularity because they allow turbines to reach higher rotor speed without visual or noise constraints. Upwind configuration might be preferred as it allows less dynamic loads and has less rhythmic noise effects. Detailed investigations need to be carried out on the blade deflections and resonant modes under turbulent wind loading.

The WindPACT rotor study [18] was designed to explore many of these configurations and attempt to determine their impact on overall turbine operation and COE. The study has found that several loads in the final two-bladed downwind machine were higher than the corresponding loads in the baseline design. The downwind two-bladed rotors also experience strong harmonic loading from the tower shadow, which may excite natural frequencies. In the future offshore wind projects detailed investigations need to be carried out on the relative advantages and disadvantages of a two-bladed upwind with the corresponding three-bladed version. The real benefit of two-bladed design with unconstrained tip speed is simplified gearbox and more optimal direct-drive drive train. Preliminary studies conducted by GE Global Research Center indicate that the rated speed of the wind turbine has a large impact on the direct-drive cost. Considering aerodynamic power only, the rated speed for a two-bladed turbine could be up to 30% higher than a three-bladed turbine. In which case a direct-drive drive train for a two-bladed turbine rated at 19.1 rpm would cost substantially less than one for a three-bladed turbine rated at 11.7 rpm for a 5 MW system. However, energy capture will likely to be less than an equivalent three-bladed wind turbine of the same diameter and hub loading needs to be investigated as both tower and hub will require reinforcement due to the two per rev loads. The suggestion in the WindPACT study to combine the rotor design with drive-train configuration studies needs to be implemented, which could ultimately contribute to reduce the COE.

6.3 Service model

A solid and viable offshore service model is extremely important for any successful offshore wind project. The basic philosophy will evolve around a global performance and product data warehouse specifically aimed for autonomous offshore operations.

The performance data from offshore wind power plants can be processed in the global data warehouse, which will feed information to different areas. The operating information could be used by future product development teams or for existing product improvements. The service information will be used for making contractual service agreements, remote monitoring and diagnostics and knowledge-based maintenance. Customers can use the availability information to understand the capacity factor and overall plant health.

Contractual services has a major emphasis on developing new technology tools to support the offshore wind business. The new technology tools are aimed at improving reliability and availability, extending parts lives and enhancing plant performance.

7 Conclusion

Offshore wind turbines need to achieve high reliability and availability at low COE, which is competitive to other energy sources. The chapter identifies innovative options for new foundation concepts, construction techniques, rotor design, drive train and electrical system while optimizing the total life cycle cost of offshore wind power plants. Turbine design will need to incorporate best technologies and practices from the land-based turbines while incorporating lessons learned from first generation offshore pilot projects to develop a new robust turbine concept optimized for offshore operations. Optimum turbine size will be determined for locations and is expected to be in the multi- megawatt to 5–7 MW range suitable for more than 25 m water depths.

References

[1] US wind resource map, DOE/NREL Wind powering America program, 2009, http://www.windpoweringamerica.gov/wind_maps.asp
[2] Musial, W. & Butterfield, S., Future for offshore wind energy in the United States (NREL/CP-500-36313), EnergyOcean 2004, Palm Beach, Florida, June 28–29, 2004
[3] Ali, M., Zheng, D., Kothnur, V. & Grimley, R., Offshore wind energy generation in the United States – Challenges and R&D needs. *Global Windpower 2004 Conference & Exhibition*, Chicago, USA, 2004.

[4] Netherlands renewable energy policy and role of offshore wind energy, 2009, http://www.offshorewindenergy.org/txt2html.php?textfile=home/dutch_ corner.txt

[5] Rules & Guidelines 2000: IV Non-marine Technology – Regulations for the Certification of (Offshore) Wind Energy Conversion Systems, Germanischer Lloyd: Hamburg, Germany, 2000.

[6] Hasselman, K., Ross, D.B., Mueller, P. & Sell, W., A parametrical wave prediction model. *Journal of Physical Oceanography*, **6**, 1976.

[7] Ewing, J.A., Some results from the joint North Sea Wave Project of the interest to engineers. *International Symposium on the Dynamics of Marine Vehicles and Structures in Waves*, University College, London, UK, 1974.

[8] Houmb, O.G. & Overvik, T., Parameterization of wave spectra and long term distribution of wave height and period. *BOSS '76 Conf.*, Trondheim, 1976.

[9] Rules for Certification of Offshore Wind Energy Conversion System, Germanischer Lloyd, Hamburg, Germany, 1995.

[10] Anastasiou, K. & Bokaris, J., Physical and numerical study of 2D wave breaking and nonlinear effects. *Proc. of the Coastal Engineering Conf.*, 2000.

[11] Kriebel, D.L., Breaking waves in intermediate depths with and without current. *Proc. of the Coastal Engineering Conf.*, pp. 203–213, 2000.

[12]. Seidel, M., Jacket substructures for the REpower 5M wind turbine. *Conf. Proc. European Offshore Wind 2007*, Berlin, Germany, 2007.

[13] Lüddecke, F., *et al.*, Load-Bearing behaviour of cast steel components in offshore wind turbines under predominantly static and non-static strain. *Stahlbau*, **77(9)**, pp. 639–646, 2008.

[14] Villalobos, F., Houlsby, G. & Byrne, B., Suction caisson foundation for offshore wind turbines, 2004, http://www-civil.eng.ox.ac.uk/people/gth/c/ c81.pdf

[15] Musial, W., Butterfield, S. & Boone, A., Feasibility of floating platform systems for wind turbines. *23rd ASME Wind Energy Symposium*, Reno, Nevada, Jan. 5–8, 2004.

[16] Offshore Wind Turbine Foundations – Current & Future Prototypes, 2009, http://offshorewind.net/Other_Pages/Turbine-Foundations.html

[17] de Goeij, W., Implementation of bending-torsion coupling in the design of a wind turbine rotor blade, Master Thesis, TU Delft, 1998.

[18] Malcolm, D.J. & Hansen, A.C., WindPACT turbine rotor design study, NREL Subcontractor Report, NREL/SR-500-32495, 2002.

[19] Butterfield, S., Musial, W., Jonkman, J. & Sclavounos, P., Engineering challenges for floating offshore wind turbines. *The 2005 Copenhagen Offshore Wind Conf.*, Copenhagen, Denmark, Oct. 26–28, 2005.

CHAPTER 12

New small turbine technologies

Hikaru Matsumiya
HIKARUWIND Lab. Ltd., Japan.

Small wind turbines (SWTs) have various long history and yet huge market potential globally. Looking back last a couple of decades, various SWTs, both horizontal-axis and vertical-axis types, have emerged, however, they have kept standing behind in the shadows of large-scale wind turbines, as if they could not play meaningful roles in energy production. Small turbines may be easily designed, manufactured, tested and operated. But somehow mostly their unit costs are not only much higher but also reliabilities are much lower when compared with those of middle to large turbines. The truth lays in technical difficulty with SWTs. Low Reynolds number is one of the problems. Today, various new technical approaches have revealed problems to be solved in delivering advanced SWTs to global markets with lower cost and higher reliability.

1 Introduction

For a long time it had been thought that large wind turbines (LWTs) are global, while small wind turbines (SWTs) are local. Actually the style of use of SWTs is local, but all of the problems, both technical and institutional, that SWTs have today are mostly global. This turned out during the IEA Annex XI's Topical Expert Meeting on "Challenges of Introducing Reliable Small Wind Turbines" in Stockholm in September 2006. What was common consensus among the experts, engineers and researchers are generally high cost and low reliability with SWTs in general. SWTs could provide us more general values in electricity or in other forms. This was the main conclusion at the expert meeting, because all of the participants had already understood that advanced technologies were coming up.

In this chapter, attentions are mainly focused on new technologies and their applications to SWTs, while there are plenty of types and intelligent applications that have already come in the markets.

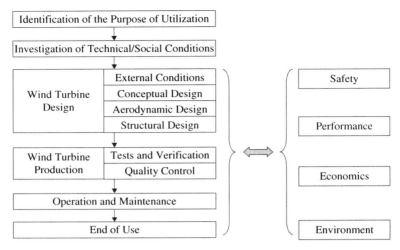

Figure 1: Lifecycle of a wind turbine.

A general life cycle of a wind turbine is illustrated in Fig. 1. Today SWT technology has evolved and is required to contribute to energy production and environment protection much more than before. Therefore, safety, performance, economics and environment are keen issues.

Discussions will be focused on those issues that are particular with SWTs in their design procedure in the following sections.

1.1 Definition of SWT

SWTs are defined as "wind turbines with a rotor swept area smaller than 200 m^2, generating at a voltage below 1000 V a.c. or 1500 V d.c." in IEC Standard [1]. If the swept area is less than 2 m^2, support system need not be included for the SWT. This is the only one definition accepted internationally.

However, basic understandings of SWTs may not be restricted within the above boundaries. On the rotor size, "Mini" and "Micro" wind turbines with much smaller rotor diameters or power outputs can be proposed. Also restriction to electricity-producing systems is not essential. Various styles of application are possible such as pumping system, some hybrid systems, etc.

The technically important features of SWTs which defer from LWTs are considered first. With high performance wind turbines of both horizontal- and vertical-axis types, they are all driven by lift force created on the rotor blades. Lift is everything and what could affect lift then? They are geometry of aerofoil section, attack angle and Reynolds number. Only the last one depends on scale of rotor. Therefore, one reasonable definition of SWT may the rotor scales that are liable to suffer from "*low Reynolds number problem*".

1.2 Low Reynolds number problem

Most of the aerofoil sections for wind turbines have the common weakness induced by the low Reynolds number problem, i.e. they have a critical Reynolds number Re_c around:

$$Re = Re_c \approx (1-2) \times 10^5 \qquad (1)$$

If $Re \leq Re_c$, lift coefficients drop considerably. An example is shown in Fig. 2 for a MEL18M31 aerofoil section. Lift and drag coefficients are shown for four Reynolds numbers, $Re = 1 \times 10^5$, 2×10^5, 5×10^5, and 1×10^6. In the small attack angle region of $a < 25°$, the lift coefficient for $Re = 1 \times 10^5$ is quite different from and much smaller than those for $Re \geq 2 \times 10^5$. The similar characteristics are found with various aerofoil sections.

What will happen with such a SWT that is often affected by low Reynolds number problem? It would not start to rotate even if the wind speed becomes sufficiently higher than cut-in wind speed, because of absence of pitch control mechanism and

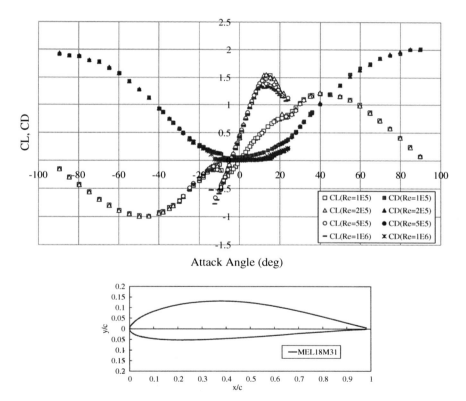

Figure 2: A sample of low Reynolds number effect (lift coefficient for various Reynolds number of two-dimensional MEL aerofoil section).

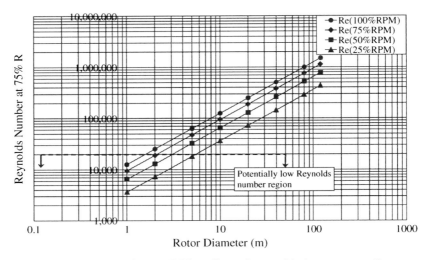

Figure 3:   Reynolds number at 75% radius of rotor blade vs. rotor diameter at
10 m/s of wind speed and different rotor speed (100%, 75%, 50% and
25% of rated value).

due to laminar flow separation. Then it will rapidly be accelerated with good wind,
however, the operation will very unstable when the wind is turbulent. One of the
worst cases, the brake system will soon shutdown the system to prevent the WT
from excessive rotation.

Figure 3 shows how SWTs suffer from low Reynolds number problem. Reynolds
number varies along blade axial location. However for simplicity, a representative
Reynolds number is defined here at 75% axial location of a blade from the rotor
centre as follows:

$$\mathrm{Re} = \frac{c_{75\%}W}{v} \qquad (2)$$

where $c_{75\%}$ is chord length of a blade at $r = 0.75R$, R is rotor radius, W is relative
flow speed to the blade aerofoil section in the cylindrical plane of rotor at $r = 0.75R$, v is kinematic viscosity.

Having simply designed a non-dimensional rotor by blade element and
momentum theory (BEM theory) with tip speed 60 m/s and tip speed ratio is 6
as a typical case, one may evaluate what size of SWT shall be potentially affected
by the low Reynolds number problem by changing the rotor diameter and rotational speed. In Fig. 3 the region where a low critical Reynolds number potentially affect on rotor aerodynamic performance is shown below dotted line.
SWTs below several meters of diameter or rotors rotating at a low rotor speed
will tend to enter into the critical region.

Figure 4 shows some typical relations between rotor diameter and output power
for three kinds of rated wind speed: 7, 10, and 13 m/s assuming 35% of system

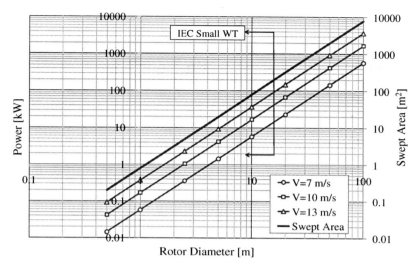

Figure 4: Relation between rotor diameter and power or rotor swept area.

Table 1: Sub-classification of SWT.

Sub-class	Rotor diameter	Technical feature
Micro wind turbine	$D < 1.6$ m*	Both size and power are very small
Mimi wind turbine	1.6 m $\leq D < 5$ m	Full or small Reynolds number problem
SWT	5 m $\leq D < 16$ m**	Less small Reynolds number problem

*In IEC 61400-2, a rotor swept area smaller than 2 m², which corresponds to 1.6 m in diameter, is specially classified in such a way that the maximum yaw rate shall be 3 rad/s.

** $\sqrt{200(4/\pi)} = 15.96 \approx 16$.

efficiency and rotor swept area. IEC definition of SWT is 200 m² of rotor swept area which corresponds to 16 m of rotor diameter and to 10–100 kW of power output depending on the rated wind speed.

Based on the above technical reason, some finer classification for SWT is desirable. Considering both Figs 3 and 4, an idea of sub-classification for small WT is shown in Table 1.

2 Other technical problems particular with SWTs

A non-dimensional design of wind turbines gives optimal power coefficient for a given tip speed ratio if the rotor performance is not affected by Reynolds number. At design point, design tip speed ratio λ_D is given as:

$$\lambda_D = \frac{R\Omega_D}{L_D} \tag{3}$$

where R, V_D and Ω_D are rotor radius, design wind speed and design rotor speed, respectively. Therefore, when design wind speed V_D is fixed, the following relation is obtained:

$$\Omega_D = \frac{\lambda_D V_D}{R} \qquad (4)$$

The smaller the turbine becomes, the faster the rotor rotates. This means a SWT is a high-speed turbo-machine and a particular attention shall be paid on control and safety even though the machine is very small. All the design issues for mechanical safety are well described in IEC 61400-2.

Safety issue relates also to the operation and environmental conditions, because SWTs are very easy to be built on domestic roofs, for example. For any kind of WT, an open area with free wind flow without obstructs nearby is the optimal for the WT and safest for human beings.

3 Purposes of use of SWTs

Among all, the objectives of the use of SWT(s) must be identified first. Some SWTs are very interesting as rotating toys or can be impressive artistic monuments in public parks or in front of buildings. There are also many wind engines pumps or mills. However, the primary important technical aspect of a SWT is its feature as an energy producing machine.

Drag-driven WTs such as Savonius rotors are also often used. These WTs are of high-torque/low-speed type and their efficiencies are not as high as lift-driven WTs of low-torque/high-speed type. Among vertical-axis wind turbines (VAWTs), Darrieus rotor is lift-driven type with an elegant shape. Straight Darrieus rotor is also lift-driven type. If carefully designed, these lift-driven VAWTs may also have not far less efficiency than horizontal-axis wind turbines (HAWTs). As technical grounds are quite same between lift-driven VAWTs and HAWTs, main efforts are made to describe the technical issues of HWATs.

Those who conduct a technical design of a SWT had better refer to IEC standard "IEC 61400-2, Wind turbines – Part 2: Design requirements for small wind turbines", which fully describes safety philosophy and engineering integrity with all technical requirements for SWTs.

However, due to various reasons, most of SWTs are technically and commercially behind LWTs.

When IEA WIND, Topical Expert Meetings "Challenges of Introducing Reliable Small Wind Turbines" was held in Stockholm in fall 2006, 17 experts from 8 countries came to such a conclusion that to create a technically and commercially reliable common market for SWTs is crucial. As a result, a new task (Task 27) entitled "Development and Deployment of Small Wind Turbine Quality Labelling" was initiated in 2008 under the cooperative activities of IEA R&D WIND (IEA Implementing Agreement for Co-operation in the Research, Development, and Deployment of Wind Energy Systems). Considering the present situation that although internationally accepted IEC standards relevant for SWTs industry already exist (IEC 61400-2: 2nd Ed:2006), its application is still not common and

Table 2: Basic parameters for SWT classes (IEC 61400-2).

	SWT class			
	I	II	III	IV	S
Reference wind speed, V_{ref} (m/s)	50	42.5	37.5	30	Values to be specified by the designer
Annual averaged wind speed, V_{ave} (m/s)	10	8.5	7.5	6	
Turbulence intensity at $V = 15$ m/s	0.18				

its scope is mostly limited within the design safety parameters of SWTs. The primary goal of the new task is to give incentives to this industrial sector to improve the technical reliability of SWTs and therefore their performance.

4 Wind conditions

4.1 External conditions

SWTs are subjected to external conditions, which are divided into wind conditions, other environmental conditions and electrical conditions [1].

4.1.1 Wind turbine class

Wind conditions are the primary external conditions and SWT classification is given in IEC standard IEC 61400-2 depending on wind conditions as shown in Table 2.

Reference wind speed is a basic parameter which is defined as an extreme 10 min averaged wind speed with a recurrence period of 50 years. For standard SWT classes I–IV, annual average wind speed is given by the formula:

$$V_{ave} = 0.2 V_{ref} \tag{5}$$

4.1.2 S class

Offshore conditions and conditions experienced in tropical storms (hurricanes, cyclones and typhoons) are excluded from Standard classes I–IV.

The SWTs that will be driven by the winds whose characteristics exceed those of standard classes in Table 2 is classified to special S classes.

Both for designers and owners of SWTs, site selection will be quite important, although mini- or micro-WTs are easily installed at any place or on a building. It is because wind conditions considerably vary depending on the site. For large WTs, the sites are mostly open free flow fields, while for SWTs the sites may often gardens, parks, roofs or rooftops of urban houses/high buildings, etc. At these sites, flow characteristics are often quite turbulent and those SWTs installed there should be also classified into S.

4.2 Normal wind conditions and external wind conditions

Wind conditions are usually divided into two wind conditions: normal wind conditions and extreme wind conditions. The former will occur frequently during normal operation of a WT through its lifetime and very much related to fatigue, while the latter will occur very rare after their definition of 1-year or 50-year recurrence period and relate to ultimate loads.

4.3 Models of wind characteristics

Wind speed distribution, wind profile (wind shear model), turbulence model, extreme wind speed model, gust model and model of wind direction change are to be used for calculations of loads.

5 Design of SWTs

A wind turbine is designed to capture maximum energy from the wind and to ensure the safety under any external and internal condition. Safety and performance are the main technical requirements in the design procedure.

Design procedure is given in IEC 16400-2 to attain the engineering and technical requirements to ensure the safety of the structural, mechanical, electrical and control systems.

5.1 Conceptual design

5.1.1 HAWT or VAWT

Wind turbines are classified into two types: HAWT and VAWT. They have some different technical features as shown in Table 3.

It is well known that lift-driven wind turbines have much higher performance than drag-driven types from the aerodynamic theory.

Table 3: Technical features of HAWT and VAWT.

Item	HAWT	VAWT
Performance	Lift-driven and high	Lift-driven and high (Darrieus rotor), drag-driven and low (Savonius rotor)
Power/speed control	Variable pitch control is possible	Usually, variable pitch control is not possible
Yaw control	Needed	Free from yaw control
Structural arrangement	Nacelle is on the top of the tower	All the components except rotor are on the ground level
Support structure	Support structure like gay wire will not induce vibration	Support structure like gay wire may induce vibration

Structural simplicity is reliability as well as beauty. From this point of view, no pitch control and/or no yaw control is desirable. However, power control/rotational speed control is vital with such a rotating machine in the free field under all weather conditions. This results in that power/speed control system including brake system is very important.

As shown in Table 3, in general, a HAWT can easily be equipped with both aerodynamic regulation system (pitch control or stall regulation) and mechanical brake. However, with SWTs, especially with micro-WTs, pitch regulation is preferably not engaged in order to avoid the structural complexity. This point is very special with SWTs and a designer must first decide the method how to control the power/speed of the WT together with the type of WT.

5.1.2 Wind characteristics

Design wind turbine class is decided according to Table 2 depending on the wind characteristics of the site where the wind turbine is to be installed.

The wind characteristics of the site are expressed in a series of mathematical models in IEC standard so that all important load cases that act on the wind turbine can be evaluated for the structural design. A full combination of aerodynamic design, structural design and design of control system will give the engineering integrity.

5.2 Aerodynamic design

5.2.1 Annual energy production

Design methods for SWTs are basically same as those for LWTs. However, variable pitch control system is often avoided with mini- and micro-WTs. Yaw system is also passive to avoid additional power sources for control. Therefore, simplified systems in structures or in control will bring some special problems to SWTs.

With a SWT with fixed pitch control system, aerodynamic design must be completed in combination with power/speed control system and brake system. Once technical solutions to start or stop the turbine as designed are found, the aerodynamic design is proceed.

Let $f(V)$ and $P(V)$ be a probability density function of wind speed at a site and a power curve of a wind turbine, respectively. Then the expected annual energy production (AEP) is:

$$\text{AEP} = \int_0^\infty f(V)P(V)\mathrm{d}V = \int_{V_{\text{in}}}^{V_{\text{out}}} f(V)(0.5\rho V^3 A C_P)\mathrm{d}V \tag{6}$$

where V_{in} and V_{out} are cut-in and cut-out wind speed respectively, and A and C_P are the rotor swept area and power coefficient.

$f(V)$ is usually well fitted by Weibull distribution function expressed in the following formula:

$$f(V) = \frac{k}{C}\left(\frac{V}{C}\right)^{k-1} \exp\left(-\left(\frac{V}{C}\right)^k\right) \tag{7}$$

where k and C are shape and scale parameters of Weibull distribution, respectively.

5.2.2 Optimization for maximum energy yield

An optimization is given by the condition that AEP will be maximum, i.e.:

$$AEP \rightarrow Max \qquad (8)$$

A rotor designer must find the optimal power curve $P(V)$ together with WT operation modes. An optimal rotor geometry which comprises of chord and twist distributions and aerofoil sections is found in combination with operation modes; cut-in, cut-out and rated wind speed in principle. Thus an optimal rotor design is site-dependent. Further, since every reliable operation mode depends on power/speed control methods, conceptual design of control methods are also needed at the initial stage.

5.2.3 One-point optimal rotor design

Simplest optimization of rotor design is "One-point design method", in which an optimal rotor geometry is determined for one operation point: one combination of a design wind speed V_D and a design rotor speed Ω_D by using BEM theory [2]. The optimization is given by the following condition for $V = V_D$ with selected rotor speed Ω_D:

$$C_P \rightarrow Max. \text{ At every axial position of rotor blade} \qquad (9)$$

Algebraic solution determines the chord and twist distributions along its axis of the blade. However, it must be noticed that "One-point design method" does not always give the maximum AEP, because a maximum AEP is obtained under the best combination of decided WT's operation mode and wind characteristics at the site.

The question is how to decide design wind speed V_D.

The simplest design by BEM theory is conducted under the following assumptions:

- Low Reynolds number effect is negligible
- Power curve is a function of tip speed ratio λ only.

Then eqn (6) is expressed as:

$$AEP = \frac{1}{2}\rho A \left\{ \int_{V_{in}}^{V_R} f(V)V^3 C_P(\lambda)dV + V_R^3 C_P(\lambda_R) \int_{V_R}^{V_{out}} f(V)dV \right\} \qquad (10)$$

where V_R is rated wind speed and

$$\lambda_R = \frac{R\Omega_R}{V_R} \qquad (11)$$

Max(AEP) is derived as follows:

In case of variable-speed operation WT, an optimal $\lambda_{opt} \approx \lambda_R$ is defined and the optimal value can be possibly in the manner:

$$\lambda \approx \lambda_{opt} \quad \text{for } V_{in} \leq V \leq V_R \qquad (12)$$

Then the optimal value of $C_{Popt} = C_P(\lambda_{opt})$ is kept and by choosing $\lambda_{opt} = \lambda_R$, AEP $\rightarrow$ Max in eqn (10). The second term of in the integral of the right side of eqn (10) is determined by the wind speed probability function $f(V)$.

5.2.4 In case of constant-speed WT

In case of fixed-speed WT, λ varies with V in proportion to $1/V$. Therefore, an optimal value of design tip-speed ratio λ_{opt} may vary from λ_R. It will be soon understood if eqn (10) is rewritten as:

$$\text{AEP} = \frac{1}{2}\rho A \left\{ \int_{V_{in}}^{V_R} f(V)V^3 C_P(V)dV + V_R^3 C_P(V_R) \int_{V_R}^{V_{out}} f(V)dV \right\} \tag{13}$$

Since C_P is a function of V, AEP is also a function of V, but depends on $C_P(V)$ and $f(V)$. This means an optimal rated wind speed V_R and an optimal design wind speed V_D will be obtained through making AEP $\rightarrow$ Max. taking into account the characteristics of WT, i.e. $C_P(V)$, and wind, i.e. $f(V)$, which are independent of each another.

An empirical formula is known as

$$V_R \approx (1.2 - 1.5)V_{mean} \tag{14}$$

where V_{mean} is the annual mean wind speed at the site and is statistically given as follows when the wind characteristics follow Weibull distribution:

$$V_{mean} = E[V] = \int_0^\infty V \cdot f(V)dV = C\Gamma\left(1 + \frac{1}{k}\right) \tag{15}$$

where Γ denotes the Gamma function.

5.2.5 Multi-point optimal rotor design

A goal is to give AEP a maximum value of the integration in eqn (10). One of the simplest methods is "Multi-points design method", with which some desirable power curve is delivered by combination of plural design points. Various original designs are possible, but most important thing is to find a solution that the operation modes together with the power curve of a SWT will most fit to the characteristics of the wind at the site where to install the WT.

It is very often experienced that under low wind speed conditions just around cut-in wind speed, a fixed-pitch SWT will hard to start rotation or generation if it stands still, however it will soon do if it is in idling state. Let assume that cut-in and rated wind speed are 2.5 and 10 m/s, respectively. The ratio of the power output is 1:64. Although the power output at cut-in wind speed is less than 2% of that at rated wind speed, fast start-up performance will result in better availability in time and higher energy production.

5.3 Selection of aerofoil sections

In general, C_P is a function of several parameters:

$$C_P = C_P(\lambda, \beta, \varepsilon, Re) \tag{16}$$

where λ, β, ε and Re tip-speed ratio, pitch angle, drag-lift ratio and Reynolds number, respectively. All these parameters relate to the flow condition on the rotor blade. Typically, λ and β define the flow angle at every location along the blade and ε and Re directly relate to the performance of the aerofoil section at the location. Thus, selection of aerofoil section is very important.

5.3.1 Mechanism of performance reduction
Wind turbines of high performance are driven by lift.

According to BEM theory, local power produced by a blade element has a expression:

$$\text{Local power: } \Delta P \infty (1 - \varepsilon \cot \varphi) \Delta r \tag{17}$$

where φ is flow angle on a blade element of length Δr and ε is drag lift ratio of the blade element. At an ideal flow angle, $\varepsilon \approx 0.01\text{–}0.02$, but if a stall occurs on the blade element $\varepsilon > 1.0$, which suggests both selection of aerofoil sections and control of relative low condition between a blade and air flow are keen to the rotor aerodynamic performance.

Thin aerofoil sections have usually smaller drag coefficients than thick sections. However, they will soon fall into stall if the angle of attack varies from the ideal condition. At such laminar separations with SWTs, the performance reduction is considerably large. Thick aerofoil sections have high structural stiffness than thin aerofoil sections, but usually have higher values of ε which induce higher energy loss.

5.3.2 High lift devices
To avoid the laminar separations, and also to regulate the power output and rotor speed, several techniques are used.

Pitch regulation is the most effective active method, but is sometimes difficult to apply to SWTs with less space or capacity for control power.

Vortex generator (VG) is simple and useful high lift device. Even a large WT benefits from this device. Longitudinal vortexes generated by VGs prevent the boundary layer from inducing large scale laminar separations, but drag coefficient will usually slightly increase.

Start-up assist method is a smart technology to improve the flow condition on rotor blades by accelerating the rotor. When a rotor condition is far from design point, such as rotor speed is very low or wind is very low, for example. An instantaneous acceleration of the rotor by additional power source for certain seconds will bring the flow angle into regular conditions capable of generation and the operation becomes more ideal condition. Although this technology does not work as a brake device, it works as an alternate method with pitch regulation.

5.3.3 Blade material

Aerodynamic performance of a rotor depends on blade structures and materials. If rotor stiffness is not enough, deformation of rotor blade will move the operation point from the designed point, thereby the power performance will decrease. As long as such an aero-elastic effect cannot be reflected in the rotor design, stiff materials shall be chosen.

On the other hand, an advanced design may utilize the aero-elastic effect. A thought-out design of a flexible rotor may obtain an ability of regulating power although it is a fixed pitch system.

5.4 Structural design

Most important works in the structural design is to verify that all the limit states are not exceeded for the WT design [1]. Therefore, all the loads that will act on the WT are determined first and then structural design of the WT and its components will follow in combination with the WT conditions.

Loads are classified as:

* Vibration, internal and gravitation loads
* Aerodynamic loads
* Operational loads caused by yawing, braking, furling, pitching, etc.
* Other loads such as wave loads, wake loads, transport, etc.

Three different methods for the determination of design loads are described in IEC Standard 61400-2. Simplified load equation methods are simple but conservative. The method using aero-elastic modelling is finer and more complicated than the first one. This method has common base with that for general wind turbines. Design loads can be derived from load measurements.

A designer of SWTs shall refer the IEC standard to understand general design methodology and conduct design by using equations for loads, table of load cases and safety factors.

6 Control strategy of SWTs

Decision of the control strategy of a SWT is also important element in the initial system design. Both of its power output control and yaw control are usually quite different if compared with those of LWTs, particularly with mini- and micro-WTs.

The reasons are:

* The representative system speed Ω_D is proportional to $1/D$ as shown in eqn (4), which means SWTs are higher rotational speed machines than LWTs and the control rate shall be faster in proportion to $1/D$.
* SWTs are more apt to be influenced by various kinds of local turbulence than LWTs.

Thus it should be understood that both speed control and yaw control performances are requested to be as high as or higher than LWTs.

Upon this nature, the first issue to decide on the control strategy is whether to adopt active control or passive control concept. The former is a concept to control the system using a yaw drive with any kind of active power source such as electric or hydraulic power actuators, while the latter is to control by natural forces generated by wind or mechanical forces such as centrifugal force without using any active power source.

The finer control system based on active control concept is desirable at one hand, however, there are another reasons that passive control concept is attractive at another. The passive control concept gives us the advantage of structural simplicity, which would give greater reliability in general, provides as smart control measures using natural force without extra control power sources and supplies us at more economical cost.

In principle both concepts exist in LWT design and large turbines, but the status of technical tendency is that more complete active control concept with LWTs and more passive control concept or less active control concept with SWTs. Should "smart" technologies be really smart, the passive control concept could be also popular with LWTs in future.

There is an evidence for this. For long, a constant-speed operation system was technically thought to be most suitable for grid connected systems simply because the electrical grid has constant frequency and the rotor should be regulated to rotate at constant speed. But wind will seldom keep blowing constantly at the rated wind speed at most of the time during the operation of wind turbine. It is much more natural for a WT to work at any given wind speed with higher efficiency but with less extra active power source. The solution was given by "variable operation systems" which employ a power convertor system such as AC–DC–AC generator system. Electronically finer and faster while mechanically less loaded, is modern technical tendency.

Decision of choosing either of constant-speed operation or variable-speed operation is also important task.

Figure 5 shows an example of a basic performance of an optimal rotor design. The rotor is designed by BEM theory at design wind speed of 10 m/s, tip-speed ratio of 7 m for 10 m of rotor diameter using a thin wing section of Illinois University. Once aerofoil section(s) selected, Reynolds number effect neglected and pitch angle fixed, then C_P is a function of $\lambda = r\Omega/V$ only.

Maximum C_P is 0.475 for $\lambda = 7$ at the design point. When neglecting the effect of low Reynolds number, as long as $\lambda = 7$ is kept, the rotor produces power with maximum aerodynamic efficiency. As shown in Fig. 5, $C_P(\lambda)$ will decrease with constant-speed system ($\Omega = \Omega_D = $ Const.) as λ departs from the design point while keeping pitch angle constant. As a result,

$$C_P \geq 0.4 \quad \text{only for } 7.5 \, \text{m/s} < V < 14 \, \text{m/s} \tag{18}$$

On the other hand, with a variable-speed operation rotor, optimal C_P can be realized if the optimal rotor speed is properly regulated. In this case, under the varying natural wind speed, it is desirable to keep $\lambda = \lambda_{\text{opt}} = 7$, or to vary rotor speed in the manner: $\Omega = (V/R)\lambda_{\text{opt}}$.

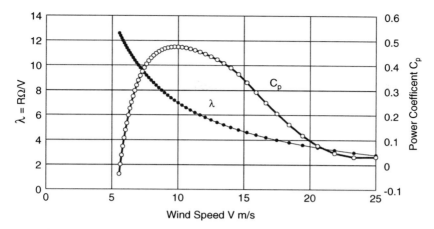

Figure 5: A sample of power coefficient obtained by BEM theory.

7 Yaw control

Yaw control is a function to face the rotor to the natural wind so that the WT can capture maximum energy from the wind. Therefore, a WT shall face perpendicular to the wind directions as much as possible at all times. Because of the dimensional effect, a SWT will much more susceptible to be caught in local smaller eddies than LWTs and as a result, yaw control is technically more complicated than LWTs. One thing most desirable is stable wind, however, the people often like to install SWTs rather close to or directly upon buildings where the wind tends to be more turbulent.

It might be better to provide a fine yaw controls system but actually almost all mini- and micro-WT have passive control systems.

Yaw motion is governed not only by yaw control device but also inertia and gyroscopic moment of the nacelle and yawing dynamic ability of the rotor itself. Some typical yaw control systems with mini- and micro-WTs are described below.

7.1 Tail wing

An upwind oriented rotor could have yaw instability. Most of LWTs except downwind type is equipped with a yaw control system driven by a hydraulic or electric power. With SWTs of mini- or micro-classes, however, a tail wing is mostly used as a passive control system. Tail wing is a simple and classic tool since the middle aged.

7.1.1 Variety of tail wings
The performance of a tail wing as an aerofoil as well as the performance of yaw control, both in static and dynamic response, is decided by the aerodynamic performance of the tail wing. Then, there are various types and shapes of tail wings.

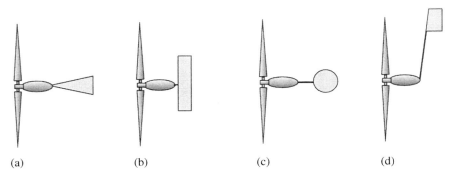

(a)   (b)   (c)   (d)

Figure 6: Various types of tail wings.

Figure 7: Modern tail wing (Courtesy of Anna Estanqueiro, IEA WIND).

Figure 6 shows a various shapes of tail wings. There is no ultimate shape of the tail wing at all, because there are many yaw control strategies at one hand, there are many wind characteristics depending on the site at another.

The yaw control moment is the product of aerodynamic force on the aerodynamic centre of the wing and the distance from yaw axis to the centre. The tail wing (a) has much small aspect ratio than the tail wing (b). Then wing type (a) has smaller lift coefficient at small attack angles, i.e. less yaw control force than type (b). However, under high turbulent wind conditions, too strong control force will sometimes bring unstable yaw motion.

7.1.2 New types of tail wings
Some unique and advanced types have been developed. The tail wing design shown in Fig. 7 has an elegant shape.

A swing ladder shown in Fig. 8 is a swinging tail wing. The ladder which works as a wing is pin-jointed to the tail-end of the nacelle free in rotation around the pin-axis. This system responds flexible to varying wind speed and directions and one of the large advantages is the regulative effect that prevents the excessive response and oscillating motion of the tail.

Figure 8: Swing rudder (Courtesy of Zephyr Corporation).

Figure 9: Yaw control system with downwind WT (Courtesy of Proven Energy Ltd.).

7.2 Passive yaw control with downwind system

With downwind WTs, the rotor functions as a tail wing and it has passive yawing-function to make WTs follow the wind direction. A typical commercial downwind wind turbine is shown in Figure 9. No power source is necessary for yaw control, but stability of yaw motion needs to be carefully investigated, because when one more freedom of the system is added, instability could happen under accelerated or decelerated conditions of rotor speeds.

8 Power/speed control

Power control/rotor speed control is an essential part of whole system, upon which both performance and safety depend. As described in the previous sessions, an optimal aerodynamic performance design can start, provided a feasible power/speed control is available.

8.1 Initial start-up control

A WT with pitch control system are capable to make start-up at lower wind speed by regulating the pitch angle of the blades. However, SWTs of fixed pitch rotor blades must wait until stronger wind starts to rotates the rotor into operation. Power assist system can catch the chance to bring the WT into operation when wind speed is near cut-in wind speed. It works just the same as variable pitch system. The power assist system is driven by a motor, which means it needs additional power source. This may lose certain energy but such energy loss would be just marginal since the start-up mechanism will not be activated during constant rotation operation, and that if improved rate of operation is considered, the gain will be larger.

8.2 Power/speed control

In the whole operation wind speed range, power/speed control must be reliable and fail-safe. Therefore, the control system provides more than two independent control or brake mechanisms.

8.2.1 Aerodynamic methods

There are two aerodynamic control methods. One is traditional stall regulation. In this method, when a gust attacks the rotor, a separation will occur on a blade which decreases the torque resulting in power/speed regulation, preventing over speed of the rotor as well. It is considered stall regulation method is more suitable in the condition of relatively stable wind of small turbulence. However, active stall system has been developed that is more reliable and available for higher turbulent wind conditions.

According to fail-safe concept, a stall regulation is usually supported by other systems such as mechanical brake. Collapsible rotor blade tip is an aerodynamic device for over-speed limitation.

Variable pitch control or pitch regulation is also aerodynamic method. The generated torque of the rotor as well as the lift on the blades is controlled by the attack angles by changing the pitch angle. Hydraulic power or motor(s) are necessary to change the pitch angle(s) by way of some movable mechanical elements such as links. Variable pitch control is reliable and works also at starting-up at cut-in state and shutdown at cut-out state.

8.2.2 Mechanical methods

All kind of mechanical brakes are possible. A mechanical method need considerable power source to shut down the system without any aerodynamic method. During the shutdown, all the kinetic energy lost from the rotor is converted to thermal energy that will be absorbed into the brake materials.

8.2.3 Electro-magnetic methods

Electro-magnetic method is also possible. The lost kinetic energy from the rotor is converted to thermal energy and temperature monitoring/control is very important.

One 1 kW SWT with such an electro-magnetic regulation has an experience that the WT operates during high wind under a typhoon attack. The turbine was generating power at 300–650 W under the wind speed range of 27–37 m/s until just before 5 s when the maximum wind speed of 47 m/s blew through the site as illustrated in Fig. 10. The turbine immediately responded to the extreme wind by falling into

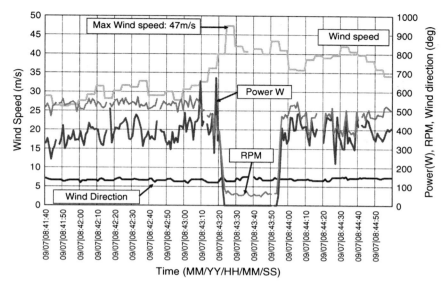

Figure 10: An operation data of a SWT under a typhoon attack.

idling motion. Thus the WT can resist against extreme winds under typhoons as well as capture energy instead of shutting down. This feature of high wind operation is especially very important for remote sites like offshore locations or high mountain uninhabited sites.

9 Tests and verification

9.1 Safety requirements

Safety requirements are described in IEC Standard 81400-2. Basic procedure of certification is also indicated. Particularly for SWTs, key items for demonstration or verification of technical integrity of a SWT are introduced. Informative Annex A gives the structure of type certification of SWTs as shown in Fig. 11.

9.2 Laboratory and field tests of a new rotor

After system design including aerodynamic design and structural design, then proto type tests and verification are required.

Regarding to rotor design, at least three methods of rotor performance tests are possible.

9.2.1 Wind tunnel test

Provided a wind tunnel facility is available in scale and in wind speed range, the rotor performance vs. arbitrary wind speed as desired can be captured at any time by wind tunnel testing. However, the stable performances obtained under stable winds

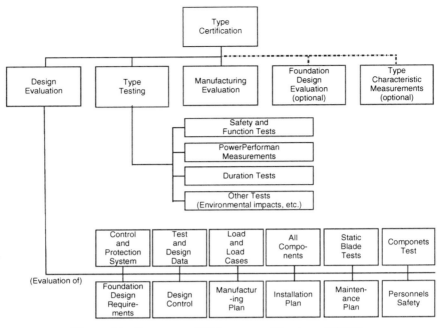

Figure 11: Structure of IEC type certification of SWTs.

in a wind tunnel seriously differ from the dynamic rotor performances obtained in the turbulent flow, sometimes gusty flow over a natural open test field.

A wind tunnel test is capable of verifying the basic stable design power performance. If proper corrections of the wind tunnel wall effect are not done, the power curve might be higher than the actual because of the channel flow effect of the wind tunnel. If the tested WT is scale-down model, Reynolds number effect shall be carefully analysed.

9.2.2 Track test

Track test is an alternative method with wind tunnel test, in which relative wind speed is realized by driving a vehicle with a test turbine on it along a test course of a vehicle. The relative wind speed is as high as that of running vehicle. If the maximum speed is 180 km/h, even an extreme wind speed of 50 m/s can be created. Since an extreme wind speed, which is a maximum 10-min averaged wind speed with 50 years recurrence period for IEC Class I turbines, is a very rare case to occur and cannot usually happen during a few years field test. Therefore, it is quite useful test method for structural safety tests and power regulation or brake tests.

Figure 12 shows a vehicle for track test. A SWT is fixed on a support structure on the vehicle. An anemometer is also mounted on the vehicle at a suitable position where flow distortion by the vehicle is the least.

Figure 12: A vehicle for track test.

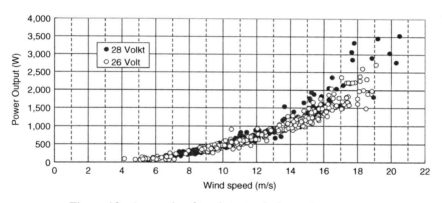

Figure 13: A sample of track test, wind speed vs. power.

If the natural wind does not affect seriously, a series of steady performance tests and structural tests including extreme conditions can be conducted. Highest attentions shall be given for personal safety and structural safety.

Another valuable task is development of control strategy, which relates power control, speed control, load regulation, break adjustment and yaw control.

Figure 13 is a sample of track test data of a Micro WT of 1.8 m rotor diameter. Power output data are shown for two values of volt of battery load. By conducting parametric tests, optimal control parameters can be found.

9.2.3 Field test

Final evaluation and validation of a SWT's power performance must be determined by field tests. It is because actual energy products from a WT are obtained under free wind flow varying with time both in speed and in direction.

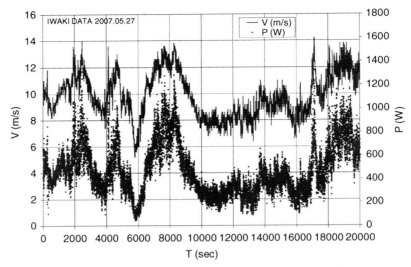

Figure 14: A sample of field test data of a 1 kW SWT.

The power performance is statistically expected values of power output and Bin method is applied.

Figure 14 shows a field test data of a 1 kW SWT. The sampling period is 1 s and the period is continuous 8 h. Wind speed is varying from 5 to 15 m/s. Due to various uncertainties in measurement, data scatter widely. Therefore, a statistically expected power curve represents the performance of a WT.

A power performance curve is decided by applying Bin-averaging method (Bin method).

In Fig. 15, 1-min averaged data are plotted with symbol ×.

For SWTs, data shall be collected continuously at a sampling rate of 1 Hz or faster and every 1-min data set gives a 1-min mean value, 1-min standard deviation, 1-min maximum value and 1-min minimum value for wind speed and power output. Then plenty of mean values are further averaged by Bin method as follows:

$$V_{Bin,i} = \frac{1}{N_i} \sum_{j=1}^{N_i} V_{i,j} \tag{19}$$

$$P_{Bin,i} = \frac{1}{N_i} \sum_{j=1}^{N_i} P_{i,j} \tag{20}$$

where i is the Bin number (when using 0.5 m/s Bins, $0.5i - 0.25 \leq V_{i,j} < 0.5i - 0.25$); $V_{i,j}$ is the normalized wind speed of data set j in Bin i; $P_{i,j}$ is the normalized power output of data set j in Bin i; $V_{Bin,j}$ is the Bin-averaged wind speed in Bin i; $P_{Bin,j}$ is the Bin-averaged power output in Bin i; N_i is the number of data sets in Bin i.

The white circles in Fig. 15 are Bin-averaged values.

A power performance curved is completed if sufficient data sets cover all the operation range of wind speed. Then an uncertainty analysis follows.

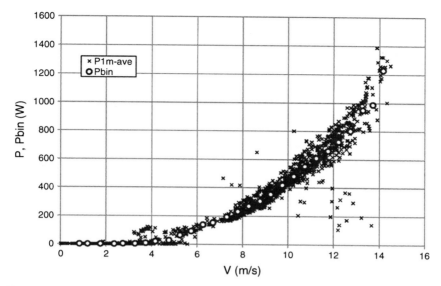

Figure 15: Sample of 1-day field test data at 1-s sampling rate and Bin-averaged data.

10 Captureability

There are various parameters that should be optimized such as power coefficient, capacity factor, cost of energy, etc.

Parameter "captureability" is a new indicator which shows how much energy has been captured from the nature in a certain period [3]. Now assume a WT has a power curve function $P(V)$ and the wind at the site has probability density function $f(V)$ vs. wind speed V.

Choosing a continuous time period T, the following formulas are obtained:

- Wind power generation G during period T:

$$G = T \int_0^\infty P(V) \cdot f(V) \, dV \qquad (21)$$

- Cumulative wind energy flux E_F past through the rotor swept area A during period T:

$$E_F = T \int_0^\infty E(V) \cdot f(V) \, dV \qquad (22)$$

where $E(V) = \rho A V^3 = \frac{1}{2} \rho V^3 \pi R^2$.

Designed maximum WT generation G_R under the assumption that WT has constantly operated at rated power P_{rated} during the period T is given as:

$$G_R = P_{rated} \cdot T \qquad (23)$$

Then capacity factor C_F is:

$$C_F = \frac{G}{G_R} \tag{24}$$

Captureability C_{PT} is defined as:

$$C_{PT} = \frac{G}{E_F} \tag{25}$$

Captureability is an absolute indicator of the wind turbine plant performance reflecting both WT performance and wind characteristics at the site. But capacity factor is not absolute indicator because rated power can be changed on purpose. For example, if a WT of very low rated wind speed is built at an excellent wind site, its capacity factor could be very high although captureability would be quite poor.

When the wind speed probability function is expressed by Weibull function and the system efficiency at rated condition is given, the following equation is obtained:

$$\frac{C_F}{C_{PT}} = \frac{\sum_V E(V) \cdot f(V)}{P_{rated}} = \frac{1}{\eta_{rated}} \left(\frac{V_{ave}}{V_{rated}} \right)^3 \frac{\Gamma(1+3/k)}{\Gamma^3(1+1/k)} \tag{26}$$

where V_{ave} is the annual mean wind speed, V_{rated} is the rated wind speed, η_{rated} is the system efficiency at rated condition and Γ is the Gamma function.

Figure 16 shows general relations between C_F/C_{PT} and V_{ave}/V_{rated} for $\eta_{rated} = 0.4$ and for different Weibull parameter k.

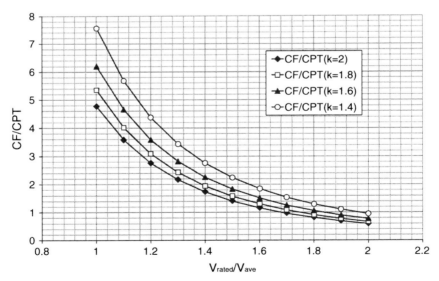

Figure 16: General relations between C_F/C_{PT} and V_{ave}/V_{rated} ($\eta_{rated} = 0.4$, Weibull shape parameter $k = 1.4$–2.0).

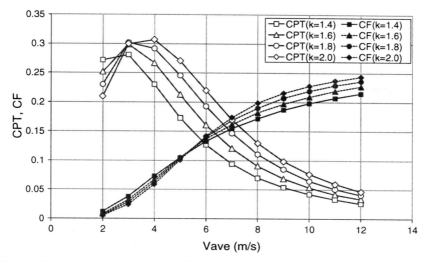

Figure 17: A sample calculation of captureability and capacity factor of a SWT with parameters annual mean wind speed and Weibull parameter k.

Using the power curve of a 1 kW SWT, expected captureability C_{PT} and capacity factor C_F are shown for different wind Weibull shape parameter k in Fig. 17.

As a conclusion, an optimal design concept is to design a WT and choose a site so that the captureability is maximized. This is the same as traditional optimal design concept to maximize the annual energy product. In spite of this fact, there are wide tendency to evaluate WT plants by capacity factor.

Therefore, a performance evaluation should be done by captureability rather than by capacity factor.

References

[1] IEC 61400-2 2nd Ed., Wind turbines – Part 2: Design requirements for small wind turbines, 2006.
[2] Hansen, Martin O.L., *Aerodynamics of Wind Turbine*, Earthscan, 2008.
[3] Matsumiya, H., *et al.*, Field Operation and Track Tests of SHWT "Airdolphin" under High Wind Conditions, EWEC2008, 2008.

PART III

DESIGN OF WIND TURBINE COMPONENTS

CHAPTER 13

Blade materials, testing methods and structural design

Bent F. Sørensen, John W. Holmes, Povl Brøndsted & Kim Branner
Risø National Laboratory for Sustainable Energy
The Technical University of Denmark.

A major trend in wind energy is the development of larger wind turbines for offshore wind farms. Since access to offshore wind turbines is difficult and costly, it is of great importance that they operate safely and reliable. The wind turbine rotor blades, which are the largest rotating component of a wind turbine, are designed for an expected lifetime of 20 years. During this period of time, the blades will be subjected to varying loads. Large wind turbine blades are made of composite materials and can develop a number of interacting failure modes. High structural reliability can be achieved by designing the blades against the development of these failure modes. This chapter provides an overview of experimental and modeling tools for the design of wind turbine blades, with particular emphasis on evolution and interaction of various failure modes. This involves knowledge of materials, testing methods and structural design.

1 Introduction

Structural design of wind turbine blades for horizontal axis wind turbines is a complicated process that requires know-how of materials, modeling and testing methods. A wind turbine blade must be designed against undesired aero-elastic phenomena and failures for a great variety of aerodynamic load cases and environmental conditions. Thus, the design process involves a number of different areas, such as knowledge of the external loads originating from wind and gravity and knowledge of the performance, the strength and the endurance of the full structure and of the basic materials used. The goal of the design process is to ensure that the wind turbine blade will function safely for its design life. The design lifetime of modern wind turbines is normally thought to be 20 years, and the corresponding number of rotations (blade-tower passings) is of the order 10^8 to 10^9, which is approximately two orders of magnitude higher than the load cycles experienced by composite materials used in other highly loaded structural applications such as helicopter blades.

The main trends in the development of wind turbine blades are towards longer and optimized blades; this is particularly the case for offshore wind turbines. The weight of a large wind turbine blade also increases the loads on the rotor input shaft and bearings as well as the wind turbine tower and mechanisms used to control yaw and pitch of the blades. Weight savings is therefore of great importance, and significant efforts are devoted by wind turbine companies in the selection of materials. To ensure that the blades can meet the required design life, the materials must have high stiffness, be fatigue resistant, and be damage tolerant.

As for other low-weight-driven designs, the material considerations thus involve the specific stiffness (i.e. the stiffness divided by density), specific strength (strength divided by density) and specific fatigue limit (fatigue limit stress divided by density).

This chapter provides an overview of the interconnection between the blade design process, material properties, materials testing and sub-component testing and full-scale blade testing. As shown in Fig. 1, the design of modern wind turbine blades involves an understanding of material behavior and failure modes at many length scales. This design process requires close collaboration between engineers involved in modeling of aerodynamic loads, structural analysis and composite materials and technicians responsible for the manufacturing process, quality control and on-site inspection and monitoring of blades. This chapter starts with a review of manufacturing processing (Section 2), followed by a description of full-scale blade testing including some results for blades tested to failure (Section 3) that leads to a classification of common failure modes (Section 4). The material properties that control the development of the various failure modes are presented in Section 5. Section 6 describes experimental methods for determination of these properties. Various examples of the use of the modern design methods that make use of these strength-controlling material properties are given in Section 7. Finally, Section 8 contains a discussion and closure.

2 Blade manufacture

The design of the wind turbine blade is a compromise between aerodynamic and structural considerations. Aerodynamic considerations usually dominate the design for the outer two-thirds of the blade, while structural considerations are more important for the design of the inner one third of the blade.

2.1 Loads on wind turbine rotor blades

The rotor blade is loaded in a combination of flapwise and edgewise loads. Basically the blades are exposed to three different load sources. One is the wind load that through the lift and drag on the aerodynamic profile loads the blade primarily

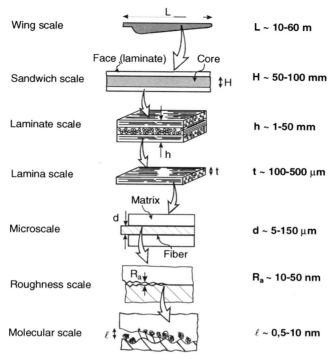

Wing scale   L ~ 10-60 m

Sandwich scale   Face (laminate)   Core   ↕H   H ~ 50-100 mm

Laminate scale   h ~ 1-50 mm

Lamina scale   ↕t   t ~ 100-500 μm

Microscale   Matrix   d   d ~ 5-150 μm
                         Fiber

Roughness scale R_a↓ R_a ~ 10-50 nm

Molecular scale   ℓ↕   ℓ ~ 0,5-10 nm

Figure 1: Modern blade design requires an understanding of how materials and structures behave at various length scales, ranging from the molecular scale (e.g. interfacial adhesion in adhesive joints) to the blade scale (e.g. the dynamic coupling of the blade and tower). Changes made at any length scale will affect the blade reliability. For instance, an increase in the bond strength of the fiber/matrix interface will increase the materials strength at the lamina scale, potentially leading to an increase in the overall blade strength.

in bending flapwise. The second load source is the gravity varying edgewise from tension/compression in leading edge and compression/tension in trailing edge. This is the main reason for the edgewise fatigue bending of the blade. Finally, the blades are exposed to centrifugal forces during the rotation. However, these longitudinal loads are relatively low and often not taken into account in the design. Furthermore, the design loads are divided into static loads and cyclic loads. International design recommendations (e.g. IEC 61400-1 [1]) specify both types of loads. Moreover, the blades will be subjected to a wide range of environmental conditions.

2.2 Blade construction

Modern wind turbine blades are structurally advanced constructions utilizing composite laminates, sandwich core materials, gelcoat films and adhesive joints.

Although there are a variety of wind turbine designs (reflecting different manufacturing processes, material selection and design philosophy), the functionality of wind turbine blades from a structural viewpoint can be understood by considering the blade as a load-carrying beam (spar) enclosed by a shell. The primary purpose of the shell is to give the blade an aerodynamic shape, creating the aerodynamic forces that make the wind turbine blade to rotate and thus extracts energy from the wind to make electrical power. The aerodynamic forces are transmitted to the wind turbine hub through a load-carrying beam within the blade. The load-carrying beam can be made as a box girder, sometimes called the main spar, or as laminates in the aeroshell supported by webs. Figure 2 shows a sketch of the cross section of a typical wind turbine blade. Steel bolts are present at the root, where the blade is to be attached to the hub of a wind turbine.

Most blade manufacturing techniques involves making the blades in several parts that are eventually joined by adhesive bonds. Figure 3 shows two common design approaches. The one design involves the use of load-carrying laminates in the aeroshells and webs for providing shear stiffness and buckling resistance. The other design constitutes of two aeroshells bonded to a load-carrying box girder. Sandwich structures are used extensively in the aeroshells and are also often used in the webs (or correspondingly, in the sides of the box girder).

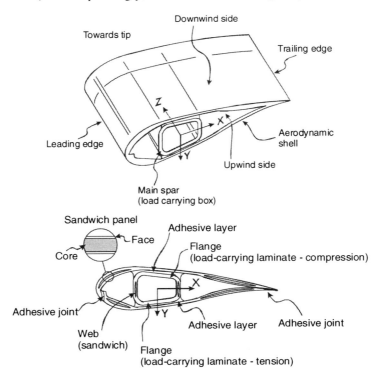

Figure 2: Terminology and definitions of coordinate system and structural parts of a wind turbine blade [2].

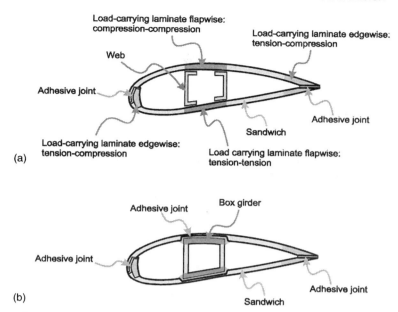

Figure 3: Two common design concepts of wind turbine blades manufactured as two half-shells joined by adhesive bonds: (a) integrating the load-carrying laminates in the aeroshell, which is supported by shear webs, and (b) use of a box girder which is glued to the inside of the aeroshells.

2.3 Materials

Materials used for wind turbine blades must be low density and possess high strength, fatigue resistance and damage tolerance. Large parts of the blades are made of composite materials, i.e. materials that consist of more constituents, e.g. long aligned fibers embedded in a continuous phase called the matrix material. The shells that define the aerodynamic blade profile are typically constructed using polymer matrix composites (PMCs) (e.g. glass fiber reinforced polyester) and sandwich structures consisting of PMC face sheets and lightweight closed-cell polymer foam or end-grain balsa wood cores. The box girder is constructed using glass fiber composites or carbon fiber composites.

Concerning *fibers*, the majority of large turbine blades use E-glass laminates for the aeroshells. Because blade mass is a critical design consideration, more expensive carbon fiber composites are also being increasingly utilized. In addition to much lower density (a factor of 2/3 lower than glass fibers), carbon fibers offer several advantages for blade applications, including a much higher elastic modulus, strength and fatigue life. Carbon fiber laminates and hybrid carbon/glass fiber laminates are currently being utilized by some blade manufacturers. For the aeroshell, carbon fibers can be used for selective reinforcement but offer great potential for innovative blade designs. Although carbon fibers have attractive modulus, density and strength, there are some drawbacks beyond higher cost. The compressive strength can be lower than that of glass fiber composites. In addition, the compressive strength of carbon fiber composites is very sensitive to the presence of manufacturing-related defects such as fiber misalignment and fiber waviness [3–5]. For certain regions of the blade, such as the aeroshell and internal sandwich structures, wood or other natural materials which typically have a much lower density than glass fiber composites can be used. For example, Vestas UK utilizes birch wood which is laminated with carbon fibers to produce the aeroshell of large (40 m and longer) wind turbine blades. Because of their low mass, these blades can be produced in thinner profiles which further reduce rotating mass. Other materials, including bamboo are also being developed as green alternatives for use in wind turbine aeroshells and for sandwich structures within the blade [6].

The composite *matrix* is also an important consideration for wind turbine blades. Currently, polyester is the most common choice as the matrix for glass fiber laminates, but epoxy resins and vinylester are also used because of their superior mechanical properties. Each matrix material has a set of cost, manufacturing and mechanical behavior tradeoffs that must be considered. For example, in comparison with vinylester, the use of epoxy resins can increase the compressive strength of glass fiber laminates by as much as 10–15%. For carbon fiber composites, epoxy is the most common matrix material. Current research efforts are aimed at the development of matrix materials with improved toughness, cure profiles and the development of improved low-viscosity infusion grades for permeating thick fiber stacks. From an environmental viewpoint, the use of thermosetting polymers like polyester and epoxy poses many problems and additional attention is being given to the use of thermoplastics such as polybutylene teraphthalate (PBT). A key environmental advantage of thermoplastics is their potential to be recycled at the end of the turbine life cycle.

Since most wind turbine blades are bonded together, the *adhesives* used to join blade segments will have a direct influence on the reliability of the blade. The adhesives utilized in blades are primarily epoxy, polyurethane and methacrylate-based adhesives. The quality and mechanical properties (e.g. strength, creep, defect tolerance) of blade adhesives are critical since they are used to bond very large areas of the blade aeroshells and box girder and are subjected to complex cyclic loading histories. Moreover, adhesives must maintain adequate bond strength over a wide range of environmental conditions. As discussed later, the defect tolerance and fracture resistance of adhesive joints is determined through fracture mechanics mechanics-based testing.

As noted earlier, sandwich structures are widely used in the blade aeroshell and shear webs in some spar designs. The *core* material is low-density materials, primarily balsa wood and polymer foams.

Finally, the surface of wind turbine blades are painted with *gelcoats* to protect the composite materials from damage originating from UV-radiation and to limit the environmental exposure to the blade, e.g. humidity which may change (decrease) the mechanical properties of the composite materials. The gelcoats should maintain their adhesion to the underlying surface despite large changes in humidity and temperature and should be sufficiently wear resistant to last for 20 years.

2.4 Processing methods

Wind turbine blades can be manufactured using procedures similar to that used for composite aircraft structures and composite boat hulls; namely lamination by the use of pre-pregs, hand lay-up and vacuum assisted resin transfer molding.

Both the manufacturing approaches shown in Fig. 3 involve the manufacturing of the aeroshells as the first step. Since the aeroshells do not carry much load themselves, they are primarily made of lightweight sandwich structures. The fibers (or pre-pregs) and sandwich materials are usually placed by hand in an open mould. The mould is closed before curing, so that fumes can be removed during the processing. Next, in the second step, the moulds are opened and the webs or box girder is placed in between the aeroshells. In the third step, the moulds with the two half-shells are closed and the two half-shells and webs (or box girder) are bonded together. Such processing methods have several advantages. First, the quality of the aeroshells and webs (or box girder) can be controlled before the final assemblage. Secondly, the adhesive layers allow a relatively large dimensional tolerance, as the adhesive layer thickness can range from a millimeter to several centimeters. This is particularly useful with increasing sizes of the blade components.

3 Testing of wind turbine blades

3.1 Purpose

Full-scale testing (Fig. 4) is mandatory for certification of large wind turbine blades. The basic purpose of these blade tests is to demonstrate that the blade type has the prescribed reliability with reference to specific limit states with a reasonable level of certainty. According to Det Norske Veritas (DNV) [7], a limit state is a defined as a state beyond which the structure no longer satisfies the requirements. The following categories of limit states are of relevance for structures: ultimate limit state (ULS), fatigue limit state (FLS), and serviceability limit state (SLS). The blade should be manufactured according to a certain set of specifications in order to ensure that the test blade is representative of the whole series of blades. In other words, the purpose of the blade tests is to verify that the specified limit states are not reached and that the type of blade possesses the projected strength and lifetime.

Normally, the full-scale tests used for certification are performed on a very limited number of samples; only one or two blades of a given design are tested so that no statistical distribution of production blade strength can be obtained. Therefore, although the tests do give information valid for the blade type, they cannot replace either a rigorous design process or the use of a quality control system for blade production.

Figure 4: Full-scale testing: a 61.5-m long wind turbine blade subjected to static
test (flapwise direction) [Courtesy of LM Glasfiber A/S].

Additionally, tests can be used to determine blade properties in order to validate
some vital design assumptions used as inputs for the design load calculations.
Finally, full-scale tests give valuable information to the designers on how the
structure behaves in the test situation and which structural details that are important
and should be included in the structural models for design. Especially, valuable
information is obtained if the blade is tested to failure.

3.2 Certification tests (static and cyclic)

According to DNV [7], it is required that the test program for a blade type shall be
composed of at least the following tests in this order:

- Mass, centre of gravity, stiffness distribution and natural frequencies
- Static tests
- Fatigue load tests
- Post fatigue static tests

All tests should be done in flapwise direction towards both the downwind (suction) and upwind (pressure) sides and in edgewise directions towards both the leading and trailing edges. If it is important for the design, also a torsion test is needed in order to determine the torsional stiffness distribution. The tests are undertaken to obtain two separate types of information. One set of information relates to the blade's ability to resist the loads that the blade has been designed for. The second set of information relates to blade properties, strains and deflections arising from the applied loads.

All tests in a given direction and in a given area of a blade shall be performed on the same blade part. The flap- and edgewise sequence of testing may be performed on two separate blades. However, if an area of the blade is critical due to the combination of flap- and edgewise loading, then the entire test sequence shall be performed on one blade.

3.3 Examples of full-scale tests used to determine deformation and failure modes

In the following, the focus is on the ultimate strength of the rotor blades. Results and findings from recent full-scale tests to failure are studied.

A 25 m blade was tested to failure in three sections in the study by Sørensen *et al.* [2]. The blade was loaded in the flapwise direction. The purpose of the test was to gain detailed information about failure mechanisms in a wind turbine blade especially with focus on failures in the compression side (downwind side) of the blade. Prior to the tests the blade was inspected by ultrasonic scanning to get an overview if any imperfections and damages were present already before starting the test. The supports and loading of the blade was changed during the test such that it was possible to use the same blade in three tests, i.e. having independent failures in three different sections of the blade. During each of the tests the behavior of the blade was recorded by means of video and photos, strain gauges, acoustic emission and deflection sensors. Two different types of deflection sensors were mounted on the blade, one giving the total deflection of the blade and another giving skin and main spar displacements, locally. The identified failure modes are presented in the next section (Section 4).

In a study by Jensen *et al.* [8], results from a full-scale test of a 34 m blade were compared with finite element (FE) analysis. The blade was loaded to catastrophic failure. Measurements supported by FE results show that detachment (delamination) of the outer skin from the box girder was the initial failure mechanism followed by delamination of the load-carrying laminate, leading to collapse.

4 Failure modes of wind turbine blades

4.1 Definition of blade failure modes

Wind turbine blades can fail by a number of failure and damage modes. Obviously, the details of damage evolution will differ from one blade design to another. However, experience shows that, irrespective of specific blade design, several types of material-related damage modes can develop in a blade. In some instances, these damage modes can lead to blade failure or require blade repair or replacement.

It is useful to define a few key concepts. The term failure is used here as a broad term covering various processes that creates damage or cracking. Blade failure indicates the critical state where the wind turbine blade loses its load-carrying capability. Failure mode describes classification of the macroscopic types of failure which can occur. Define damage as non-reversible processes that occur as distributed phenomena (e.g. multiple matrix cracking or fiber failure). Damage modes is the term used to characterize specific types of damage at the material scale. Fracture indicates damage in the form of macroscopic cracks. Fracture modes indicate specific types of cracking (e.g. cracks between different plies or cracks along interfaces between different materials). Crack initiation is defined as the process of the formation of a sharp crack from a pre-existing flaw. The precise occurrence of crack initiation depends on microstructural details, such as the size and distribution of porosity or other defects and the fracture resistance of interfaces; all of these depend on the materials and processing methods used. Crack propagation concerns the growth of sharp crack. Depending upon the crack size and load level, a crack may extend in a stable manner (i.e. in small increments), or stop (crack arrest) or rapidly (unstable).

The load-carrying laminates in the blade aeroshell and box girder are made of composite materials and adhesive joints that are damage tolerant. *Damage tolerant* behavior implies that the first mode of damage does not lead directly to failure, but propagates in a stable manner and gives detectable changes so that the damage can be detected before it reaches a critical size where it leads to failure. Therefore, failure of wind turbine blades does not occur as a direct result of crack initiation along an interface or by progressive damage to the fibers and matrix. Rather, global failure of a wind turbine blade involves the progression of several damage mechanisms that can act in series or in parallel. This hierarchical failure evolution can be thought into the blade design, creating a damage tolerant design. For example, interface debonding along an adhesive joint can cause a detectable reduction in structural stiffness while the redistribution of stresses causes corresponding higher cyclic strain amplitudes in the blade or the initiation of cracks in laminates or sandwich structures in the vicinity of the debond. However, global blade failure will not occur until a damage type (typically a crack) reaches a critical size leading to unstable fracture.

4.2 Identified blade failure modes

A considerable amount of knowledge is required to assess how damage develops in a wind turbine blade and to design a blade against failure using analytical or numerical methods. Therefore, in order to validate the design, and to provide insight into possible damage modes and their severity, blades are sometimes tested to failure by full-scale testing. Figure 5 shows sketches of the failure modes (summarized in Table 1) found in a wind turbine blade tested to failure [2].

The consequences of the various damage and failure modes listed in Table 1 are widely different. For instance, cracking of the gelcoat film is not as severe as

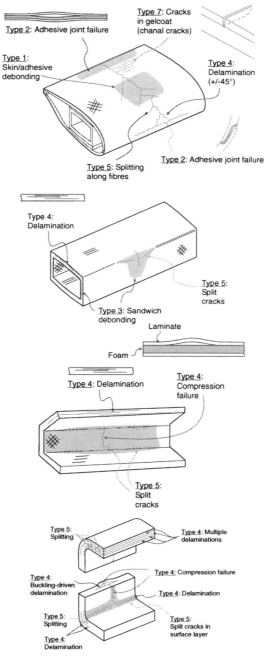

Figure 5: Sketches of observed failure modes in a wind turbine blade purposely tested to failure [2]; damages in the aeroshell (top) and box girder (below).

Table 1: Failure modes and appropriate strength and fracture concepts used for
the design and analysis of wind turbine blades.

Basic damage modes	Material property (static strength)
Adhesive joint failure	
Crack in adhesive layer	G_{Ic} (fracture energy)
Laminate/adhesive interface cracking	$G_c(\psi)$ (interface fracture energy)
Interface cracking with fiber bridging	$\sigma(\delta)$ (cohesive law)
Sandwich failure	
Interface cracking	$G_c(\psi)$ (interface fracture energy)
Interface cracking with fiber bridging	$\sigma(\delta)$ (cohesive law)
Laminate failure	
Tensile failure (fiber fracture mode) – damage zone	σ_{Lu}^+ (tensile strength in fiber direction)
Compressive failure (fiber fracture mode) – damage zone	σ_{Lu}^- (compressive strength in fiber direction)
Tensile failure (matrix fracture mode) – cracking	σ_{Tu}^+ (tensile strength perpendicular to fibers)
Shear failure	τ_{LTu} (shear strength)
Splitting crack (crack parallel with fiber direction)	$G_c(\psi)$ (mixed mode fracture energy)
Delamination crack between plies	$G_c(\psi)$ (interface fracture energy)
Gelcoat/skin delamination	
Interface cracking	$G_c(\psi)$ (interface fracture energy)
Interface cracking with fiber bridging	$\sigma(\delta)$ (cohesive law), J_c (work of separation)
Gelcoat cracking	
Thin film cracking	G_{Ic} (fracture energy)

tensile failure (fiber fracture) in a laminate. When the gelcoat cracks, the laminate
loses its protection against environmental exposure – which can lead to laminate
damage over a long time. In contrast, tensile failure in the form of fiber fractures
decreases the stiffness and residual strength of the load-carrying laminates – this
can lead to rapid failure within a short period of time. Models and criteria for
assessing the various types of damages will be described in Section 7.

5 Material properties

5.1 Elastic properties

Modern wind turbine blades are three-dimensional structures made by the use
of several different materials and the elastic properties and thermal–physical constants, such as thermal expansion coefficient, of materials influence the damage
developed in a blade. As a result, the stress field depends on the elastic properties of
the materials used. For isotropic materials, the elastic properties are the Young's modulus, E, and the Poisson's ratio, v. Orthotropic materials, such as composite laminates

Table 2: Classification of various materials used in wind turbine blades and the anisotropy level used to characterize their elastic constants.

Isotropic materials	Orthotropic materials
Adhesive	Glass fiber/polyester composites
Steel	Carbon fiber/epoxy composites
Polymer foam	Wood (e.g. birch or balsa)
Gelcoat	Bamboo

with aligned continuous fibers, have different elastic properties in different directions. Therefore, the elastic properties must be related to a coordinate system. It is convenient to use a global x–y–z coordinate system (see Fig. 2) and a local coordinate system that follows the direction of the fibers. The longitudinal direction (the fiber direction) is assigned the subscript L, the (in-plane) direction perpendicular to the longitudinal direction is called the transverse direction and given subscript T, and the out-of-plane direction, orthogonal to the L and T directions is denoted TT, i.e. T with a prime. Then, the elastic properties are specified in terms of E_L, E_T, $E_{T'}$, v_{LT}, $v_{LT'}$, $v_{TT'}$, G_{LT}, $G_{LT'}$ and $G_{TT'}$ where E_i denotes the Young's modulus in the i-direction and v_{ij} is the Poisson's ratio in direction j due to a normal stress in the i-direction, and G_{ij} is the shear modulus in the i–j-plane. Table 2 lists some common materials used in wind turbine blades and their anisotropy classification.

5.2 Strength and fracture toughness properties

Damage and failure modes are described by various parameters that may be stress-based, energy-based or length-based (e.g. critical defect length). A damage mode that involves a distributed *damage zone* is usually described in terms of a critical stress value, i.e. by a maximum stress criterion (tensile or compressive strength). Crack growth along a fracture plane is a localized phenomenon. The onset of crack growth can be described in terms of a maximum stress intensity factor (fracture toughness) or a maximum energy release rate (fracture energy). A crack experiencing fiber bridging requires modeling of the bridging fibers. This can be done by a cohesive law (a traction-separation law). The area under the traction-separation law is the work of separation. Table 1 lists parameters that are typically used to characterize common damage and failure modes. These concepts are applicable to static failure. A similar distinction can be made for cyclic damage evolution. A complete analysis that would involve the design against the failure modes listed in Table 1 will require the knowledge of all the relevant materials parameters.

A maximum stress criterion (or maximum strain criterion) can be used for materials that develop a damage zone (using appropriate safety factors, typically around 1.5–1.8). As an example, unidirectional fiber composites, loaded in uniaxial tension in the fiber direction, usually display a distributed damage zone during failure. Consequently, an appropriate strength measure is the tensile strength, σ_{Lu}^+ (here, subscript L indicates the longitudinal direction, subscript u indicates ultimate strength and superscript + indicates tension). Other failure modes that are usually characterized in terms of stress criteria are the compressive strength of fiber composites loaded in the fiber direction, σ_{Lu}^- and the composite shear strength, τ_{LTu}. As for elastic properties (Section 5.1) the strength properties of an orthotropic material, such as a unidirectional fiber composite, must be related to specific directions. The longitudinal direction is assigned the subscript L, the (in-plane) direction orthogonal to the longitudinal direction is called the transverse direction and given subscript T. For a unidirectional fiber composite, the tensile strength in the fiber direction, σ_{Lu}^+, is usually much higher than the tensile strength perpendicular to the fiber direction, σ_{Tu}^+.

Fracture by a single *sharp crack* is most often characterized by linear-elastic fracture mechanics concept such as fracture toughness (the critical stress intensity factor) or equivalently the fracture energy (the critical energy release rate G_c). The crack opening is described in 3 pure opening modes: Pure normal opening (Mode I), pure tangential crack opening/shearing (Mode II) and tearing (Mode III). The fracture toughness and fracture energy are material constants but are influenced by temperature, loading rate and environmental conditions such as humidity level. In homogenous materials, cracks tend to propagate under pure Mode I. Materials interfaces are usually weaker than the surrounding materials; therefore cracks tend to remain at interfaces. Therefore, the fracture energy, G_c, of an interface between two dissimilar materials is a function of the mode mixity ψ, where the mode mixity, ψ, is defined from the complex stress intensity factor and a characteristic length scale, see [9] (in isotropic materials, pure normal crack opening displacement corresponds to $\psi = 0°$, whereas pure tangential opening corresponds to $\psi = 90°$). As noted earlier, the interaction of aerodynamic and gravity loading during blade rotation produces multiaxial loading in wind turbine blades, which results in mixed mode loading of interfaces and cracks. Thus, the fracture energy of interface cracks in wind turbine blades must be measured for various load cases, corresponding to different mode mixities that exist in a given cross section and along the blade length. This is the case for gelcoat/laminate delamination, skin/core delamination, cracking along interfaces in adhesive joints and delamination of laminates. Because fracture energy is a material constant, it can be used for different geometries so long as the mode mixity is the same. As discussed later, this simplifies the testing requirements.

The energy required for crack initiation is less than the energy required for crack propagation. Thus, the fracture energy of an interface is typically separated into the energy required for initiation and the energy required for further crack extension; however both values are strong functions of mode mixity.

As indicated in Fig. 5, laminated fiber composites can fail by delamination, which is a cracking mode that can involve *fiber bridging between the crack faces*. If the fiber composite develops a large scale fiber bridging zone, it cannot be properly characterized by linear-elastic fracture mechanics. Instead, the mechanical behavior of a large scale fracture process zone can be characterized by non-linear fracture mechanics, in terms of the J integral [10] and a cohesive law [11]. A cohesive law is the relationship between the local crack opening, δ, and the local stress, σ, across the failure process zone. The cohesive stress is assumed to depend upon the local crack opening only, $\sigma = \sigma(\delta)$. The cohesive stresses can be normal and shear stresses (mixed mode cohesive laws). Figure 6 illustrates the concepts of stress–strain and cohesive laws.

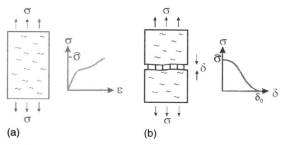

Figure 6: Schematic illustration of (a) stress–strain behavior, which describes the continuum response up to the peak stress, $\hat{\sigma}$ (onset of localization) and (b) behavior after localization which is described in terms of a traction-separation relationship called a cohesive law.

6 Materials testing methods

Because the cost of full-scale blade testing and certification is significant, various laboratory tests are being developed to better access the structural reliability of composite laminates and adhesive joints under conditions that simulate expected stress or strain states in the turbine blade. In this section, test methods that can be used for the measurement of relevant material properties are reviewed. The material properties that influence the damage development in a wind turbine blade can be divided into the elastic properties, the strength properties and fracture mechanics properties.

6.1 Test methods for strength determination

Different test methods are used for tensile and compressive strength determination. An overview of common test methods for strength measurements is given in Fig. 7.

Tensile strength is determined by uniaxial tensile tests [12, 13]. Tensile test specimens are usually long, straight-sided or have a narrow gauge section (dog-bone shape) to ensure that failure develops in the gauge section region where the stress field is uniform. Bending tests can sometimes be used to estimate tensile strength [14, 15]; but bending failure can also occur by shear or compression failure and the failure mode and test results must be carefully analyzed. Moreover, because of the non-linear stress state developed in bending, the initiation of damage on one side of a specimen will result in a shift of the neutral axis of the specimen which further complicates interpretation of test results. Therefore, tensile tests which provide a uniform volume of stressed material are preferred in order to avoid invalid determination of tensile strength.

Specimens for compressive failure are short in order to avoid buckling of the specimen [16]. Unfortunately, the stress field is not uniform in the specimens and specimens frequently fail away from the gauge section. In an effort to improve the accuracy of compressive testing, various approaches have been developed to reduce bending and transverse loading. In a test compressive test fixture designed by Bech *et al.* [17], hemispherical bearings are used to reduce transverse loading bending strains during compressive loading. The test fixture provides more accurate

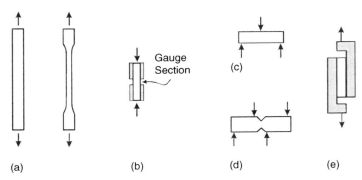

Figure 7: Schematics illustration of test specimens for the determination of strength data. (a) Specimens for determination of tensile strength are usually straight-sided for unidirectional composites or dog-bone shaped, (b) specimens for determination of compressive strength are short to prevent global buckling, (c) shear strength of laminates can be determined using the short-beam-shear test or (d) the V-notched beam (Iosipescu) test. For sandwich structures, rail shear testing is typically utilized for shear strength measurements (e).

measurements of compressive stress–strain behavior for both monotonic and fatigue loading of composites.

The shear strength of composite laminates can be measured by the use of a short beam subjected to three-point bending [18] or by the Iosipescu shear test [19]. Again, the test results must be carefully analyzed and the failure mode documented since tensile failure can occur prior to shear failure for laminates that possess a high shear strength. For sandwich structures, shear testing is commonly performed using a rail shear approach [20].

6.2 Test methods for determination of fracture mechanics properties

Fracture properties are determined from tests of specimens having an artificial crack in the form of a pre-cut notch or a thin slip foil (e.g. by use of Teflon between interfaces) introduced during specimen manufacture. However, a machined or artificial notch is not as sharp as a real crack. Thus, as discussed later, test methods that allow the initiation and arrest of cracking are preferred, since they enable the fracture properties to be determined from a truly sharp crack. An overview of commonly used specimen geometries for fracture mechanics-based testing of composite interfaces is given in Fig. 8.

First, methods for determination of the fracture energy of elastically isotropic materials are reviewed. The Mode I (pure normal opening) fracture properties of isotropic materials are often determined from the compact tension (CT) specimen. For thin (isotropic and orthotropic) laminated structures manufactured with the same material layers, such as composites and sandwich structures, double cantilever beam (DCB) specimens loaded with wedge forces are commonly employed, both for static and cyclic crack growth [21]. For both these specimens the energy release rate depends

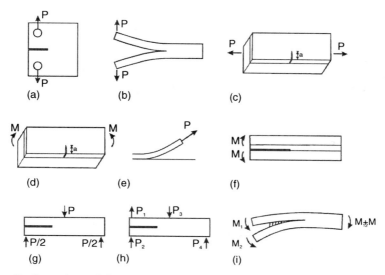

Figure 8: Overview of fracture mechanics test methods: (a) CT specimen, (b) DCB specimen loaded by wedge forces, (c) thin film on a substrate loaded in tension or bending (d), (e) the peel test; (f) DCB sandwich specimens loaded with pure bending moments, (g) the end lap shear (ENS), (h) the MMB specimen, and (i) the DCB loaded with uneven bending moments (DCB-UBM).

on the crack length; for the determination of the energy release rate, the crack length must thus be measured experimentally and correlated with the applied force.

To determine the Mode II fracture energy of composites (pure tangential crack opening displacements) the end-notched specimen (ENS) is commonly used [22]. The ENS specimen is loaded by compressive forces; part of the applied force is transmitted from one beam to the other by contact between the beams. However, this method does not provide stable crack growth, so the value of the fracture energy may be overestimated due to crack initiation and due to friction between the crack faces.

A mixed mode bending (MMB) specimen, also loaded by transverse forces, was proposed by Reeder and Crews [23]. The MMB specimen allows the entire range of mode mixities, from pure mode I to pure mode II, for the same specimen geometry. The energy release rate depends on the crack length. For Mode II dominated loading the crack propagation can be unstable. As shown in Fig. 8, another approach is to load DCB specimens with uneven bending moments (DCB-UBM) [24, 25]. For the DCB-UBM specimen configuration, the energy release rate is independent of crack length and stable crack growth occurs for all mode mixities making the specimen well suited for measuring the interface fracture behavior of various materials, including laminates and sandwich structures.

For large-scale bridging problems, such as cracking with cross-over bridging, it is of relevance to determine cohesive laws which can be used to describe the mechanical response of the bridging zone and thus represents large-scale bridging

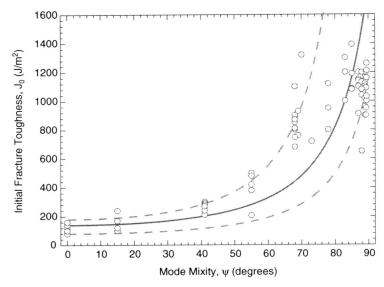

Figure 9: Initial interface fracture energy as a function of mode mixity for interlaminar cracking of a unidirectional glass fiber/polyester composite [30]. A mode mixity of 0° corresponds to a crack opening displacement normal to the cracking plane and a mode mixity of 90° corresponds to tangential crack opening displacements.

in models. Few direct measurement methods exist. For pure mode I, the cohesive law can be determined by direct tension tests [26]. Another method is to determine the bridging stresses from the crack opening profile [27]. Yet another method is to derive the bridging law from the J integral [28]. In that respect a DCB specimen loaded with pure bending moments is preferred [29], since the J integral can be expressed in closed analytical form [11].

Figure 9 shows a typical example of the dependency of the fracture energy on mode mixity for a unidirectional glass fiber composite laminate; the testing was conducted using the DCB-UBM specimen [30]. The initiation fracture energy increases significantly with increasing mode mixity (increasing amount of tangential crack opening displacement).

Very few studies address the determination of mixed mode cohesive laws. In some studies, only pure mode I and mode II cohesive laws are determined from pure mode I and mode II tests and treat them as independent cohesive laws [31, 32]. A more recent approach is to obtain mixed mode cohesive laws from results of a DCB-UBM specimen. For this specimen and loading configuration, the J integral is obtained in closed form even in the case of large-scale bridging. With the closed-form solution for the J integral, cohesive laws can be obtained by partial differentiation of the J integral with respect to the end-opening and end-sliding displacements of the crack [30, 33].

Testing of thin surface layers, such as gelcoats on wind turbine blades, offers special challenges, since standard fracture mechanics test methods cannot be applied. The determination of the fracture energy of a gelcoat can be determined by tensile [34] or bending experiments based on the concept of steady-state cracking of a channeling crack [35, 36]. The peel test can be used for measuring the fracture energy of the interfaced between a gelcoat and a substrate [37]; however, large-scale plasticity in the gelcoat may lead to erroneous results if this is not accounted for [38]. Alternatively, a DCB sandwich specimen can be made by bonding an additional beam onto the gelcoat attached to a substrate.

6.3 Failure under cyclic loads

Test methods used under static loads can in many cases also be used for the study of fatigue damage evolution. In some cases, however, special requirements mean that special concerns have to be accounted for in the selection of test specimens. Moreover, the data collection, the data analysis and materials properties used for describing fatigue are different from those used to describe strength properties under static loading.

Under cyclic loading, it is useful to distinguish between failure due to a damage zone and failure due to crack growth. For materials that fail by a damage zone, such as a unidirectional fiber reinforced composite loaded in the fiber direction, the life under cyclic loading can be described by a so-called S–N curve, which is the relationship between the maximum applied stress, σ_{max}, and the number of cycles to failure, N_f. A schematics of an S–N curve is given in Fig. 10. The application of S–N data in design is straightforward. For example, assume that a component should be designed such that it safely survives a given number of load cycles. Then, from the S–N curve one reads off the maximum applied stress, σ_{max}, corresponding to that number of cycles. The S–N curve depends on the minimum applied stress, σ_{min}; usually expressed in terms of the R-ratio, which is $R = \sigma_{min}/\sigma_{max}$. The approach can refined to predict for a small fraction, say 1/1000 of failed specimens (instead of the average fatigue life) and to account for different maximum load (load spectrums), e.g. by the use of the Palmer-Miners rule [39].

S–N curves for tension–tension and compression–compression testing of glass-epoxy unidirectional laminates for the spar beam for wind turbine blades are shown in Figs 11 and 12. The data are from the Optidat database [40]. The measured fatigue data are evaluated according to ASTM E739, and the 50% median line and the lower 95% confidence limit are shown in these diagrams. The statistical 95% lower confidence limits based on the 95% survival line for the Siemens Wind Power shell materials is also shown in the figure. This analysis is based on the work by DNV [41]. The fatigue lifetime depends on both the applied amplitude and the applied mean value and can be presented in other useful graphs [42].

Of particular importance is whether or not an endurance fatigue limit exists for a material. A fatigue limit implies that a stress limit exists, such that if the material is never loaded beyond this value, σ_{fl}, then the material will never fail due to fatigue.

In addition to buckling and failure due to static overload, perhaps the most important mode of damage that needs to be addressed in design is the cyclic growth

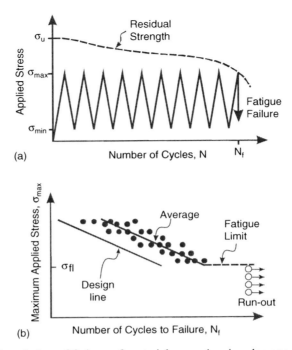

Figure 10: Description of fatigue of materials experiencing damage. (a) The number of cycles to failure, N_f, is determined by cycling specimens to a pre-selected maximum stress, σ_{max}. (b) A schematics of an S–N curve, which is the relationship between the maximum applied stress and the number of cycled to failure.

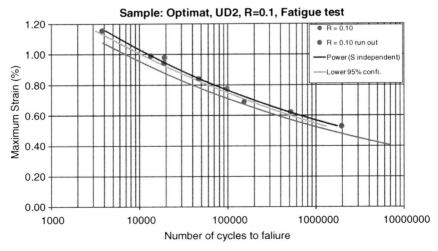

Figure 11: S–N curve for tension–tension fatigue of a glass fiber/epoxy composite [40].

rate of cracks along laminate and adhesive interfaces. If an interface is not properly designed for damage tolerance, even small initial cracks could potentially grow to a critical size over the 20-year design life of a wind turbine blade. For materials or interfaces containing cracks, the cyclic crack growth rate, da/dN, can be measured as a function of the applied energy release rate. Figure 13 shows a

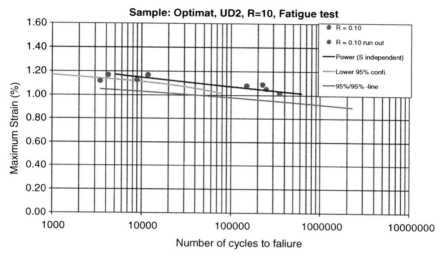

Figure 12: S–N curve for compression–compression fatigue of a glass fiber/epoxy composite [40].

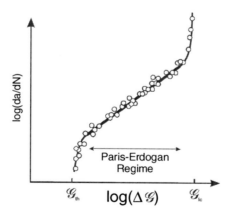

Figure 13: Schematics of a typical relationship between cyclic crack growth rate (da/dN) and maximum applied energy release rate, G_{max} for cyclic crack growth. A threshold value, G_{th}, often exists, below which no crack growth is assumed to occur. As G_{max} approaches the fracture energy, G_{Ic}, the crack growth rate increases asymptotically. In between these regimes, the crack growth rate can be described in terms of the Paris-Erdogan relationship.

schematic drawing of typical material behavior. For most materials, a threshold, G_{th}, exists, below which no crack growth occurs. For $G_{max} > G_{th}$, crack grow occurs, but the rate depends on $\Delta G = G_{max} - G_{min}$, where G_{min} is the minimum cyclic applied energy release rate. For G_{max} increasing close to G_{Ic} (the fracture energy), the crack growth rate increases rapidly. For intermediate values of ΔG the crack growth rate can be described in terms of the Paris-Erdogan relation [43]. For Mode I cracking, the Paris-Erdogan relation can be written as [44]:

$$\frac{da}{dN} = A(\Delta G)^n \tag{1}$$

where a is the crack size, N is the number of cycles, A and n are the fitting parameters. The crack growth rate, da/dN can be understood as the crack extension per load cycle and has the units mm/cycle.

As an example, results from cyclic crack growth experiments are shown in Fig. 14 [44]. The results are obtained from tests of DCB specimens loaded with wedge forces under constant load amplitudes. For this specimen configuration, the range of the energy release rate increases with increasing crack length. Thus, a single test gives data for the crack growth rate under various values of ΔG. A curve-fit, based on the Paris-Erdogan relation (1), is shown as a solid line in the figure.

For materials experiencing large-scale bridging under cyclic crack growth, the situation is more complicated. As the crack tip advances, a large-scale bridging zone develops. The bridging stresses restrain the crack opening, leading to a decreasing crack growth rate [45]. However, the cohesive laws that operate under cyclic loading are likely to be different from those present under monotonic crack opening. Thus, the cohesive laws should be characterized as a function of the number of cycles. Precisely how this should be done is not quite clear, although a few ideas have been developed [45–47].

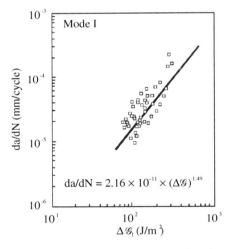

Figure 14:   The crack growth rate, da/dN, is shown as a function of the energy release rate range, ΔG, for a unidirectional glass fiber/epoxy composite [44].

7 Modeling of wind turbine blades

7.1 Modeling of structural behavior of wind turbine blades

7.1.1 Modeling of entire wind turbine blade

This section outlines basic rules for structural design of wind turbine blades. A more thorough description of the overall design of wind turbines for various onshore and offshore applications can be found in the DNV/Risø guidelines [48] and in other chapters of this book.

Some designs are constructed with a load-carrying box girder (main spar) that supports the outer aeroshell as shown in Fig. 2, which illustrates a typical structural layout for a wind turbine blade with a load-carrying girder. The purpose of the box girder is to give the blade sufficient strength and stiffness, both globally and locally. Globally, the blade should be sufficiently stiff in order not to collide with the tower under all types of loading. Locally, the webs, together with the stiffness of the outer shell, ensure that the shape of the aerodynamic profile is maintained.

The box girder or the webs usually extend from the root of the blade to a position close to the tip. The load-carrying flange of the box girder, sometimes called the cap, is usually a single skin construction (i.e. consisting of a single thick laminate, with most of the fibers aligned along the blade length, i.e. the z-direction, see Fig. 2). The webs are usually quite thin sandwich structures; the main purpose of these is to take the shear loads of the blade. The proper design of the blade requires careful analysis. For example, geometrical non-linear effects can result in higher than expected loading of the webs which may result in blade failure at a stress level that is much lower than predicted when the design is based on linear calculations. The design should ensure that the failure criteria discussed in the previous section are not exceeded anywhere in the blade during regular and extreme load situations. This section presents an overview of modeling tools used to predict the static and dynamic behavior of wind turbine blades and to determine the stress and strain distribution within a blade.

7.1.2 Beam models

The global deflection of wind turbine blades, Eigen frequencies and other global behavior can in general be analyzed with good accuracy by use of beam models. However, if greater accuracy is needed or more locally structural phenomena need to be analyzed, more detailed shell and/or solid FE models must be used.

The idea of a beam model is to describe the cross section properties in terms of suitable coefficients, such as area, moment of inertia, torsional stiffness, etc. The behavior of the beam is then described entirely in terms of one-dimensional functions, such as axial and transverse displacement and torsion. In order for such a theory to give an accurate representation of the actual behavior of a wind turbine blade, it is important that the description of the cross section parameters contain all the relevant information. This includes stiffness parameters, the center of mass, the elastic center and the center of torsion. For a typical wind turbine blade these three centers will be located fairly close to each other, but are not coincident.

While the basic properties of cylindrical beams date back to the late 19th century, consistent theories accounting for cross section variation, pre-twist of the blade and material anisotropy are much more recent and theories describing these features are not yet fully developed.

An important non-linear large-deflection effect is called the *Brazier effect* [49]. The Brazier effect is a non-linear effect resulting from curvature when bending a beam or a slender structure. Because of the curvature the longitudinal compressive and tensile stresses result in transverse stresses towards the neutral plane of the beam. This causes flattening of a hollow cylinder or suck-in deformation or a hollow box. This then result in reduction of the bending stiffness of the section.

A fully consistent representation including the three centers has been given e.g. by Krenk and co-workers [50–52]. This theory incorporates the effect of pre-twist in the form a geometric coupling of extension and twist [52]. A numerical procedure was developed for the parameters of a moderately thin-walled cell cross section often used for wind turbine blades [53]. A further development of these principles has been carried out later under the name of Variational Asymptotical Beam Section Analysis (VABS) by Hodges [54]. In this method, a beam with arbitrary cross sections consisting of different materials can be analyzed by a one-dimensional beam theory. The method provides a simply way to characterize strain in an initial curved and twisted beam and all components of cross sectional strain and stress can be accurately recovered from the one-dimensional beam analysis.

7.1.3 FE models

In FE analysis a structure is modeled with a finite number of discrete elements represented by some element nodes in which the elements are connected. Because of the blade size, and lack of symmetry, most published research on wind turbine blade design using FE analysis is done using relatively coarse meshes.

A comparison between a geometrically non-linear FE analysis and full-scale blade testing has been investigated by Jensen *et al.* [8]. In their experiments, a 34 m glass fiber PMC blade was statically loaded to catastrophic failure. Strong non-linearities in various blade responses were found. The Brazier effect was found to dominate in the inner part of the blade. The relative deflection of the box girder cap was measured during the experiments and compared with linear and non-linear FE analysis (see Fig. 15). The linear analysis was not capable of predicting these relative cap deflections, in particular at high loads. The non-linear analysis provided reasonable agreement with experimental measurements. It is clear from this that non-linear FE analysis is required for certain aspects of blade design.

The most common types of FE models used for the structural design of wind turbine blades are:

- Outer surface shell model – using shell element offset
- Mid-thickness shell model
- Combined shell/solid model

The application of these various models is described below.

The outer surface model of a blade is a shell model based on shell elements that are located on the physical outer surface of the aerodynamic shell. This approach is convenient since the outer surface is often specified from aerodynamical purposes. The material is then offset inwards in order to locate it at the correct physical position, see Fig. 16. This type of model is typically used for the practical design of wind turbine blades today.

The mid-thickness model is also created from the geometry of the aerodynamic shell. However, here the shell elements are located at the mid-plane for the different parts of the cross section. The different material thicknesses in the cross section imply that the FE shell will not have a continuous surface like the outer surface

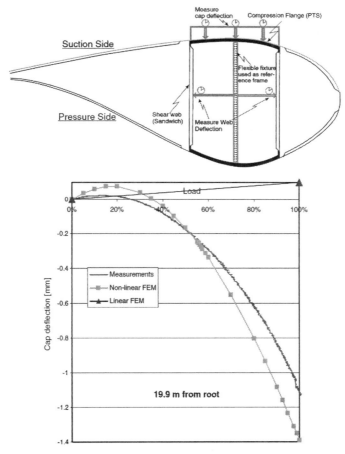

Figure 15: Measured relative deflection of the box girder cap compared with linear and non-linear FE analysis [8].

Figure 16: Outer surface shell FE model with material offset inwards from the outer surface.

model. The discontinuous surfaces are connected by rigid (fixed) elements. These rigid elements are capable of transferring all displacements and rotations from one node to another without deforming.

The combined shell/solid model is a combined shell/solid model constructed based on the following surfaces representing the blade section:

- Outer aerodynamic surface (outer sandwich skins)
- Inner sandwich skins in the leading and trailing part of the blade
- Leading and trailing edge
- Web sandwich skins
- Box girder caps

After creating the surfaces, the solids are then created from two opposing surfaces. The solids represent the following:

- Sandwich core in the leading and trailing part of the blade
- Sandwich core in the webs
- Adhesive bonds between the aerodynamic shell and the spar

Layered shell elements are then used to represent the composite laminates on both sides of the solids. An example is shown in Fig. 17. The combined shell/solid model provides the highest degree of accuracy.

7.1.4 Limitations with shell models

A number of studies have shown that there is limited correlation between the torsional response obtained by numerical structural models and measurements. Madsen [55] compared the responses of a beam model and a shell FE model. Poor correlation was found for predictions of torsional Eigen frequency and Eigen mode. Larsen [56] compared the response of the numerical models from Madsen [55] with a number of measured modal modes; the correlation related to torsional response was limited especially for the higher torsional modes. In predicting torsional behavior, problems associated with the use of offset nodes for layered shell elements in FE analysis has also been reported by Laird *et al.* [57].

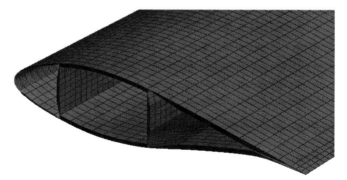

Figure 17: Meshed part of shell/solid FE model. Solid elements are used for sandwich cores and for the adhesive connection between the box girder and the aerodynamic shell. Shell elements are used for the other structural part.

Branner *et al.* [58] recently investigated how well different FE modeling techniques can predict bending and torsion behavior of a wind turbine blade. The results from the numerical investigations were directly compared with measurements obtained from experimental testing of a section of a full-scale wind turbine blade. Torsional testing was performed by locking the tip cross section in a point directly over the center of the box girder (see Fig. 18). This point is fixed, but the cross section can rotate around the z-axis and translate in the horizontal (x–z) plane, since the vertical bar, indicated in Fig. 18, is able to rotate in both ends. The movement in the horizontal plane is not entirely free since the movement is restricted to a circular arc. The numerical and experimental results are in shown in Fig. 19.

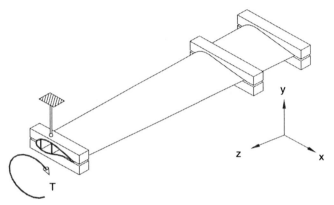

Figure 18: Locked torsion of blade section. The vertical bar is able to rotate in both ends. The cross section can rotate and is restricted to move along a circular arc.

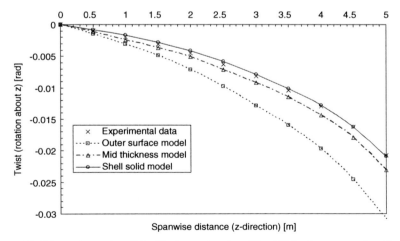

Figure 19: The twist angle of the blade section (see Fig. 18) is shown as a function of the distance from the root, z. The experimental results are shown as points and the model predictions for various FE models are shown as lines [58].

The twist angle predicted by the outer surface shell model deviates from the experimental values by as much as 32% near the loaded end of the blade section (see Fig. 19). The major reason for this disagreement is the offset configuration of the shell elements. As also found by Laird *et al.* [57], this configuration has serious problems modeling correct torsional behavior. In contrast, it was found that the outer surface shell FE model can accurately predict flapwise bending response and, to a reasonable degree, accurate edgewise bending response [58].

Mid-thickness FE models are generally not capable of modeling accurate flapwise and edgewise bending when the model includes details like ply-drops in the spar cap [58]. Using rigid elements to connect regions with different material thickness can therefore not in general be recommended.

Finally, by comparing results from experiments with the global displacements and rotations for the combined shell/solid FE models, it can be concluded that the shell/solid model provides good accuracy for predicting flapwise, edgewise and torsional behavior. Also, by comparing results from the shell/solid model of the modified blade section with experiments, it has been shown that this FE model type can also be used to model bend-twist coupling [58]. These studies will continue in order to develop more understanding regarding why shell models with material offsets have problems with modeling the torsional behavior.

The combined shell/solid model is more detailed and accurate than the other two shell models but degrees of freedom needed for the combined shell/solid model is also considerable larger and therefore more time consuming to analyze.

7.2 Models of specific failure modes

7.2.1 Criteria for laminate failure

Stress- or strain-based criteria are widely used for prediction damage and fracture of individual laminas in laminated structures [59]. Such criteria are easy to use in connection with numerical modeling of wind turbine blades, since the stress (or strain) is usually calculated as a part of the analysis. Denote the in-plane stresses as follows: σ_L is the normal stress acting in the fiber direction (the L direction defined in Section 5.1), σ_T is the normal stress acting perpendicular to the fiber direction (the T direction) and τ_{LT} is the in-plane shear stress.

The simplest type of stress-based criteria assumes that failure is governed by one stress component. Thus the criterion for tensile failure in the fiber direction (the L direction) is ($\sigma_L > 0$):

$$\frac{\sigma_L}{\sigma_{Lu}^+} \geq 1 \qquad (2)$$

where σ_{Lu}^+ is the tensile strength when the material is loaded in uniaxial tension along the fibers (the longitudinal direction), as described in Section 5.2. The criterion for compression failure in the fiber direction ($\sigma < 0$) is:

$$\frac{\sigma_L}{\sigma_{Lu}^-} \geq 1 \qquad (3)$$

The criterion for tensile failure perpendicular to the fiber direction is ($\sigma_T > 0$):

$$\frac{\sigma_T}{\sigma_{Tu}^+} \geq 1 \tag{4}$$

and the criterion for compression failure perpendicular to the fiber direction ($\sigma_T < 0$) is:

$$\frac{\sigma_T}{\sigma_{Tu}^-} \geq 1 \tag{5}$$

while the criterion for shear failure is

$$|\tau_{LT}| \geq \tau_{LTu}. \tag{6}$$

More advanced failure criteria account for the multiaxial stress state that is usually present in laminated composites. A popular criterion is the so-called Tsai-Hill criterion, which is a generalization of an orthotropic yield criterion form metallic materials. It can be written as (assuming $\sigma_L > 0$ and $\sigma_T > 0$):

$$\left(\frac{\sigma_L}{\sigma_{Lu}^+}\right)^2 - \left(\frac{\sigma_L}{\sigma_{Lu}^+}\right)\left(\frac{\sigma_T}{\sigma_{Lu}^+}\right) + \left(\frac{\sigma_T}{\sigma_{Tu}^+}\right)^2 + \left(\frac{\tau_{LT}}{\tau_{LTu}}\right)^2 \geq 1 \tag{7}$$

If $\sigma_L < 0$, σ_{Lu}^- should be used instead of σ_{Lu}^+ and σ_{Tu}^- should be replace σ_{Tu}^+ if $\sigma_T < 0$.

The Tsai-Hill criterion provides a single function for the prediction of failure, but it does not give any prediction of the *failure mode*. Other criteria are, like the simple maximum stress criteria described above, are split after various failure mode. One such example is the Hashin criteria [60], which is split into two criteria, which enables the distinction between failure that involves fiber failure (fracture in a plan perpendicular to the fiber direction) and matrix failure (failure along planes parallel to the fiber direction). The criterion for fiber failure is ($\sigma_L > 0$):

$$\left(\frac{\sigma_L}{\sigma_{Lu}^+}\right)^2 + \left(\frac{\tau_{LT}}{\tau_{LTu}}\right)^2 \geq 1 \tag{8}$$

and the criterion for fiber failure under axial compression ($\sigma_L < 0$) is:

$$\frac{\sigma_L}{\sigma_{Lu}^-} \geq 1 \tag{9}$$

Tensile failure in the matrix is predicted from ($\sigma_T > 0$):

$$\left(\frac{\sigma_T}{\sigma_{Tu}^+}\right)^2 + \left(\frac{\tau_{LT}}{\tau_{LTu}}\right)^2 \geq 1. \tag{10}$$

Similar criteria are developed by Puck and Schürmann [61].

The cracking of a $90°$ ply due to a tensile stress in the layer deserves more comments. For thin layers, where the flaw size is comparable to the layer thickness, a $90°$ layer can form transverse cracks (sometimes also called tunneling cracks). That is a cracking mode where the $90°$ layer develops a crack that spans the layer thickness and runs across the laminate parallel to the fiber direction. This is a steady-state problem: when the crack length is a few times the layer thickness, the energy release rate takes a constant value independent of the crack length. A stress criterion for the stress level in the $90°$ layer at which a tunneling crack can propagate in the $90°$ layer is [62–64]:

$$\sigma_T = \sqrt{\frac{E_{22}G_{Ic}}{fh}} \tag{11}$$

where E_{22} is the Young's modulus in the directions perpendicular to the fiber direction, G_{Ic} is the Mode I fracture energy of the $90°$ layer and h is the thickness of the $90°$ layer undergoing tunneling cracking, and f is a dimensionless parameter (of the order of unity) that depends on the elastic properties of the lamina and the surrounding layers. From (11) it follows that, for fixed material properties, the stress level at which tunneling cracking can occur decreases with increasing h. Conversely, tunneling cracking can be suppressed by decreasing the layer thickness. The development of a tunneling crack unloads the $90°$ ply only in the vicinity of the crack. Remote from the crack, the stress field is unaffected; the stress fulfils the criterion (11), so that anther tunneling crack can propagate. This leads to a characteristic damage state called multiple cracking, consisting of tunneling cracks developing with fairly regularly even crack spacing.

For cyclic loading, similar criteria can be used; the only modification is that the stress value corresponding to the failure life or the fatigue limit is used instead of the monotonic strength values.

7.2.2 Delamination of composites
Delamination is fracture along the interface between different layers. As mention, delamination can be analyzed by linear-elastic fracture mechanics concepts. The criterion for crack propagation is

$$G = G_c \tag{12}$$

where G is the energy release rate and G_c is the fracture energy. The fracture energy is a material property that can be measured by fracture mechanics testing as described in Section 6.2. The energy release rate must be calculated for the given structure, accounting for the elastic properties, the geometry (including crack size) and applied loads.

For cyclic crack growth, the aim is to determine how fast a crack can propagate during cyclic loading and to estimate the number of load cycles that remains before the blade fractures rapidly. The following procedure can be used for predicting the crack size as a function load cycles: First, the structure (a sub-structure or an entire blade) containing a crack is analyzed, e.g. by the use of a FE model. The model gives the energy release rate, G as a function of the applied load level. Next, using the Paris-Erdogan relation, eqn (1), the crack growth rate can be calculated. The crack growth rate is then used to estimate the size of the crack after some, say 1000 cycles. The structure is analyzed again with the new, larger crack size, and a new crack growth rate is calculated, etc. This is repeated until the calculated energy release rate equals the fracture energy; this corresponds to rapid crack propagation. By summing the number of load cycles, the reminder fatigue life can be calculated.

Buckling-driven delamination is a particular type of delamination fracture. Buckling-driven delamination can occur in laminates and sandwich structures used for blade aeroshells and load-carrying girders. The compressive stresses during operation of a wind turbine can drive a stable or unstable delamination. The overall structural strength will usually be reduced by the presence of delaminations [65, 66]. In general, there are two macroscopic buckling modes observed for compressively loaded panels that contain delaminations: local buckling and global buckling (Fig. 20).

When a panel with a delamination is subjected to compressive loading, the plies on one side of the delamination may buckle. This buckling will then introduce bending in the remaining plies on the other side of the delamination. The remaining plies will then be subjected to both bending and the compressive loading resulting in higher stresses than observed without the delamination [66, 67]. This type of buckling is referred to as a local buckling mode and will typically occur when the delamination is large and close to one of the surfaces.

When a delamination is small and deep within the laminate, compressive loading can cause global buckling, where the plies buckle in one direction (i.e. on the same side of the panel).

Buckling-driven delamination growth can only take place when two criteria are fulfilled: (1) the transverse constraint is sufficiently low that a layer adjacent to the delamination is able to displace in a transverse direction and (2) the energy release

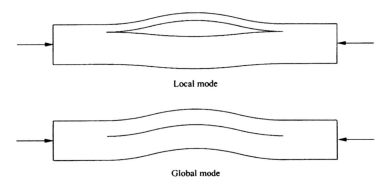

Local mode

Global mode

Figure 20: Local and global modes of delamination induced buckling. In the local buckling mode, transverse displacement of the buckled region occurs in opposing directions on either side of the delamination. Global buckling involves transverse displacement of the buckled region generally occurs in one direction (i.e. on the same side of the panel) [68].

rate equals the fracture energy of the interface. To illustrate this, consider delamination of a thin layer of a thick beam. An approximate value for the energy release rate can be calculated by an Euler model, assuming clamped ends. For a long delamination, the energy release rate approaches a value given by [9]:

$$G = \frac{(1 - v_1^2)\sigma_1^2 t}{2E_1}$$

(13)

where G is the energy release rate, σ_1 is the critical stress in the layer that undergoes buckling-driven delamination (material #1), E_1 and v_1 are the Young's modulus and Poisson's ratio of material #1 and t is the thickness of the layer. Recall, that the criterion for delamination is $G = G_c$, where G_c is the fracture energy of the interface (at the appropriate mode mixity).

Note the effect of the thickness of the thin laminate, t. A thin laminate (small t) – corresponding to a delamination lying close to the top of the original laminate – gives a low buckling stress. However, from (13) it is apparent that a small value of t gives a low energy release rate. Thus, if the delamination is positioned close to the surface (small t), the delaminated laminate can easily buckle, but the delamination crack cannot propagate since the energy release rate is lower than the fracture energy. Conversely, if the original delamination is positioned deep in the laminate, then, from (13), the energy release rate is high and the delamination can propagate if the laminate can buckle. However, the stress to cause buckling increases with increasing t. It follows that a delamination positioned deep inside the laminate will not cause local buckling – the delamination cannot propagate. Thus, neither delaminations very close to the surface or very far away from the surface will grow. However, a certain range may exist where both the buckling criterion and the crack propagation criterion are fulfilled. This simple argument illustrates that the delamination position through the thickness or a component such as a blade is of great importance when determining the impact that a delamination will have on blade failure. Additional details can be found in Karihaloo and Stang [69].

Occasionally, delamination occurs with significant crack bridging, viz., a zone behind the crack tip where fibers connect the crack faces; this is denoted crossover bridging. Since this zone can be large, the use of linear-elastic fracture mechanics is invalid. Instead, the fracture resistance can be described in terms of the J integral and cohesive laws, representing the mechanical response of the bridging fibers. An approach for determination of mixed mode cohesive laws by the use of DCB-UBM specimens and the J integral [33] was demonstrated by Sørensen and Jacobsen [30].

7.2.3 Adhesive joints

Crack growth inside an adhesive layer can occur when the energy release rate equals the fracture energy of the adhesive. However, often cracking of adhesive joints occurs as interface fracture, since the crack is subjected to both normal and shear stresses at the crack tip (mixed mode cracking) so that the crack kinks to the adhesive/laminate interface. Often, the adhesive/laminate interface is the weakest plane. Therefore, the crack tends to remain at the interface for a wide range of mode mixities. Assuming that the fracture process zone remains small, an appropriate criterion for crack propagation is

$$G(\psi) = G_c(\psi) \tag{14}$$

where G is the energy release rate, ψ is the mode mixity and $G_c(\psi)$ is the fracture energy for the specific mode mixity. The criterion should be understood as follows: an analysis of a structure with a crack along an adhesive joint gives an energy release rate and a mode mixity value. Then, the appropriate fracture energy, corresponding to the determined mode mixity should be used. This is required since the fracture energy depends strongly on the mode mixity, as shown in Fig. 9.

In case of a large scale fracture process zone, e.g. in the form of fiber bridging, the appropriate material laws are the J integral and cohesive laws, as described above. An example of this will be given in Section 7.3.

7.2.4 Sandwich failure

Delamination crack growth in sandwich structures is another example of interface fracture. As described above, the appropriate cracking criterion for interface cracking is given by eqn (14). However, some core materials (e.g. low-density polymer foams) possess low fracture energy and the crack can kink into the core material.

7.2.5 Gelcoat/skin delamination

Delamination of the gelcoat is also interface cracking. Delamination of the gelcoat can be caused by residual stresses originating from the hardening of the gelcoat. As mentioned, the peel test (Fig. 8e) is frequently used for characterizing the interfacial fracture energy. Another useful test method is the DCB-UBM test configuration. A sandwich DCB specimen can be made by gluing an addition beam onto the gelcoat layer. By changing the ratio between the applied moments, the mode mixity can be adjusted so that the crack does not kink out of the interface between the gelcoat and the skin (face sheet) layers, but remains at the interface.

7.2.6 Channeling cracking in the gelcoat

Typically, the cracking in the gelcoat takes place in the form of cracks that extends widely across the gelcoat layer, but the crack extends only to the depth of the gelcoat layer – it does not penetrate into the underlying laminate. This mode of cracking is denoted *channeling cracking*. Channeling cracking has been modeled by Nakamura and Kamath [36] who found that the energy release rate of a crack in a coating attains a steady-state value once the crack has extended a distance of a few times the coating thickness. The crack then propagates under a constant stress – therefore channeling cracks can spread widely across the gelcoat layer. We denote the Mode I fracture energy of the gelcoat with G_{1c}. The critical stress level below which no channeling cracking can occur is given by [35]:

$$\sigma_{1ch}^c = \sqrt{\frac{2}{\pi} \frac{G_{1c} E_1}{(1 - v_1^2) g t}} \tag{15}$$

where g is a non-dimensional function that depends on the stiffness mismatch. From eqn (15) it is clear that the critical stress in the gelcoat, σ_{1ch}^c, increases with decreasing t. In other words, channeling cracking can be suppressed by decreasing the thickness of the gelcoat layer.

7.3 Examples of sub-components with damage

It is of great importance to perform additional laboratory tests on larger, generic samples or on blade sub-components in order to validate the design strategy and design methods. The approach is first to predict the strength characteristics of the sub-component using measured fracture properties, next manufacture and test the sub-component. Finally, the predicted and measured strength behaviors are compared. Such case studies can validate (or invalidate) the proposed design approach.

7.3.1 Failure of adhesive joints
The first example concerns the strength of adhesively bonded joints for glass fiber composites (see Fig. 21). As described in the previous section, the proper parameter for characterizing cracking of an adhesive joint in a wind turbine blade is the fracture toughness (or fracture energy) measured by mixed mode fracture mechanics testing. In the present example [70], mixed mode fracture mechanics testing was conducted for an adhesive joint comprised of a polymer adhesive joining two beams made

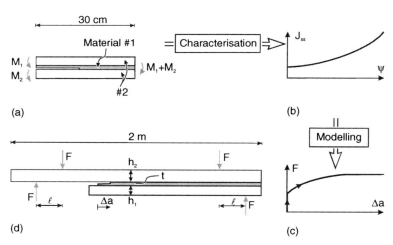

(a) (b) (c) (d)

Figure 21: A study of mixed mode cracking of adhesive/glass fiber composite: mixed mode values of the fracture resistance, J_{ss}, determined from DCB specimens loaded with uneven bending moments, were used for prediction of the load-carrying capacity of medium size specimens, and compared with experimental results obtained from flexural loading of 2 m long "medium size" beam specimens.

from glass fiber composites typical of wind turbines. For the bonding of glass fiber composites, the analysis is complicated by the development of large-scale bridging, which requires that the fracture resistance is characterized in terms of the J integral.

For the medium size specimens, the fracture load takes a steady-state value, F_{ss}, after some crack extension. Figure 22 shows the predicted steady-state fracture load, calculated from the steady-state fracture resistance data measured using DCB-UBM specimens as a function of the thickness ratio of the two beams. The measured steady-state strength values are superimposed as points. It is seen that there is a good agreement between the model predictions and experimental strength values of the medium scale specimens. The experimental results confirm the trends of the model predictions with respect to the effect of the thickness ratio of the beams: reducing the thickness of the thin beams leads to a higher fracture load. Furthermore, the predicted uncertainty of the strength of the medium size specimens (due to the variation in the measured fracture resistance of the DCB-UBM specimens) is of the same magnitude as found experimentally. Thus, the strength scaling can be successfully achieved for large-scale bridging problems if the fracture resistance data are analyzed by the J integral.

7.3.2 Cyclic crack growth of adhesive joints

Building upon the work described in Section 7.3.1 [70], Holmes *et al.* [71] investigated the cyclic crack growth behavior of adhesive joints in 1.2 m long medium size beam specimens subjected to four-point bending. For this specimen configuration, a constant load amplitude gives a constant J amplitude, ΔJ, even though the crack

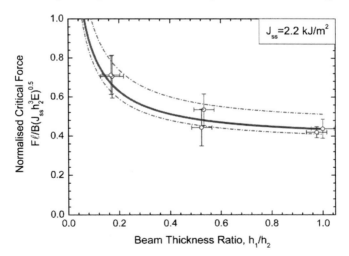

Figure 22: The load-carrying capacity of adhesive/glass fiber composite medium size specimens predicted using DCB-UBM data (solid line). Dashed lines are predictions based on upper and lower values of J_{ss}. Experimental results from tests performed on the medium size specimens subjected to four-point bending are shown as points; error bars represent the maximum and minimum values of the measured steady-state force, F_{ss} [70].

advances. Therefore, such specimen is well suited for crack growth studies involving large-scale bridging. It was found that single-cycle overloads of 20 and 30% did not significantly affect the subsequent crack growth rate. Cycling under different values of ΔJ gave difference crack growth rates. However, a given ΔJ gave, after a series of lower amplitude load cycles, a lower crack growth rate than in the earlier stages of the experiment. This behavior is attributed to crack deflection away from the adhesive/laminate interface into the laminate. A possible mechanism for the crack deflection is the additional fiber bridging which would increase the crack growth resistance along the crack path, resulting in crack deflection away from the adhesive interface into the laminate. Investigations using cyclically loaded DCB-UBM specimens have also shown a decreasing crack growth rate during constant ΔJ amplitude in glass fiber laminates [72].

7.3.3 Buckling-driven delamination of panels

In order to understand the effect of delamination on the compressive behavior of laminated composite materials used in wind turbine blades, it is instructive to examine recent results from compression tests performed on composite and sandwich panels. Short *et al.* [68] tested glass fiber reinforced polymer specimens containing artificial delaminations of various size and depth. Good agreement between FE predictions and experimental measurements were found for the entire range of delamination geometries that were tested. FE and simple closed-form models were also developed for delaminated panels with isotropic properties. This enabled a study of the effect of delamination geometry on compressive failure without the complicating effects of orthotropic material properties. A buckling mode map, allowing the buckling mode to be predicted for any combination of delamination size and through-thickness position was developed and is shown in Fig. 23. The results of this study can be used to derive a graph of non-dimensional failure load versus non-dimensional failure stress as shown in Fig. 24.

Nilsson *et al.* [73] made a numerical and experimental investigation of buckling-driven delamination and growth in carbon fiber/epoxy panels with an implanted artificial delamination. The average maximum load for the delaminated panels was approximately 10% lower than the maximum load for the undelaminated panels. The maximum load was found to be almost unaffected by delamination depth. The experimental results as well as the numerical analyses showed a strong interaction between delamination growth and global buckling load for all delamination depths. Delamination growth initiated at or slightly below the global buckling load in all specimens. In all tests, the delaminations grew more or less symmetrically and perpendicular to the direction of load. At the maximum load, the crack growth rate increased sharply. It was concluded [73] that structures where delaminations are likely to occur should be designed against the possibility of global buckling.

For wind turbine blades, delaminations will likely be present along curved surfaces. The effect of curvature on the failure load of laminates containing delaminations was studied experimentally [74]. Tests on flat and curved GFRP specimens with delaminations between plies showed that the failure load for flat specimens was the same as or higher than that for curved specimens. As shown in Fig. 25, the

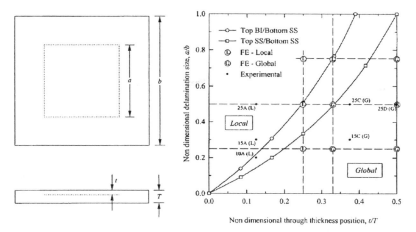

Figure 23: Buckling-driven delamination map used to determine the effect of varying delamination size and through-thickness position on buckling mode. The panels are square with width $b = 40$ mm and are loaded in compression in the vertical direction. Square delamination with $a = 10, 15$ and 25 mm are present in the position A, C, or D indicated in Fig. 25 [68].

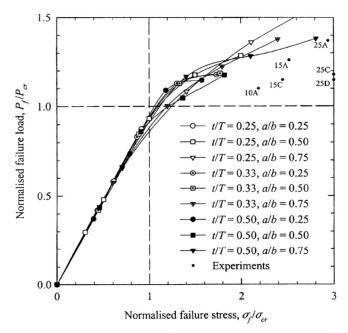

Figure 24: Normalized compressive failure stress versus normalized failure load. Labels on experimental data indicate the through-thickness positions (shown in Fig. 25) as well as the relative delamination size [68].

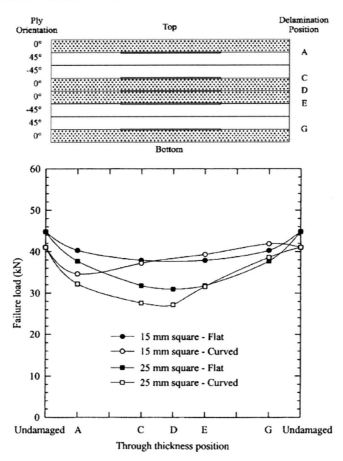

Figure 25: Comparison of failure loads for the compressive buckling of flat and curved laminates as a function of through-thickness position for 15 and 25 mm^2 delamination in square plates (width 40 mm) [74].

failure load was dependent on the size and through-thickness position of the delamination. The curved laminates exhibited an asymmetry of failure load with the through-thickness position of the delamination; when both delaminations were at the same depth, it was observed that a delamination located near the outer radius gave a greater strength reduction than a delamination near the inner radius.

Damage resulting from impacts may consist of multiple delaminations. As would be expected, depending upon size, multiple delaminations cause greater strength reduction than a single delamination. To determine structural integrity, is therefore of interest to understand how the geometric distribution in multiple delaminations affects strength reduction in laminates. Wang et al. [75] studied flat glass fiber reinforced plastic specimens containing a single or two embedded delaminations. With reference to Fig. 26, they found that that the maximum reduction in

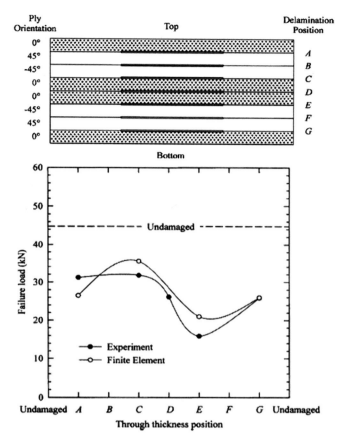

Figure 26: Comparison of experimental and FE predicted failure loads for specimens containing a delamination in the C position and one of the positions from A to G [75].

strength for two delaminations occurred when the delaminations split the laminate into sub-laminates of similar thickness.

7.3.4 Modeling of a full main spar to determine the webs influence on ultimate strength

The deformation and strength of a box girder used as a spar in a 34 m wind turbine blade was examined by Branner *et al.* [76]. The study also examined the effect of sandwich core properties on ultimate strength of the box girder. In Fig. 27, the transverse (vertical or 90°) strains for both faces of the sandwich web are compared with both linear and non-linear FE analysis. The strains presented in Fig. 27 are measured on the upper part of the web towards the leading edge at the position where the failure was observed (also indicated in the figure). The longitudinal normal strains (in the z-direction) caused by bending of the box girder lead to associated strains in the transverse direction (the y-direction) due to the Poisson's

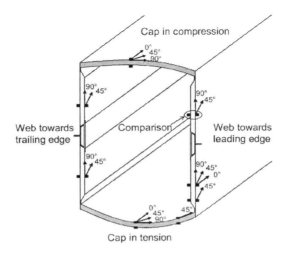

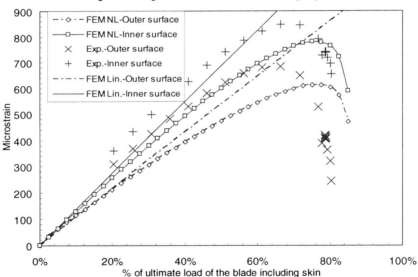

Figure 27: Comparison of FE results and measurements for back to back strain gauges on upper web part. The positions for the strain gauges are indicated in the sketch (top) [76].

effect. The transverse strains are positive (tension) in the shear webs of the upper half part of the box girder where the bending cause compression and the transverse strains are negative (compression) in the lower half part.

Initially, the transverse strain increased linearly with load, see Fig. 27. However, as the load increases, the strains become non-linear with respect to applied load and it is noted that the graphs deflect towards compressive strains.

The difference between the linear and the non-linear results is, at least in part, caused by the Brazier effect. The crushing pressure (that is the transverse stresses due to curvature described in Section 7.1.2) is flattening the cross section and introduces compressive strains into the shear webs. This crushing pressure varies with the square of the applied load as found by Brazier [49], resulting in the noted deviation from linear responses. The flattening of the cross section will probably initiate buckling which accelerates the failure evolution. Other non-linear phenomena, such as changes in geometry and loading configuration (which follows the geometry in the non-linear analysis), will also contribute to the observed non-linearities. For further discussion of this see Jensen *et al.* [77].

In the study in Branner *et al.* [76], it was found that the corner stiffness greatly influence the overall non-linear behavior of the box girder. The webs take a larger part of the overall cross section deformation when the corners are stiff. It is seen from the stiff corner models, that the core density have some influence on the ultimate strength. In contrast, the soft corner models show that the core density has almost no influence on the ultimate strength. The ultimate strength of a box girder was also studied experimentally and numerically [78]. It was found that for the shear webs, failure was initiated by shear failure of the core (see Fig. 27).

7.3.5 Testing of long sections of a main spar to determine buckling behavior

Kühlmeier [79] worked with buckling of wind turbine blades in his PhD thesis. In Kühlmeier *et al.* [80], a 9 m long airfoil blade section, designed to fail in buckling, was build and tested destructively in a four-point bending configuration. Based on numerical analysis, it was found that a linear buckling analysis will over-predict the ultimate strength of the blade. How much the strength is over-predicted depends on the size of the imperfections present in the blade and the sensitivity of the structure with respect to the imperfections. It was suggested that a bifurcation buckling analysis with a knockdown factor applied on the buckling load will give a good estimate on the ultimate strength of the blade. The value of the knockdown factor will depend on the size of the imperfections as well as the imperfection sensitivity. For the 9 m blade section a factor of 1.25 was found to assess the ultimate strength of the blade.

7.4 Full wind turbine blade models with damage

Overgaard and Lund [81] analyzed the full-scale blade collapse described in Section 4 [2]. A geometrically non-linear and linear pre-buckling analysis was performed for predicting the failure of the blade due to local buckling on the suction side of the airfoil. The geometrical imperfection sensitivity of the blade was evaluated by imposing the strain gauge measurement for the full-scale experiment as an imperfection pattern. Figure 28 displays the response of the obtained imperfection amplitude where the 23% amplitude model fits the best of the evaluated models. It is seen that the imposed imperfection pattern is directly proportional with the longitudinal strain measured in the strain gauges [82]. An important discovery is that the buckling shape is unaffected by the presence of geometrical imperfections,

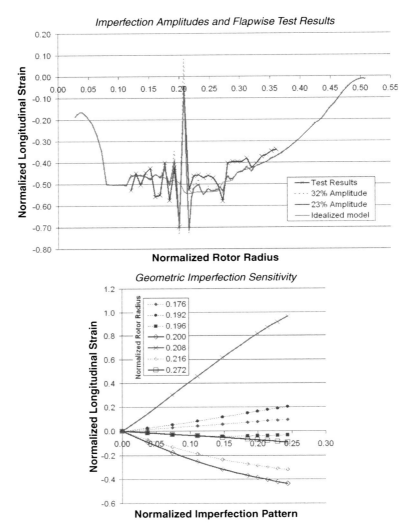

Figure 28: Response of imperfection amplitude compared with test results [81].

but the local strain at the maximum geometric imperfection amplitude is linearly depending on the imperfection amplitude. The epic centre of the buckling shape mode is at the core and flange material transition, but the structural collapse of the structure is at the geometric imperfection amplitude.

The advantages of applying a sandwich construction, as opposed to traditional a single skin laminate, in the flanges of a load-carrying box girder, was studied by Berggreen *et al.* [83] for a future, very large, 180 m wind turbine rotor. The study showed that buckling by far is the governing criterion for the single skin design. A significant weight reduction by more than 20% and an increased buckling capacity was obtained with the sandwich construction in comparison with the

Load level:0.70 Increment: 81 Damage: 0.48%   Load level:0.68 Increment: 87 Damage: 0.76%

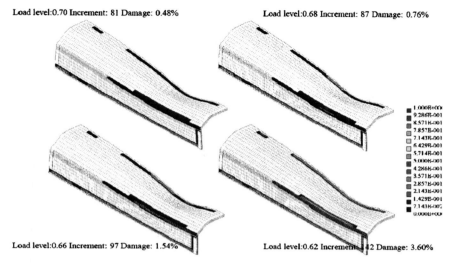

Load level:0.66 Increment: 97 Damage: 1.54%   Load level:0.62 Increment: 42 Damage: 3.60%

Figure 29: Simulation of failure of a box girder (progressive delamination growth in combination with buckling) [85].

traditional single skin composites. However, introducing a sandwich construction in the flange was found to result in a globally more flexible structure, making tower clearance the critical criterion.

Overgaard and Lund [84, 85] simulated the full-scale blade test till failure [2] by FE analysis using a cohesive zone modeling. A mixed mode bilinear cohesive law was used. The combination of a relative coarse mesh and the use of constitutive laws with softening leads to severe solution difficulties, and therefore an efficient and robust solution strategy for dealing with large three-dimensional structures was implemented.

This work shows that it is possible to predict structural behavior of a wind turbine blade based on non-linear fracture mechanics in a geometrically non-linear framework, i.e. account for buckling and delamination interaction (see Fig. 29). The numerical damage predictions were compared with a flapwise static test result. The damage simulation displayed strong geometric and material instability interaction which indeed caused a progressive collapse of the wind turbine blade as seen in the full-scale experiment [2, 82]. According to the model, the critical buckling load of the blade triggered a delamination at the corner and in middle of the flange at the point of inflection of the buckling pattern. This started a progressive chain of events that lead to a structural collapse of the complete wind turbine blade.

8 Perspectives and concluding remarks

This chapter has presented an overview of blade materials, blade design and testing methods. With more knowledge evolving, the blade design will be made more accurately and safety factors may be reduced. However, safety factors should only be reduced if quality control and damage assessment methods are improved. The reduction of safety factors could lead to an increasing amount of damage evolution in wind turbine blades in service. The safely level can nevertheless be maintained or increased if structural health monitoring (SHM) techniques are used. The perspective is that the turbine blades will be equipped with a sufficient amount of sensors that enables the detection of damages with high fidelity. The extra costs of the SHM sensors and systems may be less than the cost savings by skipping the regular service inspection, at least for large offshore wind turbines, where access is difficult and costly. The potentially earlier detection of damage by SHM will give the owner of the wind turbine the possibility to inspect, repair or replace the blades before the damages become serious. The cost savings come mainly from the fact that the vast majority of the blades are not expected to develop significant damages; with the use of SHM, these blades would not require any manual inspection.

From a knowledge point of view, it is envisioned that the future developments is towards a better understanding of the damage evolution at a smaller length scales, leading to a better, more complete and more fundamental understanding of the damage evolution in composite structures. This opens the possibility for the development of larger and better, more damage tolerant materials.

References

[1] ICE 61400-1, International Standard: Wind Turbines – Part 1: Design requirements, 2005.

[2] Sørensen, B.F., Jørgensen, E., Debel, C.P., Jensen, F.M., Jensen, H.M., Jacobsen, T.K. & Halling, K., Improved design of large wind turbine blade of fibre composites based on studies of scale effects (Phase 1), Summary report, Risø-R-1390(EN), 2004.

[3] Veers, P., et al., Trends in the design, manufacture and evaluation of wind turbine blades. Wind Energy, 6, pp. 245–259, 2003.

[4] Budiansky, B., Micromechanics. Computers & Structures, 16, pp. 3–12, 1983.

[5] Mishnaevsky, L. & Brøndsted, P., Statistical modelling of compression and fatigue damage of unidirectional fiber reinforced composites. Composite Science and Technology, 69(3–4), pp. 477–484, 2009.

[6] Holmes, J.W., Brøndsted, P., Sørensen, B.F., Jiang, Z., Sun, Z. & Chen, X., Development of a bamboo-based composite as a sustainable green material for wind turbine blades. Wind Engineering, 33, pp. 197–210, 2009.

[7] Det Norske Veritas, Design and Manufacture of Wind Turbine Blades, Offshore and Onshore Wind Turbines, Offshore Standard DNV-OS-J102m, 2006.

[8] Jensen, F.M., Falzon, B.G., Ankersen, J. & Stang, H., Structural testing and numerical simulation of a 34 m composite wind turbine blade. Composite Structures, 76, pp. 52–61, 2005.

[9] Hutchinson, J.W. & Suo, Z., Mixed mode cracking in layered materials. Advanced in Applied Mechanics, ed. J.W. Hutchinson & T.Y. Wu, Academic Press, Inc., Boston, 29, pp. 63–191, 1992.

[10] Rice, J.R., A path independent integral and the approximate analysis of strain concentrations by notches and cracks. *Journal of Applied Mechanics*, **35**, pp. 379–386, 1968.

[11] Suo, Z., Bao, G. & Fan, B., Delamination R-curve phenomena due to damage. *Journal of Mechanical Physics of Solids*, **40**, pp. 1–16, 1992.

[12] ISO 527-4: Plastics – Determination of tensile properties – Part 4: Test conditions for isotropic & orthotropic fibre-reinforced plastic composites, 1997.

[13] ISO 13003:2003(E), Fibre-reinforced plastics – Determination of fatigue properties under cyclic loading conditions, 2003.

[14] ASTM Standard D790-03, Flexural properties of unreinforced and reinforced plastics and electrical insulating materials, American Society for Testing & Materials, West Conshohocken, Pennsylvania (first issued in 1970), 2003.

[15] ASTM Standard D6272-02, Flexural properties of unreinforced and reinforced plastics and electrical insulating materials by four-point bending, American Society for Testing & Materials, West Conshohocken, Pennsylvania (first issued in 1998), 2002.

[16] ISO 14126:1999(E), Fibre-reinforced composites – Determination of compressive properties in the in-plane direction, 1999.

[17] Bech, J., Goutianos, S., Andersen, T.L., Torekov, R.K. & Brøndsted, P., A new static and fatigue compression tests method for composites. *Strain*, doi: 10.1111/j.1475-1305.2008.00521.x, 2008.

[18] ASTM Standard D2344-00, Short beam strength of polymer matrix composite materials and their laminates by short-beam method, American Society for Testing and Materials, West Conshohocken, Pennsylvania (first published in 1965), 2000.

[19] ASTM Standard D5379-98, Standard test method for shear properties of composite materials by the V-Notched beam method, American Society for Testing & Materials, West Conshohocken, Pennsylvania (first published in May 1993), 1998.

[20] ASTM D4762-08, Standard guide for testing polymer matrix composite materials, ASTM International, West Conshohocken, PA 19428-2959 USA, 2008.

[21] ASTM D5528-01, Standard test for Mode I interlaminar fracture toughness of unidirectional fiber-reinforced polymer matrix composites, American Society for Testing & Materials, West Conshohocken, Pennsylvania, 2001.

[22] Carlsson, L.A., Gillespie Jr., J.W. Jr. & Pipes, R.B., On the analysis & design of the end notch flexure (ENF) specimen for mode II testing. *Journal of Computational Materials*, **20**, pp. 594–604, 1986.

[23] Reeder, J.R. & Crews Jr., J.H., Redesign of the mixed-mode bending delamination testing. *Journal of Composites Technology and Research*, **14**, pp. 12–19, 1992.

[24] Plausinis, D. & Spelt, J.K., Application of a new constant G load-jig to creep crack growth in adhesive joints. *International of Journal of Adhesion & Adhesives*, **15**, pp. 225–232, 1995.

[25] Sørensen, B.F., Jørgensen, K., Jacobsen, T.K. & Østergaard, R.C., DCB-specimen loaded with uneven bending moments. *International Journal of Fracture*, **141**, pp. 159–172, 2006.

[26] Brenet, P., Conchin, F., Fantozzo, G., Reynaud, P., Rouby, D. & Tallaron, C., Direct measurement of the crack bridging tractions: a new approach of the fracture behavior of ceramic-matrix composites. *Computational Science and Technology*, **56**, pp. 817–823, 1996.

[27] Cox, B.N. & Marshall, D.B., The determination of crack bridging forces. *International Journal of Fracture*, **49**, pp. 159–176, 1991.

[28] Li, V.C. & Ward, R.J., A novel testing technique for post-peak tensile behaviour of cementitious materials, *Fracture Toughness and Fracture Energy – Testing Methods for Concrete and Rocks*, eds. H. Mihashi, H. Takahashi & F.H. Wittmann, A.A. Balkema Publishers: Rotterdam, pp. 183–195, 1989.

[29] Sørensen, B.F. & Jacobsen, T.K., Large scale bridging in composites: R-curve and bridging laws, *Composites, Part A*, **29**, pp. 1443–1451, 1998.

[30] Sørensen, B.F. & Jacobsen, T.K., Characterizing delamination of fibre composites by mixed mode cohesive laws. *Composite Science and Technology*, **69**, pp. 445–456, 2009.

[31] Yang, Q.D. & Thouless, M.D., Mixed-mode fracture analysis of plastically-deforming adhesive joints. *International of Journal of Fracture*, **110**, pp. 175–187, 2001.

[32] Kafkalidis, M.S. & Thouless, M.D., The effect of geometry and material properties of the fracture of single lap-shear joints. *International Journal of Solids Structures*, **39**, pp. 4367–4383, 2002.

[33] Sørensen, B.F. & Kirkegaard, P., Determination of mixed mode cohesive laws. *Engineering Fracture Mechanics*, **73**, pp. 2642–2661, 2006.

[34] Nichols, M.E., Anticipating paint cracking: the application of fracture mechanics to the study of paint weathering. *Journal of Coating Technology*, **74**, pp. 39–45, 2002.

[35] Beuth, J.L., Cracking of thin bonded films in residual tension. *International Journal of Solids and Structures*, **29**, pp. 1657–1675, 1992.

[36] Nakamura, T. & Kamath, S.M., Three dimensional effects in thin film fracture mechanics. *Mechanics of Materials*, **13**, pp. 67–77, 1992.

[37] ASTM D903-98, Standard test method for peel or stripping strength of adhesive bonds, ASTM International, 01-Apr-2004.

[38] Wei, Y. & Hutchinson, J.W., Interface strength, work of adhesion and plasticity in the peel test. *International Journal of Fracture*, **93**, pp. 315–333, 1998.

[39] Brøndsted, P., Andersen, S.I. & Lilholt, H., Fatigue performance of glass/polyester laminates and the monitoring of material degradation. *Mechanical Composite Materials*, **32**, pp. 21–29, 1996.

[40] Optidat database, http://www.wmc.eu/optimatblades_optidat.php

[41] Ronold, K.O. & Echtermeyer, A.T., Estimation of fatigue curves for design of composite laminates. *Composites, Part A*, **27**, p. 485, 1996.

[42] Brøndsted, P., Lilholt, H. & Lystrup, Aa., Composite materials for wind power turbine blades. *Annual Review of Materials Research*, **35**, pp. 505–538, 2005.

[43] Paris, P. & Erdogan, F., A critical analysis of crack propagation laws. *Journal of Basic Engineering, Series D*, **85**, pp. 528–534, 1963.

[44] Kenane, M. & Benzeggagh, M.L., Mixed-mode delamination fracture toughness of unidirectional glass/epoxy composites under fatigue loading. *Composite Science and Technology*, **57**, pp. 597–605, 1997.

[45] McMeeking, R.M. & Evans, A.G., Matrix fatigue cracking in fiber composites. *Mechanics of Materials*, **9**, pp. 217–227, 1990.

[46] Cox, B.N., Scaling for bridged cracks. *Mechanics of Materials*, **15**, pp. 87–98, 1993.

[47] Sørensen, B.F. & Jacobsen, T.K., Crack bridging in composites: connecting mechanisms, micromechanics and macroscopic models. In: *Proceedings of the International Conference on New Challenges in Mesomechanics (Mesomechanics 2002)*, eds R. Pyrz, J. Schjødt-Thomsen, J.C. Rauhe, T. Thomsen & L.R. Jensen, Aalborg University: Denmark, pp. 599–604, 2002.

[48] DNV/Risø, *Guidelines for Design of Wind Turbines*, 2nd Edition, DNV/Risø Publication: Denmark, 2002.

[49] Brazier, L.G., On the flexure of thin cylindrical shells & other thin sections. *Proceedings of the Royal Society London*, **A 116**, pp. 104–114, 1927.

[50] Krenk, S., A linear theory for pretwisted elastic beams. *Journal of Applied Mechanics*, **50**, pp. 137–142, 1983.

[51] Krenk, S. & Gunneskov, O., Pretwist and shear flexibility in vibration of turbine blades. *Journal of Applied Mechanics*, **52**, pp. 409–415, 1985.

[52] Krenk, S., The torsion-extension coupling in pretwisted elastic beams. *International Journal of Solids & Structures*, **19**, pp. 67–72, 1985.

[53] Krenk, S. & Jeppesen, B., Finite elements for beam cross sections of moderate wall thickness. *Computers and Structures*, **32**, pp. 1035–1043, 1985.

[54] Hodges, D.H., Nonlinear composite beam theory. *Progress in Astronautics & Aeronautics*, **213**, *AIAA*: Virginia, USA, 2006.

[55] Madsen, H.A. (ed), *Research in Aeroelasticity – EFP-98* (in Danish). Risø-R-1129(DA), Risø National Laboratory, Roskilde, Denmark, 1999.

[56] Larsen, G.C., Modal analysis of wind turbine blades. Risø-R-1181(EN), Risø National Laboratory, Roskilde, Denmark, 2002.

[57] Laird, D.L., Montoya, F.C. & Malcolm, D.J., Finite element modeling of wind turbine blades. *Proceedings of AIAA/ASME Wind Energy Symposium*, Reno, Nevada, USA, AIAA-2005-0195, pp. 9–17, 2005.

[58] Branner K., Berring P., Berggreen C. & Knudsen H.W., Torsional performance of wind turbine blades – part II: Numerical validation. *Proceedings of the 16th International Conference of Composite Materials*, Kyoto, Japan, 2007.

[59] Hull, D. & Clyne, T.W., *An Introduction to Composite Materials*, Cambridge University Press, 1996.

[60] Hashin, Z., Failure criteria for unidirectional fiber composites. *Journal of Applied Mechanics*, **47**, pp. 329–334, 1980.

[61] Puck, A. & Schürmann, H., Failure analysis of FRP laminates by means of physically based phenomenological models. *Composite Science and Technology*, **58**, pp. 1045–1067, 1998.

[62] Parvizi, A., Garrett, K.W. & Bailey, J.E., Constrained cracking in glass fibre-reinforced epoxy cross-ply laminates. *Journal of Materials Science*, **13**, pp. 195–201, 1978.

[63] Aveston, J. & Kelly, A., Tensile first cracking strain and strength of hybrid composites and laminates. *Philosophical Transactions of the Royal Society of London*, **A294**, pp. 519–534, 1980.

[64] Ho, S. & Suo, Z., Tunneling cracks in constrained layers. *Journal of Applied Mechanics*, **60**, pp. 890–894, 1993.

[65] Abrate, S., Impact on laminated composite materials. *Applied Mechanical Review*, **44**, pp. 155–190, 1991.

[66] Pavier, M.J. & Clarke, M.P., Experimental techniques for the investigation of the effects of impact damage on carbon fibre composites. *Composites Science & Technology*, **55**, pp. 157–169, 1995.

[67] Peck, S.O. & Springer, G.S., The behaviour of delaminations in composite plates – analytical and experimental results. *Journal of Composite Materials*, **25**, pp. 907–929, 1991.

[68] Short, G.J., Guild, F.J. & Pavier, M.J., The effect of delamination geometry on the compressive failure of composite laminates. *Composites Science and Technology*, **61**, pp. 2075–2086, 2001.

[69] Karihaloo, B.L. & Stang, H., Buckling-driven delamination growth in composite laminates: guidelines for assessing the threat posed by an interlaminar delamination. *Composites Part B*, **39**, pp. 386–395, 2008.

[70] Sørensen, B.F., Goutianos, S. & Jacobsen, T.K., Strength scaling of adhesive joints in polymer-matrix composites. *International Journal of Solids & Structures*, **46**, pp. 741–761, 2009.

[71] Holmes, J.W., Sørensen, B.F., Brøndsted, P. & Liu, L., Wind turbine blades: analysis of adhesively bonded laminates, *Global Wind Power 2008*, Beijing, China, 2008.

[72] Holmes, J.W., Rasmussen, U.R., Sørensen, B.F. & Brøndsted, P., Experimental technique for mixed-mode cyclic crack growth testing of composite laminates. *Engineering Fracture Mechanics*, manuscript in preparation, 2010.

[73] Nilsson, K.F., Asp, L.E., Alpman, J.E. & Nystedt, L., Delamination buckling & growth for delaminations at different depths in a slender composite panel. *International Journal of Solids & Structures*, **38**, pp. 3039–3071, 2001.

[74] Short, G.J., Guild, F.J. & Pavier, M.J., Delaminations in flat and curved composite laminates subjected to compressive load. *Composite Structures*, **58**, pp. 249–258, 2002.

[75] Wang, X.W., Pont-Lezica, I., Harris, J.M., Guild, F.J. & Pavier, M.J., Compressive failure of composite laminates containing multiple delaminations. *Composites Science and Technology*, **65**, pp. 191–200, 2005.

[76] Branner, K., Jensen, F.M., Berring, P., Puri, A., Morris, A. & Dear, J., Effect of sandwich core properties on ultimate strength of a wind turbine blade. *Proceedings of the 8th International Conference on Sandwich Structures,* University of Porto, Portugal, 2008.

[77] Jensen, F.M., Weaver, P.M., Cecchini, L.S., Stang, H. & Nielsen, R.F., The Brazier effect in wind-turbine blades and its influence on design. *Strain,* accepted, 2009.

[78] Branner, K., Modelling failure in cross-section of wind turbine blade. *Proceedings of the. NAFEMS-Seminar: Prediction and Modelling of Failure Using FEA,* Risø National Laboratory, Denmark, 2006.

[79] Kühlmeier, L., Buckling of wind turbine rotor blades. Analysis, design and experimental validation. PhD-Thesis. Special Report No. 58, Aalborg University, Denmark, 2007.

[80] Kühlmeier, L., Thomsen, O.T. & Lund, E., Large scale buckling experiment and validation of predictive capabilities. *15th International Conference on Composite Materials (ICCM-15),* Durban, South Africa, 2005.

[81] Overgaard, L. & Lund, E., Structural design sensitivity analysis and optimization of Vestas V52 wind turbine blade. *Proceedings of the 6th World Congress on Structural and Multidisciplinary Optimization,* 2005.

[82] Jørgensen, E.R., Borum, K.K., McGugan, M., Thomsen, C.L., Jensen, F.M., Debel, C.P. & Sørensen, B.F., Full scale testing of wind turbine blade to failure – flapwise loading. Risø-R-1392(EN), p. 28, 2004.

[83] Berggreen, C., Branner, K., Jensen, J.F. & Schultz, J.P., Application & analysis of sandwich elements in the primary structure of large wind turbine blades. *Journal of Sandwich Structures & Materials,* **9,** pp. 525–552, 2007.

[84] Overgaard, L. & Lund, E., Damage analysis of a wind turbine blade. *Proceedings of the ECCOMAS thematic Conference on Mechanical Response of Composites,* Portugal, 2007.

[85] Overgaard, L. & Lund, E., Interdisciplinary damage & stability analysis of a wind turbine blade. *Proceedings of the 16th International Conference on Composite Materials,* Japan, 2007.

CHAPTER 14

Implementation of the 'smart' rotor concept

Anton W. Hulskamp & Harald E.N. Bersee
Faculty of Aerospace Engineering, Delft University of Technology, The Netherlands.

Fatigue is one of the biggest issues in wind turbine design. Most of the sources for fatigue loads are related to changes in the airflow around the blades due to, for instance, turbulence, tower shadow or jaw misalignment. Therefore systems are proposed that counteract these load fluctuations, which are both deterministic and stochastic in nature. The systems aim at influencing the lift and drag at different stations along the blade's length. This way, the aerodynamics along the blade can be controlled and the dynamic loads and modes can be dampened. This is called the 'smart' rotor concept. Such systems are already used in aerospace, both with airplanes and helicopters, although in the latter case mostly experimental. They aim at reconfiguration of the wing or rotor, flight control or vibration reduction. The concepts are often implemented using adaptive materials such as piezoelectric materials and shape memory alloys (SMAs). Piezoelectric materials are mostly implemented for high frequent actuation, in which low strains but high actuation forces are required. SMAs can exert very high forces and can recover very large strains, but within a limited bandwidth. At the Delft University of Technology a series of experiments is conducted in which the feasibility of the concept is proven and in which several control issues are being addressed. For instance, it has been shown that the control concept can be based on the structural response of the blade to the flow disturbance and that the presence of natural modes has a large influence on the performance of the system.

1 Introduction

Currently, horizontal axis wind turbine (HAWT) manufacturers are facing several challenges related to both a fast increase in turbine size (see Fig. 1) and market growth. The trends force manufacturers to rapidly increase production capacity

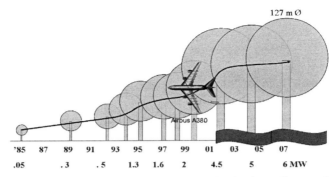

Figure 1: The increase of turbine state of the art size in time. Figure adapted from van Kuik in [1].

and to upscale existing blade designs. However, the boundaries of current technologies are being approached.

The increase in size is driven by the fact that the power conversion by a wind turbine increases with the square of the rotor blade's diameter. However, component costs will also increase. For instance, historically the mass of blades has increased with the radius to the power 2.4 [2] to 2.65 [3], depending on how the trend line is fitted. Thus the mass and with it the costs regarding materials and installation of a blade increase faster than the power it captures. This is more acceptable for offshore turbines than for their onshore counterparts because a large part of the costs, such as foundations, are related to the amount of turbines, rather than the installed power. Increasing the yield per turbine will therefore decrease the costs per kilowatthour.

Mass and size increase has several negative effects, beside higher material costs. First of all, installation cost increase because the heavy rotor has to be lifted to hub height. Secondly, at a certain point in time gravitational loading becomes a serious design drive. This fluctuating load comes on top of the already high fatigue loads that a turbine has to suffer because of fluctuating aerodynamic loading. These are caused by changes in the wind field and by the rotation of the blade through this wind field. Examples of these are wind shear, yaw misalignment, tower shadow and fluctuations in inflow. These loads will also increase with increasing size, for instance through the more severe wind shear experienced by larger machines. In fact, fatigue is already one of the biggest design drives and current blades are dimensioned for at least 10^8 cycles. Mitigating the amplitude of the fluctuations could lead to a longer service life of blades, but also to lighter blades and a reduced down time due to fatigue load induced component failure. In general mitigating fatigue loads will lead to a large reduction in the price per kilowatthour.

1.1 Current load control on wind turbines

Wind turbines have always had some system to control their aerodynamic loading. With old wind mills sails could be adjusted and with modern turbines a number of systems have been implemented to control the loads and the power output. Most of these systems comprise of a combination of variable or fixed rotational speed and pitch angle. Currently, the most widely implemented system [4] is the variable speed, variable pitch machine. These machines maintain an optimal tip speed ratio λ below rated wind speeds and the blades are pitched to regulate power above rated wind speeds. Usually pitching to feather is implemented. The controller is based on the rotational speed and the generator torque signals. The system is mainly in place to control the quality of the power output and the mean loading of the turbine, but it can also alleviate transient loading. Above rated wind speeds, an increase in loading will lead to an increase in generator torque and the controller will react to this increase by pitching the blades to feather. Pitching can be implemented with all blades being pitched collectively or with an actuator pitching each blade. If each blade is driven by its own pitch actuator, two control systems can be distinguished. In one case there exists one controller and the blades are pitched with a 120° phase difference. This so-called cyclic pitch control is used to mitigate 1P loads due to wind shear and yaw misalignment. With true individual pitch control (IPC) each blade is really controlled separately, based on local flow or load measurements. See Fig. 2 for the control loop of a variable speed, variable pitch turbine above rated wind speeds.

Here Ω stands for rotational speed, T for torque, β for the pitch angle and θ for the scheduling parameter that is tailored as the operating point of the turbine changes. The subscript g indicates that it concerns the generator and c that it is a control signal. However, numerous other concepts exist [5]. These include concepts where the loading above rated wind speeds is controlled by active or passive stall. Also passive control systems such as the FLEXHAT have been developed where a part of the blade is passively pitched as the rotational speed increases because it reacts to centrifugal forces.

Some concepts have been proposed with the goal of mitigating root bending moment fluctuations, such as teetering hinges in the hub and (active) cyclic pitch control. Also, a passive concept, called bent-twist coupling, exists. In this concept the UD laminate in the main bending load carrying spar caps is placed under an angle to the blade's longitudinal axis. This way, as the blade bends, it will twist [6].

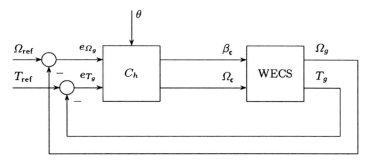

Figure 2: Variable speed, variable pitch control scheme for above rated wind speeds (figure adapted from [5]).

Both twisting toward feather and to stall is possible [7, 8]. In the first case the angle of attack is lowered and thus the aerodynamic loading. In the second case the blade is pushed into stall and is thus subjected to low lift and high drag forces. The concept is passive and it requires very accurate fiber placement and the implementation of expensive carbon fibers to be successful.

Many of the systems in use today react to the rotational speed of the rotor or generator torque. This means that that the peak loads are dampened after they occur and that the main source of fluctuations – the way in which the airflow around the blade is converted into mechanical loading – is not controlled. Nor is the loading of individual blades. Other concepts are passive. This is a disadvantage because the loading of the turbine is hard to control and stability issues may arise. Moreover, passive systems only react to the type of signals to which they are tuned (e.g. centrifugal loading for the FLEXHAT or blade bending with bend-twist coupling). Stall control seems to be a very simple way to control the loading of turbines, but it cannot mitigate load fluctuations. Blades that go dynamically in and out of stall are actually subjected to very high fluctuations in loading.

Active systems, based on feedback control, could dampen all possible load fluctuations which are picked up by sensors in the blade (e.g. strain gages), regardless of the source. The most widely implemented active load control system, variable speed and variable (individual) pitch control on turbines, also has the potential to actively mitigate load fluctuations. But controlling the loads this way puts a heavy strain on the pitch bearings and hydraulics. Furthermore, the actuation speeds are limited because the whole blade needs to be pitched. Finally, pitching the blade results in a change in angle of attack and thus loading over the whole span of the blade. Having the capability to control the aerodynamics along the span would give much greater control over the dynamics of the blade and the other components of the turbine at a much smaller effort.

1.2 The 'smart' rotor concept

Therefore the following new concept is proposed: by controlling the aerodynamics at different stages along the blade's span the way the blade is loaded can be controlled, counteracting the disturbances and mitigating fatigue loads. This, in combination with appropriate sensors that measure the loads or structural response and a controller that computes an actuation signal, is defined as the 'smart' rotor concept. Such an aerodynamic load control system has been intensively researched for helicopter blades and recently also feasibility studies for wind turbine blades have been made [9–12]. The goals of the system would be to react both to deterministic loads, such as wind shear and tower shadow, as to indeterministic loads such as gusts and turbulence. Most authors focus on deformable aerofoil geometry, but there is also an effort into microtabs or Gurney flaps by van Dam and co-workers [13–15]. These are small tabs of a height that is in the order of 1% of the cord, that are placed near the trailing edge, perpendicular to the flow. The tabs jets the flow in the boundary layer away from the blade's surface, effectively changing the circulation around it and thus the lift and drag characteristics. Depending on whether they are placed on the top or bottom side of the aerofoil the lift can be reduced or increased and thus the loading of the blade can be controlled. van Dam has made several studies into the topic, including CFD analysis and wind tunnel experiments, determining the optimal configuration and testing the performance of the device. Another 'add on' device under investigation is the synthetic jet [16]. With this, air is sucked out of, and jetted back into the boundary layer at a very high rate. This can be implemented to control flow separation, but when placed near the trailing edge, the device has an effect similar to the microtab at low angles of attack [17].

Here the possibilities will be addressed to implement the spanwise distributed devices into the structure of rotor blades. For wind turbine blades a solution that leads to an integrated adaptive structure is sought because of maintenance and weight considerations. Ideally a construction would be obtained that can deform to the designer's wishes by dissolved actuators. This deformation can be, e.g. a flapping motion of the trailing edge, blade twist or activating a control surface like the microtab. The focus will be on adaptive materials for actuation purposes because they are not subjected to wear and require no lubrication. Therefore their implementation will lead to a structure that can stay in service with little inspection and maintenance. Moreover, adaptive materials offer a very high power/weight ratio [18, 19]. In this section first aerodynamic control concepts found in other structures that are aimed at controlling aerodynamic forces will be addressed. These are aircraft wings and helicopter rotors. Secondly the most widely applied and most suitable types of adaptive materials will be discussed, namely piezoelectrics and SMAs. And finally some considerations regarding control and structural integration are addressed.

2 Adaptive wings and rotor blades

One research field in which adaptive structures are heavily researched, is aerospace. SMAs are already used to deploy space antenna's or as fasteners that do not require bolting [19, 20]. In addition, both for fixed wing and rotor craft many concepts have been considered and build to control the aeroelastic behavior of the machines. In this section the 'smart' wing and helicopter rotor concepts that proof to be a good benchmark and example for wind turbine rotors will be discussed.

2.1 Adaptive aerofoils and smart wings

A very comprehensive explanation of adaptive aerofoils is presented by Campanile [21]. Campanile presents in a historic overview that an adaptive wing was firstly aimed at in man's first attempts to fly and that even the Wright's brother's aircraft had morphing wings for flight control. Campanile asserts that because of the ever more severe demands for load carrying capability, maneuverability and speed (thrust), the tasks were separated in the wing box, flaps or ailerons and engines. But according the Campanile reintegration should be sought and it will lead to lighter, aerodynamically more efficient structures. In the end that is also the goal of the 'smart' rotor research. However, there is one mayor difference between aircraft wings and wind turbine rotor blades regarding flow control. Aircraft wings have had a control function from the first day they have been around, whereas for wind turbine blades actively controlling the aerodynamic loading is relatively new and spanwise aerodynamic load control has not been implemented on any commercial turbine. And in general wind turbine blades are not designed to house any systems – they are static fiber reinforced plastic components that are bolted to the hub, which possibly houses a pitch system. Still, the research into smart wings provides interesting insights since both structures have the same task – generating lift.

The first goal of adaptive wings is the replacement of current control features into the wing to form an integrated structure. These control surfaces comprise of flaps, slats and ailerons. Their common denominator is that actuating them induces a change in camber. This is also what is mostly aimed for in the research into spanwise distributed devices for wind turbine blades [9]. Monner *et al.* [22, 23] presents a system where linked plate-like elements form deformable ribs, so-called "fingers". The fingers have to be actuated only at one point. A kinematic mechanism distributes the work over the finger and thus the finger is gradually deformed (see Fig. 3). Monner allows for deformability of the outer skin by a flexible, sliding metallic skin.

Campanile presents a concept called the belt-rib design [24] (see Fig. 4). The principle of the belt-rib design is based on the idea that for camber control on a

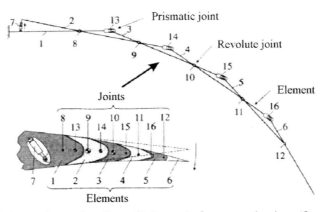

Figure 3: Kinematics and outline of Monner's finger mechanism (figure adapted from [23]).

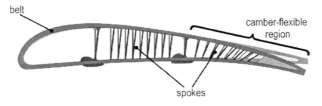

Figure 4: The Belt-rib-concept by Campanile.

continuous body, of the pressure and suction side one must lengthen and the other shorten depending on the direction of defection. Instead of straining both surfaces, it can also be conceived that transporting material from the suction to the pressure side, or vice versa attains the same effect. In the belt-rib design this is attained by implementing the aerofoil outer surface as a 'belt': circulation of the belt around the leading edge leads to a flapping motion of the trailing edge. Ribs inside the belt assure that actuation of the belt only leads to a change in camber and only little change in aerofoil shape and thickness.

A disadvantage of this concept is that the whole aerofoil needs to be deformed. With other trailing edge concepts like Monner's finger concept or phase two of the DARPA project Smart Wing [25], the central, load carrying part of the wing is often left unchanged. But combining full deformability in cordwise direction with stiffness in the flapwise direction, is not that straight forward.

Another concept that aims as camber control is presented by Inthra et al. Here genetic algorithms (GAs) [26] are implemented to determine the optimal distribution of piezoelectric stack actuators to induce a distributed deformation in the trailing edge. GAs are also a powerful tool in designing compliant mechanisms. Many concepts for adaptive wings are compliant structures in the sense that they comply to an actuator signal in a predetermined way and that they rely on the compliance of the material of which the structure is made. In general the ultimate goal is to distribute or to lever the control action of a discrete actuator through the body's volume in a predetermined manner[27]. Or to optimally distribute an array of actuators for a certain control action, like in [26]. GAs are very suitable for designing these structures because the user only supplies the input and the parameters that should be optimized. If the algorithm is correctly constructed it will determine the optimal material distribution of the structure to achieve, for instance, maximal control action for minimal weight.

See Fig. 5 for an example of a compliant mechanism concept by Saggere and Kota [28]. They elaborate the concept for a deformable leading edge, a so-called drooped nose. Trease et al. [29] presents a similar concept which just aims at deforming the trailing edge, but in this case for bio-mimenic propulsion in water. Trease does not elaborate on the means of actuation, but calls it smart actuator driven concepts. Kota also reports commercial implementation by the Flexsys company [30]. Recently they have also moved into the wind turbine engineering field as an application of their product.

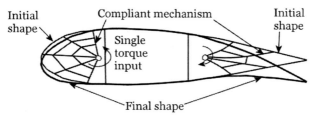

Figure 5: Schematics of compliant mechanisms for load control by deformable leading and trailing edge (figure adapted from [28]).

Two types of compliance mechanisms can be distinguished: lumbed and continuous compliance. The disadvantage of lumped compliance – where the deformability is concentrated in discrete points – is that stress concentrations occur around the flexible points, whereas with distributed compliance the strain is distributed over the whole mechanism.

A second form of adaptive wings aims at flight control trough wing twist. Gern *et al.* [31] mentions that wing twist can be employed to reduce roll reversal due to flap actuation. Wing twist requires a large amount of distributed actuators because the entire wing needs to be deformed to get a significant change in angle of attack. Stanewski [32] also mentions the suitability of 'smart' materials for this task because of their high power density and the possibility of distributed actuation, which is needed when deforming a wing though gradual torsional deformation, either on the wing itself, like Krakers [33] suggests, or by a added torsion tube, like Jardine *et al.* [34] has implemented.

Shape change of an aerofoil is also used for reconfiguration. In this case the aerofoil is deformed to set an optimal shape for different flight modes, e.g. from sub- to transonic flight or from a landing to cruising modes and vice versa. Reconfiguration concepts [35, 36] are often similar to other adaptive aerofoil concepts, but the needed actuation speeds are much lower. Therefore it opens up new possibilities for low frequent control concepts and adaptive materials such as SMAs and shape memory polymers (SMPs) [34, 37–39].

A final application of adaptive materials in aircrafts is vibration control. On the subject of 'smart' materials for vibration reduction in general numerous publications exist, but here the focus will be on vibration reduction of aerodynamically excited panels and wings. Vibration suppression then usually involves increasing aeroelastic damping. With increasing size, flutter might also become an issue for wind turbine blades [40] because the first bending and torsion mode move closer to each other, so it is an interesting topic for the wind turbine community as well. Guo *et al.* [41] reports increased stability of a panel by activating embedded SMA wires. Wu *et al.* [42] achieves vibration suppression by adhering PZT patches to a panel of a F15 fighter that suffers from aeroelastic excitation. These patches are connected to a shunt circuit which is tuned for maximum energy dissipation at specific frequencies. Hopkins *et al.* [43] reduces the vibration due to twin tail buffeting by adhering PZT patches to the tails and controlling these using feedback control based on accelerometers and strain gages. An increase in aeroelastic stability cannot only be attained by actively or passively increasing the structural damping, but also by increasing the aerodynamic damping, e.g. though control of a flap [44, 45].

A program that encompassed many of these concepts is the DARPA Smart Wing program. In the first phase of this program wing twist and SMA-activated trailing edge flaps were implemented on a scaled wing of a fighter aircraft [46]. Wing twist was induced by a SMA torque tube [34, 47] and the trailing edge flaps were designed as flexible glass-epoxy plates, covered with aluminum honeycomb and room temperature vulcanized (RTV) sheets to give the trailing edge its shape and smoothness respectively. In the second phase flaps, driven by piezoelectric motors, were implemented on a unmanned aerial vehicle (UAV) [48]. This concept is allowed for a high rate, continuously deforming trailing edge in both cordwise and spanwise direction. As a base structure the same construction as the trailing edge flaps, but aramid instead of aluminum honeycomb were used. The piezoelectric motors were chosen because of their high power-to-weight-ratio. Interesting conclusions from the program were that active surfaces can indeed be employed but that the actuation bandwidth of SMA material is very low and that the technology readiness level [49] was around five [25]. This means that it is at the level of component and/or breadboard validation in a relevant environment. This point is also addressed by Boller [50]: adaptive structures have shown great potential over the last 20 years, but only little real structural implementations have been achieved.

The overview of all these concepts of 'smart' wings leads to some reflections. First of all, a distinguish between different amounts of required deformation can be made: vibration control requires the littlest deformation of the structure since it means controlling the stiffness of a structure through stressing, or actively counteracting the vibration with a force. For wing twist medium strains are required because the resulting twist results from the accumulated strain along the blade's span. For integrated control surfaces relatively large strains are needed. This is also illustrated by the fact that many studies focus on the use of servo or servo-like actuators and that often silicone or latex skins are employed to allow for large strains in the skin, e.g., [48]. Classical concepts mentioned by Campanile [21] often employ surfaces that slide over each other.

Another distinction can be made in the different speeds required for actuation. Active vibration control requires very high actuation frequencies, whereas reconfiguration is quasi-static. Concepts for flight control require medium actuation speeds which are similar to the current control surfaces. Thirdly, for all adaptive concepts, but especially with wing twist, a trade-off must me made between on one hand the possible weight reduction due to the integration of several functions in one structure and on the other the added actuator mass. A final consideration is the readiness of the technology.

2.2 Smart helicopter rotor blades

A great deal of research on adaptive (aero)elastic structures has been conducted in the field of helicopter rotors. The research has been into different features such as torsion tubes, active twist control, trailing edge flaps, etc. Straub presents a good overview of early concepts [51]. In the following they are grouped and discussed.

2.2.1 Trailing edge flaps
Although helicopter rotors are much smaller than wind turbine blades and operate at much higher rotational speeds, they still pose an interesting benchmark as adaptive structures primarily because the structure under consideration is also a rotor. Secondly, because the intended effect, obtaining vibration reduction through load control, is usually the same. And finally because the flap deflections that are aimed at are roughly the same [9, 52], viz. several decrees for a flap size of ~10% of the cord.

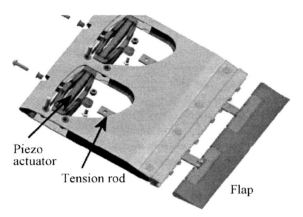

Figure 6: The piezoelectric stack driven concept by Enenkl (figure adapted from [54]).

There are several ways at which the flap can be controlled by adaptive materials. The following are elaborated in literature. Lee and Chopra [53] proposes a so-called piezo-based 'double L' amplifier. This mechanism drives the flap. Enenkl [54] proposed a similar system which consists of a piezo stack between two leaf springs. The Enenkl concept (see Fig. 6) has been tested on a full-scale helicopter rotor.

Bothwell [52] elaborates on a magnetostrictive driven torsion tube but he also mentions SMA wires for twisting the tube. Other torsion tube options [55] use pure piezo tubes by evoking twist in the piezo ceramic elements. Another way to drive the flap would be to employ the piezo ceramic in a bender and to use that bender to actuate the flap close to its hinge, thus amplifying displacement induced by the bender [56, 57]. A final application is to use SMA wires to pull the flap in either direction [58].

All these helicopter rotor trailing edge concepts are based on hinged flaps. This is not wanted in wind turbine rotor trailing edge flaps because of maintainability issues. Helicopters are already subjected to mandatory, regular inspections and rotor blades are replaced regularly.

2.2.2 Active twist

With helicopter blades another option has been researched, although less thoroughly than trailing edge flaps. This is to twist a section of the blade to influence the aerodynamics. Such a system could theoretically replace the cyclic pitch system, but the deflections are not large enough, so the goal remains vibration control [59]. With active twist a piece of the blade at the root can be twisted. Two technical implications can be distinguished. One is to apply adaptive materials under an angle to induce shear and thus torque. The second one is to apply an angled laminate with longitudinal applied adaptive materials. This induces the same type of shear. Possibilities are mentioned by Chopra [56], Strehlow and Strehlow [60] and Barrett et al. [59]. Barrett makes a comparison for different twist inducing configurations for a fin [61] and later to a small helicopter rotor [59]. But according to Boller [50] the solution is only applicable to small rotors.

As with concepts for smart wings, some general considerations can be deducted from this. The first one is that generally the goal of adding smart features to helicopter blades is vibration control and sometimes quasi-static blade tracking. But not replacing the current flight control system. For vibration control most research is aimed at active twist or trailing edge flaps, actuated by piezoelectric driven mechanisms. With piezoelectrics high actuation frequencies are attainable.

3 Adaptive materials

In the previous section on aerodynamic features, two types of adaptive materials are repetitively implemented: piezoelectrics and SMA material. Others can also be discarded for adaptive features on wind turbine blades because of the need for heavy coils (magnetostrictives, magneto-rheological fluids), too low bandwidth (electroactive and SMPs).

3.1 Piezoelectrics

Piezoelectric materials exhibit a coupling between mechanical deformations and dielectric effects. Actually, the term piezo is derived from the Greek word for squeeze or press. Moreover, the piezoelectric effect was firstly discovered in crystals which gave of a surface charge when strained. This is known as the direct effect. But the effect also works in the inverse manner: the material will strain under the application of an electric field. This is known as the converse effect.

These effects, as well as the governing equations have been documented by many [62–67]. They can be written in many forms [62, 66], depending on the electric and mechanical boundary conditions, but here they have written in two forms which are relevant to the use of actuator and strain sensor and the appropriate constants that describe the material will be discussed:

$$\varepsilon_i = S_{ij}^E \sigma_j + d_{mi} E_m \tag{1}$$

$$D_m = d_{mi}\sigma_i + e_{ik}^\sigma E_k \tag{2}$$

with: $i,j = 1,2...,6$ and $m,k = 1,2,3$ The subscripts are related to the material's coordinate system and the superscripts denote under which constant boundary condition the parameter is assessed – σ for constant stress and E for constant electric field. In eqs (1) and (2) the following variables define the electric and mechanical state of the piezoelectric materials.

ε_i: Strain component.

σ_i: Stress component.

D_i: Electric displacement component. Electric displacement is related to the amount of charge q on the electrodes: $q = \iint DdA$

E_i: Applied electric field component.

The other symbols are the piezoelectric constants and they are defined as follows:

e: Permittivity. It is the electric field per unit applied electric displacement. e is often related to the permitivity of vacuum: 8.85×10^{-12} F/m

d: matrix of piezoelectric charge constants. It is the mechanical strain per unit applied electric field or the electrical polarization per unit mechanical stress applied.

S: Elastic compliance constant which is defined as the amount of strain in the material per unit applied stress.

There are many ways to rewrite these equations into other forms. The IEEE [62] and Moulson and Herbert [66] present a good overview. These are chosen here because eqn (1) is very useful in describing the behavior of piezo-electrics as actuator and eqn (2) as sensor. Moheimani and Fleming [63] describes very well how these equations can be applied to patches and Waanders [64] to stacks. Another very important piezoelectric parameter is the effective coupling coefficient, k, which is a measure for the ability of the material to convert mechanical energy in electrical energy, or vice versa:

$$k^2 = \frac{\text{converted energy}}{\text{input energy}} \tag{3}$$

k can also be expressed in terms of piezoelectric constants described above:

$$k_{ij}^2 = \frac{d_{ij}}{s_{ij}^E \varepsilon_{ij}^S} \tag{4}$$

However, the total efficiency of a piece of piezoelectric material is not only defined by k, but also by the way it is incorporated into a mechanical system. Giurgiutiu and Rogers [68] define r as the ratio between the (internal) stiffness of the piezoelectric material and the (external) stiffness of the structure against which it acts, and then derives that the total energy conversion coefficient equals:

$$\eta = \frac{1}{4}\left(\frac{k^2}{1 - k^2(r/r+1)}\right) \tag{5}$$

Note that this equation holds for low frequent and quasi-static applications. Many dynamic analyses [64, 66, 68] of piezoelectric materials in electro-mechanical systems exist.

3.1.1 PZT

PZT, or lead zirconium titanate, is the most widely used piezoelectric material. It is a very brittle ceramic. Below the Curry temperature the ceramic crystal exhibits a lattice structure with a dipole because of tetragonal symmetry. Crystals with adjoining dipoles are grouped in domains, called Weiss domains, which are randomly oriented in the material. Therefore the material has no net dipole.

However, applying a high electric field just below the Curie temperature lets the domains which lie in the field direction grow at expense of others. This operation is called poling and causes a net dipole in the material. The matrix of piezoelectric strain constants is of the form:

$$\begin{bmatrix} 0 & 0 & 0 & 0 & d_{15} & 0 \\ 0 & 0 & 0 & d_{15} & 0 & 0 \\ d_{31} & d_{31} & d_{33} & 0 & 0 & 0 \end{bmatrix} \tag{6}$$

In the case of a patch, the three-direction denotes the out of plain and poling direction and d_{33} is much larger then d_{31}. 4–6 denote shearing. Specific forms and applications of PZT are discussed later.

3.1.2 PVDF

PVDF, or polyvenyldiphosphate, is one of the many polymers that can exhibit piezoelectric effects. It is discussed here because it has the highest coupling coefficient and already applications exist. Still, the coupling is much lower than the PZT, but a big advantage of PVDF is that it is very ductile. Making PVDF piezoelectric requires two steps: obtaining the right crystal structure and secondly, obtaining the net dipole. There are numerous ways to get to a piezoelectric form, but since most available materials consist of so-called phase II form, the following procedure is suggested [69].

The first step, obtaining the right crystal structure, is performed by stretching the material at elevated temperature after which it recrystallizes in phase I. This form has a non-centrosymmetric crystal, and thus a dipole. However, the crystals are still randomly orientated. Subsequent poling can be achieved by applying a high electric field from a corona at room temperature or a relative low field at elevated temperature. When poling, the PVDF's constitutive units are rotated around their chain bonds and thus a net dipole is attained. Hundred percent alignment with the field is not possible because some of the chains my have a vector component in the poling direction.

For piezoelectric PVDF sheets the charge constant d which links a field in the out of plane poling direction to in-plane strain, is different for both in-plane directions: $d_{31} \neq d_{32}$ and $d_{51} \neq d_{42}$, because of the uni-axial stretching. Therefore the charge constant matrix has the following form:

$$\begin{bmatrix} 0 & 0 & 0 & 0 & d_{15} & 0 \\ 0 & 0 & 0 & d_{24} & 0 & 0 \\ d_{31} & d_{32} & d_{33} & 0 & 0 & 0 \end{bmatrix} \tag{7}$$

The charge constants, the coupling coefficient and the compliance of PVFD are much lower than those of PZT, but it is much easier to handle and less brittle. In addition, PVDF with PZT granules dissolved in them have been proposed already years back [70]. This boosts the piezoelectric coupling, while maintaining the ductility of PVDF. However, Furukawa shows that the piezoelectric functionality of the composite can mainly be attributed to the PZT part of the composite.

3.1.3 Single crystal piezoelectrics

A final possible alternative is called Langasite [33]. This is a piezoelectric crystal that is grown under very high pressure. The crystal has a lower coupling coefficient than PZT, but a very high Curie temperature and depolarization field. Thus, when high voltages are applied, similar performances can be attained. It is also much less brittle and lighter than PZT. However, the crystal is very limitedly available and more expensive than PZT. The structure of the d-matrix is as follows:

$$\begin{bmatrix} d_{11} & d_{12} & 0 & d_{14} & 0 & 0 \\ 0 & 0 & 0 & 0 & d_{25} & d_{26} \\ 0 & 0 & 0 & 0 & 0 & 0 \end{bmatrix} \tag{8}$$

Bohm *et al.* [71] reports that due to the symmetry of the crystal, only two parameters are independent. Usually d_{11} and d_{14} are mentioned.

3.1.4 Piezoelectrics as actuators

Because the coupling between a field in the poling direction is the strongest, it is most obvious to employ this effect. But since the mutual distance between electrodes must remain small to attain a high electric field for a given voltage, special configurations have been developed for employing the d_{33}-effect. First of all, one can stack a series of disks or small patches on top of each other, each patch with its own set of electrodes. This is called a 'piezo stack' (see Fig. 7). By applying the same voltage to each patch, a high field in the poling direction is attained and because the dimension of the piezoelectric material in its poling direction is increased, so is its displacement. However, the displacements are still very small; a typical stack of several centimeters can only attain displacements in the order of magnitude of 10^2 μm. But large forces can be exerted this way.

Another concept in which the d_{33}-effect is implemented, is the embedding of fibers that are poled in their longitudinal direction in a polymer with the application of interdigitated electrodes on the surface. These materials are called active fiber composite [73] (see Fig. 8). Because these electrodes are placed at small instances, a high electrical field in the poling direction per unit applied voltage is attained. These plies can in turn be embedded into fiber reinforced polymers.

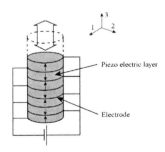

Figure 7: Piezo stack configuration (figure adapted from [72]).

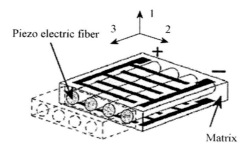

Figure 8: The principle of active fiber composites (figure adapted from [72]).

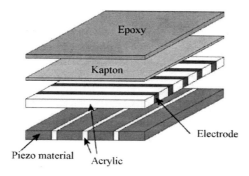

Figure 9: The build-up of a MFC (figure adapted from [74]).

A special type of these active fiber composites are called macro fiber composites (MFC's) [74] (Macro Fiber Composites), developed by NASA. MFCs are produced by sawing very fine strips from a patch and embedding those in epoxy between Kapton film (see Fig. 9).

Another way to amplify the displacement of a piezo-based actuator is to apply sheet material in a bender, either as a unimorph, with a patch on one side, or as a bimorph with a patch on each side. This way, the small strain of the piezo patch can be used to induce relatively large deflections of the bender. The behavior of such a bender is quite accurately described by extending the classical laminate theory (CLT) with piezoelectric effects analogous to thermal expansion. But other models which incorporate out-of-plane shear have also been developed [72]. When applying piezo patches in laminates, one must be aware that a piezo patch strains in all in-plane directions when a field is applied. However, usually straining the construction or substrate in one direction is wanted. Barrett *et al.* [75] therefore introduced the enhanced directionally attached piezoelectric (EDAP). With the concepts, narrow strips or special adhesion strategies are used to exploit the induced strain in only one direction.

There are also ways of increasing the deflection of these benders. This is done by introducing a geometrical non-linearity by axially compressing the bender [76] or using the thermal mismatch between the plies to introduce an internal stress state. Two types of benders that make use of the thermal mismatch are called Thunders [77, 78] and lightweight piezo ceramic actuators (LiPCA's) [79–81]. The Thunder actuators consist of a piezo ceramic sheet that is laminated between a thin steel plate and an even thinner aluminum foil by means of an adhesive. This laminate is consolidated at about 300°C and when it cools, the difference in coefficient of thermal expansion (CTE) causes a bend, and even slightly domed shape of the bender, depending on the aspect ratios and the presence of tabs. Aimmannee presents a model [78] which accurately predicts the shape of the Thunder. LiPCAs are based on the same principle, but they are produced by laminating a piezo patch between glass-epoxy and carbon-epoxy plies.

Mulling et al. [82] makes an analysis of the Thunder under external mechanical loading with different end conditions (e.g., clamping or simply supported). Hyer and Jilani [83] focus on the geometrically non-linear deformation of the actuator that results from the production method. But Haertling and co-workers [84] mention another effect that might enhance the deflection of Thunder-type actuators under actuation. The residual stresses that occur after cooling from processing at elevated temperature cause additional domain orientation. This increases the piezoelectric charge constant d and thus the coupling between an applied field and the strain of the PZT in the actuator.

Haertling and co-workers [84] focuses mainly on this effect in the analysis of Rainbow actuators. The concept of this actuator is similar to that of Thunders and LiPCA's, but with Rainbow actuators the layered constitution is obtained by 'reducing' a PZT patch to a certain depth. In this procedure [85], the oxygen from the PZT reacts with a carbon plate on which the PZT is placed at high temperature. This reduced layer has no piezoelectric coupling and has different mechanical properties and CTE than the unreduced layer. In addition, a volume change occurs in the reduction process. This leads to residual stresses after cooling. A Rainbow actuator is therefore not laminated but monolithic.

There are also other ways to mechanically amplify the deflection of piezoelectric actuators. Niezrecki et al. [86] presents an overview including concentric cylinder telescopic actuators, leave springs and lever systems. These mechanical amplifiers are not feasible for integration into active surfaces, but should be regarded as optional, low-wear, stand-alone actuators that could drive other mechanisms in turn.

3.2 Shape memory alloys

3.2.1 Material characteristics
SMAs derive their name from the fact that large, plastic deformations to which the material is subjected can be recovered. At low temperatures the material is in martensitic. The martensitic lattice structure has two variants, or twins. Loading the martensite above a certain level will lead to 'detwinning'. In this process one variant flips to the other over a certain stress range and the material can be

deformed heavily – up to 8% – under a small increase in loading. Upon unloading the deformation remains. This can be observed in Fig. 10. Here no elaboration into metallurgy is made, but Otsuka and Ren [20] offer a comprehensive review.

The nearly flat part of the tensile ε, σ-curve is called the stress plateau. The deformation is plastic, but recoverable by raising the temperature above a certain threshold. Above this threshold the material is austenitic. Austenite has a cubic lattice structure and therefore no variants. The material 'remembers' its undeformed shape because the net shape of the austenite is the same as of the undeformed variants of the martensite state. When the material is cooled down again the lattice structure becomes multi variant martensite again. This deformation and heating cycle can be observed in Fig. 11 and is called the shape memory effect (SME).

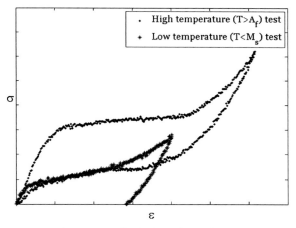

Figure 10: Stress–strain curves of a SMA wire at low and high temperatures.

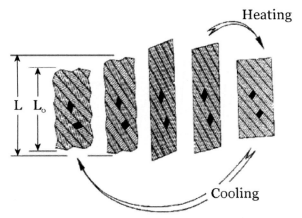

Figure 11: The SME cycle in SMA (figure adapted from [87]).

The transitions from martensite to austenite and vice versa do not occur suddenly at a certain temperature but they rather occur gradually over a certain temperature range. The temperatures at which the martensite-to-austenite (MA) transition begins (A_s) and finishes (A_f) are different then those of the reverse AM transition. The latter transition occurs at lower temperatures (see Fig. 12). Something else that can be observed in Fig. 12 is that the MA transition is postponed by tensile stresses.

When the temperature is raised until $T > A_f$, this causes the material to also exhibit pseudo-elasticity: stretching a piece of austenitic SMA causes the formation of martensite, so-called 'stress induced martensite' or SIM. The deformation that is obtained under the formation of martensite is recovered as the material transformed back into austenite when the tension is released (see Fig. 10). The behavior is apparently elastic since all the deformation is recovered, but the physical behind this process is a reversible change in lattice structure, not atomic bond stretching. Moreover, the stress–strain loop shows a considerable amount of hysteresis.

Therefore the SMA is said to have two observable effects during thermal and mechanical load cycles: super- or pseudo-elasticity when A_f is below operating temperature and the SME when the material is deformed at $T < M_f$ and then reheated, or a combination of both in between M_f and A_f. A third effect that is reported [19, 20] is the rubber-like effect, exhibited by some alloys. With this, the material shows recovery of deformation below M_f. This is usually regarded as an anomaly.

3.2.2 SMA behavior modeling

Usually the mechanical properties of a SMA are described as an ε, σ, T-behavior, but actually the underlying, connecting parameter is the martensite fraction ξ. Three types of models can be distinguished to model this behavior: the thermodynamical,

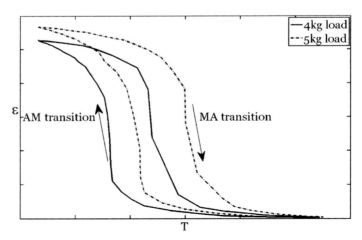

Figure 12: The displacement characteristics of an SMA wire as a function of the temperature, under constant loading.

the phenomenological and curve fitting models. The Young's modulus of the low temperature martensite is lower than that of austenite. In all models it is assumed to be linearly decreasing with increasing martensite fraction (ξ):

$$E(\xi) = (1-\xi)E_A + \xi E_M \tag{9}$$

Thermodynamic models are based on potential energy functions. In these models, the possible states are mathematically represented by 'wells', being local minima of potential energy with respect to the shear length. The transformation dynamics are described by the probability of a crystal being in one well to overcome the energy barrier to jump to the next using Boltzmann statistics. One of the earlier models for SMA behavior by Achenbach [88] is such a model. Others are by Seelecke [87] and Massad et al. [89].

Phenomenological models, like those by Tanaka [90], Liang and Rogers [91] and Brinson [92–94] are also based on thermodynamic potential formulations, but in these models often Gibs and Helmholtz free energy functions are employed because they do not rely on entropy as an internal parameter [19]. So-called hardening functions are assumed to describe the transformation dynamics. With these models, the martensite fraction of the material is determined by using the σ, T-state of the material. The models differ in the way that transition areas are modeled. Tanaka derives the following constitutive relation from the Helmholtz free energy:

$$d\sigma = E d\varepsilon + \Theta dT + \Omega d\xi \tag{10}$$

In this equation E refers to the modulus of elasticity, Θ is related to the CTE and Ω is called the 'transformation tensor'. Equation (10) can be written in integral form, with constant material properties:

$$\sigma - \sigma_0 = E(\varepsilon - \varepsilon_0) + \Theta(T - T_0) + \Omega(\xi - \xi_0) \tag{11}$$

Tanaka only distinguishes between austenite and martensite and models the stress-temperature dependency of the martensite fraction ξ with an exponential function. For the AM transition:

$$\xi = 1 - \exp^{a_M(M_s - T) + b_M\sigma} \quad \text{for} \quad \sigma \geq \frac{a_M}{b_M}(T - M_s) \tag{12}$$

and for the MA transition:

$$\xi = \exp^{a_A(A_s - T) + b_A\sigma} \quad \text{for} \quad \sigma \leq \frac{a_A}{b_A}(T - A_s) \tag{13}$$

The coefficients a_M, b_M, a_A and b_A are dependent on the transitions' start and finish temperatures and the stress dependency of these temperatures, the Clausius Clapeyron constants C_A and C_M. Assuming that the transition is complete with 99% conversion:

$$a_A = \frac{2\ln 10}{A_f - A_s}, \quad b_A = \frac{a_A}{C_A} \tag{14}$$

$$a_{\mathrm{M}} = \frac{-2\ln 10}{M_{\mathrm{s}} - M_{\mathrm{f}}}, \quad b_{\mathrm{M}} = \frac{a_{\mathrm{M}}}{C_{\mathrm{M}}} \tag{15}$$

The shape functions for phase transition are valid on certain stress-dependent temperature domains which can thus be plotted on the T, σ-plane (see Fig. 13).

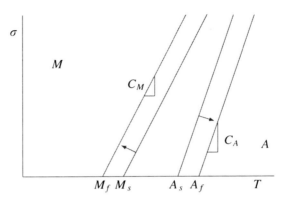

Figure 13: T, σ-phase diagram from the Tanaka model. The arrows indicate in which direction of the T, σ-path the phase change occurs.

Here the graphical representation of the Clausius Clapeyron constant can also be seen. Tanaka actually defines b_{A} and b_{M} in terms of the height of the transition band:

$$\Delta\sigma_{\mathrm{A}} = \frac{2\ln 10}{b_{\mathrm{A}}} \tag{16}$$

$$\Delta\sigma_{\mathrm{M}} = \frac{-2\ln 10}{b_{\mathrm{M}}} \tag{17}$$

This constitutes the same as eqns (14) and (15) because of the definition of the Clausius Clapeyron constants:

$$C_{\mathrm{A}} = \frac{\Delta\sigma_{\mathrm{A}}}{A_{\mathrm{f}} - A_{\mathrm{s}}} \tag{18}$$

$$C_{\mathrm{M}} = \frac{\Delta\sigma_{\mathrm{M}}}{M_{\mathrm{s}} - M_{\mathrm{f}}} \tag{19}$$

Liang and Rogers propose a similar model, but with a cosine-shaped dependency of ξ on T and σ.

For $C_{\mathrm{M}}(T - M_{\mathrm{f}}) - (\pi / |b_{\mathrm{M}}|) \leq \sigma \leq C_{\mathrm{M}}(T - M_{\mathrm{f}})$:

$$\xi = \frac{1 - \xi_{\mathrm{A}}}{2} \cos(a_{\mathrm{M}}(T - M_{\mathrm{f}}) + b_{\mathrm{M}}\sigma) + \frac{1 + \xi_{\mathrm{A}}}{2} \tag{20}$$

for the AM transition. For the MA transition the following holds.

For $C_A(T - A_s) - (\pi/|b_A|) \leq \sigma \leq C_A(T - A_s)$:

$$\xi = \frac{\xi_M}{2}\left[\cos(a_A(T - A_s) + b_A\sigma) + 1\right] \tag{21}$$

with:

$$a_A = \frac{\pi}{A_f - A_s}, \quad b_A = \frac{-a_A}{C_A} \tag{22}$$

$$a_M = \frac{\pi}{M_s - M_f}, \quad b_M = \frac{-a_M}{C_M} \tag{23}$$

and ξ_M and ξ_A are the start martensite fractions at the beginning of the respective transformations. a_M, b_M, a_A and b_M are slightly differently defined than in the Tanaka model, but they constitute the same physical meaning.

Brinson makes a distinction between temperature induced, multi variant ('twinned') martensite and stress induced, single variant ('detwinned') martensite. The constitutive is relation is then rewritten, also taking into account non-constant material properties:

$$\sigma - \sigma_0 = E(\xi)\varepsilon - E(\xi_0)\varepsilon_0 + \Omega(\xi)\xi_s - \Omega(\xi_0)\xi_{s0} + \Theta(T - T_0) \tag{24}$$

where the subscript 's' denotes the stress induces, detwinned martensite. Brinson also explains that:

$$\Omega(\xi) = \varepsilon_L E(\xi) \tag{25}$$

In which ε_L is the maximal recoverable strain. Thus eqn (24) reduces to:

$$\sigma = E(\xi)(\varepsilon - \varepsilon_L\xi_s) + \Theta(T - T_0) + K_0 \tag{26}$$

where K_0 is a collection of terms that represent the initial conditions:

$$K_0 = \sigma_0 - E(\xi_0)(\varepsilon_0 - \xi_{s0}\varepsilon_L) \tag{27}$$

This parameter is dependent on the loading history of the material. Brinson, like Liang and Rogers, also assumes a cosine-shaped transition path. Unlike Liang and Rogers, Brinson makes no distinction between the fraction at the start of the AM or MA transition and denotes the state at the beginning of the transition with the subscript '0'. Because of the distinction between twinned and detwinned martensite, below M_s another transition is introduced for the formation of detwinned martensite, also following a cosine-shaped path. For $T < M_s$ and $\sigma_s^{cr} < \sigma < \sigma_f^{cr}$:

$$\xi_s = \frac{1 - \xi_{s0}}{2}\cos\left[\frac{\pi}{\sigma_s^{cr} - \sigma_f^{cr}}(\sigma - \sigma_f^{cr})\right] + \frac{1 + \xi_s 0}{2} \tag{28}$$

$$\xi_s = \xi_{t0} - \frac{\xi_{t0}}{1 - \xi_{s0}}(\xi_s - \xi_{s0}) + \Delta_{t\xi} \tag{29}$$

with, if $M_f < T < M_s$ and $T < T_0$:

$$\Delta_{T\xi} = \frac{1-\xi_{t0}}{2}\cos(a_M(T - M_f)) + 1 \tag{30}$$

Else:

$$\Delta_{T\xi} = 0 \tag{31}$$

Subscript 't' denotes temperature induced martensite. If the temperature is below M_f the detwinning is only stress-dependent. But if the temperature is between M_f and M_s, the model takes into account the formation of detwinned martensite due to cooling through the AM transition zone. This is captured in the $\Delta_{T\xi}$ parameter. If the stress is below σ_s^{cr}, only twinned martensite is formed. For the formation of detwinned martensite above M_s Brinson derives the following.

For $T > M_s$ and $\sigma_s^{cr} + C_M(T - M_s) < \sigma < \sigma s_f^{cr} + C_M(T - M_s)$:

$$\xi_s = \frac{1-\xi_{s0}}{2}\cos\left[\frac{\pi}{\sigma_s^{cr} - \sigma_f^{cr}}(\sigma - \sigma_f^{cr} - C_M(T - M_s))\right] + \frac{1+\xi_{s0}}{2} \tag{32}$$

The function for the formation of austenite above A_s is the same as with Liang and Rogers, but a function for the split in stress and temperature induced martensite is added:

$$\xi = \frac{\xi_0}{2}\left[\cos(a_A(T - A_s) - \sigma/C_A) + 1\right] \tag{33}$$

$$\xi_s = \xi_{s0} - \frac{\xi_{s0}}{\xi_0}(\xi_0 - \xi) \tag{34}$$

$$\xi_t = \xi_{t0} - \frac{\xi_{t0}}{\xi_0}(\xi_0 - \xi) \tag{35}$$

Like with the model of Tanaka, the different phase regions can be plotted on the T, σ-plane (see Fig. 14). In a later publication Bekker and Brinson [95] introduce so-called switching points. At these switching points, the phase transition is either complete or the σ, T-path reverses. In the model, the start fractions ξ_{s0} and ξ_{t0} are then reset. This way, uncompleted transitions and embedded loops can be modeled.

These models are very insightful in understanding the underlying mechanisms of the SME and superelasticity because they map the martensite fraction based on the actual parameters on which it is actually depending: stress and temperature. And more importantly: they seem to predict the SMA behavior well [96]. However, the models provide the strain as a function of temperature, stress and load history. Inverting the model is not possible and a solution must be found iteratively. Leo presents a similar model in [97].

With the curve fitting models, like those by Spies [98] and van der Wijst [99], the force–displacement behavior is derived directly from the stress–strain path. The temperature dependency of this path is taken into account by linearizing the ε,

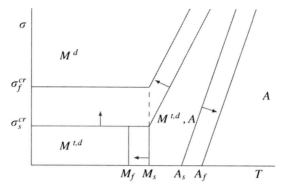

Figure 14: T, σ-phase diagram for the Brinson model. The arrows indicate in which direction of the T, σ-path the phase change occurs.

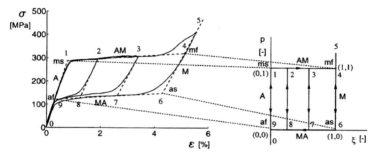

Figure 15: ε, σ-space to ξ, p-space mapping in the van der Wijst model.

σ-paths between transition points and shifting these points with the temperature. van der Wijst does that by mapping the stress–strain envelope onto a ξ, p-plane, where p is the elastic load parameter and ξ is again the martensite fraction (see Fig. 15).

It is then stipulated that changes in elastic stress and in martensite fraction cannot occur simultaneously:

$$\dot{p}(t)\dot{\xi}(t) = 0 \;\forall\; t \tag{36}$$

van der Wijst then uses the ξ, p-map in conjunction with a set of bilinear equations for stress and strain to determine the state of the SMA material and calculate the corresponding stress and strain state:

$$\varepsilon = \varepsilon_1 + \varepsilon_2 p + \varepsilon_3 \xi + \varepsilon_4 p\xi \tag{37}$$

$$\sigma = \sigma_1 + \sigma_2 p + \sigma_3 \xi + \sigma_4 p\xi \tag{38}$$

In these equations, the coefficients are linearly dependent on the temperature:

$$\varepsilon_1 = \varepsilon_{1a} T + \varepsilon_{1b}$$
$$\cdots \tag{39}$$
$$\sigma_4 = \sigma_{4a} T + \sigma_{4b}$$

To calculate the state from time step to time step the equations are split over the two states and differentiated with respect to time. For the elastic regime (with constant martensite fraction):

$$\dot{\xi} = 0 \tag{40}$$

$$\dot{p} = \frac{1}{\varepsilon_{,p}}\dot{\varepsilon} - \frac{\varepsilon_{,T}}{\varepsilon_{,p}}\dot{T} \tag{41}$$

$$\dot{\sigma} = \frac{\sigma_{,p}}{\varepsilon_{,p}}\dot{\varepsilon} + \left(\sigma_{,T} - \frac{\varepsilon_{,T}}{\varepsilon_{,p}}\sigma_{,p}\right)\dot{T} \tag{42}$$

for the transformation state (with constant elastic load parameter):

$$\dot{\xi} = \frac{1}{\varepsilon_{,\xi}}\dot{\varepsilon} - \frac{\varepsilon_{,T}}{\varepsilon_{,\xi}}\dot{T} \tag{43}$$

$$\dot{p} = 0 \tag{44}$$

$$\dot{\sigma} = \frac{\sigma_{,\xi}}{\varepsilon_{,\xi}}\dot{\varepsilon} + \left(\sigma_{,T} - \frac{\varepsilon_{,T}}{\varepsilon_{,\xi}}\sigma_{,\xi}\right)\dot{T} \tag{45}$$

In these equations a comma denotes a derivative to the subsequent parameter. Coupled with a thermal model for the temperature of the wire and a model for the external forces on the wire, the wire's behavior can be predicted each time step. The curve fitting models are not based on the thermodynamics behind the material behavior and a linear path is fitted between the transition points, but van der Wijst has shown that they can be a powerful tool for trajectory control, both with feedforward and feedback controllers. However, without feedback on the position of the actuator, good trajectory control is not possible. This is both due to the difficulties in modeling as in uncertainties in the thermal balance of the system.

3.2.3 Applications

SMAs are mainly employed in the form of wires and ribbons. Lagoudas [19] and Prahlad and Chopra [96] describe the procedure to characterize the material for its application. This implies determining the borders of the transition zones in the σ, T-phase diagram. The following tests are proposed:

- Differential scanning calorimetry (DSC) measurements to determine the stress-free transition temperatures, M_s, M_f, A_s and A_f
- Tensile test below M_f to determine σ_s^{cr} and σ_f^{cr}
- Tensile test above A_f to determine the stress dependency of the transitions' start and finish temperatures. Alternatively, it is possible to do recovery experiments under constant loading (isobaric tests). The first gives vital information on the

superelastic behavior and the other about the ability of the materials to exert work. Both test will provide points another set of points on the transitions' borders (the first being the result of the DSC measurements).

Lagoudas further mentions it is also important to determine the stabilizing behavior under cyclic loading, especially if more than one cycle are part of the functionality. Prahlad compares the results from a model, based on the characterization experiments with experimental results for restrained recovery.

To be applied as an actuator, the SMA material must be prestrained and prestressed and attached to, or embedded in the structure. When the SMA material is heated it will start to recover its deformation. The structure will resist to the deformation and the resulting stresses will postpone the formation of austenite. The structure or a bias force (spring, mass) will also have to force it back to its original position because typically the SMA is employed with one-way behavior. In addition, two wires can be set to act against each other.

If the structure is stiff enough, the behavior of a SMA can be described as restrained recovery: the strain remains negligible in comparison to the maximal recoverable strain, and it reduces to a σ, T-behavior. This still shows a considerable amount of hysteresis and the behavior is non-linear. The restrained recovery force can be used to determine the deflection of structures.

Restrained SMA wires can exert high forces, up to several hundred MPa. The force that can be exerted increases linearly for moderate amounts of prestraining, but it flattens off for high rates of prestraining [100]. Practical functionalities of (embedded) SMA wires and ribbons include tuning of dynamic behavior [101] and increasing aeroelastic stability [41], increasing critical buckling loads [102] and increased impact resistance [103]. Practical applications are mentioned by [19] and [20]. They mention pipe couplings that do not require fasteners and deforming cheyfrons on jets in order to change the jet outlet from low noise configuration during landing and take off to optimal performance while cruising. SMA material is also often implemented in bio-mechanical engineering, because of its good bio-compatibility. Use of the SME in bio-mechanical engineering can be found in stents to open arteries and in minimally invasive surgical equipment. See also [104].

SMA materials seem very suitable for application for control surfaces on MW-sized turbines because of their high power density, high actuation force and/or strain capability and because their bandwidth is in the required range. However, several drawbacks exist:

- Like all conductors, they are susceptible to lightning strike.
- The goal of the control system of which the actuator is a part is to alleviate fatigue loads on the blade. However, SMA material itself shows poor fatigue properties. Several options exist to increase the fatigue life:
  - Only subject the material to partial cycles
  - Implement materials that exhibit the R-phase transition
  - Use special high fatigue alloys

- The bandwidth that is mentioned in literature [19] is only attainable in laboratory conditions. In applications the bandwidth is limited by the cooling rate that the system can impose on the SMA material.
- Typically, the heat that is put in, is not recovered and therefore energy loss. This makes the power consumption of SMA materials relatively high as compared to, for instance, piezoelectrics.

4 Structural layout of smart rotor blades

The most promising concept until now has been camber control and the trailing edge flap design. The flow will stay attached and the boundary layer is not disturbed by the presence or actuation of the device. The difference between camber control and a continuously deformable trailing edge flap, is only in which amount of the cord is deformable and the distinction between the two is arbitrary.

However, in order to introduce this concept, the aft part of the cord over the part of the span of the blade where the flaps are to be integrated, will have to be flexible. In current, rigid blades, usually a sandwich construction is applied in these regions of the blade to assure the shape stability of the shell and to provide resistance against buckling [105]. A thin monolithic laminate is favorable for actuation by adaptive materials or to house a mechanism that is deformable in cordwise direction (e.g. a compliant mechanism or Monner's "finger" concept).

If the design relies on the trailing edge for its edgewise properties, for instance by the presence of UD strands there, the design will need reinforcements in sections where the trailing edge is flexible. But these features are mainly implemented in the outboard region of the blades. The trailing edge reinforcement, if implemented, is placed in the inboard section of the blade. Another issue resulting from the removal of slots where the actuators are placed in, is the occurrence of stress concentrations. This can be tackled by the introduction of reinforcing elements (e.g. ribs, additional spar) to locally strengthen the blade.

Figure 16: Topology optimization of the internal outline of a blade by Joncas *et al.* [106].

A rib-spar structure has also been proven to be the optimal topology for load transfer through the blade [106] (see Fig. 16). The rib-spar concept can therefore also be applied throughout the whole blade, in conjunction with a thermoplastic composite (TPC) material [107]. TPC materials are more feasible for the multi-component rib-spar concept because TPC parts can be assembled by means of welding, which is much faster than adhering and – if done well – leads to a stronger bond.

Alternatively, the trailing edge can be extended with a flat morphing surface as was done by Bak et al. [108] in their load control experiment. Structurally it is a very favorable solution because only minor adaptations to the blade are required. The active surface is simply added. The flat surface could be activated by piezo-electrics or SMA wires. However, an aerofoil with flat trailing edge will have to be developed and a transition to parts of the aerofoil with non-flat trailing edge will have to be made. The load carrying part of the cord at sections with control surfaces is also reduced, assuming that the total cord length must remain the same.

5 Control and dynamics

An important aspect of the smart rotor is the sensor and control strategy for the load control features. In Fig. 17 the possible control possibilities can be observed. Implementation of sensors that measure the structural response is most straight-forward. They can be embedded in or attached to the structure. The problem with flow measurements techniques is that they add complexity to the system and some, like Lidar or pressure taps, are not reliable enough yet. However, Lidars have been shown to show great correspondence with cup anemometer data [109] and a nacelle mounted Lidar that measured turbulence in the inflow has been reported [110]. On the other hand, Lidars do not work under all atmospheric conditions. Pitot tubes may be more feasible and are already suggested for control purposes by Larsen *et al.* [111]. Larsen also mentions that the drawback of measuring the structural response is the phase difference between the load fluctuation and the blade response.

With sensors that measure the blade's mechanical loading, such as strain gages, already some implementations have been seen on wind turbine blades, but not for control purposes. Here the goal was to measure loads to validate the load assumptions

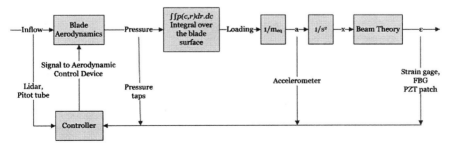

Figure 17: Various sensor concepts for feedback and feedforward control.

in design or for monitoring. For control purposes, measuring the structural response is especially useful if the harmonics of the blade play a large role. If the blade is excited close to a resonance peak, suppressing these dynamics already poses a significant load reduction potential. In this case, flow measurement might still be needed, but primarily for controlling the performance of the aerodynamic load control device, not for the load control itself.

This has also been shown in a load control experiment by the Delft University of Technology's wind energy research institute DUWind. Here, a series of experiments was conducted to research the (dynamic) load reduction potential of the 'smart' rotor concept. The primary goal of this experiment was showing the feasibility of the concept and to have a test set-up to test new control algorithms and actuator designs. Recently non-rotating experiments have been conducted and plans on a scaled turbine are planned.

5.1 Load alleviation experiments

A first approach these experiments were performed on a non-rotating blade. In these experiments the blade operates as a cantilever beam with uniform cross-section – the DU96 W180 aerofoil profile. The blade is mounted onto a pitch system at the wind tunnel's top wall and free to deflect over a table at the bottom side. The pitch system can be used to change the mean angle of attack, as well as inducing the dynamic disturbances that are to be mitigated. This way rotational effects are not taken into account and the blade has no twist or taper and constant thickness, unlike actual HAWT rotor blades. The table ensures that there are no tip-effects, because only 2D aerodynamic analyses were made. Thus, quasi-2D flow would be obtained in the static case. However, additionally experiments without table were also performed. See Fig. 18 for a picture of the set-up.

For controlling the aerodynamic loads it was chosen to implement partial camber control: the aft half of the cord at certain stations in the outboard section of the blade was made deformable therefore allowing for a change in camber of that part of the span. Such aerodynamic load control systems were also suggested for wind turbine blades by Buhl et al. [12] and Joncas et al. [10] and intensively discussed before. The actuator is based on a piezoelectric Thunder™ actuator, already elaborated on in section 3.1.4. The actuator is covered with a soft polyether foam which in turn is covered with a latex skin to provide a smooth surface. See Fig. 19 for the actuator design.

5.2 Control

In order to control the actuators and read the signals from the sensors, a dSpace™ system was employed for both feed forward as feedback experiments. With these systems sensors signals are converted to a digital signal and sampled. These signals can be recorded as well as fed to a feedback control algorithm. The output of the controller (whether it is feedforward or feedback) is converted to an analogue signal and send to the different actuators. The system of processing signals as well as the feedback controller is designed in Simulink™ and compiled onto

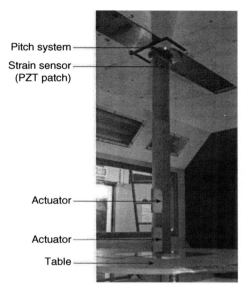

Figure 18: Wind tunnel set-up for load alleviation experiments. The airflow goes from right-to-left in this picture.

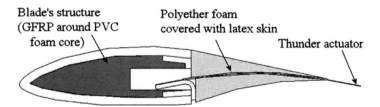

Figure 19: Design of the active control surface.

the dSpace™ system. Inputs and outputs for, e.g. setting values and plotting and recording signals can also be incorporated and linked to Control Desk™, a graphical user interface (GUI) (Fig. 20). From the dSpace™ hardware, one signal goes to the pitch system, which consists of a linear motor and two signals go to the high voltage amplifier which drives both sets of piezoelectric benders.

Inputs to dSpace™ include: the actual pitch displacement (feedback from the pitch system), the actual voltage on the piezoelectric benders (output of the amplifier), strain at the root of the blade and acceleration of the tip.

A critical part of the blade's design is the dynamics. The first natural vibration mode should be scaled with respect to two parameters:

1. The frequencies of the disturbances on the blade. HAWT blades are mainly subjected to loads associated with its rotational frequency or multiples of that – 1P, 2P and 3P. Proximity of vibrational modes will influence the dynamic response under

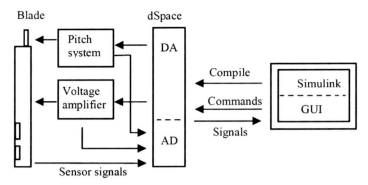

Figure 20: Set-up of the control system.

loading. However, this can also be tuned by changing the frequency spectrum of the disturbances which is controlled by the pitch system.

2. The second type of dynamic effects to take into account, is the unsteadiness of the aerodynamics. This is expressed by the parameter k, called the reduced frequency:

$$k = \frac{\omega b}{V} \tag{46}$$

in which Ω is the frequency of the disturbances, V the undisturbed airspeed and b the half cord of the aerofoil. With it, frequencies of disturbances can be scaled to the dimensions of the blade and the wind speed. The aerodynamic delay, the phase between a sine on the flap and the resulting lift forces, is dependent on this reduced frequency [112].

The blade is designed to match the frequencies that were derived from these considerations. The target first flapping frequency was determined to be 19.2 Hz and in the actual blade the eigenfrequency was 12.5 Hz. This was easily compensated for by changing the airspeed and the frequencies of the disturbances to which the blade is subjected. The blade was tuned by changing the internal structure, viz. the number of glass-epoxy plies and the presence of a spar. A spar was added in the tip because here the actuator slots were cut out. The spar adds additional stiffness and strength and can be used as mounting point for the actuators.

The blade was produced using vacuum infusion in a closed mould and after assembly of the sensors and actuators, it was mounted on the pitch system and connected to the control hardware. Several tests were conducted:

• Feedforward on disturbances with a sinusoid signal.
• Feedback control on a sinusoid signal with a strain sensor at the root.
• Feedback control on a spectrum angle of attack disturbances that resembles the turbulence that an actual blade experiences.
• Feedback control on a step on the angle of attack (simulating gusts or tower shadow).

5.3 Results and Discussion

The results of the first set are promising. Here the focus will be on the results on the step experiments. See van Wingerden *et al.* [113] for details. In these experiments a step, simulating a gust, was put on the pitch of the blade which triggered a sudden change in lift. This was firstly done without controlling the flaps and secondly with feedback control. The results in two cases, $a = 6°$ (around maximum C_L/C_D) and $a = 10°$ (higher than the static stall angle), can be seen in Figs. 21 and 22. A significant reduction in the vibration behavior, as well as a reduction of the first peak can be observed.

Observing Figs. 21 and 22 and an important conclusion can be drawn: the control system based on structural response is only partially able to mitigate the

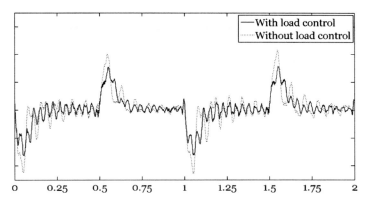

Figure 21: Strain signal at the root as a result of a step on the pitch at 6° angle of attack (close to maximum C_L/C_D, desired operating point for the DU-96 aerofoil).

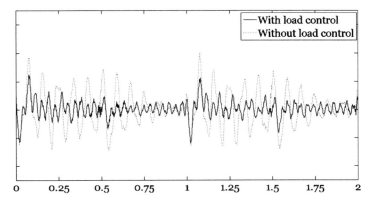

Figure 22: Strain signal at the root as a result of a step on the pitch at 10° angle of attack (higher than the static stall angle).

response to a gust. But it is possible to considerably increase the aerodynamic damping and to decrease the peak load.

The measuring of inflow could increase the load alleviation potential of the concept. Then so-called collocated control [114] is possible, where a local flow sensor directly coupled to a local control surface keeps the local aerodynamics constant. However, a global control system must be installed too to make sure that the ultimate goal on the system, reducing the load fluctuations, is assured. Keeping the aerodynamic load at certain stations constant does not assure that, because not all stations can be controlled and because non-aerodynamic loading on the blade exist, e.g. wave loading on the tower for offshore turbines. Including inflow sensors will complicate the control system and it can be questioned whether acting solely on the structural response is not already sufficient to attain a satisfying level of load reduction.

To answer this question, it must be researched which part of the load spectrum is dominant: are the loads dominated by quasi-static components or is the turbulence exciting the dynamics of the blade? MacMartin [114] makes the same distinction: he calls the mitigation of load fluctuation due to forced excitation isolation, whereas the suppression of dynamic modes is called active damping. In the latter case, basing the control system on the structural response, possibly with feedforward control on deterministic components of turbulence, will suffice. For issues concerning the control algorithms, please refer to [115].

5.4 Rotating experiments

These control issues are also being addressed in a second series of experiments in which an actual blades equipped with flaps is tested on a small turbine. This will allow for the study of the effect of rotationally induced disturbances, such as yaw misalignment, as well as the interaction between multiple blades and between the rotor and the wake. In addition, some design enhancements are made.

Blade design and manufacturing was done similar to the non-rotating experiment, except that the blade was made out of one piece and the dynamic scaling was performed with respect to the ratio between the rotational and the eigenfrequency, not the reduced frequency. This was done because from the non-rotating experiment it was concluded that the proximity of natural modes to parts of the disturbance spectrum has an influence on the dynamics loading of the blade. However, the reduced frequency for the model is higher than with the reference blade, and thus is the aerodynamic delay. A photo of the finished blades can be seen in Fig. 23. As you can see, the blade has twist and tapper, but a straight tip. This is to facilitate the installation of the actuators.

In this experiment, more attention was given to the dynamic behavior of the blade. In harmonic analyses, the transfer between flap excitation and stresses at different points on the blade was calculated in order to determine the right placement of sensors. The sensors were placed at locations where the normal stresses for a given excitation were relatively high and a safe phase margin existed. Both piezoelectric MFCs as well as strain gages are adhered to the blade. At the center of the flaps accelerometers are build in. The accelerometers record both in and out of plane acceleration and can be used for so-called collocated control as discussed above. The sensor array measures the structural response of the blade. However, flow measurement devices could be considered in the future.

Also, the actuator design was updated. The Thunder^{TM} actuator was placed at the suction side instead of the center of the profile to form a hard. In addition, stiff dissected foam was used to fill the remainder of the profile. High modulus foam was used to prevent the indentation under the dynamic pressure during operation. The dissected foam was covered with a thin layer of soft polyether foam, which in turn was covered with a polypropylene skin (see Fig. 24).

Wind tunnel experiments with the 'smart' rotor are scheduled for Fall 2009.

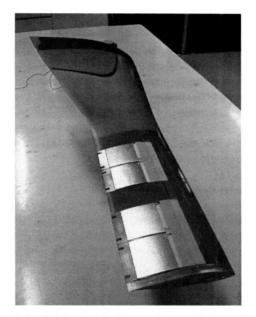

Figure 23: Photo of the finished scaled 'smart' rotor blade with two actuator slots in the tip.

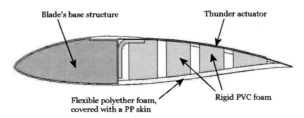

Figure 24: Design of the enhanced actuator for the rotating experiments with the 'smart' rotor.

6 Conclusions and discussion

6.1 Conclusions on adaptive aerospace structures

The adaptiveness of aerospace structures is advocated with many purposes in mind; from vibration control to reconfiguration. But the underlying idea is the same: to obtain lighter, better performing structures. Both the smart rotor blade research for helicopters, as the smart wing research for aeroplanes pose interesting benchmarks. In rotor blades aerodynamic load control is mostly pursued for vibration control. The concepts are usually based on hinged flaps where adaptive materials are implemented for their high power to weight ratio. Moreover, with piezoelectrics very high actuation frequencies are attainable. For vibration control or flight control for helicopters mostly piezoelectrics are proposed. For quasi-static blade tracking SMAs are referred to.

In smart wing research most research into aerodynamic load control is aimed at replacing the current control systems or for reconfiguration from one flight mode to another. Most concepts for control surfaces are still aimed at mechanisms rather than integrated structures. In addition compliant mechanisms have been proposed for morphing surfaces. Actuation of these surfaces does not necessarily have to be done by means of adaptive materials, but they are mentioned because of their high power-to-weight ratio. For the smart wind turbine blades, the morphing flap or aileron concepts are mostly interesting.

6.2 Conclusions on adaptive materials

Two adaptive materials are of most interest: piezoelectrics and SMAs. The challenge with piezoelectrics is to sufficiently amplify their strain and to take precautions for their brittleness, in the case of PZT. Precompression and applying it in Thunder-type benders or in mechanisms are good solutions.

The challenges with SMAs are actuation speed and controllability. The bandwidth of an SMA actuator could be increased by active cooling and for its controllability models exist, but the material behavior is highly non-linear and dependent on the load history. An issue with SMA material is the fatigue properties of the material. The advantage of SMA material is that the theoretically attainable bandwidth and force–displacement characteristics are very well suited for actuation in MW-sized HAWT blades.

With all electrically controlled actuators, whether they are electro-mechanical actuators (EMA) or adaptive materials, an issue is lightning strike.

6.3 Conclusions for wind turbine blades

From a stiffness point of view a rigorous alteration of the blade design probably not needed, depending on the blade design. Reinforcing and stiffening elements such as ribs or an additional spar could be placed around the actuator slots. Thermoplastic materials are favorable for such a blade concept because they allow for the easy assembly of multi-component designs. In addition, TPCs are to be preferred for deformable surfaces because many are tougher than thermosets. Finally it was mentioned that a rib-spar design is also the optimal topology for transferring loads through the blade.

6.4 Control issues

Load alleviation experiments at the TU Delft, using strain measurements as a feedback signal have shown that a significant reduction of the fatigue loads is possible. But many other signals, including inflow measurements are possible. Measuring the inflow could increase the load alleviation performance of any control system because then the largest source of fluctuating loads is known and feedforward control can be applied to it.

However, before aerodynamic load control on wind turbines can become a reality many hurdles have to be taken. Although the 'smart' structures from aerospace pose an interesting benchmark, the demands for wind turbines are different. Secondly, there has been a large effort into aero-servo-elastic modeling over the last few years, but the structural implementation of the spanwise distributed devices has been relatively ignored. Here some light on the matter has been shed, as well as on some of the control issues involved.

References

[1] Hanjalic, K., Krol, R. & Lekic, A., (eds.) *Sustainable energy technologies options and prospects.* Springer, 2008.
[2] Griffin, D., Windpact turbine design scaling studies technical area 1. composite blades for 80- to 120-meter rotor. Technical report, Sandia, 2001.
[3] Brøndsted, P., Lilholt, H. & Lystrup, A., Composite materials for wind power turbine blades. *Annual Review of Materials Research*, **35**, pp. 505–538, 2005.
[4] Hansen, A. & Hansen, L., Wind turbine concept market penetration over 10 years (1995-2004). *Wind Energy*, **10**, pp. 81–97, 2007.
[5] Bianchi, F., Battista, H. D. & Mantz, R., (eds.) *Wind Turbine Control Systems - Principles, Modelling and Gain Scheduling Design.* Springer, 2007.
[6] Goeij, W., Tooren, M. & Beukers, A., Implementation of bending-torsion coupling in the design of a wind-turbine rotor-blade. *Applied Energy*, **63(3)**, pp. 505–538, 1999.
[7] Lobtiz, D. & et al., P. V., The use of twist-coupled blades to enhance the performance of horizontal axis wind turbines. Technical report, Sandia, 2001.
[8] Lobitz, D. & Laino, D., Load mitigation with twist-coupled HAWT blades. *Proc. of the ASME Wind Energy Symposium*, 1999.
[9] Andersen, P., Gauna, M., Bak, C. & Buhl, T., Load alleviation on wind turbine blades using variable airfoil geometry. *Proc. of European Wind Energy Conf. and Exhibition*, European Wind Energy Association, Brussels, 2006.

[10] Joncas, S., Bergsma, O. & Beukers, A., Power regulation and optimization of offshore wind turbines through trailing edge flap control. *Proc. of the 43rdAIAAAerospace Science Meeting and Exhibit*, 2005.

[11] Basualdo, S., Load alleviation on wind turbine blades using variable airfoil geometry. *Wind Engineering*, **29(2)**, 2005.

[12] Buhl, T., Gauna, M. & Bak, C., Potential load reduction using airfoils with variable trailing edge geometry. *Journal of Solar Engineering*, **127**, 2005.

[13] Standish, K. & van Dam, C., Computational analysis of a microtab-based aerodynamic load control system for rotor blades. *Journal of the American Helicopter Society*, **50(3)**, 2005.

[14] Mayda, E., van Dam, C. & Nakafuji, D. Y., Computational investigation of finite width microtabs for aerodynamic load control. *Proc. of the 43rd AIAAAerospace Science Meeting and Exhibit*, 2005.

[15] Nakafuji, D. Y., van Dam, C., Michel, J. & Morrison, P., Load control for turbine blades: a non traditional microtab approach. *Collection of Technical Papers of the 40th ASME Wind Energy Symposium; AIAA Aerospace Sciences Meeting and Exhibit*, 2002.

[16] Glezer, A. & Amitay, M., Synthetic jets. *Anual Review of Fluid Mechanics*, **34**, 2002.

[17] Traub, L., Miller, A. & Rediniotis, O., Comparisons of a gurney and jet flap for hingeless control. *Jounral of Aircraft*, **41(2)**, 2004.

[18] Barrett, R. & Stutts, J., Design and testing of a 1/12th-scale solid state adaptive rotor. *Smart Materials and Structures*, **6**, 1997.

[19] Lagoudas, D., (ed.) *Shape Memory Alloys - Modeling and Engineering Applications*. Springer, 2008.

[20] Otsuka, K. & Ren, X., Recent development in the research of shape memory alloys. *Intermetallics*, **7**, 1999.

[21] Campanile, L., *Adaptive Structures, Engineering Applications*, J. Wiley and Sons, chapter Chapter 4, Light Shape-Adaptable Airfoils: a New Challenge for an Old Dream, 2007.

[22] Monner, H., Bein, T., Hanselka, H. & Breitbach, E., Design aspects of the adaptive wing - the elastic trailing edge and spoiler bump. *Multidisciplinary Design and Optimization: Proceedings*, Royal Aeronautical Society, 1998.

[23] Monner, H. & et al., Design aspects of the elastic trailing edge for an adaptive wing. *Proc. of the RTO AVT Specialists Meeting on "Structural Aspects of Flexible Aircraft Control"*, 1999.

[24] Campanile, L. & Sachau, D., The belt-rib concept: A structronic approach to variable camber. *Journal of Intelligent material systems and structures*, **11**, pp. 215–224, 2000.

[25] Kudva, J., Overview of the darpa smart wing project. *Journal of Intelligent material systems and structures*, **15**, pp. 261–267, 2004.

[26] Inthra, P., Sarjeant, R., Frecker, M. & Gandhi, F., Design of a conformable rotor airfoil using distributed piezoelectric actuators. *AIAA Journal*, **43(8)**, pp. 1684–1695, 2005.

[27] Lu, K.-J. & Kota, S., Design of compliant mechanisms for morphing structural shapes. *Journal of intelligent material systems and structures*, **14(6)**, pp. 379–391, 2003.

[28] Saggere, L. & Kota, S., Static shape control of smart structures using compliant mechanisms. *AIAA journal*, **37(5)**, pp. 572–578, 1999.

[29] Trease, B., Lu, K. & Kota, S., Biomemetic compliant system for smart actuator-driven aquatic propulsion: Preliminary results. *Proc. of IMECE03*, ASME, 2003, pp. 1–10.

[30] Flexsys. http://www.flxsys.com/.

[31] Gern, F., Inman, D. & Kapania, R., Computation of actuation power requirements for smart wings with morphing airfoils. *Proc. of the 43rd AIAA/ASME/ASCE/AHS/ASC Structures, Structural Dynamics, and Materials Conference*, 2002.

[32] Stanewski, E., Adaptive wing and flow control technology. *Progress in Aerospace Sciences*, **37(7)**, pp. 583–667, 2001.

[33] Krakers, L., Ductile piezo-electric actuator materials. Technical report, Netherlands Institute for Metal Research, 2006.

[34] Jardine, A., Bartley-Cho, J. & Flanigan, J., Improved design and performance of the SMA torque tube for the darpa smart wing program. *Proc. of SPIE Conference on Industrial and Commercial Applications of Smart Structures Technologies*, 1999.

[35] Love, M., Zink, P., Stroud, R., Bye, D. & Chase, C., Impact of actuation concepts on morphing aircraft structures. *Proc. of the 45th AIAA/ASME/ASCE/AHS/ASC Structures, Structural Dynamics and Material Conf.*, 2004.

[36] Black, The changing shape of future aircraft. *High Performance Composites*, pp. 52–54, 2006.

[37] Strelec, J., Lagoudas, D., Khan, M. & Yen, J., Design and implementation of a shape memory alloy actuated reconfigurable airfoil. *Journal of Intelligent material systems and structures*, **14(4-5)**, pp. 257–273, 2003.

[38] Perkins, D., J.L. Reed, J. & Havens, E., Morphing wing structures for loitering air vehicles. *Proc. of the 45th AIAA/ASME/ASCE/AHS/ASC Structures, Structural Dynamics and Material Conf.*, 2004.

[39] Dietsch, B. & Tong, T., A review - features and benefits of shape memory polymers. *Journal of Advanced Materials*, **39(2)**, pp. 3–12, 2007.

[40] Berring, P., Branner, K., Berggreen, C. & Knudsen, H., Torsional performance of wind turbine blades - part I: experimental investigation. *Proc. of the 16th International Conference on Composite Materials*, Japan Society for Composite Materials, 2006.

[41] Guo, X., Przekop, A. & Mei, C., Supersonic nonlinear panel flutter suppression using aeroelastic modes and shape memory alloys. *Proc. of the 46th AIAA/ASME/ASCE/AHS/ASC Structures Structural Dynamics and Materials Conf.*, 2005.

[42] Wu, S.-Y., Turner, T. & Rizzi, S., Piezoelectric shunt vibration damping of F-15 panel under high acoustic excitation. *Proc. of SPIE Conf. on Smart Structures and Materials 2000: Damping and Isolation*, SPIE, 2000, volume 3989.

[43] Hopkins, M., Henderson, D., Moses, R., Ryall, T., Zimcik, D. & Spangler, R., Active vibration suppression systems applied to twin tail buffeting. *Proc. of SPIE Conf. on Smart Structures and Materials 1998: Industrial and Commercial Applications of Smart Structures Technologies*, SPIE, 1998, volume 3326.

[44] Heinze, S. & Karpel, M., Analysis and wind tunnel testing of a piezoelectric tab for aeroelastic control applications. *Journal of Aircraft*, **43(6)**, pp. 1799–1804, 2006.

[45] Raja, S. & Upadhya, A., Active control of wing flutter using piezoactuated surface. *Journal of Aircraft*, **44**, pp. 71–80, 2007.

[46] Kudva, J. & et al., Overview of the DARPA/AFRL/NASA Smart Wing program. *Proc. of the SPIE Conf. on Industrial and Commercial Applications of Smart Structures Technologies*, 1999.

[47] Martin, C., Bartley-Cho, J., Flanigan, J. & Carpenter, B., Design and fabrication of smart wing tunnel model and sma control surfaces. *Proc. of the SPIE Conf. on Industrial and Commercial Applications of Smart Structures Technologies*, 1999.

[48] Bartley-Cho, J. & *et al.*, Development of high rate, adaptive trailing edge control surface for the smart wing phase 2 wind tunnel model. *Journal of Intelligent material systems and structures*, **15**, pp. 261–267, 2004.

[49] Moorhouse, D., Detailed definition and guidance for application of technology readiness levels. *Journal of Aircraft - Engineering notes*, **39**, pp. 190–192, 2002.

[50] Boller, C., *Adaptive Structures, Engineering Applications*, J. Wiley and Sons, chapter Chapter 6, Adaptive Aerospace Structures with Smart Technology - A retrospective and Future View, 2007.

[51] Straub, F., A feasibility study of using smart materials for rotor control. *Smart Materials and Structures*, 5, 1996.

[52] Bothwell, M., Chandra, R. & Chopra, I., Torsion actuation with extension-torsion composite coupling and a megnetostrictive actuator. *AIAA Journal*, **33(4)**, 1995.

[53] Lee, T. & Chopra, I., Design of piezostack-driven trailing-edge flap actuator for helicopter rotors. *Smart materials and structures*, **10**, pp. 15–24, 2001.

[54] Enenkl, B., Kloppel, V., Preiler, D. & Jänker, P., Full scale rotor with piezoelectric actuated blade flaps. *Proc. of the 28th European Rotorcraft Forum*, 2002.

[55] Centolanza, L., Smith, E. & Munsky, B., Induced-shear piezoelectric actuator for rotor blade trailing edge flaps. *Smart materials and structures*, **11**, pp. 24–35, 2002.

[56] Chopra, I., Recent progress on the development of a smart rotor system. *Proc. of the 26th European Rotorcraft Forum*, 2000.

[57] Hall, S. & Prechtl, E., Development of a piezoelectric servoflap for helicopter rotor control. *Smart Materials and Structures*, 5, pp. 26–34, 1996.

[58] Singh, K., Sirohi, J. & Chopra, I., An improved shape memory alloy actuator for rotor blade tracking. *Journal of Intelligent material systems and structures*, **14(12)**, pp. 767–786, 2003.

[59] Barrett, R., Frye, P. & Schliesman, M., Design, construction and characterization of a flightworthy piezoelectric solid state adaptive rotor. *Smart Materials and Structures*, **7**, 1998.

[60] Strehlow, H. & Rapp, H., Smart materials for helicopter active control. *75th Meeting of the AGARD Structures and Materials Panel, AGARD Conf. Proc. 531*, 1993.

[61] Barrett, R., Aeroservoelastic dap missile fin development. *Smart Materials and Structures*, **7**, 1993.

[62] Meitzler, A., Belincourt, D., Coquin, G., F.S. Welsh, I., Tiersten, H. & Warner, A., Ieee standard on piezoelectricity. Technical report, IEEE, 1988.

[63] Moheimani, S. & Fleming, A., *Piezoelectric Transducers for Vibration Control and Damping*. Springer Verlag, 2006.

[64] Waanders, J., *Piezoelectric Ceramics - Properties and Applications*. Philips Components, 1991.

[65] Sihora, J. & Chopra, I., Fundamental behavior of piezoceramic sheet actuators. *Journal of Intelligent Material Systems and Structures*, **11**, 2000.

[66] Moulson, A. & Herbert, J., (eds.) *Electroceramics: Materials, Properties, Applications*. John Wiley and Sons, Ltd, 2003.

[67] Leo, D., (ed.), *Engineering Analysis of Smart Material Systems*, John Wiley and Sons, Ltd, chapter 4, 2007.

[68] Giurgiutiu, V. & Rogers, C., Power and energy characteristics of solid-state induced-strain actuators for static and dynamic applications. *Journal of Intelligent Material Systems and Structures*, **8**, 1997.

[69] Sessler, G., Piezoelectricity in polyvinylidenefluoride. *Journal of the Acoustical Society of America*, **70(6)**, 1981.

[70] Furukawa, T., Ishida, K. & Fukada, E., Piezoelectric properties in the composite systems of polymers and pzt ceramics. *Journal of Applied Physics*, **50**, 179.

[71] Bohm, J. & et al., Czochralski growth and characterization of piezoelectric single crystals with langasite structure: $La_3Ga_5SiO_{14}$ (LGS), $La_3Ga_{5.5}Nb_{0.5}O_{14}$ (LGN) and $La_3Ga_{5.5}Ta_{0.5}O_{14}$ (LGT), Part II, Piezoelectric and elastic properties. *Journal of Crystal Growth*, **216**, 2000.

[72] Chopra, I., Review of state of art of smart structures and integrated systems. *AIAA Journal*, **40(11)**, 2002.

[73] Zhang, H. & Shen, Y., Three-dimensional analysis for rectangular 1-3 piezoelectric fiber-reinforced composite laminates with the interdigitated electrodes under electromechanical loadings. *Composites: Part B*, **37**, 2006.

[74] Sodano, H., Park, G. & Inman, D., An investigation into the performance of macro-fiber composites for sensing and structural vibration applications. *Mechanical Systems and Signal Processing*, **18**, 2004.

[75] Barrett, R. & et al., Active plate and wing research using edap elements. *Smart Matererials and Structures*, **1**, 1992.

[76] Barrett, R. & *et al.*, Post-buckled precompressed piezoelectric flight control actuator design, development and demonstration. *Smart Materials and Structures*, **15**, 2006.

[77] Aimmannee, S. & Hyer, M., Deformation and blocking force characteristics of rectangular thunder-type actuators. *Proc. of the International Conf. for Emerging System and Technology (ICEST 2005)*, 2005.

[78] Aimmannee, S. & Hyer, M., Analysis of the manufactured shape of rectangular thunder-type actuators. *Smart Materials and Structures*, **13**, 2004.

[79] Yoon, K. & *et al.*, Design and manufacture of a lightweight piezo-composite curved actuator. *Smart Materials and Structures*, **11**, 2002.

[80] Yoon, K. & *et al.*, Analytical design model for a piezo-composite unimorph actuator and its verification using lightweight piezo-composite curved actuators. *Smart Materials and Structures*, **13**, 2004.

[81] Kim, K. & *et al.*, Performance evaluation of lightweight piezo-composite actuators. *Sensors and Actuators A*, **120**, 2005.

[82] Mulling, J. & *et al.*, Load characterization of high displacement piezoelectric actuators with various end conditions. *Sensors and Actuators A*, **94**, 2001.

[83] Hyer, M. & Jilani, A., Deformation characteristics of circular rainbow actuators. *Smart Materials and Structures*, **11**, 2002.

[84] Li, G., Furman, E. & Haertling, G., Stress-enhanced displacements in plzt rainbow actuators. *Journal of the American Ceramics Society*, **80**, 1997.

[85] Wang, Q. & Cross, L., Analysis of high temperature reduction processing of rainbow actuator. *Materials Chemistry and Physics*, **58**, 1999.

[86] Niezrecki, C., Diann, B., Balakrishnan, S. & Moskalik, A., Piezoelectric actuation: State of the art. *The Shock and vibration digest*, **33**, 2001.

[87] Seelecke, S., Shape memory alloy actuators in smart structures: Modeling and simulation. *Appl Mech Rev*, **57**(**1**), 2004.

[88] Achenbach, M., A model for an alloy with shape memory. *International Journal of Plasticity*, **5**, 1989.

[89] Massad, J., Smith, R. & Garman, G., A free energy model for thin-film shape memory alloys. *Proc. of the SPIE, Smart Structures and Materials 2003: Modeling, Signal Processing, and Control*, 2003.

[90] Tanaka, K., A thermomechanical sketch of shape memory effect. *Res Mechanica*, **18**, 1986.

[91] Liang, C. & Rogers, C., One-dimensional thermo mechanical constitutive relations for shape memory alloys. *Journal of Intelligent Material Systems and Structures*, **1**, 1990.

[92] Brinson, L., One dimensional consttitutive behavior of shape memory alloys: Thermomechanical derivation with non-constant material functions and redefined martensite internal variable. *Journal of Intelligent Material Systems and Structures*, **4**(**2**), 1993.

[93] Brinson, L., Deformation of shape memory alloys due to thermo-induced transformations. *Journal of Intelligent Material Systems and Structures*, 7, 1996.

[94] Brinson, L. & Huang, M., Simplifications and comparisons of shape memory alloy constitutive models. *Journal of Intelligent Material Systems and Structures*, **7**(1), 1996.

[95] Bekker & Brinson, L., Phase diagram based description of the hysteresis behavior of shape memory alloys. *Acta Materialia*, **46**(10), 1998.

[96] Prahlad, H. & Chopra, I., Comparative evaluation of shape memory alloy constitutive models with experimental data. *Journal of Intelligent Material Systems and Structures*, **12**, 2001.

[97] Leo, D., (ed.), *Engineering Analysis of Smart Material Systems*, John Wiley and Sons, Ltd, chapter 6, 2007.

[98] Spies, R., An algorithm for simulating the isothermal hysteresis in the stress-strain laws of shape memory alloys. *Journal of Materials Science*, **31**, 1996.

[99] van der Wijst, M., *Shape control of structures and materials with shape memory alloys*. Ph.D. thesis, University of Eindhoven, 1998.

[100] Rogers & Liang, One dimensional constitutive relations of shape memory materials. *Journal of Intelligent Material Systems and Structures*, **1**(2), 1990.

[101] Epps, J. & Chandra, R., Shape memory alloy actuation for active tuning of composite beams. *Smart Materials and Structures*, **6**, 1997.

[102] Choi, S., Lee, J., Seo, D. & Choi, S., The active buckling control of laminated composite beams with embedded shape memory alloy wires. *Composite Structures*, **47**, 1999.

[103] Jia, H., *Impact Damage Resistance of Shape Memory Alloy Hybrid Composite Structures*. Ph.D. thesis, Virginia Polytechnical Institute and State University, 1998.

[104] Langelaar, M., *Design optimization of Shape Memory Alloy Structures*. Ph.D. thesis, Delft Technical University, 2006.

[105] Burton, T., Sharpe, D., Jenkins, N. & Bossanyi, E., *Wind Energy Handbook*. John Wiley and Sons, 2001.

[106] Joncas, S., Ruiter, M. & Keulen, F., Preliminary design of large wind turbine blades using layout optimization techniques. *Proc. of the 10th AIAA/ISSMO Multidisciplinary Analysis and Optimization Conf.*, 2004.

[107] Rijswijk, K., Joncas, S., Bersee, H. & Bergsma, O., Vacuum infused fiber-reinforced thermoplastic MW-size turbine blades: A cost-effective innovation? *Proc. of the 43rd AIAA Aerospace Science Meeting and Exhibit*, 2005.

[108] Bak, C., Gaunna, M. & Andersen, P., Load alleviation through adaptive trailing edge control surfaces: Adapwing overview. *Proc. of the European Wind Energy Conf. and Exhibition*, 2007.

[109] Smith, D., Harris, M., Coffey, A., Mikkelsen, T., Jørgensen, H., J.Mann & Danielian, R., Wind lidar evaluation at the danish wind test site in Høvsøre. *Wind Energy*, **9**, pp. 87–93, 2006.

[110] Harris, M., Bryce, D., Coffey, A., Smith, D., Birkemeyer, J. & Knopf, U., Advanced measurements of gusts by laser anemometry. *Journal of Wind Engineering and Industrial Aerodynamics*, **95**, pp. 1637–1647, 2007.

[111] Larsen, T., Madsen, H. & Thomsen, K., Active load reduction using individual pitch, based on local blade flow measurements. *Wind Energy*, **8**, pp. 67–80, 2005.

[112] Leishman, J., Unsteady lift of a flapped airfoil by indicial concepts. *Jounral of Aircraft*, **31**, 1994.

[113] van Wingerden, J., Hulskamp, A., Barlas, T., Marrant, B., van Kuik, G., Molenaar, D.-P. & Verhaegen, M., On the proof of concept of a smart wind turbine rotor blade for load alleviation. *Wind Energy*, **11**, 2008.

[114] MacMartin, D., Collocated structural control: motivation and methodology. *Proc. of the 4th IEEE Conf. on Control Applications*, 1995.

[115] van Wingerden, J., *Control of Wind Turbines with Smart Rotors: Proof of Concept & LPV Subspace Identification*. Ph.D. thesis, Delft University of Technology, 2008.

CHAPTER 15

Optimized gearbox design

Ray Hicks
Ray Hicks Limited, UK.

Superficially, gearboxes for wind turbines are required for a low technology, low speed and relatively low power application. However, their very high torque and speed increasing ratio requirements coupled with the capricious nature of the power source have created many problems which have had a detrimental effect on reliability. In reality therefore, they have had to be manufactured to the highest possible quality with corrections to gear teeth, etc. to compensate for the parasitic loads and deflections to which they are subjected. This chapter explains the basics of gear design criteria and offers solutions to the various problems.

1 Introduction

Wind turbines in common with virtually all other rotary machinery are subject to speed limits such that the product of rotor diameter and rotational speed is a constant, i.e. blade tip diameter is inversely proportional to speed. Since the proportions of the blade length and chord section tend to be constant, then given similar materials, wind speeds, etc., the rotor weight and torque are proportional to the linear dimension cubed.

Because of the inverse relationship of diameter and speed, the product of torque and speed, i.e. power, is directly proportional to the rotor diameter squared and therefore, the swept area. Thus, the power to weight ratio diminishes as power increases.

For example, if power is increased from 750 to 3000 kW, i.e. by a factor of 4, the rotor diameter is doubled and its rotational speed is halved. It follows that the weight and torque of the rotor are increased by a factor of 8. Incidentally the moment of inertia (related to the 5th power of the diameter) is increased by a factor of 32.

Unlike other methods of power generation such as gas turbines, the input energy source of wind turbines is of an uncontrolled stochastic nature. Its velocity, direction and pressure distribution over the swept area are all subject to sudden changes which require complementary changes in the rotor speed, blade pitch and nacelle orientation. However, because of inertia effects such changes cannot be made within a compatible time scale during which, the rotor hub is transiently required to sustain whatever loads this might entail.

Since power is proportional to wind speed cubed, a transient speed increase of only 50% will more than double the torque and treble the power. Even if the wind speed remains constant, a change of its direction with respect to the axis of rotation means that the rotor will run yawed such that the angle of attack on the individual blades will vary continuously as they rotate. Since it is virtually impossible to keep moving the nacelle in step with every transient it is only practicable to respond to a sustained change of direction. Thus, the turbine could spend a significant amount of its time running yawed.

In any case, most wind turbines face upwind with their rotor axes tilted some $5°$ up at the front to reduce overhang from the nacelle and the danger of blades colliding with the tower. The rotor therefore, will always be yawed even if in other respects it is perfectly aligned to the wind. Over the large swept area of a turbine there are significant variations in wind speeds, angles of attack and blade deflections which inevitably promote angular fluctuations at the rotor hub and consequentially, large cyclic torque and electrical power variations in the generator. While the electrical fluctuations may be dealt with electronically, the associated mechanical torque fluctuations due to the referred inertia of the generator rotor can only be absorbed by strain energy deflections in the drive train and/or an active form of torque control.

Assuming similar density materials, a direct drive generator will have 100 times the torque and weight of with a step up ratio of 100/1 the geared version and if their respective generator rotor lengths are approximately the same, it would have the same polar moment of inertia as that of the high speed generator whose inertia is multiplied by gear ratio squared when referred to the turbine rotor. The power to weight ratio of the direct drive generator, like the turbine will be subject to the same disproportionate decrease in its power to weight ratio, whereas that of a geared generator is constant.

Due to the universal application of constant frequency grid systems and cheap standardised high speed generators produced in large numbers, the cost of direct drive generators produced in relatively small numbers is inevitably much greater.

As power increases, the input shaft of a gearbox is subject not only to the same disproportionate increase in turbine torque but also a bigger step up ratio. However, this incurs a much smaller increase in overall weight and cost of the nacelle/tower assembly compared with the direct drive alternative. Thus, despite their reliability problems, geared generators have generally been the preferred option for the vast majority of wind turbines.

2 Basic gear tooth design

Toothed gearing is historically the most effective and efficient mechanism for coupling machines having different optimum speeds. Its development has therefore been driven by purely economic considerations.

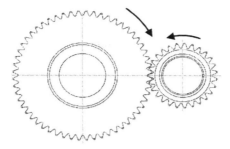

Figure 1: Simple wheel and pinion.

In its simplest form, a fixed ratio gear comprises a pinion with a smaller number of teeth meshing with a wheel having a larger number of teeth whose respective axes are parallel.

The difference in tooth numbers then determines the ratio between the respective speeds of pinion and wheel; e.g. a 100-tooth wheel will drive a 20-tooth pinion at five times its own speed. As shown in Fig. 1, the wheel and pinion rotate in opposite directions.

To provide a constant velocity ratio, the respective teeth must have the same precise circular pitch and a geometric shape which enables the torque to be transmitted from one tooth to the next by a slide/roll mechanism which ensures a constant circumferential velocity. The universally chosen tooth form is an involute whose properties are clearly described in any gearing text book. While toothed gearing is very simple in principle, it is very difficult to implement in practice. Torque is transmitted as a normal load between the mating teeth .but even if they are geometrically perfect, this load creates surface and bending deflections which in effect create pitch errors that vary with torque. In addition, misalignments occur due to associated deflections in the shafts, bearings, mountings, casings, etc. which support the gears. It becomes even more difficult when the gearbox is subjected to externally generated forces due to the variable nature of the wind. All these effects create unacceptable mal-distribution of tooth load across the face width of the gears.

Figure 2 shows the pitch circles of a pinion and wheel which contact one another at a pitch point on the line joining their respective centres. The pitch line passing through this point is tangential to the pitch circles and therefore, crosses the centre line at right angles. The circumference of the respective pitch circles is equal to their tooth numbers multiplied by the common circular pitch. As shown, the path of contact between the mating gears is a straight line common tangent to their respective base circles from which the involute tooth flanks are generated. This passes through the pitch point at an angle to the pitch line known as the pressure angle (usually 20°). Its length is determined by the distance between the two points where the respective tooth tip diameters cut across the common tangent. For continuity of transmission the normal distance between successive tooth flanks (the base pitch) has to be less than this length by a factor known as the contact ratio. For most standard gears this varies between 1.4 and 1.7. Thus, at the beginning and

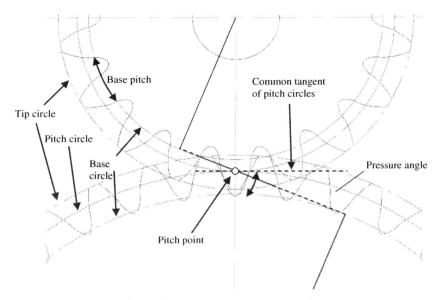

Figure 2: Base tangent contact path.

end of the contact path there are two pairs of teeth engaged, whereas in the centre, there is only one. Considering the meshing sequence: just as an unloaded pair of teeth are about to enter the double tooth contact zone at the beginning of contact there is only one pair of teeth transmitting the load at one base pitch from the beginning of the contact path. These loaded teeth will therefore, have a combined deflection which creates a relative pitch error with respect to the unloaded teeth. It is standard practise to modify the involute profiles of mating gears by tip relief designed to ensure that the tooth load increases progressively from zero to its nominal value as it passes through the double tooth contact zone at the beginning of the contact path into the single tooth contact zone, with a complementary decrease as it subsequently passes through the double contact zone at the exit.

Gear tooth design is required to satisfy two basic fatigue stress criteria, i.e. tooth root tensile bending and surface compressive stresses. The critical area therefore, for both is in this single tooth contact zone.

Surface contact stress is the criterion which effectively determines the pitch cylinder volumes of a pair of gears, i.e. their respective diameters squared multiplied by their face width. The compressive stress generated by the normal force between the teeth is determined by dividing this force by the meshing face width and the relative radius of curvature at the contact point which varies as it progresses from the beginning to the end of the path of contact. This is because relative radius is the product of the respective tangent lengths to the contact point divided by their sum, i.e. the constant length of the common tangent. For a given common tangent length, the product of respective pinion and wheel tangent lengths would be a maximum if they were equal. Clearly, this would only happen if the pinion and wheel were of the same size. It follows that relative radius of curvature is minimum at the lowest point of contact between the wheel tip diameter in the root of the pinion. However, this is in the double tooth contact zone and thus the chosen load point for calculating the highest surface stress is at the lowest point of single tooth contact on the pinion flank.

The criterion calculated as above is known as the Sc factor whose value is directly proportional to torque. It is therefore, valid for directly comparing load capacity taking into account any linear application and service factors (factors of ignorance!). While superficially, it has the dimensions of stress, in fact it is necessary to take the square root of Sc (after it has been multiplied by the various factors) then further multiplying this by a constant (190 for N/mm^2) or (2290 for lb/in^2) to get the "actual" compressive stress. The reason for the non-linear relationship between load and stress is that the contact area increases as it flattens so that if load is increased by a factor of 4, stress is only doubled. Most international design standards use this as their surface stress criterion. This leads to the anomaly that an acceptable surface safety factor based on stress is the square root of the associated Sc and bending safety factors directly related to load.

Historically, a simplified surface criterion known as the "K" factor, has been universally used for gear design. In effect, it is similar to Sc but as an approximation, it takes the pitch point as the chosen load point and further simplifies calculation by treating the sine and cosine of pressure angle as constants. Arbitrary limits for K may then be used as appropriate, for different applications, gear materials, pressure angles, etc. Using this approach, it is much easier to relate the volume of gears directly to the torque and ratio in a particular application viz.

$$fd^2 = \left(1+\frac{1}{n}\right)\frac{T}{K} \tag{1}$$

$$fd_w^2 = (n+1)\frac{T_w}{K} \tag{2}$$

where K is the surface criterion, n the wheel/pinion ratio, f the face width, d the pinion pitch diameter, d_w the wheel pitch diameter, T the pinion torque and T_w is the wheel torque.

The chosen load point for calculating bending stress in both pinion and wheel, is their respective highest point of single tooth contact, i.e. one base pitch from either end of the contact path as appropriate. Figure 3 shows the angle at which the normal tooth load at this point crosses the centre line of the tooth.

This load is then resolved into its tangential and radial components which respectively, create bending and direct compressive stresses in the tooth root. The resultant maximum tensile and compressive root fillet stresses, in particular the tensile, may be determined for the actual pitch, face width, etc. by an iteration process which takes account of the precise geometric shape of the fillet and the associated stress concentration factor. The results are then compared with the permissible tensile fatigue limit suggested by the Goodman diagram as shown in Fig. 4. This shows that when the load is unidirectional the mean stress is half the tensile

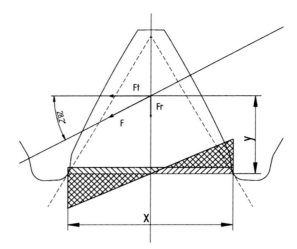

Figure 3: Highest point of single tooth contact.

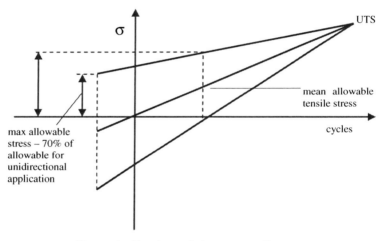

Figure 4: Goodman fatigue stress diagram.

maximum with an alternating range from zero to the maximum. In the case of an idler such as an epicyclic planet as shown later, tooth load reverses as it alternately meshes with the sun and annulus. The mean root stress obtained by taking the algebraic sum of the tensile and compressive stresses divided by 2 is therefore, negative and while this leads to a greater permissible alternating range about the mean, the allowable tensile maximum stress is only some 70% of that of the limit for unidirectional application. As a first approximation, for a given gear volume, fillet stresses are inversely proportional to pitch, i.e. if pitch is doubled, stress is halved.

Again, historically, this has led to a simple criterion for tooth bending known as the "C" factor. As for the "K" factor this can be arranged as a volumetric expression viz.

$$C = \frac{T}{fdm} \tag{3}$$

where m = module = d/N. Then,

$$fd^2 = \frac{TN}{C} \tag{4}$$

where f, d and T are as before and N is the number of teeth.

By equating the surface and bending volumes derived as above, it is possible to obtain a non-dimensional "optimum" number of pinion teeth based on the balance of bending to surface criteria and the gear ratio viz. By comparing eqns (1) and (4), it yields,

$$N = \left(1 + \frac{1}{n}\right)\frac{C}{K} \tag{5}$$

Since tooth number is unaffected by face width to diameter ratio or torque, it is only necessary to choose a rounded down number compatible with the nearest standard pitch and the required face width, diameter and ratio. Thus, root fillet stress is not usually a limiting criterion because the pitch is easily increased by reducing the number of teeth. Nonetheless, there are big incentives for making pitch as fine as possible.

1. A smaller pitch with bigger tooth numbers has a somewhat greater contact ratio but a shorter path of contact and commensurately lower tooth sliding velocities. This improves efficiency and reduces sliding losses and associated surface related problems such as scuffing. It also reduces surface stress slightly by increasing the relative radius at the chosen load point.
2. For a given load, the reduced bending moment on the root of a shorter tooth means a thinner rim is required for its support. This is very important in epicyclic gears as described later.

For simplicity, the foregoing consideration of gear geometry is confined to spur rather than helical gears. The latter also embody tip relief to mitigate pitch error problems as the teeth enter and leave the contact zone. However, experience suggests that although they are generally quieter than spurs, they are both subject to the same problems associated with the effects of parasitic loads and deflections and the helix corrections required to compensate for them. Unfortunately, such corrections only work for one condition. Attempts to cater for varying conditions by crowning the teeth, inevitably lead to higher stresses.

3 Geartrains

It is not practical to have a single stage gearbox to provide a step-up ratio of 100:1. In practice, such an overall ratio invariably requires three stages. To minimise size

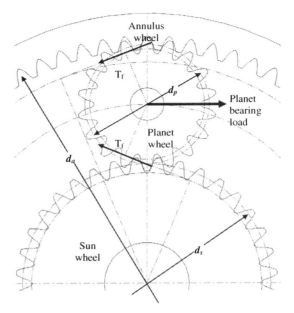

Figure 5: Half section epicyclic gear.

and weight, particularly in the first two, high torque low speed stages, it is usual to employ epicyclic gearing in which load is shared via three or more parallel load paths. As shown in Fig. 5, such gears have the further advantage of having co-axial input and output shafts rather than the offset parallel axes of a simple wheel and pinion.

The simplest form of epicyclic gear comprises three co-axial elements; a sun wheel, a planet carrier, which provides a straddle mounting for a number of equi-spaced planet wheels and an internally toothed ring gear or annulus. The figure shows that the planet wheels serve as idlers (no residual torque) between the sun and annulus wheels. If the planet carrier is fixed, the sun and annulus rotate in opposite directions, with the sun rotating at $-R$ times the speed of the annulus where

$$R = \frac{N_a}{N_s} = \frac{d_a}{d_s}$$

(6)

where N_a and N_s are the teeth numbers, and d_a and d_s are the pitch diameters of the annulus (ring) and sun wheel, respectively.

It can be seen that the carrier has a torque reaction equal and opposite to the sum of the sun and annulus torques. From this, it can be inferred that if the annulus is fixed then the sun will rotate at $+(R + 1)$ times the speed of the carrier and in the same direction.

Conventionally, most simple epicyclic gears have three planets and to ensure equal load sharing the sun is allowed to float so that it can find an axis which ensures its equilibrium and compensates for the collective errors in the concentricity of the respective axes of the sun wheel, planet carrier and annulus. This therefore requires a suitable flexible coupling to transmit the sun wheel torque.

Various solutions have been used to provide load sharing for epicyclic gears having more than three planets. The most widely used have employed a flexible annulus ring which subject to its tooth forces deflects as shown in Fig. 6. However, the maximum number of planets is usually limited to 6, because with greater numbers, load sharing becomes less effective as the deflections decrease. Even though more planets enable the ring thickness and weight to be appreciably reduced, it is not enough to give the required deflections without excessive stresses. In addition, the planet spindles are straddle mounted in a carrier which requires rigid webs between the planets to try and minimise its torsional wind up and the mal-distribution across the meshing faces of the planet wheel teeth.

The main problem associated with flexible annulus rings is that even with constant torque, they are subject to fully reversed cyclic bending stresses due to the outward and inward deflections, with the passage of each planet (see Fig. 7).

The most logical location for flexibility is in fact the planet spindle. Because it serves as a mounting for an idler with zero torque, the relative load on the spindle is always in a constant direction, whether or not the carrier is rotating. It follows therefore that subject to constant torque, deflection is static, and not subject to a

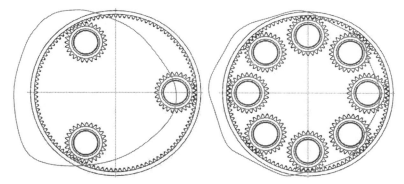

Figure 6: Annulus ring bending deflections. The deflection curves should not be offset laterally but located symmetrically so that they show the radial inward and outward distortions of the respective 3 and 8 planet annulus rings from their circular shapes.

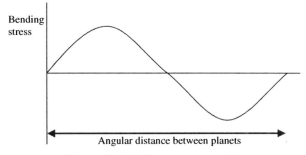

Figure 7: Cyclic stress reversals.

primary fatigue condition. However, when torque is variable, there is clearly an associated secondary unidirectional fatigue condition in the planet spindles as well as the gear teeth. In any case, even with constant torque, sun and annulus gear teeth are subject to unidirectional fatigue as they rotate, in and out of mesh with the planet wheel whose teeth are subject to full fatigue load reversals as they alternately mesh on opposite flanks with the sun and annulus. Therefore, all gears whether epicyclic or otherwise, have to be designed to accept primary fatigue loads as well as the secondary effects of torque fluctuation.

As stated previously, conventional planet carriers cannot be made completely rigid so that inevitably, the webs joining the two flanges which support either end of the planet spindle are subject to shear and bending deflections that create a torsional deflection of one flange with respect to the other to misalign the planet wheel. While it is feasible to calculate this deflection and compensate for it either by boring the carrier skewed or by helix corrections on the mating gears this only helps at one nominal torque. It is therefore, usual to crown the face widths of the gear teeth to avoid edge contacts on either end which would otherwise occur at different loads. This reduces the contact area and increases the local stresses. Furthermore, the planet bearing load is no longer on the centre of its spindle which can also be a source of bearing problems.

Figure 8 shows the principles of the compound cantilever flexible planet spindle comprising a flexible inner member and a comparatively rigid co-axial outer sleeve. Central tooth loads at the planets sun and annulus mesh points create equal and opposite moments at either end of the inner pin with a point of inflection at the centroid of load, where the bending moment is zero. The spindle is very soft in an angular sense to such an effect that it cannot sustain any unequal loading across either gear face, e.g. if a planet wheel has a helix error which could lead to heavier loads at opposite ends of its respective face contacts with the sun and annulus, then the

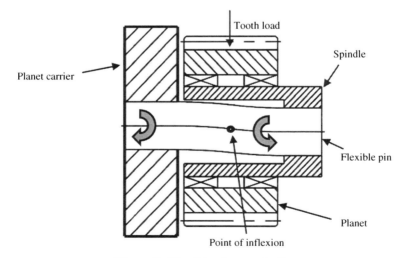

Figure 8: Flexible planet spindle.

effective centroid of its tangential loads would still be at the midpoint of the spindle. However, the associated radial components of tooth load due to pressure angle create a tilt in the radial plane, which reduces the tipping couple and the mal-distribution of load by which it is generated, so that the planet adopts a skewed equilibrium attitude with a low mal-distribution commensurate with its very low angular rigidity. The crossed helix effect created by having non-parallel axes leads to a notional point contact rather than line contact on the tooth faces. Since the crossed helix angle is very small it relieves the tendency for edge contacts in a manner analogous to crowning. The flexible spindle has proved conclusively that it can compensate for helix errors of different magnitude and hand in sun, planet and annulus by tilting in a complex way to a position of minimum strain energy to enable the planet wheel to avoid the load mal-distributions that are imposed by a more rigid support. In simple terms, the planet dictates where it wants the spindle to be rather than vice versa. Unlike a flexible annulus, the planet spindles are all independent of one another so they are all free to do their own thing and because they are cantilevers, the only limit on the number of planets is the clearance of their adjacent tip diameters and the annulus to sun ratio. As this ratio varies from 2.15 to 5.2 the number of planets reduces from 8 to 4. For bigger ratios than 5.2 only 3 can be accommodated.

A larger number of equally loaded planets directly reduce the overall volume of an epicyclic geartrain. This is shown by deriving a similar volumetric expression as that shown above for a simple parallel shaft pinion and wheel viz.

It can be seen in Fig. 5 that the relationship of three pitch diameters can be expressed as

$$d_p = \frac{d_a - d_s}{2} \tag{7}$$

Epicyclic analogy gives

$$n = \frac{d_p}{d_s} = \frac{R-1}{2} \tag{8}$$

Noting that

$$1 + \frac{1}{n} = 1 + \frac{2}{R-1} = \frac{R+1}{R-1} \tag{9}$$

Thus,

$$N_s = \left(\frac{R+1}{R-1}\right)\frac{C}{K} \tag{10}$$

$$N_a = N_s R = R\left(\frac{R+1}{R-1}\right)\frac{C}{K} \tag{11}$$

$$f d_s^2 = \frac{T_s}{QK}\left(\frac{R+1}{R-1}\right) = \frac{T_c}{QK(R-1)} \tag{12}$$

and

$$fd_a^2 = fd_s^2 R^2 = \frac{T_c R^2}{QK(R-1)} \tag{13}$$

where d_p is the planet wheel pitch diameter, d_s the sun wheel pitch diameter, d_a the annulus pitch diameter, f the sun wheel face width, N_s the sun wheel tooth number, C the planet tooth bending criterion, K the sun wheel surface criterion, T_s the sun wheel torque, T_c the planet carrier torque and Q is the number of planets.

From Fig. 5 it can be seen that in effect, an annulus has a negative diameter exemplified by the concave flanks on internal teeth. This means that given the same pressure angles, the product of the annulus and planet base tangent lengths is $-R$ times that of the planet and sun whereas the sum of the respective base tangent lengths are equal and opposite so that algebraically, the relative radius of curvature at the planet/annulus mesh point is precisely R times that of the planet/sun. Given the same face widths its K value is reduced accordingly by the reciprocal of R. The internal tooth root thickness is also somewhat thicker due to its concavity so that lower grade material and/or a smaller face width may be used.

The significance of the above is illustrated by comparing the annulus volumes of two planetary gears having the same carrier torque and sun wheel surface stress but with R equal to either 2 or 3, i.e. planetary ratios of 3 and 4 having either 8 or 5 planets respectively viz.

$$fd_a^2 = T_c(4/8) = 0.5T_c \tag{14a}$$

or

$$fd_a^2 = T_c(9/10) = 0.9T_c \tag{14b}$$

The larger ratio annulus is therefore, 1.8 times the volume!

The comparable volumes of a simple wheel subject to the same torques, surface stress and ratios are viz.

$$fd_w^2 = T_c(3+1) = 4T_c \tag{15a}$$

or

$$fd_w^2 = T_c(4+1) = 5T_c \tag{15b}$$

Without considering the pinion offset, the first is 8 times and the second 5.56 times the volumes of the annuli of the alternative planetary gears.

Even with only three planet wheels, the volume of the annuli is always 30% of an equivalent parallel shaft wheel for any ratio from 3 to 12.

4 Bearings

Rolling element bearings are the type most commonly used in wind turbines for both parallel shaft and epicyclic gears. Generally, the design criteria for such bearings leads to a finite life which takes account of the total number of hours at varying loads. The most heavily loaded are in the high torque low speed primary trains and in particular the planet spindles which sustain the double tooth loads on the planet meshes with the sun and annulus. The most successful arrangement has been a pair of preloaded taper roller bearings which ensure that at light loads there is no risk of skidding.

To maximise the bearing space available between the bore of the planet and the spindle especially for low annulus/sun ratios it helps to have fine pitch teeth to increase the root diameter, reduce rim thickness and increase the bore. It also helps if roller outer races are embodied in the planet bores. Timken have gone further by also integrating the inner races in the planet spindle and using full complement preloaded tapered rollers. All planet bearings together with all other lower loaded higher speed bearings in the secondary trains require a pressurised supply of lubricant. No bearing should be subjected to misalignment and self-aligning bearings should be avoided. They cannot be effectively preloaded because they have clearances which may lead to skidding on low loads.

In this context, the flexible planet spindle ensures that however much the torque may transiently vary, the bearing load always stays in the same place, i.e. the plane of the face width centres so that it is equally shared when two or more bearings are required to carry the load.

For smaller gears it is quite possible to have fully floating suns and annuli whose dead weight can, without detriment, be supported on their gear tooth meshes but generally not for planet carriers. As power increases, the tooth force to component weight diminishes and there comes a point where annulus rings and even sun wheels have a significant effect on load sharing and need support.

5 Gear arrangements

As shown in Fig. 9 the most commonly used arrangement employs two planetary step-up gear stages (with fixed annuli) coupled in series with the secondary sun wheel driving a parallel shaft wheel via a double tooth type coupling. This wheel meshes with a pinion having a parallel offset determined by the required location of the generator which it drives via a proprietary spacer type coupling. The primary reason for the offset is to provide a co-axial access to the turbine rotor from the rear of the gearbox for pitch control purposes, e.g. electrical slip rings.

Figure 10 shows an arrangement of the epicyclic stages featuring a star/planetary differential with its input torque divided between the annulus of a primary star stage and the planet carrier of a secondary differential stage whose annulus is coupled to the primary sun wheel. Thus the primary planet carrier is the sole static torque reaction member of the combined trains, while the secondary differential sun wheel is the output coupled to the parallel shaft wheel. The significance of this is that the torque reaction is no longer transmitted to the gear case via a live gear such as an annulus. This reduces structure-borne vibrations particularly when flexible planet spindles are used.

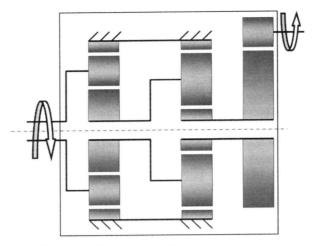

Figure 9: Conventional 3 train arrangement.

Fully floating torque reaction arm

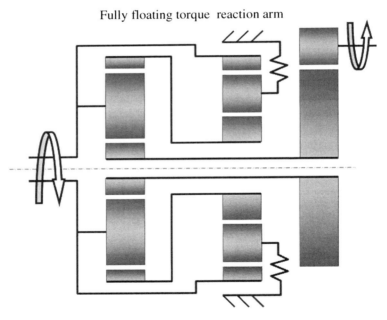

Figure 10: Star/planetary differential arrangement.

It is usual to mount a brake disc on the output shaft of the gear. This has two functions, first as a parking brake and secondly to stop the turbine in an emergency. The second function generally imposes up to three times the nominal full power torque on the drive train. However, in the light of earlier comments, this is quite probably no worse than the torque fluctuations it experiences in normal operation.

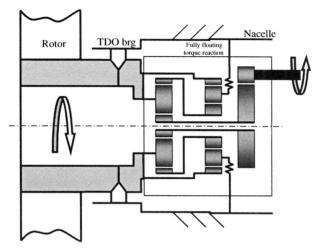

Figure 11: Independent rotor support arrangement.

The most critical aspects affecting reliability are the mounting of the gearbox and the coupling of the turbine shaft to its input. Hitherto, a majority have had the turbine shaft supported at its front by a single bearing mounted on the nacelle bed plate while its rear end has been supported by the gearbox input shaft to which it is rigidly coupled via a shrink fit coupling. The rear end of the turbine shaft is therefore, supported by the gear case and its resilient mounts via the input shaft bearings. This has created detrimental parasitic loads on the gearbox due to the pitching and yawing couples and associated shaft bending deflections plus deflections in the mounts due to torque fluctuations. In the light of the problems that have arisen from this situation, as powers have increased, most recent designs have featured large back to back taper roller bearings in TDO configuration to independently support the turbine rotor in a mounting frame. The gearbox input shaft is rigidly coupled to the rotor hub while the gear case torque reaction is supported by a suitable mechanism designed to impose only pure torque (see Fig. 11). In effect the rotor hub supports the gearbox, not vice versa.

6 Torque limitation

In its simplest form, the differential properties of an epicyclic gear can be exploited by allowing what would otherwise be its fixed reaction member to rotate in the direction imposed by its torque. This is effected by gearing it to a fixed stroke positive displacement pump with its delivery bypassed to its inlet via a pressure relief valve to give a limited slip at a controlled torque. Such a gear is best used on the final low torque high speed stage where the component sizes are much smaller and more manageable. This has been used very successfully by the Windflow company in New Zealand for 500 kW turbines driving synchronous generators. They have

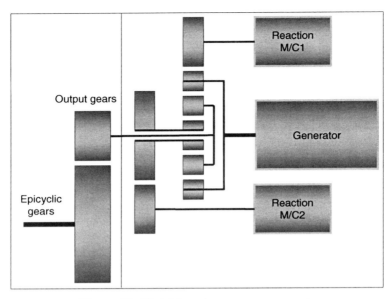

Figure 12: Variable ratio gear arrangement.

found that at this size, it has worked quite successfully without heat dissipation problems with a limited slip of up to 5%.

For larger powers, to provide better control and a bigger speed range with greater energy capture without excessive losses, it is necessary to control the reaction with a closed loop bypass branch comprising either a hydraulic pump and motor or an electrical equivalent to recover the power which would otherwise be lost. With such a system, torque may be monitored to enable transient referred inertia effects to be eliminated. Variable ratio gears using this principle have been successfully developed for powers up to 3.6 MW with synchronous generators driven by turbines with speeds ranging from 60 to 100%. In effect, such gears allow turbine speeds to increase when subject to a transient torque increase so that the excess torque is absorbed by the increased kinetic energy in the rotor while the excess speed is absorbed by the reaction member. Conversely, when the turbine torque has a transient decrease its speed can be reduced by a ratio change to recover the kinetic energy. For more sustained changes the gear ratio is changed accordingly (see Fig. 12).

7 Conclusions

The purpose of this chapter is to show how the transient torque/speed characteristics of a wind turbine affects the volume/weight of the drive train and the benefits that accrue due to the use of epicyclic gears not only for reducing weight and increasing compliance but also for their differential torque limiting properties.

It also emphasizes the importance of isolating the gearbox from the parasitic forces imposed by the turbine on its rotor support.

The volumetric concept facilitates the synthesis of the initial design of gears rather than using an analytical/iterative approach. It helps to optimise the overall size and weight of gears by showing the value of using lower ratios in the high torque low speed stages of high ratio applications, particularly when epicyclic trains are involved. Ultimately, all stress criteria are subject to arbitrary limits embodying a string of "factors of ignorance" which tend to be treated as virtual constants. Ten such factors, with a 5% increase in each, would reduce permissible load by 40%!

CHAPTER 16

Tower design and analysis

Biswajit Basu
Trinity College Dublin, Ireland.

This chapter addresses some of the design and analysis issues of interest to structural and wind engineers involved in ensuring the serviceability and survivability of wind turbine towers. Wind turbine towers are flexible multi-body entities consisting of rotor blades which collect the energy contained within the wind, and the tower which supports the weight of the rotor system and nacelle and transfers all gravity and environment loading to the foundation. Two themes on the design and analysis aspects of the tower have been presented. The first is the mathematical representation of the behaviour of wind turbine towers when subjected to wind loading and the second is the suppression of the vibrations caused by this wind action. The first theme focuses on a series of mathematical models representing the rotor blades, the tower with the added mass of the nacelle, and the coupled rotor blade and tower system which are used to determine the free and forced vibration characteristics of the structure. Response estimation for the rotating blades includes the effects of centrifugal stiffening, dynamic gravity effects due to rotation and rotationally sampled turbulence. A gust factor approach is also presented for design of the wind turbine towers. The second theme considers the mitigation of vibrations under dynamic wind action by adding energy dampers to the system, and finding the optimal properties of these dampers in order to maximise the reduction of vibration. Modelling and analysis of offshore towers have also been discussed.

1 Introduction

With the exponential growth in the wind energy market, turbines with larger rotor diameter and hence taller towers are becoming more common. This has a crucial impact on the design and analysis of wind turbine towers. The primary function of the wind turbine tower is to elevate the turbine rotor for a horizontal axis wind

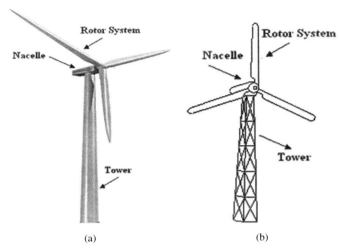

(a)   (b)

Figure 1: (a) Free standing tubular wind turbine tower; (b) lattice wind turbine
           tower.

turbine (HAWT) and support the mechanical and electrical system housed in the
nacelle. Wind speed increases with altitude and also tends to become less turbulent. As a result more energy can be extracted with taller towers. However, this
comes at a price of higher cost of construction and installation. Choice of tower
height is based on a tradeoff between increased energy production at a particular
site and the increase in the cost of construction.

   The principal types of towers currently in use are the free standing type using
steel tubes (Fig. 1a), lattice (or truss) towers (Fig. 1b) and concrete towers. For
smaller turbines, guyed towers are also used. Tower height is typically 1–1.5
times the rotor diameter. Tower selection is greatly influenced by the characteristics of the site. The stiffness of the tower is a major factor in wind turbine
system dynamics because of the possibility of coupled vibrations between the
rotor and tower. In addition, there are several other factors which affect the selection of the type of tower and its design, such as the mode of erection and fabrication, sizes of crane required for construction, noise, impact on avian population
and aesthetics. Among the different type of towers, tubular towers are more common and they are also preferable due to aesthetics and in minimizing impact on
avian population.

   One of the primary considerations in the tower design is the overall tower stiffness, which in turn affects its natural frequency. From a structural dynamics point
of view, a stiff tower whose fundamental natural frequency is higher than that of
the blade passing frequency (rotor's rotational speed times the number of blades)
is preferable. This type of tower has the advantage of being relatively unaffected
by the motions of the rotor-turbine itself. However, the cost may be prohibitive due
to a larger mass and hence more material requirement.

Towers are usually classified based on the relative natural frequencies of the tower and the rotor blades. Opposite to the stiff towers, soft towers are those whose fundamental natural frequency is lower than the blade passing frequency. A further subdivision differentiates a soft and a soft–soft tower. A soft tower's natural frequency is above the rotor frequency but below the blade passing frequency while a soft–soft tower has its natural frequency below both the rotor frequency and the blade passing frequency. These kinds of towers (soft and soft–soft) are generally less expensive than the stiffer ones, since they are lighter. However, they require particular attention and need careful dynamic analysis of the entire system to ensure that no resonances are excited by any motions in the rest of the turbine.

2 Analysis of towers

2.1 Tower blade coupling

Design engineers are interested in understanding and analyzing the coupled dynamics of wind turbine towers with associated components, especially with proliferation of such systems worldwide for renewable energy production. As wind turbines are becoming larger in size and are being placed in varying global wind environments, knowledge of the dynamic behaviour is important. The behaviour of the subcomponents of the system (the tower and rotor blades) as well as the dynamic interaction of those components with each other is vital to ensure the serviceability and survivability of such expensive power generating infrastructure. Following a conventional and simplified design analysis, the mass of the components (nacelle and rotor blades) can be simply lumped at the top of the tower, and as long as the fundamental frequencies of the tower and blades are far apart, a stochastic forced vibration analysis could be carried out. While the simplicity of this is attractive, the flexibility of large rotor systems may result in either economically inefficient design due to the conservatism required to accommodate the uncertainties of component interaction or an unsafe design due to ignoring the coupling effects.

Published literature available regarding the dynamic interaction of wind turbine components, especially from the point of view of the structural design of the tower with the interaction of the mechanical rotor blade system is growing. Harrison *et al.* [1] state that the motion of the tower is strongly connected to the motion of the blades, as the blades transfer an axial force onto the low speed drive shaft which is ultimately transferred into the nacelle base plate at the top of the tower.

The dynamic characteristics of a multi-body system have traditionally been determined by the substructure synthesis or component mode synthesis method [2, 3]. In coupled analyses, it is first necessary to obtain the free vibration characteristics of all sub-entities, prior to dynamic coupling. The free vibration properties of a tower carrying a rigid nacelle mass at the top may be evaluated by techniques such as the discrete parameter method, the finite element method or by using closed form solutions. The discrete parameter method was used by Wu and Yang [4] in a study on the control of transmission towers under the action of stochastic wind loading. Lavassas *et al.* [5] also used this technique to assess the accuracy and reliability of more computationally expensive finite element analyses of wind turbine tower. Recent studies using the finite element technique for free vibration analyses of structures in wind engineering include Bazeos *et al.* [6] and Dutta *et al.* [7]. Murtagh *et al.* [8] derived an expression in closed form to yield the eigenvalues and eigenvectors of a tower-nacelle system comprising of a prismatic cantilever beam with a rigid mass at its free end.

2.2 Rotating blades

The free vibration properties of realistic wind turbine blades are computationally more difficult to obtain, and models are usually mathematically complicated due to the complex geometry of the blade and the effects of blade rotation. Baumgart [9] used a combination of finite elements and virtual work, accounting for the complex geometry of the blade to obtain the modal parameters. Naguleswaran [10] proposed an approach to determine the free vibration characteristics of a spanwise rotating beam subjected to centrifugal stiffening. This model [10] can be used in many industrial fields, such as wind turbine blades, aircraft rotor blades and turbine rotor blades. Naguleswaran [10] and Banerjee [11] both used the Frobenius method to obtain the natural frequencies of spanwise rotating uniform beams for several cases of boundary conditions. Chung and Yoo [12] used the finite element method to obtain the dynamic properties of a rotating cantilever, whereas Lee *et al.* [13] carried out experimental studies on the same. All studies indicate that the natural frequencies rise as the rotational frequency of the blade increases. Various software codes have been developed by engineers to dynamically analyse the various components of a wind turbine tower. Buhl [14] presented guidelines for the use of the software code ADAMS in free and forced vibrations of wind turbine towers.

Under the action of rotation, the free vibration parameters of the blades are affected by two axial phenomena. The first is centrifugal stiffening and the second is blade gravity (self weight) effects. In order to find the free vibration properties of the blades, each blade can be discretized into a lumped parameter system comprising of 'n' degrees of freedom. The eigenvalues of a blade undergoing flapping motion may be obtained from the eigenvalue analysis:

$$\left| [K_{B}{}'] - \omega_{B}{}^2 [M_{B}] \right| = 0 \tag{1}$$

where $[K_{B}{}'] = [K_{B} + K_{BG}]$ represents the modified stiffness matrix due to the geometric stiffness matrix $[K_{BG}]$, accounting for the effect of axial load, ω_{B} is the natural frequency, $[K_{B}]$ is the flexural stiffness matrix and $[M_{B}]$ is the mass matrix. The mass matrix may be formulated as a diagonal matrix with the mass m_i at each discrete node i.

The geometric stiffness matrix contains force contributions due to blade rotation which are always tensile, and contributions from the self weight of the blade, which may be either tensile or compressive, depending on blade position. The geometric stiffness matrix is

$$[K_{BG}] = \begin{bmatrix} \dfrac{N_1}{l_1} & \dfrac{-N_1}{l_1} & \cdots & & 0 \\[2ex] \dfrac{-N_1}{l_1} & \dfrac{N_1}{l_1} + \dfrac{N_2}{l_2} & \cdots & & 0 \\[2ex] \vdots & \vdots & \ddots & \dfrac{-N_{n-1}}{l_{n-1}} & \\[2ex] 0 & 0 & \dfrac{-N_{n-1}}{l_{n-1}} & \dfrac{N_{n-1}}{l_{n-1}} + \dfrac{N_n}{l_n} \end{bmatrix} \qquad (2)$$

where N_i is the axial force at node 'i' and l_i is the length of beam segment between the nodes 'i' and '$i + 1$'. The magnitude of the tensile centrifugal axial force, $CT(x)$, along the axis of a continuous blade, may be found from the expression given by Naguleswaran [10] as

$$CT(x) = 0.5\bar{m}_B \Omega^2 (L_B + 2L_B R_H - 2R_H x - x^2) \qquad (3)$$

where $\bar{m}_B$ represents the mass per unit length of the blade, Ω is the rotational frequency of the blade, and x is the distance along the blade from the hub. This continuous force distribution is discretized into nodal values (CT_i) and used to form the geometric stiffness matrix. The component of nodal blade gravity force (self weight), G_i, acting axially may be obtained from geometry and depends on the angle θ that the longitudinal axis of the blade makes with the horizontal global axis, in the plane of rotation. Values of N_i are obtained from the expression:

$$N_i = CT_i \pm G_i \qquad (4)$$

with the sign convention that tensile forces are positive and compressive forces are negative.

2.3 Forced vibration analysis

Forced vibration analyses of structures may either be carried out in the time or frequency domain, with each having its own distinct merits. Analysis through the time domain allows for the inclusion of behavioural non-linearity and response coupling. Due to limited availability of actual input time-histories as measured in the field, the designer has to generate relevant artificial time-histories using widely published spectral density functions. The method for generating the artificial time-histories can be divided into three categories, the first based on a fast Fourier transform (FFT) algorithm, the second based on wavelets and other time–frequency algorithms and the third based on time-series techniques such as Auto-Regressive Moving Average (ARMA) method. Suresh Kumar and Stathopoulos [15] simulated both Gaussian and non-Gaussian wind pressure time-histories based on the FFT algorithm. Kitagawa and Nomura [16] recently used wavelet theory to generate wind velocity time-histories by assuming that eddies of varying scale and strength may be represented on the time axis by wavelets of corresponding scales.

In an investigation on the buffeting of long-span bridges, Minh *et al.* [17] used the digital filtering ARMA method to numerically generate time-histories of wind turbulence.

In simulating drag force time-histories on the tower, information on spatial correlation, or coherence is necessary to be included. Coherence relates the similarity of signals measured over a spatial distance within a random field. Coherence is of great importance, especially if gust eddies are smaller than the height of a structure. Some of the earliest investigations into the spatial correlation of wind forces were carried out by Panofsky and Singer [18] and Davenport [19] and later augmented by Vickery [20] and Brook [21]. Recent publications involving lateral coherence in wind engineering include Højstrup [22], Sørensen *et al.* [23] and Minh *et al.* [17].

2.4 Rotationally sampled spectra

In order to simulate the drag force time-histories on the rotating blades, a special type of wind velocity spectrum is needed. Connell [24] reported that a rotating blade is subjected to an atypical fluctuating wind velocity spectrum, known as a rotationally sampled spectrum. Due to the rotation of the blades, the spectral energy distribution is altered, with variance shifting from the lower frequencies to peaks located at integer multiples of the rotational frequency. Kristensen and Frandsen [25], following on from work by Rosenbrock [26], developed a simple model to predict the power spectrum associated with a rotating blade, and this was significantly different to a spectrum without the rotation considered. Though literature on this topic is limited, Madsen and Frandsen [27], Verholek [28], Hardesty *et al.* [29] and Sørensen *et al.* [23] are some relevant references on this topic.

Rotationally sampled spectra are used to quantify the energy as a function of frequency for rotor blades within a turbulent wind flow for representing the redistribution of spectral energy due to rotation. The required redistribution of spectral energy can be achieved by identifying the specific frequencies 1Ω, 2Ω, 3Ω, and 4Ω (Ω being the rotational frequency of the blades), and then deriving the Fourier coefficients for those frequencies according to specific standard deviation values. These values can be obtained based on some measurements or assumption related to the rotational turbulence spectra. Madsen and Frandsen [27] observed that the peaks of redistributed spectral energy in a rotationally sampled spectrum tend to become more pronounced as distance increases along the blade, away from the hub.

The typical rotationally sampled turbulence spectra are shown in Fig. 2 [30]. It has been assumed for the spectra that the variance values increase by an arbitrary value of 10%, for each successive blade node radiating out from the hub. It is also assumed that 30% of the total variance at each node is localized into peaks at 1Ω, 2Ω, 3Ω, and 4Ω (15%, 7.5%, 4.5% and 3% of the total energy is allocated to the different peaks). Nodal fluctuating velocity time-histories with specific energy–frequency relationships can be simulated from the spectra in Fig. 2 using a discrete Fourier transform (DFT) technique.

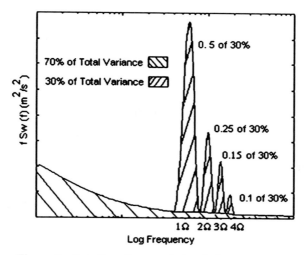

Figure 2: Rotationally sampled turbulence spectra.

Using the loading from the rotationally sampled spectra of turbulence and using a mode-acceleration method, Murtagh *et al.* [31] estimated the wind-induced dynamic time-history response of tapered rotating wind turbine blades. The mode-acceleration method was initially implemented by Williams [32] and Craig [33] reported that it has superior convergence characteristics compared to the mode-displacement method. Singh [34] presented a method for obtaining the spectral response of a non-classically damped system, based on the mode-acceleration technique. Akgun [35] presented an augmented algorithm based on the mode-acceleration method which has improved convergence for computation of stresses in large models.

2.5 Loading on tower-nacelle

The tower can be modelled as a lumped mass multi-degree-of-freedom (MDOF) flexible entity, which includes a lumped mass at the top of the tower, to represent the mass of the nacelle and the effect of the blades. An eigenvalue analysis can be performed to obtain the natural frequencies and mode shapes. As the tower-nacelle is a MDOF system, it is convenient to obtain modal force time-histories associated with each mode for analysis. This allows the spatial correlation or coherence of drag forces along the height of the tower to be included. Nigam and Narayanan [36] presented an expression for the modal fluctuating drag force power spectrum, for a continuous line-like structure, which can be used following modification for a discretized MDOF system [30].

The wind velocity auto and cross power spectral density (PSD) terms may be evaluated as

$$S_{V_{k}V_{l}}(f) = \sqrt{S_{V_{k}V_{k}}(f)S_{V_{l}V_{l}}(f)}\,\mathrm{coh}(k,l;f) \tag{5}$$

with $S_{vkvk}(f)$ and $S_{vkvk}(f)$ being the velocity PSD functions at nodes k and l respectively and $\text{coh}(k,l;f)$ is the spatial coherence function between nodes k and l. The terms $S_{vkvk}(f)$ and $S_{vkvk}(f)$ are functions of frequency f and may be calculated using the Kaimal spectra [37]. A coherence function suggested by Davenport [19], $\text{coh}(k,l;f)$, which relates the frequency dependent spatial correlation between nodes k and l, is represented as

$$\text{coh}(k,l;f) = \exp\left(-\frac{|k-l|}{L_s}\right) \tag{6}$$

where $|k-l|$ is the spatial separation and L_S is a length scale given by

$$L_S = \frac{\hat{v}}{fD} \tag{7}$$

with

$$\hat{v} = 0.5(\overline{v}_k + \overline{v}_l) \tag{8}$$

and D is a decay constant. The fluctuating component of the modal force acting on the tower may be obtained by employing the DFT technique. The mean nodal drag force component is obtained by transforming the nodal mean drag force time-histories into modal force time-histories using the modal matrix. The mean modal drag force is added to the modal fluctuating component to obtain the total modal drag force time-history.

2.6 Response of tower including blade–tower interaction

In order to couple the tower and rotating blades, equations of motion for the tower that includes the blade shear forces is necessary to be considered. This is represented by

$$[M_T]\{\ddot{x}(t)\}+[C_T]\{\dot{x}(t)\}+[K_T]\{x(t)\} = \{F_T(t)\}+\{V_B'(t)\} \tag{9}$$

where $[M_T]$, $[K_T]$ and $[C_T]$ are the mass, stiffness and damping matrices of the tower-nacelle respectively, $\{x(t)\},\{\dot{x}(t)\},\{\ddot{x}(t)\}$ are the displacement, velocity and acceleration vectors respectively, $\{F_T(t)\}$ is the total wind drag loading vector acting on the tower and $\{V_B'(t)\}$ is the effective blade base shear vector transmitted from the root of the rotating blades and acting at the top of the tower. The set of equations cannot be solved directly in time domain as the base shear is dependent on the motion of the tower (due to coupling) and hence is not known explicitly. An alternative way to solve the equations is to convert the set into a set of algebraic equations by FFT and subsequently solve by inverse FFT [30].

A numerical example [30] is presented for a steel wind turbine tower of height 60 m with three blades of rotor radius 30 m. The total mass of the nacelle and rotor system is 19,876 kg. The average wind speed at the top of the tower is 20 m/s. Figure 3 shows the displacement response time-history at the top of tower when

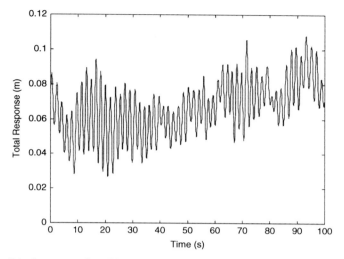

Figure 3: Displacement time-history at the top of the tower ignoring blade rotation.

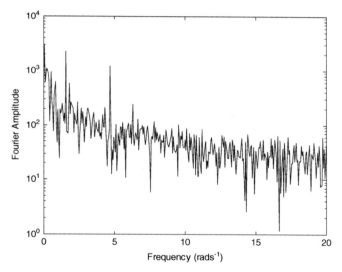

Figure 4: Fourier transform amplitude of wind velocity.

the blades masses are lumped on the top of the tower thus, ignoring the tower–blade interaction. The maximum observed tower tip response is 0.108 m.

The forced vibration response of the coupled tower–blade model is also calculated for a rotational frequency of 1.57 rad/s. Figure 4 presents a Fourier transform of the simulated fluctuating wind velocity acting at the tip of the blade. An increase in energy is clearly observable at integer products of the rotational frequency.

Figure 5 illustrates the computed blade tip displacement time-history. The maximum observed displacement is approximately 0.75 m. Figure 6 presents the total

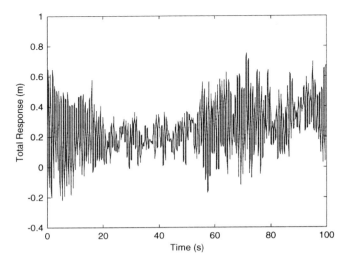

Figure 5: Blade tip displacement time-history.

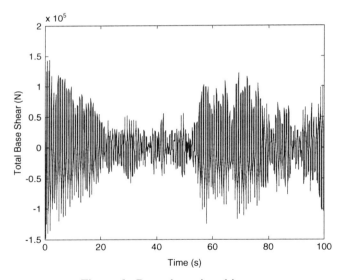

Figure 6: Base shear time-history.

base shear time-history due to the forced vibration of the three rotating blades. A maximum base shear force of nearly 150 kN is observed. The three rotating blades are now coupled to the tower-nacelle and the maximum tower tip displacement response is found to be 0.385 m, as presented in the displacement time-history in Fig. 7. Thus, inclusion of blade–tower interaction results in a 256% increase in peak tip displacement of the tower compared to the case excluding blade–tower interaction.

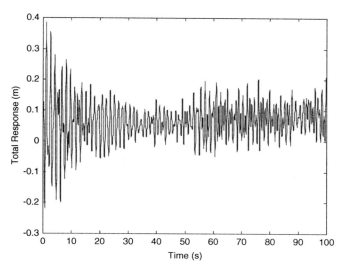

Figure 7: Displacement time-history at the top of the tower with blade interaction.

In the approach by Murtagh *et al.* [30], the coupled system equation of motion is primarily cast in the frequency domain via Fourier transform. This allows the coupling of the tower and the blades. The time domain along-wind response of the coupled assembly is ultimately obtained by inverse Fourier transform. There are a number of merits behind this type of approach. The technique is relatively simple, especially compared with a more computationally expensive finite element formulation. The approach may be used in a preliminary quantitative design, which may subsequently be validated by a more rigorous analysis. The dynamic properties of the coupled system are available using the dynamic properties of each of the two sub-systems, which is an extension of the substructure synthesis approach.

3 Design of tower

A complete dynamic analysis of the tower taking into account the effect of the rotation of the blades (rotors) and the nacelle mounted at the top is necessary for ensuring the safety and operational serviceability. However, such a detailed dynamic analysis may be time consuming and rigorous at a preliminary design stage when the initial configuration has to be chosen based on the design forces and displacements. Hence, for an initial assessment it may be more attractive to use an approximate simplified approach while taking account of the stochasticity in the wind loading (and hence in the response of the tower) and the rotor–tower interaction.

Gust response factor (GRF) approach is a simple technique used by structural engineers in the along-wind design of flexible structures and incorporates the stochastic and dynamic effects. This technique is now well developed due to the contributions of Davenport [38] and Velozzi and Cohen [39]. GRF is the ratio of the maximum or peak response quantity to the mean response quantity.

Hence, when this factor is applied to the responses from the mean wind loading, it yields the maximum design values.

The methodology developed by Davenport and, Velozzi and Cohen calculated the GRF using a ratio of displacements, and while this yielded accurate maximum expected response for displacement, it was found to fall short in providing estimates of other response parameters, such as bending moment and shear force. Following the work by [38, 39] several new models of the GRF have been proposed by Holmes [40] and Zhou and Kareem [41], with the latter being based on base bending moment, rather than displacement. The GRF methodology has also become the basis of most modern design codes worldwide [42].

3.1 Gust factor approach

The traditional Davenport-type GRF assumes that the flexible structure may be represented by a single degree-of-freedom (SDOF) representing the fundamental mode of vibration, and this is usually sufficient. However, if a structural system like a wind turbine tower (with coupled tower–rotor interaction) has more than one mode contributing to the response, the traditional GRF methodology may yield inaccurate representations of the energy contained in the response. Thus an extension of the traditional GRF methodology to include the effects of higher modes in the derivation of the GRF is required for application in the case of a wind turbine tower.

A GRF for evaluating the along-wind response of wind turbine towers has been proposed by Murtagh et al. [43]. The approach presented differs from the conventional GRF methods as the GRF contains contributions from two resonant modes, mainly due to rotor blade–tower interaction effects. The wind turbine tower model considered contains two inter-connected flexible sub-systems, representing the tower and a three-bladed rotor system. It is assumed that all the blades vibrate identically in the flapwise mode (out-of-plane) coupled with the tower. Each component is initially modelled as a separate degree-of-freedom (DOF) and these are coupled together to form an equivalent reduced order model of the coupled tower–rotor system considering the first two dominant modes. Thus, the resonant component of the response contains energy output from the two modes of the coupled system. This is an approximate way to account for the effect of the blades fed back to the tower including the coupled tower–blade interaction. The GRF is obtained for both tower tip displacement and base bending moment through numerical integration, with a closed form expression included for the former.

3.2 Displacement GRF

The displacement GRF [43], G_{DISP}, is obtained as a ratio of the expected maximum displacement response, $X_{MAX}(t)$ divided by the mean displacement, $\overline{x}$, with the latter being represented by the equation:

$$\overline{x} = \frac{\Phi_{CS,1\text{-}TT}\,\overline{f}_{D,1}}{\overline{K}_{CS,1}} + \frac{\Phi_{CS,2\text{-}TT}\,\overline{f}_{D,2}}{\overline{K}_{CS,2}} \tag{10}$$

with $\Phi_{cs,j-TT}$ ($j = 1, 2$) being the jth coupled system (CS) mode shape component at the top of the tower, $K_{CS,j}$ is the jth modal stiffness of the coupled system and $\overline{f}_{D,j}$ is the jth modal mean drag force. Because modal/generalized quantities are used in eqn (10), it is assumed that the free vibration parameters obtained from the tower–rotor system are from a classically damped one. The modal mean drag force on a structure (i.e. the tower or the blade) is obtained as

$$\overline{f}_{D,j} = \int_0^H \left(\frac{1}{2} \rho C_D(z) B(z) \overline{v}(z)^2 \right) \Phi_{CS,j}(z) dz \tag{11}$$

where H is the length over which drag is to be calculated (i.e. the total height of the tower or the length of the blade), $C_D(z)$ is the drag coefficient, $B(z)$ is the width of the tower (or blades), and $\overline{v}(z)$ is the mean wind velocity and $\Phi_{CS,j}(z)$ is the jth mode shape component of the coupled system, all as a function of the spatial variable z. The expected maximum displacement may be obtained as the product of a peak factor, Ψ (using first passage analysis, as in [44]) and the root mean square (RMS) of the displacement response at the top of the tower, σ_X. This RMS displacement response, which includes a second mode of vibration, may be obtained by taking the square root of the area under the displacement response PSD function, $S_{XX}(f)$ The PSD function $S_{XX}(f)$ is found as the sum of the products of the modal wind drag force PSD functions with their appropriate squared amplitude of the modal mechanical admittance functions [43].

The modal drag force PSD function may be obtained from the expression:

$$S_{MFjMFj}(f) = S_{VV}(f) \rho^2 \int_0^H \int_0^H C_D(z_1) C_D(z_2) B(z_1) B(z_2) \overline{v}(z_1) \overline{v}(z_2)$$
$$\Phi_{CS,j}(z_1) \Phi_{CS,j}(z_2) R(z_1, z_2; f) dz_1 dz_2 \tag{12}$$

where $S_{VV}(f)$ denotes the wind velocity PSD function at the top of the tower [37], ρ is the density of air, and $R(z_1, z_2; f)$ is the spatial coherence function between elevations z_1 and z_2 [19]. The mechanical admittance function at the top of the tower due to a unit force at that point for the jth mode may be obtained as

$$H_{D,j}(f) = \frac{\Phi_{CS,j-TT} F_{CS,j}}{4\pi^2 f_{CS,j}^2 \overline{M}_{CS,j} \left[1 - (f/f_{CS,j})^2 + 2i\xi_{CS,j}(f/f_{CS,j}) \right]} \tag{13}$$

where $F_{CS,j}$ is the jth modal force due to a unit force placed at the top of the tower, $f_{CS,j}$ is the jth natural frequency, $\overline{M}_{CS,j}$ is the jth modal mass ($\overline{M}_{CS,j} = \int_0^H m(z) \Phi_{CS,j}^2(z) dz$) with $m(z)$ as the mass distribution of the structure and, $\xi_{CS,j}$ is the jth modal damping ratio.

Two procedures have been proposed by Murtagh [43] based on how the value of σ_X may be calculated. It may be computed by numerically evaluating an integral or it may also be obtained in closed form based on some approximation. For the closed form calculation, a method of decomposition can be employed, in which it is assumed that the variance of the displacement response PSD function may be separated into two components: a background component and a resonant component. Contrary to the conventional GRF approach, in the proposed methodology [43], there are two contributions for the resonant component. The square of the non-dimensionalized form of background component of the gust factor G_B^2 can be expressed as

$$G_B^2 = \frac{\Phi_{CS,1-TT}^2 F_{CS,1}^2 \Psi^2}{16\pi^4 f_{CS,1}^4 \overline{M}_{CS,1}^2 \overline{x}^2} \int_0^\infty S_{MF1MF1}(f)df + \frac{\Phi_{CS,2-TT}^2 F_{CS,2}^2 \Psi_{CS,2}^2}{16\pi^4 f_{CS,2}^4 \overline{M}_{CS,2}^2 \overline{x}^2} \int_0^\infty S_{MF2MF2}(f)df \quad (14)$$

The integral in eqn (14) may be evaluated numerically, or by assuming the integrand to be a white noise, or from a known value of turbulence intensity.

The resonant component of the gust factor comprises of two non-dimensionalized terms representing contributions of the first and second modes of vibration, $G_{R,1}^2$ and $G_{R,2}^2$, respectively. These terms are given by the expressions:

$$G_{R,j}^2 = \frac{\Phi_{CS,j-TT}^2 \Phi_{CS,j}^2 S_{MFjMFj}(f_{CS,j})\Psi_j^2}{64p^3 f_{CS,j}^3 \overline{M}_{CS,j}^2 x_{CS,j} \overline{x}^2}, \quad j = 1,2 \quad (15)$$

where Ψ_j is the peak factor associated with mode 'j'. Thus, the closed form solution for the displacement GRF, $G_{DISP-CF}$, is obtained as

$$G_{DISP-CF} = 1 + \sqrt{G_B^2 + G_{R,1}^2 + G_{R,2}^2} \quad (16)$$

where G_B and $G_{R,j}$ represent the background and resonant components of the displacement GRF, respectively.

3.3 Bending moment GRF

A GRF also has been derived based on the bending moment GRF [41] at the tower base, G_{BM} by [43] which is presented for comparison. Similar to the displacement GRF, G_{BM} will contain contributions from two modes of vibration and is obtained as the ratio of the expected maximum base bending moment, $Y_{MAX}(t)$ $(=\Psi\sigma_{BM})$, to the mean base bending moment, $\overline{y}(\int_0^H 0.5\rho C_D(z)B(z)\overline{v}(z)^2 zdz)$. The RMS of the base bending moment, σ_{BM}, is obtained from the equation:

$$\sigma_{BM} = \left(\sum_{j=1}^2 \Gamma_j^2 \int_0^\infty S_{MFjMFj}(f)\left|H_{D,j}(f)\right|^2 df \right)^{1/2} \quad (17)$$

where Γ_j is given by

$$\Gamma_j = (2\pi f_{CS,j})^2 \int_0^H m(z)\Phi_{CS,j}(z)zdz \quad (18)$$

The base bending moment GRF, G_{BM} may be obtained as

$$G_{BM} = 1 + \Psi\frac{\sigma_{BM}}{\overline{y}} \quad (19)$$

Table 1: GRFs for SDOF lumped mass model.

$G_{DISP-NI}$	2.275
$G_{DISP-CF}$	2.291
G_B	1.019
$G_{R,1}$	0.792
G_{BM-NI}	2.429

Table 2: GRFs for coupled model with blade–tower interaction.

Ω (rad/s)	$G_{DISP-NI}$	$G_{DISP-CF}$	G_B	$G_{R,1}$	$G_{R,2}$	G_{BM-NI}
0.000	2.507	2.356	1.032	0.850	0.268	2.633
0.785	2.509	2.370	1.044	0.837	0.266	2.599
1.570	2.503	2.392	1.070	0.833	0.257	2.506
3.140	2.381	2.327	1.059	0.753	0.170	2.225

A series of numerical examples are presented from [43] to investigate the magnitude of GRFs obtained for the model which allows for blade–tower interaction, and these are compared with GRF values obtained from an equivalent SDOF model which ignores blade–tower interaction by lumping the mass of the blades in with that of the nacelle. A tower (steel) of height 50 m with rotor (GFR epoxy) diameter of 60 m is considered with the details available in [43]. Four different rotational frequencies of the rotor blades were considered. As rotational frequency of the blades increases, the fundamental frequency of the blades also increases, and this leads to increase in the natural frequencies of the coupled systems.

Tables 1 and 2 show the GRFs obtained for the lumped mass equivalent SDOF and two DOF tower–blade interaction models for a mean wind velocity of 20 m/s at the top of the tower. A time of 600 s was used to obtain the GRFs, as used in Eurocode 1 (CEN 2004) [45]. Included in these tables are the displacement GRFs obtained by numerical integration and in closed form, $G_{DISP-NI}$ and $G_{DISP-CF}$, respectively, and the base bending moment GRF obtained using numerical integration, G_{BM-NI}. It may be noted that the second mode affects the background and the resonant components and changes the response obtained from the classical gust factor approach.

It is evident from Tables 1 and 2 that the choice of modelling strategy, i.e. lumped mass SDOF or two DOF blade/tower interaction, has a bearing on the magnitudes of both the displacement and base bending moment GRFs obtained. When the blades are stationary ($\Omega = 0$ rad/s) in the two DOF case, the values of $G_{DISP-NI}$ and G_{BM-NI} obtained differ from the SDOF model values of $G_{DISP-NI}$ and G_{BM-NI} by over 10 and 8%, respectively. These differences remain nearly constant until the case of $\Omega = 3.14$ rad/s where they are equal to 5 and 8%, respectively.

The values of $G_{DISP-NI}$ and $G_{DISP-CF}$ showed a close match in most cases, though it was observed that when the two modes were closest together ($\Omega = 0$ rad/s),

$G_{\text{DISP-CF}}$ yielded a difference of 6% from $G_{\text{DISP-NI}}$. The difference in the value dropped to less than 1% when the modes move further apart at $\Omega = 3.14$ rad/s. It was also observed from Tables 1 and 2 that the displacement and bending moment GRFs obtained showed some disagreement, with the values of $G_{\text{BM-NI}}$ being higher than those of $G_{\text{DISP-NI}}$. The largest disagreements were observed at the single DOF model and the two DOF model case of $\Omega = 0$ rad/s, where differences of 7 and 5% were observed.

4 Vibration control of tower

As the wind turbines grow bigger in size and become flexible with the increase in rotor diameter, it is not only enough to estimate the design forces and ensure the safety of the wind turbine. Additionally, it is necessary to control the vibration response of the flexible wind turbine tower. It has been observed that wind-induced accelerations may be the reason for the unavailability of wind turbine with increased downtime and may cause damage to the acceleration sensitive subcomponents and devices in a wind turbine [46]. Hence, it is important to consider structural vibration control strategies for wind turbine towers for operational reliability of wind turbines.

Vibration control strategies for flexible and tall structures susceptible to large wind-induced oscillations in general are becoming increasingly important, particularly with the current tendency to build higher and lighter. HAWTs are no exception, having experienced a dramatic increase in scale in the past decade. This is particularly evident in offshore wind turbines, with rotor diameter measuring over 120 m. As the design approach is based on strength considerations, stiffness does not increase proportionally with increase in height and these flexible turbines may experience large-scale blade and tower deformations having non-linear characteristics, which may prove detrimental to the functioning of the turbine. Thus, there is distinct merit in investigating the vibratory control of both wind turbine blades, e.g. using blade pitch [47, 48] and towers [49], using an external energy damper.

Among the several structural vibration controllers available, tuned mass damper (TMD) as a passive vibration control device has become popular. It suppresses vibration by acting as an energy dissipator. Considerable amount of literature now exists on the use of TMDs for flexible structures [50–52]. Use of a TMD for suppression of vibration in a wind turbine tower including blade–tower interaction has been studied by Murtagh et al. [49]. They provided a simple analytical framework in order to qualitatively investigate the effect of a TMD on the fore-aft response of a wind turbine tower.

4.1 Response of tower with a TMD

The displacement response of a wind turbine tower including blade–tower interaction and rotationally sampled turbulence acting on the rotor blades, and with an attached TMD may be expressed as [49]:

$$[M_{\text{T}}]\{\ddot{x}(t)\} + [C_{\text{T}}]\{\dot{x}(t)\} + [K_{\text{T}}]\{x(t)\} = \{F_{\text{T}}(t)\} + \{V_{\text{B}}'(t)\} + \{F_{\text{DAMP}}(t)\} \quad (20)$$

where $[M_T]$, $[K_T]$ and $[C_T]$ are the mass, stiffness and damping matrices of the tower/nacelle, respectively, $\{x(t)\}, \{\dot{x}(t)\}, \{\ddot{x}(t)\}$ are the time-dependent displacement, velocity and acceleration vectors respectively, $\{F_T(t)\}$ is the total wind drag loading acting on the tower, $\{V_B{}'(t)\}$ is the effective blade base shear acting at the top of the tower and $\{F_{DAMP}(t)\}$ is the damping force brought about by the action of the TMD. Details on how to calculate the effective blade base shear time-histories and total wind drag loadings may be found in Murtagh et al. [30].

The response time-histories of the tower can be obtained following a modal decomposition of the tower response, transforming the set of equations in eqn (20) in a Fourier domain and subsequently applying an inverse FFT [49].

4.2 Design of TMD

For designing a TMD two important parameters need to be considered, the damping ratio and the tuning ratio. For an efficient performance of a TMD these two ratios need to be optimized.

A number of approximate and empirical expressions are available for the evaluation of the optimum damping ratio of the TMD. Given below is the simple expression by Luft [51] for the optimum damping ratio of the TMD:

$$\xi_{D,opt} = \frac{\sqrt{\mu}}{2} \tag{21}$$

where μ is the mass ratio of the damper (i.e. mass of the damper to the entire mass of the assembly). In order to tune the damper, its natural frequency is obtained as the product of a tuning ratio v, times the natural frequency of the coupled tower–blades system, i.e.:

$$v = \frac{\omega_D}{\omega_{CS,1}} \tag{22}$$

where $\omega_{CS,1}$ is the fundamental frequency of the coupled tower-rotating blades assembly. It is possible to derive a closed form expression for the optimum tuning ratio of the TMD attached to a damped structure based on the "fixed- point" theory of Den Hartog [53] which had been proposed for the case of undamped structural systems subjected to sinusoidal excitation. In the optimal design of a TMD attached to an undamped structural system subjected to sinusoidal excitation [53, 54], two "fixed-point" frequencies were obtained at which the transmissibility of vibration is independent of the damping in the TMD. It was also observed that the amplitude of the response transfer functions at the two fixed points was unequal and had a contrasting effect with the change in the tuning ratio. For a structure subjected to an external force which has wide banded energy content or which has dominant energy at the natural period of the structure, the maximum response reduction is achieved when the area under the transfer function curve is at a minimum. This implies that the values of the transfer function at the fixed points should be equal and the value of the tuning ratio for which this occurs is the

Table 3: Properties of the TMD.

	Rotational frequency (rev/min)	
	15	30
Mass ratio (%)	1	1
Tuning ratio	0.99	0.99
Natural frequency (rad/s)	4.45	4.55
Mass (kg)	997	997
Stiffness constant (kN/m)	20.64	19.74
Damping constant (kNs/m)	0.45	0.44
Damping ratio (%)	5	5

optimal tuning ratio of the TMD. Ghosh and Basu [55] extended the theory based on "fixed-points" to obtain closed form expression for optimal tuning ratio in case of a damped structure. This was used by Murtagh *et al.* [49] designing an optimal TMD for a wind turbine tower. The expression for the optimal tuning parameter v_{opt} for a wind turbine tower with damping ratio ξ_n in the fundamental mode of vibration is [49, 55]:

$$v_{opt} = \sqrt{\frac{1 - 4\xi_n^2 - \mu(2\xi_n^2 - 1)}{(1 + \mu)^3}} \tag{23}$$

The optimal tuning ratio together with an optimal damping ratio in the TMD will minimize the maxima of the displacement transfer function of a wind turbine tower.

Murtagh *et al.* [49] considered a tower of hub height 60 m and blades with radius 30 m for a three-bladed wind turbine and designed a TMD for suppression of the tip displacement. The mean wind speed at the top of the tower was assumed to be 20 m/s. The first three modal damping ratios of the tower were assumed to be 1% of the critical. A mass ratio of 1% was assumed for the TMD, giving the damper a damping ratio of 5% of critical. Thus, when used in conjunction with eqn (23), an optimal tuning ratio of 0.99 is obtained. The forced vibration responses of the coupled tower–blades model including and excluding the TMD were calculated and compared. Two rotational frequencies of the rotor system were considered, and the blades are perturbed under the action of rotationally sampled wind turbulence [30]. The design parameters of the dampers designed for the two cases are presented in Table 3.

Figure 8 presents the tip displacement transfer function amplitudes obtained for the coupled tower and rotating blades model ($\Omega = 15$ rev/min) with and without the damper. When contrasting the two transfer functions obtained, it is evident that the presence of the damper causes the peak to split and decrease substantially in magnitude. Figure 9 presents the simulated wind-induced response of the coupled blade–tower model, at the top of the tower, including and excluding the damper. From this figure, it is evident that the damper has been effective in suppressing the vibrations, particularly in the earliest portion of the time-history, where the

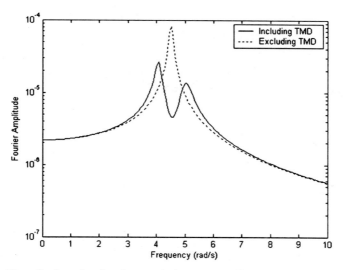

Figure 8: Transfer function for the coupled tower-nacelle and rotating blades model.

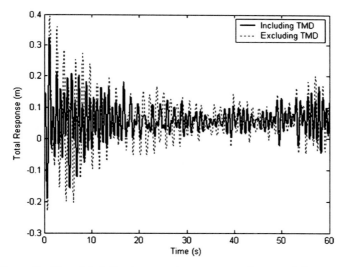

Figure 9: Simulated displacement response at the top of the tower.

maximum tower tip displacement observed without the damper of about 0.4 m, reduced to approximately 0.32 m when the damper was included.

5 Wind tunnel testing

Wind tunnel testing of scaled model in order to experimentally investigate aero-elastic and aerodynamic phenomena associated with structures has proved to be a

Figure 10: Wind turbine tower model installed in test section of wind tunnel.

valuable approach for wind engineers. Ever since the first major building study in a boundary layer wind tunnel was conducted by Cermak and Davenport in the 1960s, engineers have been able to inexpensively investigate turbulence-induced phenomena. The results provide vital information necessary to ensure the serviceability and survivability of flexible structures like a wind turbine.

Considerable experimental literature now exists regarding wind tunnel testing of structures in general. Aerodynamic studies are primarily focused on evaluation of drag and lift coefficients, such as those by Carril *et al.* [56] and Gioffrè *et al.* [57]. Aeroelastic scale model studies, similar to those by Ruscheweyh [58] and Kim and You [59], examine the link between structural geometrical form and aeroelastic phenomena, such as vortex shedding. Passive and active dampers are also proving to be valuable devices in the mitigation of wind-induced structural vibration, and the wind tunnel provides an excellent means to develop and test control strategies [60, 61]. While there is very limited literature available on wind tunnel testing of wind turbines, this kind of testing can be very useful for system identification [62], design, and analysis of wind turbines and associated vibration control systems.

Figure 10 shows a model assembly of wind turbine constructed at the Department of Civil Engineering, Trinity College Dublin, Ireland being tested in the wind tunnel facility at National University of Ireland, Galway [63]. The model assembly was composed of three main components: the tower, the nacelle and motor, and the rotor system. The model was designed so that the fundamental frequencies of the rotor blades and the tower were close to each other, ensuring significant dynamic coupling between the two subcomponents. The model was immersed in a turbulent wind flow and the responses were recorded. The recorded bending strain at the base of the tower and the corresponding Fourier amplitude spectrum are shown in Figs 11 and 12 for the case of a stationary wind turbine.

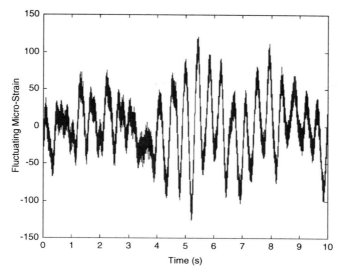

Figure 11: Strain time-history recorded at the tower base point for rotational speed of 0 rad/s.

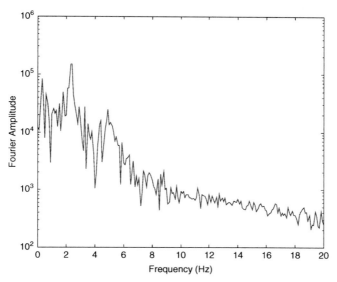

Figure 12: Fourier amplitude of strain response at tower base point for rotational speed of 0 rad/s.

6 Offshore towers

Recent expansion in the wind energy sector has seen an associated growth in energy production from offshore wind farms. Hence, turbines are becoming larger with taller towers and are being moved further out to sea. As a result the wind

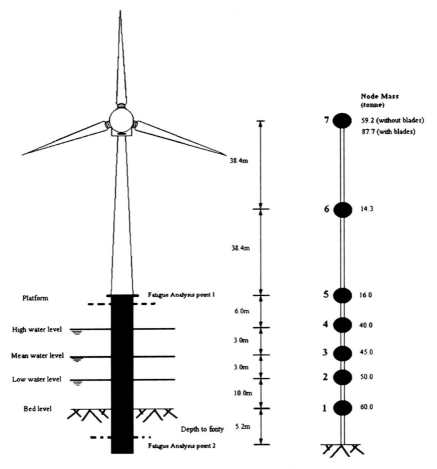

Figure 13: Structural model.

turbine towers are subjected to ever greater wind and wave forces. Thus, it is necessary to analyse the dynamics and minimize the response of wind turbine towers to simultaneous actions of joint wind and wave loadings, instead of just the wind loading as in the onshore case.

6.1 Simple model for offshore towers

A model for analysis of an offshore wind turbine tower can in general be represented by a discrete MDOF system [64]. A simple schematic model of an offshore tower is shown in Fig. 13 [65]. The response of such an MDOF system under joint wind and wave loading subjected at the nodes can be calculated by a time-history integration using a standard technique like Runge-Kutta of suitable order. A fatigue analysis can be performed using the rainflow counting method and Miner's rule in accordance with [66] following Colwell and Basu [65].

6.2 Wave loading

Following the collection of data and analysis carried out under the Joint North Sea Wave Observation Project (JONSWAP) [67], it was found that the wave spectrum continues to develop through non-linear, wave–wave interactions even for very long times and distances compared to the Pierson–Moskowitz spectrum. The wave excitation for an offshore wind turbine tower can be modelled using the JONSWAP spectrum which takes into account the higher peak of the energy spectrum in a storm. Also, for the same total energy as compared with the Pierson–Moskowitz wave energy spectra, it takes into account the occurrence of frequency shift of the spectra maximum. The spectrum takes the form

$$S_{\eta\eta}(\omega) = \frac{\alpha g^2}{\omega^5} \exp\left[-\frac{5}{4}\left(\frac{\omega_m}{\omega}\right)^4 \right] \gamma^{\exp[(\omega-\omega_m)^2/2\sigma^2\omega_m^2]} \tag{24}$$

where η is the function of water surface elevation. Equation (24) defines a stationary Gaussian process of standard deviation equal to 1. In eqn (24), γ is the peak enhancement factor (3.3 for the North sea), g is the acceleration of gravity and ω is the circular wave frequency. The wave data from the JONSWAP project was used to calculate the values of the constants in eqn (24) as follows:

$$\alpha = 0.076\left(\frac{U_{10}^2}{Fg}\right)^{0.22} \tag{25}$$

$$\omega_m = 22\left[\frac{g^2}{U_{10}F}\right]^{1/3} \tag{26}$$

and

$$\sigma = \begin{cases} 0.07, & \omega \leq \omega_m \\ 0.09, & \omega > \omega_m \end{cases} \tag{27}$$

where U_{10} is the mean wind speed 10 m from the sea surface, F (fetch) is the uninterrupted distance over which the wind blows (measured in the direction of the wind) without a significant change of direction. The fetch varies in its non-dimensional form as follows [68]:

$$10^{-1} < \frac{gF}{U_{10}^2} < 10^4 \tag{28}$$

The wave force acting on the offshore wind turbine structure can be calculated by using the linearized Morison equation [69] and from the wave surface elevation time-history calculated based on the wave spectrum (for details see [65]).

6.3 Joint distribution of wind and waves

The JONSWAP spectrum defined in the previous section is a stationary Gaussian process and can be mapped into the process of the sea state defined by the significant wave height and mean zero-crossing wave period (H_s, T_z) by letting the dimensionless time be t/T_z and the dimensionless process be $X/(\lambda_0)^{1/2} = 4X/H_s$, [68]. The wind speed at 10 m, U_{10}, and the significant wave height, H_s, from the JONSWAP spectrum can be related through the integral of eqn (24):

$$\lambda_0 = \int_0^\infty S_{\eta\eta}(\omega)\,d\omega \tag{29}$$

where $(\lambda_0)^{1/2}$ is the standard deviation of surface displacement. If a sea contains a narrow range of wave frequencies, H_s is related to the standard deviation of the sea surface displacement [70]:

$$H_s = 4\sqrt{\lambda_0} \tag{30}$$

The time-histories used for analysis in the joint distribution of wave period and height are approximated by the linear combination of trigonometric polynomials [71].

Simulated wave surface elevation time-history for 'moderate' wave excitation with target and simulated PSD have been presented for the purpose of illustration in Figs 14 and 15 which have been taken from the investigation carried out by [65]. The wave surface elevation time-history has been simulated with a joint dependence on

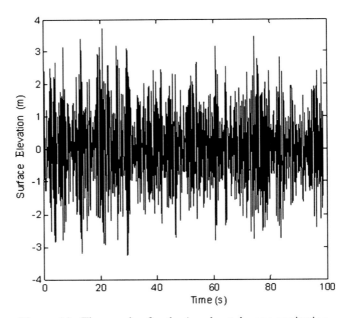

Figure 14: Time-series for the 'moderate' wave excitation.

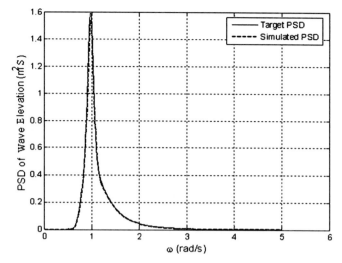

Figure 15: PSD function of 'moderate' wave elevation.

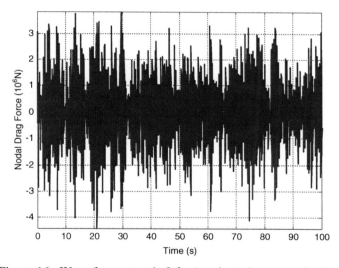

Figure 16: Wave force at node 3 for 'moderate' wave excitation.

a wind loading with mean wind speed of 18 m/s at a level of 10 m. Also, shown in Fig. 16 is the wave force time-history at node 3 of the structural model in Fig. 13 for 'moderate' wave excitation [71].

6.4 Vibration control of offshore towers

As in the case with onshore wind turbine towers several structural vibration control strategies could be adopted. The use of passive control devices such as TMD,

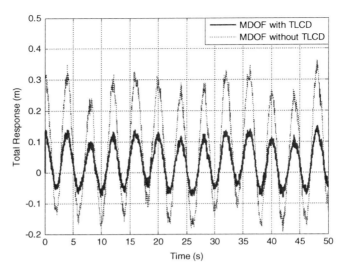

Figure 17: Time-history response of a rotating offshore turbine under wind and 'moderate' wave excitation with and without TLCD.

TLD or tuned liquid column damper (TLCD) can be useful to suppress undesirable structural vibrations. The use of a TLCD for the vibration control of offshore towers has been investigated by Colwell and Basu [65]. The simple model in the previous section was used to analyse the dynamic response of the system under joint wind and wave loading. The assumed wind velocity at a height of 10 m was 18 m/s. The blades were 60 m in length and individually weighed 9.5 tonnes. The blades rotated at a frequency equal to the fundamental natural frequency of the MDOF system. The system was subjected to joint wind and wave loading with the wind turbulence simulated from Kaimal spectra and the wave excitation simulated from JONSWAP spectra. The details on the damper parameters are available in [65]. Figure 17 compares the total displacement response at the top of the tower with and without the TLCD. It has been concluded that the passive damper has a significant beneficial impact in suppressing the structural vibrations by about 60%. The maximum design bending moment for the theoretical simulation at the base of the structure is reduced from 6.2607×10^4 to 4.0101×10^4 kNm.

7 Conclusions

Several aspects of dynamic analysis and design of wind turbine towers have been discussed in this chapter. With the increase in rotor diameter and the height of the towers, the analysis of wind turbine towers becomes crucial and needs special attention particularly in the view of tower–rotor coupling. The important physical behaviour and phenomena to be accounted for are centrifugal stiffening of blades, gravity effects, rotationally sampled turbulence and the tower–blade coupled dynamic interaction. Numerical results have indicated that responses of the tower may be severely underestimated if the tower–blade coupling is not taken into consideration properly. A simplified gust factor approach for calculating the response of wind turbine towers has been discussed. Two gust factors based on displacement and bending moment response of the tower have been presented. They have again highlighted the importance of tower–rotor coupling with the need for response specific gust factors. The importance of wind tunnel testing has been discussed and emphasized. Since vibration poses a critical challenge in wind turbine towers, vibration control strategies with TMD and other passive dampers have been discussed. Mathematical models have been presented for both onshore and offshore wind turbine towers. Joint wind and wave loading have been modelled for analyzing the offshore towers. It has been observed that vibration control dampers significantly reduce the motions in a wind turbine tower with possible design, maintenance and operational benefits.

References

[1] Harrison, R., Hau, E. & Snel, H., *Large Wind Turbines: Design and Economics*, John Wiley and Sons Ltd: New York, 2000.
[2] Jen, C.W., Johnson, D.A. & Dubois, F., Numerical modal analysis of structures based on a revised substructure synthesis approach. *Journal of Sound and Vibration*, **180**(2), pp. 185–203, 1995.
[3] Scheble, M., Strizzolo, C.N. & Converti, S.C., Rayleigh-Ritz substructure synthesis method in physical co-ordinates for dynamic analysis of structures. *Journal of Sound and Vibration*, **213**(1), pp. 193–200, 1998.
[4] Wu, J.C. & Yang, J.N., Active control of transmission tower under stochastic wind. *Journal of Structural Engineering ASCE*, **124**(11), pp. 1302–1312, 1998.
[5] Lavassas, I., Nikolaidis, G., Zervas, P., Efthimiou, E., Doudoumis, I.N. & Baniotopoulos, C.C., Analysis and design of the prototype of a steel 1-MW wind turbine tower. *Engineering Structures*, **25**(8), pp. 1097–1106, 2003.
[6] Bazeos, N., Hatzigeorgiou, G.D., Hondros, I.D., Karamaneas, H., Karabalis, D.L. & Beskos, D.E., Static, seismic and stability analyses of a prototype wind turbine steel tower. *Engineering Structures*, **24**, pp. 1015–1025, 2002.
[7] Dutta, P.K., Ghosh, A.K. & Agarwal, B.L., Dynamic response of structures subjected to tornado loads by FEM. *J. Wind Engineering and Industrial Aerodynamics*, **90**, pp. 55–69, 2002.
[8] Murtagh, P.J., Basu, B. & Broderick, B.M., Simple models for the natural frequencies and mode shapes of towers supporting utilities. *Computers and Structures*, **82**(20-21), pp. 1745–1750, 2004.
[9] Baumgart, A., A mathematical model for wind turbine blades. *Journal of Sound and Vibration*, **251**(1), pp. 1–12, 2002.
[10] Naguleswaran, S., Lateral vibration of a centrifugally tensioned uniform Euler Bernoulli beam. *Journal of Sound and Vibration*, **176**(5), pp. 613–624, 1994.
[11] Banerjee, J.R., Free vibration of centrifugal stiffened uniform and tapered beams using the dynamic stiffness method. *Journal of Sound and Vibration*, **233**(5), pp. 857–875, 2000.

[12] Chung, J. & Yoo, H.H., Dynamic analysis of a rotating cantilever beam by using the finite element method. *Journal of Sound and Vibration*, **249**(1), pp. 147–164, 2002.

[13] Lee, C.I., Al-Salem, M.F. & Woehrle, T.G., Natural frequency measurements for rotating spanwise uniform cantilever beams. *Journal of Sound and Vibration*, **240**(5), pp. 957–961, 2001.

[14] Buhl, M.L., Data preparation requirements for modelling wind turbine with ADAMS, National Renewable Energy Laboratory report for the U.S. Department of Energy, Colorado, USA, 1994.

[15] Suresh Kumar, K. & Stathopoulos, T., Computer simulation of fluctuating wind pressures on low building roofs. *Journal of Wind Engineering and Industrial Aerodynamics*, **69-71**, pp. 485–495, 1997.

[16] Kitagawa, T. & Nomura, T., A wavelet-based method to generate artificial wind fluctuation data. *Journal of Wind Engineering and Industrial Aerodynamics*, **91**, pp. 943–964, 2003.

[17] Minh, N.N., Miyata, T., Yamada, H. & Sanada, Y., Numerical simulation of wind turbulence and buffeting analysis of long-span bridges. *Journal of Wind Engineering and Industrial Aerodynamics*, **83**, pp. 301–315, 1999.

[18] Panofsky, H.A. & Singer, I.A., Vertical structure of turbulence. *Journal of Royal Meteorological Society*, **91**, pp. 339–344, 1965.

[19] Davenport, A.G., The dependence of wind load upon meteorological parameters. *Proc. of the International Research Seminar on Wind Effects on Buildings and Structures*, University of Toronto Press Toronto, pp. 19–82, 1968.

[20] Vickery, B., On the reliability of gust factors. *Proc. of the Technical Meeting Concerning Wind Loads on Buildings and Structures* Building Science Series 30, Washington, DC, pp. 93–104, 1970.

[21] Brook, R.R., A note on vertical coherence of wind measured in an urban boundary layer. *Boundary Layer Meteorology*, **9**, pp. 247, 1975.

[22] Højstrup, J., Spectral coherence in wind turbine wakes. *Journal of Wind Engineering and Industrial Aerodynamics*, **80**, pp. 137–146, 1990.

[23] Sørensen, P., Hansen, A.D. & Rosas, P.A.C., Wind models for simulation of power fluctuations from wind farms. *Journal of Wind Engineering and Industrial Aerodynamics*, **90**, pp. 1381–1402, 2002.

[24] Connell, J.R., Turbulence spectrum observed by a fast-rotating wind turbine blade. *Rep. PNL-3426, Battelle Pacific Northwest Laboratory*, Richland, WA 99352, 1980.

[25] Kristensen, L. & Frandsen, S., Model for power spectra of the blade of a wind turbine measured from the moving frame of reference. *Journal of Wind Engineering and Industrial Aerodynamics*, **10**, pp. 249–262, 1982.

[26] Rosenbrock, H.H., Vibration and stability problems in large turbines having hinged blades. *Rep. C/T 113, ERA Technology Ltd*, Surry, Great Britain, 1955.

[27] Madsen, P.H. & Frandsen, S., Wind-induced failure of wind turbines. *Engineering Structures*, **6**(4), pp. 281–287, 1984.

[28] Verholek, M.G., Preliminary results of a field experiment to characterise wind flow through a vertical plane. *Rep. PNL-2518, Battelle Pacific Northwest Laboratory*, Richland, WA 99352, 1978.

[29] Hardesty, R.M., Korrel, J.A. & Hall, F.F., Lidar measurement of wind velocity spectra encountered by a rotating turbine blade. *NOAA Technical Memorandum*, Washington DC, USA, 1981.

[30] Murtagh, P.J., Basu, B. & Broderick, B., Along wind response of wind turbine tower with blade coupling subjected to rotationally sampled wind loading. *Engineering Structures*, **27**(8), pp. 1209–1219, 2005.

[31] Murtagh, P.J., Basu, B. & Broderick, B.M., Mode acceleration approach for rotating wind turbine blades. *Proc. of the Institution of Mechanical Engineers: Part K: Journal of Multi-body Dynamics*, **218**(3), pp. 159–167, 2004.

[32] Williams, D., Dynamics loads in aeroplanes under given impulsive loads with particular reference to landing and gust loads on a flying boat. *Royal Aircraft Establishment, Farnborough, UK, Reports SMR 3309 and 3316*, 1945.

[33] Craig, R.R., *Structural Dynamics*, John Wiley and sons: New York, 1981.

[34] Singh, M.P., Mode-acceleration based response spectrum approach for non-classically damped structures. *Soil Dynamics and Earthquake Engineering*, **5**, pp. 226–233, 1986.

[35] Akgun, M.A., A new family of mode-superposition methods for response calculations. *Journal of Sound and Vibration*, **167**(2), pp. 289–302, 1993.

[36] Nigam, N.C. & Narayanan, S., *Applications of Random Vibrations*, Springer Verlag: Delhi, 1994.

[37] Kaimal, J.C., Wyngaard, J.C., Izumi, Y. & Cote, O.R., Spectral characteristics of surface-layer turbulence. *Journal of the Royal Meteorological Society*, **98**, pp. 563–589, 1972.

[38] Davenport, A.G., Gust Loading Factors. *Journal of Structural Division, ASCE*, **93**(3), pp. 11–34, 1967.

[39] Velozzi, J. & Cohen, E., Gust response factors. *Journal of Structural Division, ASCE*, **94**(6), pp. 1295–1313, 1968.

[40] Holmes, J.D., Along-wind response of lattice towers: part II – Aerodynamic damping and deflections. *Engineering Structures*, **18**(7), pp. 483–488, 1996.

[41] Zhou, Y. & Kareem, A., Gust loading factor: new model. *Journal of Structural Engineering, ASCE*, **127**(2), pp. 168–175, 2001.

[42] Zhou, Y., Kijewski, T. & Kareem, A., Along-wind load effects on tall buildings: comparative study of major international codes and standards. *Journal of Structural Engineering, ASCE*, **128**(6), pp. 788–796, 2002.

[43] Murtagh, P.J., Basu, B. & Broderick, B.M., Gust response factor methodology for wind turbine tower assemblies. *Journal of Structural Engineering, ASCE*, **133**(1), pp. 139–144, 2007.

[44] Crandall, S.H., First crossing probabilities of linear oscillator. *Journal of Sound and Vibration*, **12**(3), pp. 285–289, 1970.

[45] CEN (2004). Eurocode 1 Basis for design and actions on structures – part 2- 4: actions on structures – wind actions. *European Prestandard Env. 1991-2- 4, European Committee for Standardization, Brussels*, 2004.

[46] Dueñas-Osorio, L. & Basu, B., Unavailability of wind turbines from wind induced accelerations. *Engineering Structures*, **30**(4), pp. 885–893, 2008.

[47] Wright, A.D. & Balas, M.J., Design of control to attenuate loads in the controls advanced wind turbine. *Journal of Solar Energy Engineering, ASME*, **126**, pp. 1083–1091, 2004.

[48] Kallesøe, B.S., A low-order model for analysing effects of blade fatigue load model. *Wind Energy*, **9**(5), pp. 421–436, 2006.

[49] Murtagh, P.J., Ghosh, A., Basu, B. & Broderick, B., Passive control of wind turbine vibrations including blade/tower interaction and rotationally sampled turbulence. *Wind Energy*, **11**(4), pp. 305–317, 2008.

[50] McNamara, R.J., Tuned mass dampers for buildings. *Journal of the Structural Division, ASCE*, **103**(9), pp. 1785–1798, 1977.

[51] Luft, R.W., Optimal tuned mass dampers for buildings. *Journal of the Structural Division, ASCE*, **105**(12), pp. 2766–2772, 1979.

[52] Gerges, R.R. & Vickrey, B.J., Wind tunnel study of the across-wind response of a slender tower with a nonlinear tuned mass damper. *Journal of Wind Engineering and Industrial Aerodynamics*, **91**(8), pp. 1069–1092, 2003.

[53] Den Hartog, J.P., *Mechanical Vibrations*, McGraw-Hill Book Company Inc.: New York, 1947.

[54] Mallik, A.K., *Principles of Vibration Control*, Affiliated East West Press Pvt. Ltd.: New Delhi, India, 1990.

[55] Ghosh, A. & Basu, B., A closed form optimal tuning criterion for TMD in damped structures. *Structural Control Health Monitoring*, **14**, pp. 681–692, 2007.

[56] Carril, C.F., Isyumov, N. & Brasil, R., Experimental study of the wind forces on rectangular latticed communication towers with antennas. *Journal of Wind Engineering and Industrial Aerodynamics,*, **91**, pp. 1007–1022, 2003.

[57] Gioffrè, M., Gusella, V., Materazzi, A.L. & Venanzi, I., Removable guyed mast for mobile phone networks: load modelling and structural response. *Journal of Wind Engineering and Industrial Aerodynamics*, **92**, pp. 467–475, 2004.

[58] Ruscheweyh, H., Vortex-induced vibration of a water tank tower with small aspect ratio. *Journal of Wind Engineering and Industrial Aerodynamics*, **89**, pp. 1579–1589, 2001.

[59] Kim, Y.-M. & You, K.-P., Dynamic response of a tapered tall building to wind loads. *Journal of Wind Engineering and Industrial Aerodynamics*, **90**, 1771–1782, 2002.

[60] Cho, K.-P., Cermak, J.E., Lai, M.-L. & Nielsen, E.J., Viscoelastic damping for wind-excited motion of a five-story building frame. *Journal of Wind Engineering and Industrial Aerodynamics*, **77**(8), pp. 269–281, 1998.

[61] Wu, J.-C. & Pan, B.-C., Wind tunnel verification of actively controlled high-rise building in along-wind motion. *Journal of Wind Engineering and Industrial Aerodynamics*, **90**, pp. 1933–1950, 2002.

[62] Murtagh, P.J. & Basu, B., Identification of modal viscous damping ratios for a simplified wind turbine tower using Fourier and wavelet analysis.

Proceedings of the Institution of Mechanical Engineers, Part K: Journal of Multi-body Dynamics, **221**(4), pp. 577–589, 2007.

[63] Murtagh, P.J., Dynamic analysis of wind turbine tower assemblies, Ph.D. thesis, University of Dublin, Trinity College Dublin, Ireland, 2004.

[64] Rogers, N., Structural dynamics of offshore wind turbines subject to extreme wave loading. *Proc. of the 20th BWEA Annual Conf.*, UK, 1998.

[65] Colwell, S. & Basu, B., Tuned liquid column dampers in offshore wind turbines for structural control. *Engineering Structures*, **31**(2), pp. 358–368, 2009.

[66] DNV, *Design of Offshore Wind Turbine Structures.* Offshore Standard DNV-OS-J101, 2004.

[67] Hasselmann, K., Barnett, T.P., Bouws, E., Carlson, H., Cartwright, D.E., Enke, K., Ewing, J.A., Gienapp, H., Hasselmann, D.E., Kruseman, P., Meerburg, A., Muller, P., Olbers, D.J., Richter, K., Sell, W. & Walden, H., Measurement of Wind-Wave Growth and Swell Decay During the Joint North Sea Wave Project (JONSWAP). *Deutsche Hydrogr*, **A8**(12), 1973.

[68] Ditlevsen, O., Stochastic model for joint wind and wave loads on offshore structures. *Structural Safety*, **24**, pp. 139–163, 2002.

[69] Sarpkaya, T. & Isaacson, M., *Mechanics of Wave Forces on Offshore Structures*, Reinhold, V.N.: New York, 1981.

[70] Hoffman, D. & Karst, O.J. , The theory of Rayleigh distribution and some of its applications. *Journal of Ship research*, **19**(3), pp. 172–191, 1975.

[71] Colwell, S. & Basu, B., Simulation of joint wind and wave loading time histories. *Proc. of the Irish Signals and Systems Conf., Dublin*, paper no. 120, pp. 1–5, 2006.

CHAPTER 17

Design of support structures for offshore wind turbines

J. van der Tempel, N.F.B. Diepeveen, D.J. Cerda Salzmann
& W.E. de Vries
*Department of Offshore Engineering, Delft University of Technology,
The Netherlands.*

Offshore wind is the logical next step in the development of wind energy. With higher wind speeds offshore and the fact that turbines can be placed out of sight, offshore wind helps increase the amount of renewable energy significantly. Offshore wind has been developed through pilot projects in the 1990s and has seen commercial development over the last decade. This chapter shows the development of offshore wind and then focuses on the design of support structures. It briefly describes all fundamental steps that need to be taken to come to a proper support structure design, incorporating all turbine loads and the impact of waves and soil. Furthermore, the chapter gives an overview of the different types of structures and how they are fabricated and installed.

1 Introduction

As wind energy developed on land, the locations with a favourable wind climate became scarce, but the demand for clean energy still grew. The solution for many countries lies at their doorstep: offshore. Around the world, many densely populated areas are close to the sea. Offshore, the wind blows stronger and more constant, unhindered by obstacles. This led to the development of offshore wind farms: turbines placed at sea.

This chapter describes the design of offshore wind turbines. Turbine design for offshore follows the general design approach for onshore, although typical load cases are somewhat different and the turbine needs to withstand the more severe environment: salt. The structures on which the turbines are placed are significantly different, though. Different from their onshore counterparts but also from other structures found at sea. The design is not only dependent on turbine loads and associated overturning moment, the wave and currents add significant loads too. For the design of these structures wind, wave and current loads need to be assessed as acting on the offshore wind turbine system as a whole.

The development of offshore wind began in the 1970s and 1980s with studies and assessment of the potential wind resource offshore. In the 1990s several pilot offshore wind farms were constructed in the European waters, which helped develop knowledge and new technology. In 2002, the first large offshore wind farm Horns Rev was constructed in the North Sea off the Danish coast: 80 turbines with an installed capacity of 160 MW. In the years that followed, other countries followed with the construction of these large, commercial offshore wind farms. The EU target for 2020 is to have 40 GW of installed capacity.

2   History of offshore, wind and offshore wind development of offshore structures

2.1   The origin of "integrated design" in offshore wind energy

During the 1970s, 1980s and early 1990s, a number of studies were conducted in the field of offshore wind energy. Offshore and shipbuilding as well as renewable energy groups drafted reports on how to effectively harness the offshore wind energy potential. The first designs were mainly based on the multi-megawatt prototype turbines built in the 1970s: 3 MW and more. The structures were large, heavy and stiff: based on the accumulated experience of offshore construction in the North Sea for oil & gas exploitation. Figure 1 shows examples of a design from the British RES study and a Heerema tripod design [1, 2].

The design did incorporate combined wind and wave loading, but only on a basic level for extreme load case calculations. The stiffness of the structure prevented heavy dynamic response, so fatigue was not a big issue. For the subject operation and maintenance a direct copy of offshore platforms was made: the addition of a complete helicopter deck.

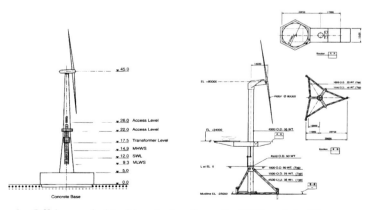

Figure 1:   Offshore wind turbine design from the RES and the Heerema study.

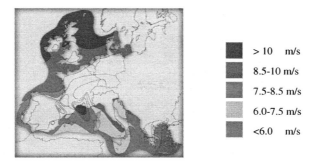

	> 10 m/s
	8.5-10 m/s
	7.5-8.5 m/s
	6.0-7.5 m/s
	<6.0 m/s

Figure 2: Yearly average wind speed at 100 m height for the European Seas.

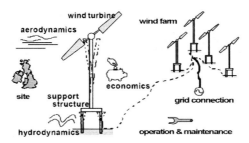

Figure 3: Subjects covered in the integrated design approach of the Opti-OWECS study.

In 1995 the Joule I "Study of Offshore Wind Energy in the EC" was published [3]. The study gave an overview of the wind potential offshore as shown in Fig. 2. The study described the design of offshore wind turbines in a more generic way with example designs for different types of offshore wind turbines. It was found that for one turbine wave loads could be dominant while for the other wind was the dominant load source. One of the main issues found was the benefit of aerodynamic damping on the dynamic behaviour of the structure when the turbine is in operation. It was also stated that a softer support structure would further enhance the aerodynamic damping effect, but at the cost of increased tower motion.

The Joule III Opti-OWECS [1] report finally made a complete design focusing on the integrated dynamic features of flexible offshore wind turbines. The design incorporated the entire offshore wind farm with all its features from turbines to operation and maintenance philosophy to cost modelling. Figure 3 gives an overview of all subjects covered in this integrated design scheme.

The Opti-OWECS study further explored the possibilities of flexible dynamic design. Although several types of support structures were reviewed, it was decided to make a full design of a soft monopile structure to benefit in full from the aerodynamic damping and assess the potential negative consequences of large structural motion. It was found that a structure could be designed with a natural frequency below both the rotation and the blade passing frequency of the turbine, a so-called soft–soft structure. The frequency distributions are shown in Fig. 4.

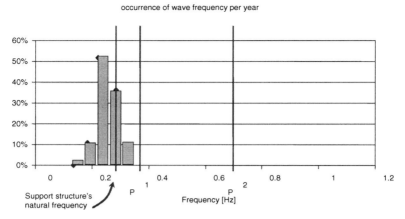

Figure 4: Rotation (1P) and blade passing frequency (2P) of the Opti-OWECS turbine with the structure's natural frequency and a histogram of the occurring wave frequencies.

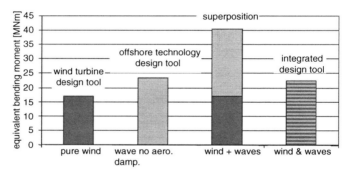

Figure 5: Comparison of fatigue calculations for wind only, wave only, wind and wave combines from separate and simultaneous analyses.

The fact that the structure's natural frequency coincided with a large portion of wave frequencies was further investigated. The aerodynamic damping of the turbine was found to reduce fatigue significantly, doubling the structures fatigue life when taken into account. To enable the analysis of this feature, full non-linear time domain simulations were found to be necessary of simultaneous wind and wave loading. Should wind and wave loads be analysed separately, the effect will not become visible by just adding the separate analyses as can be seen in Fig. 5.

Next to the detailed investigation of the dynamic behaviour in the design, a large number of practical issues were addressed in an integrated way. For installation it was found that onshore pre-installation would cause large cost reductions. For the correction of misalignment of the driven foundation pile, a transition piece was proposed. Installation of fully operational turbines and the misalignment correction are shown in Fig. 6. It was concluded that large-scale offshore wind

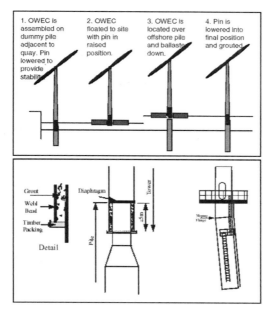

Figure 6: Installation of fully operational turbine and connection details between foundation pile and tower with misalignment correction.

energy application would require purpose-built vessels because existing vessel were either too large (offshore cranes) or too small.

2.2 From theory to practice: Horns Rev

The installation of Horns Rev in 2002 was the largest practical test of all theoretical findings. The installation of the foundation pile was done on a rather traditional manner: a small jack-up with a crane. For the installation of the turbines however two ships were entirely converted to purpose-built turbine installation vessels. Choosing a normal ship would ensure high sailing speed from and to port. A jacking system was added which only pre-stressed the legs without lifting the entire vessel out of the water. Two blades were already connected to the nacelle before placing it on the deck of the installation vessel. The method was christened "bunny ears" for obvious reasons. The installation of the tower and turbine was reduced to four lifts; two tower sections, nacelle with two blades and the final blade.

All appurtenances were pre-fitted in port to the transition piece: boat landing, J-tube, platform and the transition piece was grouted to the foundation pile. Figure 7 shows the "bunny ears", the A2Sea installation vessel, the transition piece being pre-fitted with a J-tube and the installation of the transition piece.

The design for the support structures on Horns Rev was fully covered by the owner of the wind farm: Elsam supplied all contractors with a complete pre-design, which was to be prized and for which an installation method was to be drafted.

Figure 7: Bunny ears, the pre-fitting of two blades, purpose-converted installation vessels, pre-fitting of J-tube to the transition piece and the installation of the transition piece.

Figure 8: Heli-hoist platforms are installed on turbines to lower mechanics for maintenance.

The design was well documented and integrated. The contractors were also invited to give their own alternative design. The amount of information for this part however was much less: the support structure was to end at 9 m above the mean sea level and the only interaction from the turbine was a static load and moment at this 9 m level. This did not improve integrated design but it can be argued that no contractor at that time would have any time for more detailed integrated turbine–foundation interaction analysis as all engineering went into "getting the things there".

For maintenance all nacelles are equipped with a heli-hoist platform onto which mechanics can be lowered even when boat access is not possible due to high waves (see Fig. 8).

The Horns Rev project proved that many practical issues addressed in the paper studies were applicable in real offshore wind. The amount of overall integration, or even the need for it is not crystal clear: many individual optimisations could be done without affecting the entire system.

2.3 Theory behind practice

The installation of the two turbines offshore of Blyth in the UK was part of a large EU-funded project to study Offshore Wind Turbines at Exposed Sites (OWTES).

One of the turbines is fitted with a complete measurement system to record external conditions and structural response. A picture of the turbines and the measurement systems is shown in Fig. 9.

The measurements were used to validate the current design tools for offshore wind turbines. It was found that present-day tools are very able to model the offshore wind turbine behaviour induced by wind and waves simulations. Figure 10 shows the comparison of measured and modelled mudline bending moment per wind speed.

It was found that offshore wind turbine design is very dependent on site-specific features like the wind and wave climate. At Blyth the local bathymetry is such that

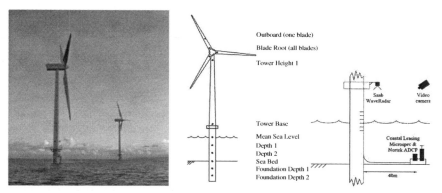

Figure 9: Turbines at Blyth with complete measurement system for external loads and responses.

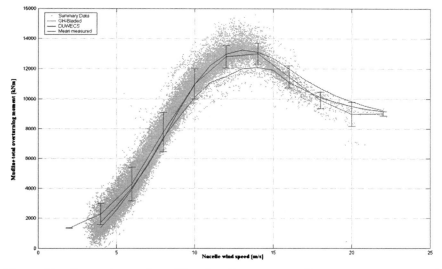

Figure 10: Comparison of mudline bending moment form measurements and modelling.

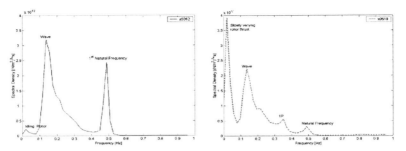

Figure 11: Response spectrum for mudline bending stress for idling (left) and operating (right) turbine.

near the turbines breaking waves are a common phenomenon. Although their influence did not affect the design dramatically in this particular case, they prove the importance of taking all details of a site into account.

Although the natural frequency of the structure is rather high at 0.48 Hz, the effect of both wind and wave loading on resonance is significant, as is the aerodynamic damping. Figure 11 shows the response spectrum for the mudline bending stress for equal environmental condition with an idling rotor (left) and a turbine in operation (right). The significant resonance peak in the wave-only case is damped dramatically when the turbine is operating.

From the measurements at Blyth it can be concluded that current modelling techniques are able to represent the critical features of offshore wind turbines properly, especially when on hindsight all structural and environmental parameters are known. It has also been shown that monopile structures are very dynamically sensitive, even in this case with relatively high natural frequency and that therefore proper analysis of resonant behaviour and aerodynamic damping deserve special attention.

3 Support structure concepts

3.1 Basic functions

The basic function of the support structure is to keep the wind turbine in place. This means that it has to be built to withstand loads originating from sea currents, waves and wind – acting on both the support structure and the turbine in operation.

A variety of wind turbines is available on the market, designed by different turbine manufacturers, in a range of power ratings. Each wind turbine has different characteristics. The offshore environmental conditions may also vary from site to site. Therefore, support structures are designed specifically for each case. It is not uncommon for one offshore site to have several variations of one type of support structure for one type of turbine.

The cost of the support structure on average amounts to around 25% of the total offshore wind turbine cost [1].

Typically the support structure is divided in two main parts:

1. The turbine tower
2. The foundation

The turbine tower normally consists of two or three sections. The design of the tower is usually provided by the turbine manufacturer. The tower is often installed in the same shift as the nacelle and the rotor.

The term foundation here refers to the turbine support structure, excluding the tower. It is essentially located below and at the water level. The function of the foundation is to direct the loads on the support structure into the seabed.

Many types of foundation for offshore wind turbines already exist. Essentially, the manner in which they are connected to the seabed determines how they are classified. The choice of foundation type depends primarily on the local water depth at the proposed site.

3.2 Foundation types

3.2.1 Monopile

The most frequently used foundation type is the monopile. It commonly consists of a foundation pile and a transition piece, on top of which the turbine tower is placed, as shown in Fig. 12.

Foundation piles are made from steel plates which are rolled and welded together to form a cylindrical section. The conventional method of installation (see Fig. 13)

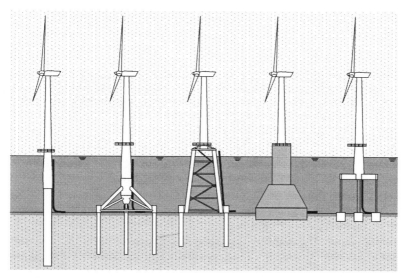

Figure 12: Foundation types, from left to right: monopole, tripod, jacket, GBS, floating.

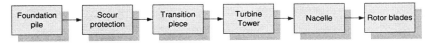

Figure 13: Installation sequence of main components for a monopile foundation.

Figure 14: The J-tube through which the power cable is directed to the seabed.

is by pile driving, whereby the foundation pile is driven into the seabed using hydraulic hammers.

If a foundation pile is designed to be driven into the seabed, a transition piece is required for the secondary structures such as J-tubes, boat landings and platforms as shown in Fig. 14.

If the seabed consists of rock, a borehole is prepared (drilled) in which the foundation pile is inserted. Since the foundation does not have to be designed for the impact forces of pile driving, the secondary structures can be attached directly to the foundation pile. Hence, no transition piece is needed.

Water currents flowing around the pile can, through erosion, create a depression in the seabed around the base of the pile, known as scour. The effect of scour on the foundation pile is as if it is positioned in deeper water with reduced soil-penetration depth. The depth of a scour hole depends on local currents and soil conditions, which is why it cannot be predicted accurately. Furthermore, an increased section of the pile is exposed to hydrodynamic loads. The increased length of the unsupported structure above the seabed may also result in a more dynamic behaviour.

To avoid scour, monopiles are provided with *scour protection*. Protection against scour is usually done by placing a "filter" layer of small stones around the pile. On top of that a layer of larger stones is positioned. The small stones keep the sand around the pile in place and the large stones keep the filter layer in place.

The relatively simple production and installation, together with the large range of exploitable water depths, have made the monopile the most widely used support structure concept. Its popularity has led the monopile to be developed for increasingly deeper waters. Monopiles are therefore likely to remain the most popular foundation type in the near future.

3.2.2 Tripod

A tripod foundation is a structure with three legs which diverge from a single node to their respective positions on the seabed. A foundation pile is driven into the ground at the base of each leg of the tripod section. On top of the tripod section, the turbine tower is placed. The procedure is visualised in Fig. 15.

Complications with production and installation make it relatively expensive. The main transition node where the three legs meet the central column is sensitive to fatigue. Stiffness benefits are only interesting in large water depths, but then the base becomes restrictively large.

The conventional installation method is to load several tripods onto a barge which is towed to the offshore site as depicted in Fig. 16. At a predetermined location, a structure is lifted of the barge, using a large crane (on the barge). Simultaneously, a smaller crane guides the tripod to its final position. The loads on the tripod

Figure 15: Installation sequence of main components for a tripod foundation.

Figure 16: Tripods on a barge, on their way to the Alpha Ventus wind farm.

structure will be mainly in axial direction. Therefore, scour protection is generally not required.

3.2.3 Jacket

Jackets, also known as space-frames or truss-towers, are relatively complex steel structures. Despite a reduction in construction materials (and hence weight), jackets are relatively expensive.

At the water depth where monopiles become uneconomical, jackets take over. However, due to ongoing developments, the monopile concept is used for increasingly deeper waters and the application of jackets is shifted to even deeper waters. For installation, a method similar to the one for tripods is applied (see Fig. 17).

3.2.4 Gravity-based structures

As the name implies, a gravity-based structure (GBS) utilizes the earth's gravitational force to stabilize its position.

GBSs usually have reinforced concrete foundations, referred to as caissons, in which a tower is placed. An example of GBSs for offshore wind turbines is shown in Fig. 18. Such GBSs are a proven technology for shallow waters. Occasionally they are used in deeper waters. The offshore wind farm Thornton Bank off the coast of Belgium applied reinforced concrete GBSs for a water depth of approximately 28 m.

Deeper waters require constructions with larger footprints in order to absorb greater moments. The increase of mass with water depth follows an approximately

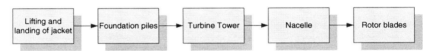

Figure 17: Installation sequence of main components for a jacket foundation.

Figure 18: GBSs for Thornton Bank, off the coast of Belgium.

quadratic relation. So even though building with concrete appears an economic choice, the large amounts required will make the structure relatively expensive.

An advantage with respect to installation is that a GBS can be floated out to its offshore location. A disadvantage is the necessary preparation of the seabed, which is done to provide the structure with a stable, horizontal floor. Another drawback is that, due to the great weight of the GBS, heavy lifting vessels are needed to perform the installation. The conventional installation sequence is shown in Fig. 19.

3.2.5 Floaters
The maximum water depth of wind farms has been steadily increasing over the last decade. Although monopiles will likely continue to be the most applied support structure for years to come, deeper waters appear to favour jacket structures.

Floating structures are seen by many as the solution to place wind farms in deeper waters (>70 μ). To keep it in place, the floating substructure is attached to the seabed through cables. In terms of installation costs, the question is whether such a system will require new installation procedures and dedicated vessels, or if it can simply be pre-assembled and transported by standard tugs (see Fig. 20).

4 Environmental loads

4.1 Waves

When calculating wave loads different wave categories can be distinguished, regular waves and irregular waves. Regular waves are periodic in nature and are usually associated with extreme load events. Irregular waves have a random appearance and are related to normal sea conditions and as such are to be adopted for fatigue evaluations.

For both regular and irregular waves several wave theories exist that allow the calculation of wave particle kinematics: the orbital motion, velocity and acceleration of infinitesimal quantities of water beneath the surface of the waves. Linear wave theory is valid for waves with infinitely small amplitudes, whereas non-linear wave theories are required for finite amplitude waves. Non-linear waves have a different surface profile compared to linear waves, with sharper, higher crests and longer and shallower troughs. Figure 21 shows which wave theory applies under

Figure 19: Installation sequence of main components for a GBS foundation.

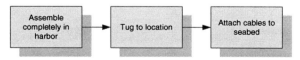

Figure 20: Proposed installation sequence for floating turbines.

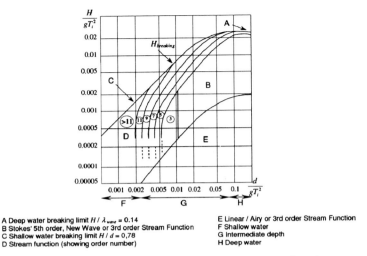

Figure 21: Range of application of various wave theories.

certain depth and wave steepness conditions. It can be seen that linear Airy wave theory can be applied in deep water waves with small steepness. Beyond this region non-linear wave theories such as Stokes' 5th order and stream function waves apply. This region in turn is limited by the wave breaking limit. In shallow water waves cannot grow higher than 0.78 times the water depth, while in deep water a wave will break if it grows too steep, with the wave height exceeding 0.14 times the wave length.

Linear Airy wave theory considers the surface elevation to be described by a harmonic wave:

$$\eta(x,t) = a\sin(\omega t - kx) \qquad (1)$$

Using potential theory and boundary conditions at the seabed and at the free surface a velocity potential Φ can be formulated corresponding to the surface elevation described as in the following equation:

$$\Phi(x,z,t) = \frac{\omega a}{k}\frac{\cosh k(h+z)}{\sinh kh}\cos(\omega t - kx) \qquad (2)$$

In this equation the term $\cosh k(h+z)/\sinh kh$ is the exponential decay function that describes the decrease of the intensity of the kinematics with increasing depth. By differentiating the velocity potential with respect to x and z the horizontal velocity u and the vertical velocity w can be derived, respectively, as follows:

$$u = \omega a\frac{\cosh k(h+z)}{\sinh kh}\sin(\omega t - kx), \quad w = \omega a\frac{\sinh k(h+z)}{\sinh kh}\cos(\omega t - kx) \qquad (3)$$

The accelerations can be determined by differentiation of the horizontal and vertical velocities with respect to t.

$$\dot{u} = \omega^2 a \frac{\cosh k(h+z)}{\sinh kh} \cos(\omega t - kx), \quad \dot{w} = \omega^2 a \frac{\sinh k(h+z)}{\sinh kh} \sin(\omega t - kx) \quad (4)$$

As the above formulations are based on linear wave theory, assuming small amplitude waves, the kinematics can only be calculated up to the still water surface. To allow for the calculation of the kinematics up to the instantaneous water surface elevation some kind of extrapolation is required. Several methods exist of which Wheeler stretching is the most common. Up till now the origin was assumed to be in the still water line, with the negative x-axis directed downward. By applying Wheeler stretching, the negative x-axis is stretched or compressed such that the origin is in the instantaneous water surface, yet intersects the seabed at the same z coordinate as the original z-axis. To this end a computational vertical coordinate z' is used that modifies the original coordinate z with the use of the dimensionless ratio q, which is dependent on the water depth h and the surface elevation ζ. Using Wheeler stretching therefore implies that the kinematics are calculated at an elevation z as if it is at an elevation z' :

$$z' = qz + h(q-1), \quad \text{with } q = h/(h+\zeta) \quad (5)$$

Using the formulations for the wave kinematics the wave loads on a structure can be computed. This can be done with the help of Morison's equation. This equation assumes the wave load to be composed of a drag load term and of an inertia load term. The drag term is dependent on the water particle velocity whereas the inertia term is induced by the accelerations of the fluid. Equation (6) shows how the Morison equation can be used to calculate the wave force on a cylindrical segment of unit height and a diameter D:

$$F(t) = \tfrac{1}{4} \pi \cdot \rho \cdot C_M \cdot D^2 \cdot \dot{u}(t) + \tfrac{1}{2} \cdot D \cdot C_D \cdot u(t) \cdot |u(t)| \quad (6)$$

From this equation it can be seen that the drag term is non-linear. Furthermore, due to the fact that the drag term is dependent on the velocity while the inertia term depends on the acceleration, the occurrence of the maximum drag force and the maximum inertia force are separated by a phase shift of $90°$.

Apart from the velocity and the acceleration of the water particle kinematics, the total wave force is dependent on a number of other parameters: the density of the surrounding water ρ and the hydrodynamic coefficients C_D and C_M. The drag coefficient C_D varies from 0.6 to 1.6, depending on the roughness of the cylinder and the Keulegan Carpenter number KC, a measure for the ratio between the wave height and the cylinder diameter. The inertia coefficient C_M can attain values ranging from 1.5 to 2.15, again depending on roughness and Keulegan Carpenter number. It should be noted that C_D increases with increasing roughness, whereas C_M decreases with increasing roughness. Finally the water depth, the water level above the still water surface and scour depth also influence the total wave load. Finally, marine organisms will accumulate on the structure below the water surface, thereby creating a layer of marine growth on the structure. This leads to an effective increase of the diameter, resulting in higher loads on the structure. This effect can be taken into account by adding twice the thickness of marine growth to the diameter of the member under consideration, without an increase in mass.

4.2 Currents

Sea currents may originate from a variety of sources. Friction of the wind with the water surface may lead to wind-driven currents. Tides also contribute to currents. Further sources of currents are density differences, due to temperature or salinity gradients, wind surge and waves.

Depending on the origin of the currents, the current is most pronounced at different depths. Wind-driven currents, for instance are felt strongest near the surface, while tidal currents may be stronger over the entire depth. Friction with the sea bed will result in a near-zero current velocity at the bottom. These effects require the use of different current profiles in different circumstances. While measurements may lead to accurate descriptions of the local current profile, in the absence of data standard current profile expressions can be used.

For subsurface currents the profile can be described by an exponential profile, which describes the decrease of the current velocity with increasing depth d from the current velocity $U_{c,sub}$ at the surface to zero at the seabed [2]:

$$U_{c,sub}(z) = U_{c,sub}\left(\frac{d+z}{d}\right)^{1/7} \tag{7}$$

For wind induced currents the following description can be adopted:

$$U_{c,wind}(z) = U_{c,wind}\left(\frac{d_0+z}{d_0}\right) \tag{8}$$

In this equation $U_{c,wind}$ is the wind induced current at the still water line and d_0 is a fixed depth at which the current is zero. If the local water depth is less than d_0 the current profile is cut off at the seabed. For water depths larger than d_0 the wind induced current is assumed to be zero for depths larger than d_0. Commonly used values for d_0 are 20 [2] and 50 [3].

For evaluation of the current loads only the drag term of the Morison equation is relevant, as the accelerations due to the variations in current velocity over can be neglected. Due to the non-linearity of the drag term, the current load cannot be evaluated separately from the wave load. For a correct evaluation of the total hydrodynamic load on a structure, the current velocity must be added to the wave particle velocity. As the direction of the wave particle velocity and the current velocity is opposite for half the wave cycle it is important to calculate the term u^2 as the velocity $(u_{wave} + u_{current})$ times its absolute value as shown in the following equation:

$$F(t) = \tfrac{1}{4}\pi \times \rho \times C_M \times D^2 \times \dot{u}_{wave}(t) + \tfrac{1}{2}\rho \times D \times C_D \times (u_{wave}(t) + u_{current}) \times |u_{wave}(t) + u_{current}| \tag{9}$$

4.3 Wind

Figure 22 shows a wind speed profile for a certain point in time. From this figure a number of characteristics can be deduced. First, that the mean wind is stronger at higher altitudes than near the surface of the earth. This is caused by friction of the moving air with the terrain. The effect becomes less pronounced as the altitude increases. The resulting difference in wind speed over altitude is called wind shear. Secondly, it is evident that the actual wind profile is very irregular. The actual wind speed deviates from the mean wind speed and direction as a result of turbulence. These two phenomena will be discussed briefly.

There are two commonly used models to describe wind shear: the logarithmic profile and the power law. The logarithmic profile is given by eqn (10), while eqn (11) describes the power law [4]:

$$\overline{V}(z) = \overline{V}_r \, \frac{\ln\!\left(\dfrac{z}{z_0}\right)}{\ln\!\left(\dfrac{z_r}{z_0}\right)} \tag{10}$$

$$\overline{V}(z) = \overline{V}_r \left(\frac{z}{z_r}\right)^a \tag{11}$$

In Fig. 23 both the logarithmic profile and the power law are shown. It clearly shows the difference between both models.

While the above gives a description for the mean wind speed, in reality the wind is never a steady flow of air that can be described with only one parameter. Local disturbances in the airflow called eddies cause the instantaneous wind speed to fluctuate around a mean value. This phenomenon is called turbulence. A measure

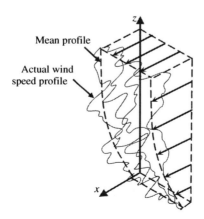

Figure 22: 3D turbulent wind velocity profile.

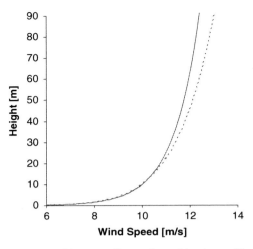

Figure 23: Wind shear profile according to logarithmic profile and power law.

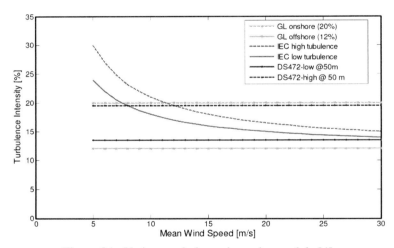

Figure 24: Various turbulence intensity models [4].

for the turbulence is given by the turbulence intensity I, which is defined as a function of the standard deviation and the mean wind speed as shown in the following equation:

$$I = \frac{\sigma}{\overline{V}} \cdot 100 \, [\%] \tag{12}$$

Recommended values for the turbulence intensity are given by various design standards. Figure 24 shows some turbulence intensity descriptions. It shows that the turbulence intensity is much higher onshore than offshore.

The wind loads on an offshore wind turbine can be split into operational loads on the turbine and loads on the structure. A description of the operational loads on the turbine and the load cases that should be considered can be found elsewhere in this work. The operational loads result in bending moments, normal forces and shear forces on the tower top.

The wind load on the tower structure itself results from drag forces only. To determine the total load on the tower structure the instantaneous wind speed should be evaluated at several elevations to account for wind shear. Subsequently, eqn (13) can be used to determine the drag force on each segment:

$$F_{tower}(t) = \frac{1}{2} \rho_{air} \cdot C_w \cdot D_{av} \cdot u_{wind}^2(t) \tag{13}$$

4.4 Soil

The soil contributes to the loading of the structure by providing the support reactions. In the case of piled foundations, these reactions are dependent on the lateral and axial pile–soil interaction. For GBSs the support reactions are generated by the vertical bearing capacity and the resistance against sliding.

Soil is generally a granular material, either cohesive such as clay, or non-cohesive such as sand. Other soil types that may be encountered are gravel, silt and peat. Soil originates either through erosion of rocks or through accumulation of organic material. Due to its geological history soil is highly inhomogeneous. The inter-particle voids are filled with water which may prevent or slow deformations [5].

The characterisation of loose to dense sand and soft to hard clay only gives a first indication of the ability of the soil to carry load. For design, more detailed knowledge is required. This is usually gathered through in-situ sampling and analysis of drilled samples in the laboratory. The first property measured for all types is the density ρ_{soil} (kg/m^3), usually for submerged soil, which is the dry density minus the density of water. A typical value is between 400 and 1000 kg/m^3. For clay, the undrained shear strength s_u and the strain at 50% of the maximum stress ε_{50} are measured. Table 1 gives an overview of typical values when no reliable soil data is available [4].

For sand the friction angle ϕ' and the relative density of sand D_r are derived directly from in-situ measurements. The initial modulus of horizontal subgrade reaction, k_s, can then be found with the graph in Fig. 25 [6].

Table 1: Characteristic parameters for clay.

Clay type	s_u (kPa)	ε_{50} (%)
Soft	0–25	1.5
Firm	25–50	1.5
Stiff	50–100	1.0
Very stiff	100–200	0.5
Hard	>200	0.5

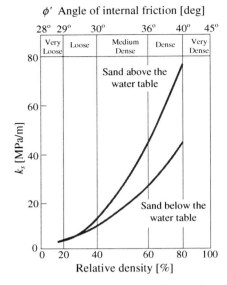

Figure 25: Initial modulus of subgrade reaction k_s as a function of friction angle ϕ' [6].

Due to its discontinuous nature soil particles can move with respect to the surrounding particles, thereby altering the structure of the soil. This creates a significantly non-linear behaviour which is usually described in terms of load displacement diagrams.

For a single pile in soil the pile–soil interaction can be described in terms of lateral resistance, shaft friction and end bearing. This behaviour is commonly modelled as non-linear load displacement curves: $P–y$ curves for the lateral resistance and $t–z$ and $Q–z$ curves for the shaft friction and the end bearing respectively. Figure 26 shows $t–z$ curves for sand and clay [6].

To model the soil reaction loads a set of soil springs is used. Figure 27 shows the springs for the horizontal and vertical direction as well as for the pile plug [4].

5 Support structure design

5.1 Design steps

The design of the support structure is an iteration between tuning the dynamic properties, optimising the amount of steel needed to resist all load cases and recalculating the loads on the optimised structure. Figure 28 shows the design steps that are typically required to come to a complete design of a support structure.

The different design steps have a strong interdependence and several iteration steps are normally required to come to an optimal design. For an entire offshore wind farm, some design details can be fixed. For instance, the hammer for installing

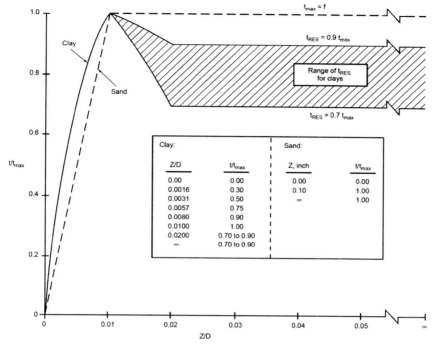

Figure 26: Load-displacement curves [6].

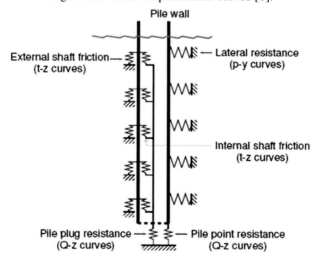

Figure 27: Modelling of pile–soil interaction [4].

the foundation piles can be of a single diameter, giving the designer less parameters to optimize.

In the previous chapter, the determination of design loads was treated. This chapter describes the steps to process this data and the turbine characteristics to come to a design of the structure.

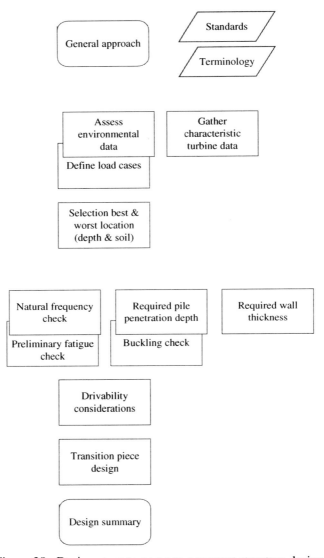

Figure 28: Design steps to come to a support structure design.

5.2 Turbine characteristics

The support structure has one main purpose: to keep the turbine up in the wind, where it produces energy. Wind turbines are fatigue machines by principle: with a rotation every 3 s on a desired availability above 98% per year for 20 years, a total of 200 million cycles. It is therefore key to design the support structure in such a way that the turbine dynamics and the support structure dynamics to not coincide.

Table 2: Turbine data.

	SWT 3.6 MW	V90 3.0 MW
Rotor		
Type	Three-bladed	Three-bladed
Diameter (m)	107	90
Swept area (m^2)	9000	6362
Turbine		
Minimum rotor speed (rpm)	5	8.6
Maximum rotor speed (rpm)	13	18.4
Blades		
Blade length (m)	52.0	44.0
Generator		
Nominal power (kW)	3,600	3,000
Tower		
Tower diameter (m)	3.051–5.000	2.300–4.200
Tower wall thickness (mm)	21–30	14–26
Operational data		
Cut-in wind speed (m/s)	4.0	4.0
Nominal power at approximate wind speed (m/s)	13.0	15.0
Cut-out wind speed (m/s)	25.0	25.0
Masses		
Nacelle + rotor mass (ton)	225	111

From the publically available turbine data, the required properties for support structure design can usually be gathered:

- turbine rotation speed range
- number of blades
- tower height
- turbine mass

Table 2 shows details for two commonly used turbine types.

5.3 Natural frequency check

From the turbine characteristics, the frequency ranges for the design of the support structure can be determined. The natural frequency of the structure should not coincide with the rotor speed range (1P) and blade passing speed range (3P for three-bladed turbines).

A first-order calculation of the natural frequency of a structure can be performed with the following simplified model. When the support structure is modelled as a mass on pole with the mass being the turbine mass and the pole a single diameter and wall thickness steel pile as depicted in Fig. 29.

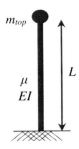

Figure 29: Structural model of a flexible wind turbine system.

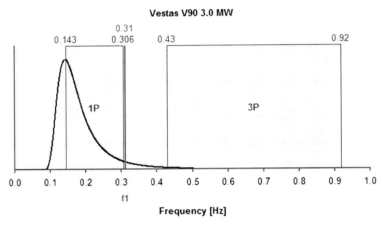

Figure 30: Frequency areas of 1P and 3P for the V90, with the designed natural frequency at 0.31 Hz between 1P and 3P to prevent dynamic interaction.

For this model consisting of a uniform beam with a top mass and a fixed base, the following approximation for the calculation of the first natural frequency is valid:

$$f_{nat}^2 \cong \frac{3.04}{4\pi^2} \frac{EI}{(m_{top} + 0.227\mu L)L^3} \qquad (14)$$

with f_{nat} is the first natural frequency (Hz), m_{top} the top mass (kg), μ the tower mass per meter (kg/m), L the tower height (m) and EI is the tower bending stiffness (N m^2).

The 1P and 3P areas can be plotted in a figure to visualize the zones in which the support structure natural frequency should not lie. In Fig. 30 this is shown for the V90 from Table 2 in the previous section.

The natural frequency will change through the next steps of design. It will need to be checked against this diagram to make sure it falls within the area between 1P and 3P.

For more detailed determination, the natural frequency will of course be calculated using a finite element model of the structure.

5.4 Extreme load cases

The main parameter resulting from the natural frequency check is the pile diameter. For the part of the support structure that is submerged, the diameter determines the hydrodynamic loads: waves and currents. The extreme load cases on an offshore wind turbine and the soil reaction to support those loads are shown in Fig. 31.

The rest of the loads are aerodynamic loads on the turbine and the tower. The combinations of these loads under different conditions are prescribed in the design standards. To take the probability of occurrence into account, several load combinations are prescribed: maximum 50-year wave load combined with a reduced 50-year gust event and the reduced maximum 50-year wave load combined with the full 50-year gust. An overview of load combinations is shown in Table 3.

5.5 Foundation design

Now that the global structural dimension and the design load cases are known, the foundation design can be detailed.

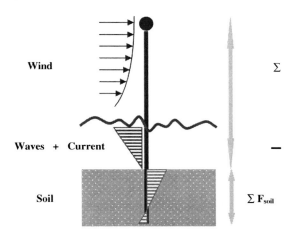

Figure 31: Extreme load cases on the offshore wind turbine and the soil reaction to support those loads.

Table 3: Load combinations [3].

Limit state	Load combination	Environmental load type and return period to define characteristic value of corresponding load effect				
		Wind	Waves	Current	Ice	Water level
ULS	1	50 years	5 years	5 years		50 years
	2	5 years	50 years	5 years		50 years
	3	5 years	5 years	50 years		50 years
	4	5 years		5 years	50 years	
	5	50 years		5 years	50 years	

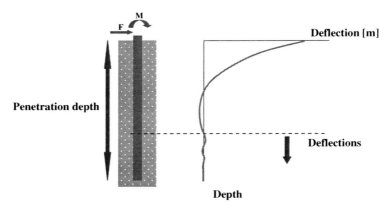

Figure 32: Pile deflection.

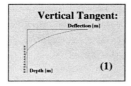
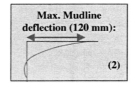
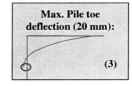

Figure 33: The three checks for foundation pile length design: vertical tangent at the pile toe, maximum deflection at the mudline, and maximum toe deflection.

The foundation pile can be modelled separately in a program incorporating detail soil models following the p–y method described in the previous chapter. The deflection of the pile is shown in Fig. 32. To start the pile penetration depth design, a pile length of seven times the pile diameter is chose to be sure the pile tip deflection is negligible.

With the foundation pile modelled in the finite element program, the pile length can be reduced while the following three checks are monitored after each step (Fig. 33):

1. the pile tip should reach a vertical tangent
2. the deflection of the pile at mudline is less than 120 mm
3. the deflection at the toe is less than 20 mm

5.6 Buckling & shear check

Now that the foundation pile has been modelled, also the pile–soil interaction is known and the deflection of the structure under different loads. The structural steel can now be checked for integrity under extreme load cases.

5.7 Fatigue check

The biggest step in optimising the design of the support structure is checking the fatigue. Fatigue is the phenomenon of slow deterioration of the steel due to continuous varying loads over time. For a fatigue check it is therefore vital to know the following details:

- environmental loads over the lifetime of the structure
- steel properties at the most severely loaded sections (typically: welds)
- fatigue resistance of the details of these welds: empirical S–N curves.

The long-term environmental loads are usually gathered in tables listing the simultaneous occurrence of wave height, period and direction, wind speed and direction and potentially several other parameters. For the fatigue check of the support structure the amount of data for a 20-year lifetime can accumulate to over 1000 load cases. To reduce these for the initial design stages, scatter diagrams are used. Table 4 shows a typical scatter diagram for wave height and period.

Such a wave scatter diagram is available for every wind speed bin (0–2 m/s wind, 2–4 m/s, 4–6). We then have the simultaneous occurrence of wave height, period and wind speed in a 3D scatter diagram.

To calculate the fatigue of the structure, we ideally need to run all these load cases through a wind turbine simulation program such as Bladed or Flex to incorporate all wind, wave and structure interactions an find the stress variations in each critical point of the structure. In the preliminary design stages, a reduced amount of data can be used that represents the most commonly occurring wind and wave conditions. Typically, the amount of data is reduced to 15 or 20 of these environmental states. An example for the Blyth turbines is shown in Table 5.

Table 4: Wave scatter diagram for H_s and T_z with occurrence in parts per thousand for the OWEZ location.

H_s (m)	T_z (s)								Sum
	0–1	1–2	2–3	3–4	4–5	5–6	6–7	7–8	
6.5–7.0									0.0
6.0–6.5								0.1	0.1
5.5–6.0							0.1	0.1	0.2
5.0–5.5							0.1	0.1	0.2
4.5–5.0							1		1.0
4.0–4.5							4		4.0
3.5–4.0						4	5		9.0
3.0–3.5						19	0.1		19.1
2.5–3.0					0.1	38			38.1
2.0–2.5					27	43			70.0
1.5–2.0				0.1	115	5			120.1
1.0–1.5				6	220	1			227.0
0.5–1.0				236	145	1			382.0
0.0–0.5	1		1	113	14	0.1			129.2
Sum	1.0	0.0	1.0	355.1	521.1	111.1	10.4	0.3	1000

Table 5: Summary of 15 environmental states for Blyth.

	H_s (m)	T_z (s)	V_w (m/s)	% of occurrence
1	0.25	2.0	5.0	20.47
2	0.25	5.2	4.9	3.73
3	0.25	4.0	11.8	21.76
4	0.25	5.6	15.7	3.85
5	0.25	5.8	20.6	1.00
6	0.75	3.4	6.7	8.62
7	0.75	5.3	5.8	13.25
8	0.75	5.5	11.7	5.58
9	1.25	5.2	8.8	10.66
10	1.25	8.0	8.5	1.25
11	1.75	6.0	9.9	4.83
12	1.75	6.7	16.2	0.55
13	2.4	6.8	12.8	3.54
14	3.4	7.8	14.5	0.77
15	3.3	9.7	18.7	0.14
				100%

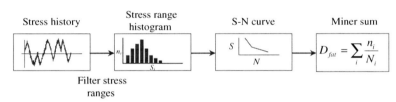

Figure 34: Flowchart of fatigue calculation due to variable stress ranges using S–N curve and Miner sum.

When the stress signal is determined for each location that needs to be checked, the fatigue calculation can be performed. Figure 34 shows the calculation steps: the stress history is converted to stress ranges via the rainflow counting method. The stress ranges are then checked against the S–N curve for the detail under consideration and the fatigue damage due to the load case is calculated using the Miner sum.

When the Miner sum is determined for each load case, it is multiplied by the percentage of occurrence during the design life of 20 years. The total fatigue damage is then found by adding the damage of all individual load cases together.

Should a detail not pass the fatigue check, changing the wall thickness will reduce the amount of stress and a re-calculation of the fatigue can be performed. To check the full fatigue of a monopile design requires several hours of computation time with current industry standard software. New methods of calculating the fatigue in the frequency domain show promising results and they have found their way to preliminary design calculations.

5.8 Optimising

The design steps described in this section have been treated on a high level only. The design process will involve several repetitions in which structural properties change in one step and require checking in all other steps again. Furthermore, some steps have not been treated here: the design of secondary steel and its impact on support structure loads and stress concentrations; the drivability analysis of the pile and the associated fatigue (the pile loses 25–30% of its fatigue resistance during installation); the impact of scour, corrosion and marine growth, etcetera.

6 Design considerations

6.1 Offshore access

The majority of the maintenance activities that are required during the entire life-time of an offshore wind farm consist of simple repairs rather than the replacement of turbine parts. Therefore, the accessibility to be treated here will involve personnel and light equipment only. The accessibility of a wind turbine depends first of all on the chosen access method. In the offshore industry there are two means of transportation used to reach offshore structures: helicopters and vessels.

6.1.1 Helicopters

Helicopters are used regularly to gain access to various offshore installations since they provide a fast means of transportation for personnel and light equipment at cruise speeds up to 250 km/h. Another big advantage of using helicopters is that both travel and access operations are not limited by wave conditions. If an offshore structure is equipped with a helicopter landing deck, the helicopter can land on this deck and passengers can safely board or exit the helicopter. However, mounting a landing deck on an offshore wind turbine would be unpractical. Instead, a hoisting platform can be placed on the turbine nacelle. The transfer of personnel from helicopter to turbine is then achieved by having the helicopter hovering above the turbine and hoisting people from the helicopter down to the platform on top of the turbine. Although this method is fast, disadvantages are the high costs of operation and the fact that a hoisting platform is required on each turbine. In addition, most exploiting parties are not eager to use this method due to the risks involved using helicopters: in case of a crash, the risk of casualties is high. In fact, the Horns Rev wind farm, located in the North Sea 14 km west of Denmark, is the only wind farm where helicopter hoisting is applied as a means of access. Furthermore, this method only allows transferring personnel with a very limited amount of tools and safe flying can be hampered by limited visibility and too large wind speeds. The accessibility of a helicopter is therefore determined by the percentage of the time that both wind speed and visibility are outside the restricted values.

6.1.2 Vessels

The use of vessels is a more cost-efficient and probably safer way of accessing offshore wind turbines than using helicopters. Currently, the most commonly used

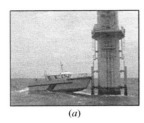  

(a) (b) (c)

Figure 35: Ship-based access to offshore wind turbines. Ships used: (a) WindCat; (b) Aaryan; (c) Moidart.

vessels for wind farm support are small vessels with lengths between 14 and 20 m, with either a single or a twin hull shape, and a bow section that is designed for access. Safe access is provided by intentionally creating frictional contact between the vessel and a boat landing structure on and in order to have no vertical vessel motions at the point of contact. A rubber bumper on the vessel bow forms this contact point; the thrusters push the boat against the structure to create the friction. The boat now pivots around the bumper and personnel can step from the vessel bow onto the turbine ladder safely. This method is generally being used for maintenance visits and applied by different types of vessels as shown in Fig. 35.

This ship-based access to offshore wind turbines is limited by wave conditions. As wave conditions get rougher, ship motions will become lager and there is a possibility that the vessel loses its contact with the boat landing. As a result, the vessel can start moving relative to the offshore structure. In this situation, the safety of the person accessing the turbine can no longer be guaranteed: the access procedure is no longer safe. For this reason, access operations are limited to certain wave conditions. The general way of describing the limiting wave conditions for access is by giving the limiting significant wave height for an access method. In wave conditions exceeding this limiting significant wave height, the access operation is considered too dangerous and will therefore not be performed.

6.1.3 Motion compensation systems: Ampelmann and OAS

The core of the problem when transferring people from a ship to a structure is that the vessel moves with the waves and the structure is stationary. The development of offshore wind sparked new innovations in this field. Several systems have been developed that compensate the wave motions partially or fully to remove the relative motion problem. The Offshore Access System is a hydraulic gangway that compensates the heave motion while connecting to the offshore structure. The offshore structure needs to be equipped with a landing station where the OAS grabs onto. As soon as the contact is made, the active compensation is switched off and the gangway hinges passively on both ends, as shown in Fig. 36.

The Ampelmann follows a different approach: it cancels all motions in the 6 degrees of freedom (surge, sway, heave, roll, pitch, yaw) to achieve a completely stationary platform. A gangway is then extended that is lightly pressed against the structure to allow quick and safe access. The additional advantage is that no landing station is

Figure 36: Access to offshore structure with OAS.

Figure 37: Ampelmann system for accessing offshore wind turbines.

needed, saving steel on each single support structure. The Ampelmann is shown in Fig. 37.

Both systems allow safe transfer in sea states up to $H_s = 2.5$ m, enabling maintenance crews to access the turbines over 90% of the year.

6.2 Offshore wind farm aspects

In this chapter the design of support structures was the main theme. The support structures and their turbines make up the offshore wind farm. The design of the offshore wind farm as a whole also has impact on the single support structures. The most pronounced items are summarised here.

6.2.1 Wind farm layout

The offshore wind farm layout is first and foremost determined by the consented stretch of seabed available for the farm. But within this area, optimisation on farm level is possible. The first design goal is to place the turbines close to each other to limit cable length, and as far apart as possible to increase power capture.

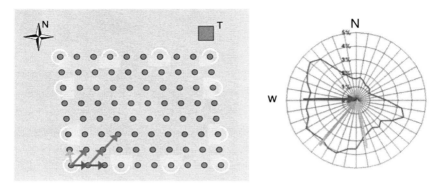

Figure 38: Wind farm layout with the wind rose on site.

Because the wind flow offshore is less turbulent, the mixing of the air behind the turbine (which has been slowed down due to the extraction of energy) takes longer to regain strength from the undisturbed wind around it. This means that turbines need to be placed further apart in offshore wind farms compared to onshore where turbulence intensity is higher. The rule of thumb is to place turbines at least seven times the rotor diameter apart. Smart layout configurations help increase the distance between the turbine in the governing wind direction to 10D or more, while maintaining the 7D minimum distance to make sure the cable length does not increase too much. Figure 38 shows such a layout with the wind rose on site.

6.2.2 Electrical infrastructure

The offshore wind power needs to be transported to the electricity grid. Within the offshore wind farm, in-field cables connect the turbines to each other. For most larger offshore wind farms (>100 MW) a transformer station is normally used. Here all in-field string cables come together. The total power is gathered and the voltage is increased to reduce losses to shore along the "shore connection cable". Typical voltages in fields are order of magnitude 33 kV, to shore 50 kV.

The in-field cable routing has influence on the design of the support structures: to most turbines two J-tubes will guide the incoming and outgoing cable, requiring secondary steel attachments that can add significant hydrodynamic loads. For installation purposes it can be very beneficial to choose a simple grid type of layout. Should a "star grid" or other, non-regular pattern be chosen, maintenance vessels may damage them in later years when there is confusion about the exact location.

6.2.3 The support structure in the offshore wind farm

Offshore wind farms are complex systems placed in a harsh environment, designed to operate on their own for large periods of time. The support structure is an integral part of these systems and should be treated as such. The design steps depicted in this chapter give a first rough guide to finding an optimised solution. The detailed design of the support structure requires intense cooperation between the different disciplines involved in offshore wind farm design and construction. Only then can the true potential be unleashed of the force we know as "offshore wind".

References

[1] Ferguson, M.C., et al. (ed), Opti-OWECS final report Vol. 4: a typical design solution for an offshore wind energy conversion system, Institute for Wind Energy, Delft University of Technology, 1998.
[2] Germanischer Lloyd, *Rules and Guidelines for the Design of Offshore Wind turbines*, Hamburg, Germany, 2004.
[3] DNV, *Design of offshore wind turbine structures*, Det Norske Veritas, DNV-OS-J101, 2004.
[4] Tempel, J. van der, Design of support structures for offshore wind turbines, PhD. Thesis, Delft University of Technology, Section Offshore Engineering, 2006.
[5] Verruijt, A., *Offshore Soil Mechanics*, Delft University of Technology, 1998.
[6] API, *Recommended Practice for Planning, Design and Constructing Fixed Offshore Platforms – Working Stress Design*, American Petroleum Institute, 21st edition, 2000.

PART IV

IMPORTANT ISSUES IN WIND TURBINE DESIGN

CHAPTER 18

Power curves for wind turbines

Patrick Milan, Matthias Wächter, Stephan Barth & Joachim Peinke
*ForWind Center for Wind Energy Research of the Universities
of Oldenburg, Bremen and Hannover, Oldenburg, Germany.*

The concept of a power curve is introduced, as well as the principles of the power conversion performed by a wind turbine. As an appropriate approach for the estimation of the annual power production of a wind turbine, the procedure to determine the power curve after the international standard IEC 61400 of the International Electrotechnical Commission (IEC) is discussed. As another approach is introduced a stochastic definition of a power curve which is based on high frequency measurement data and on the dynamic response of the wind turbine to wind fluctuations. The latter approach should be seen as a completion to the IEC definition which provides further insight into the dynamic performance of a wind turbine and may be used as a monitoring tool for wind turbines.

1 Introduction

The overall purpose of a wind turbine is to produce electrical power from wind. Quantifying this power output is necessary, on the one hand, for the financial planning of any wind energy project. On the other hand, besides the pure amount of energy production, also the dynamics of the power conversion contains essential information about, e.g. mechanical and electrical performance of the turbine and power quality. Following the turbulent behavior of the wind, the power production of a wind turbine fluctuates on short-time scales [1]. While exploiting the free, uncontrolled input that is the wind, it is of primary importance to control the stability of the power output of wind turbines. A large integration into energy networks supposes a good command of the power production, in terms of quantity, quality and availability.

To achieve such control, it is necessary to understand the behavior of wind turbines and quantify it. This is the scope of power performance techniques. This chapter introduces such methods, so as to estimate the performance of a wind turbine. The procedure applies to single wind turbines, while ongoing developments lead towards integration of entire wind parks, and possibly large networks.

The approach is restricted to large-scale horizontal-axis wind turbines here. It is also assumed that the produced electrical energy is directly fed into the grid. This is not an essential restriction but facilitates putting the presented work in a relevant context. For a detailed overview of different types of wind turbines and corresponding modes of operation see, e.g. [2].

2 Power performance of wind turbines

2.1 Introduction to power performance

In the past 30 years, recommendations and standards were defined to determine the power performance of a wind turbine. Permanently developed, the International Electrotechnical Commission (IEC) set the international standard IEC 61400-12 and its revised version IEC 61400-12-1 in 2005 [3]. These common guidelines defined the power performance characteristic of a wind turbine by the so-called power curve and its corresponding estimated annual energy production (AEP). The IEC standard gives a good estimation of long-time power production (through the AEP), which is of primary importance for an economical approach to wind energy.

Regarding actual power performance, the power curve is a powerful tool to estimate the power extraction process, as it quantifies the relation between incoming wind and power output of a wind turbine. Simultaneous measurements of wind speed $u(t)$ and power output $P(t)$ must be performed for the wind turbine concerned. Here the power output $P(t)$ is the net power released by the wind turbine into the electrical grid. From this data collection, a functional relation $P(u)$ can be defined, and a two-dimensional curve of power output vs. incoming wind speed can be derived. This is what power performance refers to. Such procedure can then be applied on single wind turbines in order to characterize their power performance, monitor their behavior over time as well as predict their power production. While this prediction is well described by the IEC definition of AEP, monitoring methods can be defined based on a dynamical approach to wind energy conversion.

In order to test power performance, a measurement of the wind velocity must be performed. As a wind turbine distorts the incoming wind field, a measurement in the rotor plane or closeby is not useful, at least not without further corrections. Instead the incoming upstream velocity is generally chosen as representative of the wind field, and measured from a meteorological mast at turbine hub height, a certain distance in front of the turbine. Based on these considerations, it becomes possible to quantify the power performance of a wind turbine in simple ways.

2.2 Theoretical considerations

The purpose of this section is to give a simple understanding of fluid mechanics applied to wind turbines. A detailed description of the formulas and derivations presented here can be found in [2]. This theoretical approach sets ground for the further power curve analysis.

In the following derivation, the complexity of turbulence will be set aside so as to understand the fundamental behavior of a wind turbine. Atmospheric wind has finite time and space structures, more commonly referred to as turbulent structures. Its statistics display complex properties like unstationarity or intermittency (such as gusts), whose effects will not be discussed in this section. They represent active research topics whose detailed analysis is outside the scope of this introduction, cf. [4]. In this section, a uniform flow at steady-state is considered.

Based on the fact that a wind turbine converts the wind power into available electrical power, one can assume the following relation:

$$P(u) = c_{\mathrm{p}}(u)P_{\mathrm{wind}}(u) \tag{1}$$

where $P_{\mathrm{wind}}(u)$ is the power contained in the wind passing with speed u through the wind turbine, and $P(u)$ is the electrical power extracted. The power coefficient $c_{\mathrm{p}}(u)$ represents the amount of power converted by the wind turbine. Because the input $P_{\mathrm{wind}}(u)$ cannot be controlled, improvements in wind power performance involve increasing the power coefficient $c_{\mathrm{p}}(u)$. Momentum theory can now be applied to determine this coefficient.

Consider a volume of air moving towards the wind turbine, which is modeled as an actuator disc of diameter D. A stream-tube is defined here as the volume of air that interacts with the turbine (see Fig. 1). The wind is affected by the wind turbine when crossing its swept area as the turbine extracts part of its energy. The extraction of kinetic energy accounts for a drop in the wind speed from upstream to downstream, as shown in Fig. 1. To ensure mass conservation, the stream-tube has to expand in area downstream, as shown in Fig. 1 [2].

Following this simple analysis, one can estimate the amount of kinetic energy available for extraction. The wind power $P_{\mathrm{wind}}(u)$ is derived from momentum theory for the wind passing with speed u through the rotor of area $\pi D^2/4$:

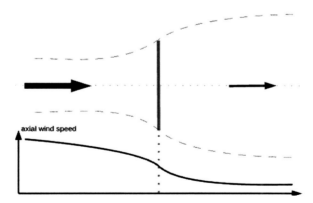

axial wind speed

Figure 1: A visual representation of an airflow on a wind turbine. The stream-tube is affected by the presence of the wind turbine that extracts part of its energy.

$$P_{\text{wind}}(u) = \frac{\rho}{2}\frac{\pi D^2}{4}u^3 \tag{2}$$

where ρ is the air density.

This stream-tube expansion shows that $c_p(u)$ has a physical limit called Betz limit such that $c_p(u) \leq 16/27 \approx 0.593$ [2, 5]. Regardless of its design, a wind turbine can thus extract at most 59.3% of the wind energy. Figure 2 shows the power coefficient as a function of $a = (1 - u_{\text{downstream}}/u_{\text{upstream}})$, the axial flow induction factor a gives the ratio of speed lost by the wind. The Betz limit corresponds to the maximum power a wind turbine can extract, when $a = 1/3$ [2]. This result is obtained for an actuator disc. The more complex shape of a real wind turbine certainly brings a lower limit for c_p. This physical limit is due to the stream-tube expansion induced by the presence of the turbine, i.e. by distorting the wind field, a wind turbine sets a limit for the energy availability. Criticism of this approach is given in [6, 7], leading to a less well defined upper limit of c_p.

Although it is based on a simplified approach, the Betz limit is a widely used and accepted value. The power coefficients of modern commercial wind turbines reach values of 0.45 and more. Physical aspects that limit the value of the power coefficient are, e.g. the finite number of blades and losses due to the drag and stall effects of the blades [2, 8].

Joining eqns (1) and (2), the theoretical power curve reads

$$P_{\text{theoretical}}(u) = c_p(u)P_{\text{wind}}(u) = c_p(u)\frac{\rho}{2}\frac{\pi D^2}{4}u^3 \tag{3}$$

$P_{\text{theoretical}}(u)$, or more simply $P(u)$ is the electrical power output and u is the input wind speed (u_{upstream}), i.e. a power curve is roughly characterized by a cubic

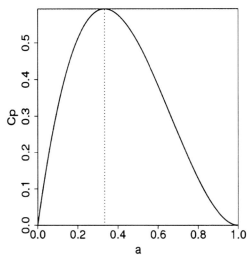

Figure 2: Power coefficient c_p as a function of the axial flow induction factor a.

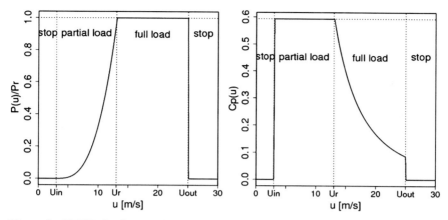

Figure 3: LEFT: Static (steady-state) power curve $P(u)$ of an active stall controlled wind turbine showing the different power operation states: stop, partial and full load. RIGHT: Corresponding power coefficient $c_p(u)$.

increase of the power output with the wind speed. The functional behavior of the power coefficient $c_p(u)$ is the result of certain control strategies as well as of Betz physical limit.

In the mechanical power extraction the usual way to control power production is achieved by stall effects on the rotor blades. Stall effects happen when the critical angle of attack for an airfoil is exceeded, resulting in a sudden reduction in the lift forces generated by the airfoil. In modern wind turbines this is achieved by so-called active stall control or pitch control [2]. This consists of a rotation of the blades into the plane of rotation and the blade cross-section. The blade rotation angle is known as blade pitch angle θ. The power coefficient c_p is in this case a function of the blade pitch angle θ and the tip-speed ratio $\lambda = \omega R/u$ (where ω is the angular velocity of the rotor, R the rotor radius, i.e. blade length and u the wind speed), i.e. $c_p = c_p[\lambda(u),\theta]$. Thus, the power extraction of wind turbines is optimized via $c_p[\lambda(u),\theta]$ to a desired power production. In particular for high wind conditions c_p is lowered to protect the turbine machinery and avoid overshoots in power production.

To achieve an efficient pitch control during wind energy conversion the wind turbine is equipped by a power controller system. This is generally composed of several composite mechanical–electrical components that, depending on the type of design, operate actively for the optimum power performance. As a consequence the power output operation for active stall wind turbine systems can be separated into two states: partial load, with maximum c_p value, and full load, with reduced c_p values. A complete detail of the overall structure of the power operation system for different wind turbine types is described in [2]. Numerical wind turbine simulation can be found, e.g. in [9]. The theoretical power curve $P(u)$ together with the corresponding $c_p(u)$ is represented in Fig. 3.

In partial load $u > u_{cutin}$, where u_{cutin} is the minimum wind speed for power production, the wind turbine yields the maximum wind energy extraction by power optimization operation. This is achieved by an effective power control system which adjusts to the desired pitch angle θ, at a given wind speed u, in order to optimize the power coefficient $c_p(u)$ and hence the power production. In practice a simple lookup table is the most used method for this operation [9]. The partial load area of the power curve is limited to the range $u_{cutin} < u < u_r$ where u_r is the rated wind speed.

In full load $u_r < u < u_{cutout}$, where u_{cutout} is the maximum wind speed (or shutdown wind speed) for power production, the wind turbine power output is limited to nominal or rated power. In this power setting typically the pitch angle θ is adjusted to control the power output to its rated power value P_r.

For $u > u_{cutout}$ the pitch angle θ is maximized (minimizing the angle of attack) to the feathered position in order to eliminate the lift forces on the rotor blades. As a consequence power generation is switched off (stopped).

The main properties of wind turbine power curves have been introduced so far. However, the theoretical power curve is derived from a laminar wind flow, which never occurs in real situations. The complexity of the wind, i.e. the turbulence needs more complex models to analyze power performance. Following the path of turbulence research, statistical models to deal with complexity will now be introduced.

2.3 Standard power curves

The power performance procedure for wind turbines defined by the IEC in 2005, and labeled IEC 61400-12-1 is now introduced. For a detailed description of the procedure, please refer to the complete proceeding [3]. This procedure provides a common methodology to ensure consistency, accuracy and reproducibility in the measurement and the analysis of power performance of wind turbines. It consists of the minimum requirements for a power performance test, as well as a procedure to analyze the measured data that can be applied without extensive knowledge.

The standard procedure first describes the necessary preparations for the performance test, such as criteria for the test equipments, guidance for the location and setup of the meteorological mast that will be used to measure the wind speed and other parameters like the wind direction, the temperature and the air pressure. The measurement sector is also described, as the range of wind directions that are valid for a representative measurement. Wind directions in the wake of the wind turbine must be excluded. A more detailed assessment of the terrain at the test site is provided in the optional site calibration procedure that reports for additional obstacles in addition to the wind turbine itself. The measurement procedure must be performed for the different variables, so that the data collection displays a sufficient quantity and quality to estimate accurately the power performance characteristics of the wind turbine.

The measured data is then averaged over periods of 10 min. These averaged values are used for the analysis, together with their corresponding standard errors (based on the standard deviations). A normalization must then be applied to the measurement data. Depending on the type of turbine, either the means of wind speed (for turbines with active power control) or of power output (for stall-regulated turbines) must be normalized to a reference air density.

The IEC power curve is then derived from the normalized values using the so-called method of bins, i.e. the data is split into wind speed intervals of a width of 0.5 m/s each. For each interval i, bin averages of wind speed u_i and power output P_i are calculated according to

$$u_i = \frac{1}{N_i} \sum_{i=1}^{N_i} u_{\text{norm},i,j} \quad \text{and} \quad P_i = \frac{1}{N_i} \sum_{i=1}^{N_i} P_{\text{norm},i,j} \tag{4}$$

where $u_{\text{norm},i,j}$ and $P_{\text{norm},i,j}$ are the normalized values of wind speed and power averaged over 10 min, and N_i is the number of 10 min data sets in the ith bin.

For the power curve to be complete, or reliable, each bin must include at least 30 min of sampled data and the entire measurement must cover a minimum period of 180 h of data sampling. The range of wind speeds shall extend from 1 m/s below cutin wind speed to 1.5 times the wind speed at 85% of the rated power P_r of the wind turbine. Such power curve was represented in Fig. 4 for a multi-MW wind turbine. Error bars were included following the recommendations below.

The standard also provides a description of the evaluation of uncertainty in the power performance measurement [3]. In a first step, the respective uncertainties are obtained from the measurement as the standard error of the normalized power data. Additional uncertainties are related to the instruments, the data acquisition system and the surrounding terrain.

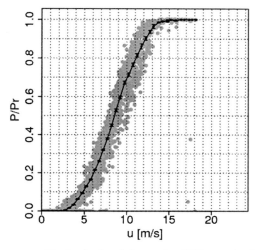

Figure 4: Power curve (line) obtained from the IEC standard procedure for a multi-MW wind turbine. Corresponding error bars were displayed. The grey dots represent 10-min averages. The power output is normalized by its rated value P_r.

Based on the IEC procedure, the AEP can be derived by integrating the measured power curve to a reference distribution of wind speed for the test site, assuming a given availability of the wind turbine [3]. The AEP is a central feature for economical considerations.

The standard procedure defined by the IEC offers some interesting insights. It is a great advance as it sets a common ground for wind power performance. As the wind industry develops, common standards help building a general understanding between manufacturers, scientists and end-users. The IEC procedure serves this purpose as the most widely used method to estimate power performance.

A detailed analysis of this standard is of great importance for anyone who wishes to test power performance. The procedure defines a set of important parameters, such as the wind direction, terrain corrections and requirements for wind speed measurements. These parameters are relevant for performance measurements, regardless of the final method used to handle data. The main strength of the method lies in the definition of these important parameters.

Unfortunately, the standard procedure presents important limits. In contrast to a good definition of the requirements above, the way the measured data is analyzed suffers mathematical imperfections. In order to deal with the complexity of the conversion process, the data is systematically averaged. A statistical averaging is indeed necessary to extract the main features of the complex process, and the central question is how to perform this averaging. The IEC method applies the averaging over 10-min intervals, which lack physical meaning. The wind fluctuates on various time scales, down to seconds (and less). A systematic averaging over such time scales as 10 min neglects all high frequency fluctuations present in the wind dynamics, but also in the dynamics of the extraction process. In combination with the fundamental non-linearity characteristics of the power curve, i.e. $P(u) \propto u^3$, the resulting power curve is spoiled by mathematical errors, as derived below [see eqn (7)].

One can split the wind speed $u(t)$ into its mean value and the fluctuations around this mean value:

$$u(t) = \langle u(t) \rangle + v(t) = V + v(t). \tag{5}$$

where $\langle u(t) \rangle$ represents the average (arithmetic mean) value of $u(t)$. Applying a Taylor expansion to $P(u)$ gives [10]:

$$P(u) = P(V) + \frac{\partial P(V)}{\partial u} v + \frac{1}{2} \frac{\partial^2 P(V)}{\partial u^2} v^2 + \frac{1}{6} \frac{\partial^3 P(V)}{\partial u^3} v^3 + o(v^4) \tag{6}$$

$$\langle P(u) \rangle \neq P(V) = P\big(\langle u \rangle\big) \tag{7}$$

It appears that the average of the power is not the power of the average, due to the non-linear relation $P(u) \propto u^3$ and the high frequency turbulent fluctuations. The IEC procedure gives $P(V)$ exactly $P(\langle\langle u(t)\rangle_{10min}\rangle_{bin})$, which neglects the high-order terms in the Taylor expansion. The resulting IEC power curve should be corrected by the second- and third-order terms.

As a consequence of this mathematical over-simplification, the result depends on the turbulence intensity $I = \sigma/V$ (where $\sigma^2 = \langle u^2(t) \rangle$) and on the wind condition

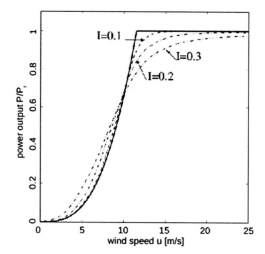

Figure 5: The effects of non-linearity of the power curve for turbulence intensities $I = 0.1, 0.2, 0.3$. The full line is the theoretical power curve $P(u)$ and the dotted line is the standard power curve given by the IEC procedure. The data has been obtained from numerical model simulations [10].

during the measurement [10]. Figure 5 illustrates this mathematical limit. The IEC power curve fails to characterize the wind turbine only, as the final result also depends on the wind condition during the measurement. For this reason, the IEC procedure cannot be fully satisfactory as a power performance procedure. The requirements for a measurement of power performance are well defined, but it is necessary to introduce a new method to process the measured data $u(t)$ and $P(t)$.

2.4 Dynamical or Langevin power curve

The averaging procedures within the IEC standard [3] induce the problem of systematic errors because of the non-linear dependence of the power P on the wind speed u in a wide range of u. Thus the standard power curve will depend not only on the characteristics of a turbine, but also on the wind situation, and on the conversion dynamics of a turbine. On the other hand, if no averaging is performed, the power conversion is discovered to be a highly dynamical system even on very short-time scales, as it can be seen in Fig. 6. Recently it could be shown that the statistics of the electrical power output of a wind turbine is close to the intermittent, non-Gaussian statistics of the wind speed [1].

2.4.1 Obtaining the Langevin power curve

To derive the power characteristic of a wind turbine from high-frequency measurements without the use of temporal averaging, one can regard the power conversion as a relaxation process which is driven by the turbulently fluctuating wind

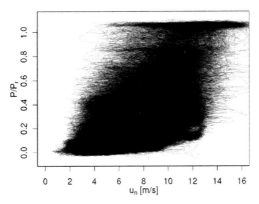

Figure 6: One hertz measurements of wind speed and electrical power output of a multi-MW wind turbine. Wind speed u_n is normalized to equal power at standard conditions, and power is normalized to rated power P_r, both according to IEC 61400-12-1 [3]. For a description of the measurements, see [11, 12].

speed, see also [13, 14]. For the (hypothetical) case of a constant wind speed u, the electrical power output would relax to a fixed value $P_s(u)$. Mathematically, these power values $P_s(u)$ are called *stable fixed points* of the power conversion process.

It is possible to derive them even from strongly fluctuating data as shown in Fig. 6. To this end the wind speed measurements are divided into bins u_i of 0.5 m/s width, as it is done in [3]. It is thus possible to account to some degree for the non-stationary nature of the wind, and obtain quasi-stationary segments $P_i(t)$ for those times t with $u(t) \in u_i$. The following mathematical considerations will be restricted to those segments $P_i(t)$. For simplicity, the subscript i will be omitted and the term $P(t)$ will refer to the quasi-stationary segments $P_i(t)$.

The power conversion process is now modeled by a first order stochastic differential equation, the Langevin equation (which is also the reason for the name Langevin Power Curve):

$$\frac{d}{dt} P(t) = D^{(1)}(P) + \sqrt{D^{(2)}(P)} \cdot \Gamma(t). \tag{8}$$

Using this model, the evolution of the power signal is described by two terms. The first one, $D^{(1)}(P)$, represents the deterministic relaxation of the turbine, which leads the power towards the fixed point of the system. According to this effect, $D^{(1)}(P)$ is commonly denoted as *drift function*. The second term involving $D^{(2)}(P)$, serves as a simplified model of the turbulent wind which drives the system out of its equilibrium. The function $\Gamma_{(t)}$ denotes Gaussian distributed, uncorrelated noise with variance 2 and mean value 0. $D^{(2)}(P)$ is commonly denoted as *diffusion function*. More details on the Langevin equation can be found in [15, 16]. For the power curve, only the deterministic term $D^{(1)}(P)$ is of interest. The stable fixed points of the system are those values of P where $D^{(1)}(P) = 0$. If the system is

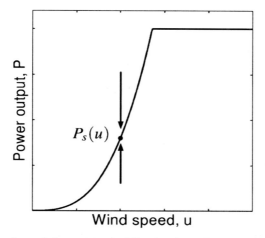

Figure 7: Illustration of the concept of fixed points. For constant wind speed u, the power would relax to the stable value $P_s(u)$. The deterministic drift $D^{(1)}(P)$, denoted by vertical arrows, drives the system towards this fixed point (see text). The sketch was taken from [17].

in such a state, this means that no deterministic drift will occur (see also Fig. 7). (To separate stable from unstable fixed points, also the slope of $D^{(1)}(P)$ has to be considered [16].) $D^{(1)}(P)$ can be interpreted as an average slope of the power signal $P(t)$, depending on the power value. For the stable fixed points this drift function vanishes because for constant u the power would also be constant, and thus the average slope of the power signal would be zero. A simple functional ansatz for $D^{(1)}$ would be $D^{(1)}(P) = \kappa[P_s(u) - P(t)]$, where $P_s(u)$ is a point on the ideal power curve (see the explanation of P_s below).

Using their definition [15], the drift and diffusion functions can be derived directly from measurement data as conditional moments (called Kramers-Moyal coefficients):

$$D^{(n)}(P) = \lim_{\tau \to 0} \frac{1}{n\tau} \left\langle \left(P(t+\tau) - P(t) \right)^n \middle| P(t) = P \right\rangle, \qquad (9)$$

where $n = 1$ for the drift and $n = 2$ for the diffusion function. The average $\langle \cdot \rangle$ is performed over t. The condition inside the brackets means that the difference $[P(t + \tau) - P(t)]$ is only considered for those times for which $P(t) = P$. This ensures that averaging is done separately not only for each wind speed bin u_i but also for each level of the power P. If one considers the state of the power conversion system as defined by u and P, one could speak of a "state-based" averaging in contrast to the temporal averaging performed in [3].

Using the mathematical framework of eqns (8) and (9), also uncertainty estimations can be performed for the fixed points. For details of the derivation, the reader is kindly referred to [16, 17]. Figure 8 presents the dynamical power characteristic

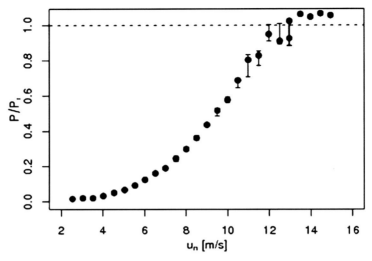

Figure 8: Dynamical power characteristic of a multi-MW wind turbine. Wind speed u_n is normalized to equal power at standard conditions, and power is normalized to rated power, both according to IEC 61400-12-1 [3]. For a description of the measurements, see [11, 12].

of a multi-MW turbine [11, 12] which has been derived following the procedure outlined above, including error bars. It can be seen that for most wind speeds the power characteristic has very little uncertainty. Nevertheless, in the region where rated power is approached larger uncertainties occur. Here it can be assumed that the state of the power conversion is close to stability over a range of power values. These larger uncertainties of the fixed points can thus be interpreted as a consequence of the changing control strategy of the wind turbine from partial to full load range. It is of great interest how different turbines behave here, and may be more or less power efficient.

It is important to note that in [11] dynamical power characteristics have been calculated using wind measurements taken by cup, ultrasonic, and LIDAR anemometers. All three power characteristics were identical within measurement uncertainty, showing that this approach appears quite robust concerning the wind measurements.

2.4.2 Summary

The Langevin equation (8) clearly is a simplifying model of the complex power conversion process. On the other hand, the drift function $D^{(1)}(P)$, see eqn (9), is well defined for a large class of stochastic processes and not restricted to those which obey the Langevin equation.

A central feature of the dynamical approach is the use of high frequency measurement data as shown in Fig. 6, which enables the analysis of the short-time dynamics of the power conversion process. The usage of the drift function eliminates systematical errors caused by temporal averaging combined with the non-linear

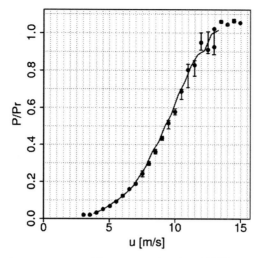

Figure 9: Langevin power curve (dots + error bars) and IEC power curve (line) for a multi-MW wind turbine (same as Fig. 8). The power output is normalized by its rated value P_r.

dependency of the power on the wind speed. The results are therefore site independent because effects of turbulence have no influence on the dynamical power curve. An interesting feature of this approach is the ability to show also additional characteristics of the investigated system. Examples are regions where the system is close to stability, as mentioned above, or multiple stable states, see also [17, 18]. Because of these features the dynamical power curve is a promising tool for monitoring the power output of wind turbines.

3 Perspectives

Different tools have been defined in the previous section to estimate power performance. The IEC power curve, in spite of being a good introduction to the topic, cannot characterize the conversion process of a wind turbine objectively, i.e. the result depends on the wind condition. The Langevin power curve, on the other hand, provides robust results that can be applied to determine and monitor the dynamical behavior of a wind turbine. Rather than competing against each other, these two power curves, when plotted together, can quickly bring useful insights on the health of any (horizontal-axis) wind turbine.

An overview of the available applications will be presented in this section.

3.1 Characterizing wind turbines

A striking feature is that the Langevin power curve offers new, complementary information to the IEC power curve. The two power curves are presented together in Fig. 9.

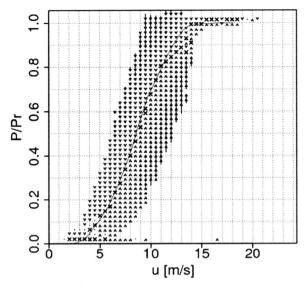

Figure 10: Drift field $D^{(1)}(P;u)$ (arrows), Langevin power curve (crosses) and IEC power curve (background line) for a multi-MW wind turbine. The intensity of the drift field is proportional to the length of the arrows. The power output is normalized by its rated value P_r. This figure represents the same wind turbine as Fig. 4.

Indeed, a striking limitation of the IEC method lies in the way it discretizes the domain $\{u, P\}$. As detailed in the Section 2.3, the IEC method averages all data in speed bins of size $\delta u = 0.5$ m/s. The domain is discretized only for the wind speed, resulting in a unique point every 0.5 m/s for the IEC power curve. The IEC power curve is one-dimensional, it is the line $P_{IEC}(u)$ (as represented in Fig. 9).

The dynamical method, however, discretizes the domain $\{u, P\}$ on both wind speed and power output. The resulting power curve, derived from the drift field $D^{(1)}(P;u)$, is a two-dimensional quantity. As shown in Fig. 10, each point in the domain displays a vector indicating how fast (length of the vectors) and in which direction the system wants to evolve.

Obviously in low power but high wind regions, the vectors point upwards to higher power values. Correspondingly, high power but low wind regions vectors point downwards to smaller power values. This is shown in Fig. 10.

This mathematical framework is necessary to observe the dynamics of a wind turbine. This point is crucial to characterize power performance. Thanks to the dynamical method, multi-stable behaviors can be observed, and a greater insight can be reached. This is easily seen in Fig. 10, where multiple fixed points appear in several speed bins. Multi-stable behavior appears in the slow region ($u \approx u_{cutin} = 4$ m/s) and in the fast region ($u \approx u_r = 13$ m/s), where complex dynamics take place. In the slow region, the turbine transits between the rest and the partial load modes.

In the fast region, the transition happens between the partial load and the full load modes. The dynamics of these transitions can be observed in Fig. 10. In both regions, the two different modes of operation are clearly separated by the Langevin power curve, while the IEC power curve averages both modes into an intermediate value.

The two different wind turbines represented in Figs 9 and 10 were very well characterized by the Langevin power curve. The method applies to all (horizontal-axis) wind turbines, even when presenting complex dynamics. Wind turbines equipped with multiple gears were characterized successfully using this method, when the IEC method revealed its limitations. The Langevin power curve is a powerful tool to visualize and quantify power performance.

3.2 Monitoring wind turbines

Monitoring is closely related to the idea of characterization (introduced in the previous section), as it follows the evolution in time of the power characteristics. Once a machine was characterized using the dynamical approach, it becomes possible to compare and monitor power performance on a regular basis. Dynamical anomalies can be rapidly brought to light when deviations appear on two consecutive Langevin power curves. The precision of the method allows localizing the anomaly in the domain $\{u; P\}$, giving more insight towards the deficient component of the wind turbine. Applied on a monthly (or even weekly) basis, such monitoring can prevent anomalies from limiting the power production, or worse, damaging other components of the wind turbine.

3.3 Power modeling and prediction

Once a machine was characterized using the dynamical approach, basically once the drift field $D^{(1)}(P; u)$ and the diffusion coefficient $D^{(2)}(P; u)$ were computed, it becomes possible to model the power output $P(t)$ from any input wind speed time series $u(t)$. The Langevin equation [see eqn (8)] can be solved knowing $D^{(1)}(P; u)$ and $D^{(2)}(P; u)$ to generate a realistic power output $P(t)$.

A simple, artificial case was created in Fig. 11. In this figure, one can see the real power output $P(t)$ of a wind turbine, then the same quantity modeled in good running condition, and finally modeled with an artificial anomaly (that limits the power extraction to roughly 45% of the rated power P_r). A comparison of the first and second graph shows that through a simple model, it is possible to estimate the power output $P(t)$ of a wind turbine knowing only $D^{(1)}(P; u)$ and $D^{(2)}(P; u)$. This model can be applied to any wind situation $u(t)$, as an effective way to study the behavior of a wind turbine in different wind conditions.

This model shows great potential in the continuing evolution of current methods, principally in the prediction of power production. When coupled with a meteorological wind forecast, the model could be used to generate the power output of a wind turbine (whose power performance has been characterized). In addition to providing quantitative power production estimates, power quality, i.e. fluctuations in power, stability, and regularity of the high frequency power output $P(t)$ too will

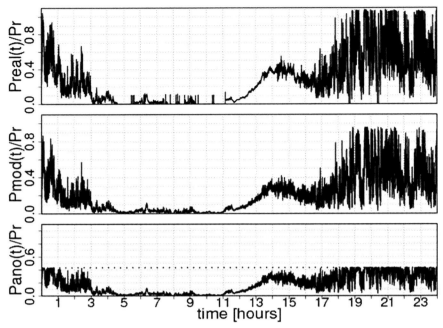

Figure 11: Time series for the power output $P(t)$. The upper graph shows the real power output measured on a multi-MW wind turbine. The graph in the middle represents the power output $P_{model}(t)$ modeled for the same turbine. The lower graph shows the power output $P_{anomaly}(t)$ modeled for the same turbine, but spoiled by an artificial anomaly. A horizontal line represents the artificial limitation in power due to the anomaly. The power output is normalized by its rated value P_r. The three graphs are given for the same time window of 24 h.

be assessable. Predictions of power quantity, and especially power quality are of major importance for a large integration of wind energy into the electrical networks. However, important developments towards a high frequency wind forecast $u(t)$ need to be performed first. Efforts towards such developments are being made.

4 Conclusions

Power curves for wind turbines establish a relation between wind speed and electrical power output. This relation is essential for project planning and operation of wind turbines. Also for the monitoring of turbines, concerning proper operation and detection of possible misconfiguration or failures, the power curve is an important tool. It can therefore be considered as a central characteristic of a wind turbine.

The current industry standard IEC 61400-12-1 [3] defines, among others, a uniform procedure for the measurement of power curves. This definition relies on temporal averaging of wind speed and power output. Due to the turbulent nature of the wind and the non-linear dependency of power on wind speed, this power curve combines the characteristic of the turbine together with the statistical features of the wind at the special site under investigation. This combination makes the estimation of the annual energy yield at a certain site especially easy. On the other hand, systematic averaging errors are introduced through the mentioned non-linearity, and the power characteristic of the turbine cannot be separated from the site effects. These weaknesses are well known, and several corrections have been proposed, e.g. [19].

As an alternative, recently a different approach has been proposed to obtain the power characteristic of wind turbines [16, 17], the Langevin power curve, which relies on high frequency measurement data (approximately 1 Hz). Inspired from dynamical systems theory, the power conversion process is regarded as a relaxation process, driven by the turbulently fluctuating wind speed. The power characteristic can then be obtained for every wind speed as the stable fixed points of this process. Averaging errors and influence of turbulence are thus avoided. Possible multiple stable states are also captured, allowing deeper insight in the dynamics of the power conversion. These features make the dynamical power characteristic especially interesting as a monitoring tool for wind turbines.

As a work in progress, the simulation of high frequency power output signals based on eqn (8) is currently developed. One application of this procedure will be the prediction of energy yields for specific wind turbines under specific wind conditions.

References

[1] Gottschall, J. & Peinke, J., Stochastic modelling of a wind turbine's power output with special respect to turbulent dynamics. *J. Phys: Conf Ser*, **75**, pp. 012045, 2007.

[2] Burton, T., Sharpe, D., Jenkins, N. & Bossanyi, E., *Wind Energy Handbook*, Wiley: New York, 2001.

[3] IEC. Wind turbine generator systems, Part 12: Wind turbine power performance testing, International Standard 61400-12-1, International Electrotechnical Commission, 2005.

[4] Böttcher, F., Barth, S. & Peinke, J., Small and large fluctuations in atmospheric wind speeds. *Stochastic Environmental Research and Risk Assessment*, **21**, pp. 299–308, 2007.

[5] Betz, A., Die Windmühlen im Lichte neuerer Forschung. *Die Naturwissenschaften*, **15**, pp. 46, 1927.

[6] Rauh, A. & Seelert, W., The Betz optimum efficiency for windmills. *Applied Energy*, **17**, pp. 15–23, 1984.

[7] Rauh, A., On the relevance of basic hydrodynamics to wind energy technology. *Nonlinear Phenomena in Complex Systems*, **11(2)**, pp. 158–163, 2008.

[8] Bianchi, F., De Battista, H. & Mantz, R., *Wind Turbine Control Systems*, 2nd ed., Springer: Berlin, 2006.

[9] Hanse, A., Jauch, C., Soerense, P., Iov, F. & Blaabjerg, F., Dynamic wind turbine models in power system simulation tool DIgSILENT, Risø Report Risø-R-1400(EN), Risø National Laboratory, 2003.

[10] Böttcher, F., Peinke, J., Kleinhans, D. & Friedrich, R., Handling systems driven by different noise sources – Implications for power estimations. *Wind Energy*, Springer: Berlin, pp. 179–182, 2007.

[11] Wächter, M., Gottschall, J., Rettenmeier, A. & Peinke, J., Dynamical power curve estimation using different anemometer types. *Proc. of DEWEK*, Bremen, Germany, 2008.

[12] Rettenmeier, A., Kühn, M., Wächter, M., Rahm, S., Mellinghoff, H., Siegmeier, B. & Reeder, L., Development of LiDAR measurements for the German offshore test site. IOP Conference Series: *Earth and Environmental Science*, **1**, pp. 012063 (6 pages), 2008.

[13] Rosen, A. & Sheinman, Y., The average power output of a wind turbine in turbulent wind. *Journal of Wind Engineering and Industrial Aerodynamics*, **51**, pp. 287–302, 1994.

[14] Rauh, A. & Peinke, J., A phenomenological model for the dynamic response of wind turbines to turbulent wind. *Journal of Wind Engineering and Industrial Aerodynamics*, **92(2)**, pp. 159–183, 2004.

[15] Risken, H., *The Fokker-Planck Equation*, Springer: Berlin, 1984.

[16] Gottschall, J. & Peinke, J., How to improve the estimation of power curves for wind turbines. *Environmental Research Letters*, **3(1)**, pp. 015005 (7 pages), 2008.

[17] Anahua, E., Barth, S. & Peinke, J., Markovian power curves for wind turbines. *Wind Energy*, **11(3)**, pp. 219–232, 2008.

[18] Gottschall, J. & Peinke, J., Power curves for wind turbines – a dynamical approach. *Proc. of EWEC 2008*, Brussels, Belgium, 2008.

[19] Albers, A., Jakobi, T., Rohden, R. & Stoltenjohannes, J., Influence of meteorological variables on measured wind turbine power curves. *Proc. of EWEC 2007*, Milan, Italy, 2007.

CHAPTER 19

Wind turbine cooling technologies

Yanlong Jiang

Department of Man-Machine - Environment Engineering,
Nanjing University of Aeronautics and Astronautics, China.

With the increase of the unit capacity of wind turbines, the heat produced by different components rise significantly. Effective cooling methods should be adopted in developing larger power wind turbine. In this chapter, the operating principle and main structure of wind turbines are firstly described, following with the analysis of heat production mechanisms for different components. On this basis, current cooling methods in wind turbines are presented. Also, optimal design of a liquid cooling system for 1 MW range wind turbine is conducted. Finally, some novel cooling systems are introduced and discussed.

1 Operating principle and structure of wind turbines

In brief, the operating principle of a wind turbine is that rotation of impellors driven by wind power converts the kinetic energy of wind into mechanical energy of the impellor shaft, which drives the generator. There are mainly two types of wind turbine operating modes. One is the independent power-supply system, which is usually used in the remote areas, where electric network is not available. The terminal electrical equipments are powered by alternating current, which is converted by a DC–AC converter from the electricity in a storage battery charged by small scale wind turbines. Generally, the unit capacity is from 100 W to 10 kW. Or a hybrid power-supply system comprising a middle scale wind turbine and a diesel generator or solar cells with capacity, range from 10 to 200 kW, is adequate to meet the need of a small community. In another wind turbine operating mode, the wind turbines are used as a power resource of an ordinary power network, paralleling in the electricity grid system. It is the most economic way to utilize wind power in a large scale. This mode can synchronize and close with a unit independently and also can be made of multiple, or even thousands of wind turbines, called wind farm [1–3].

As shown in Fig. 1, a wind turbine working in a parallel operation is mainly comprised of an impeller, a nacelle, a pylon, a foundation and an electric transformer. Among these components, the impeller is wind collecting device, including blades and hub. It can convert wind power at a certain height to mechanical energy, representing as shaft rotation at a low speed but with high torque. The nacelle, comprised of a gearbox, a generator and control systems, is the core component of the wind turbine where the mechanical movement is accelerated, then converted to electric energy with modulated frequency to meet the demands of parallel operation. The pylon and foundation are mostly used to support the nacelle and impeller to a certain height and ensure the safe operation. The function of the electric transformer is to perform the voltage regulation to the output electricity so as to transfer power efficiently.

To sum up, the operating procedure of a wind turbine is as follows: the impeller rotating under the wind force action drives the main shaft in the nacelle to rotate simultaneously. This movement is then accelerated in the gearbox, and supplies the high-speed revolution for the generator rotor by connecting with high-speed shaft. The rotor cuts the magnetic lines of force, and thus produces electric energy. With the increasing unit capacity of wind turbines, the length of impeller blades and the height of pylon are gradually increased for the purpose of capturing more wind energy.

2 Heat dissipating components and analysis

It is well known from the operation principle mentioned in Section 1, the nacelle is the core component for a wind generating set and also the concentrated area of heat production in the operating process. The configuration of the nacelle is shown in Fig. 2, and the mechanisms of heat production for different components are explained as follows.

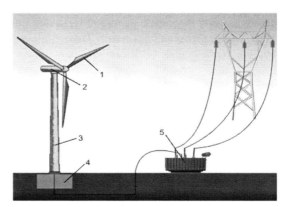

Figure 1: Sketch of a wind turbine generator connecting to power system: 1, impeller; 2, nacelle; 3, pylon; 4, foundation; 5, transformer.

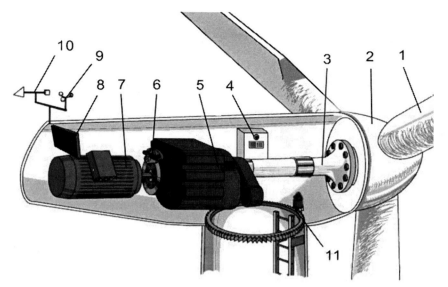

Figure 2: Sketch of a wind generating set [4]: 1, impeller blades; 2, hub; 3, main shaft; 4, controller; 5, gearbox; 6, mechanical brake; 7, generator; 8, cooling system; 9, anemoscope; 10, wind vane; 11, yawing motor and yawing bearing.

2.1 Gearbox

The gearbox is the bridge connecting the impeller and the generator. Since the rotational speed of an impeller is between 20 and 30 rpm, and the rated speed of a generating rotor is from 1500 to 3000 rpm or even higher, therefore a gearbox has to be installed between the impeller and the generator to accelerate the low-speed shaft. The running gearbox causes some power loss, most of which transfers into heat and is absorbed by the lubricating oil and, thus, causes temperature rising in the gearbox. If this temperature becomes too high, it will deteriorate the performance of lubricating oil, causing lower viscosity and shorter drain period. Moreover, it also increases the possibility of damage to the lubricating film under load pressure, which leads to impairment of the gear meshing or the bearing surface and, eventually, the equipment accident. Therefore, restriction of temperature rise in the gearbox is a key prerequisite for its endurable and reliable operation [5]. On the other hand, in winter, when the ambient temperature is below 0°C, heating measure for the lubricating oil in gearbox should also be taken into consideration in order to avoid lubricating oil from failing to splash onto the bearing surface due to high viscosity in low temperature, and, therefore, prevent impairment of the bearing from short of lubrication. Normally, every large-scale wind turbine gearbox contains a compelling cooling system and a heater for lubricating oil. However, in some regions where the temperature seldom drops below 0°C, such as the coastal areas in Guangdong Province, China, heaters can be an exemption [6].

2.2 Generator

The generator rotor is connected to the high-speed shaft of the gearbox. It drives the generator to rotate at a high speed and to cut the magnetic lines of force, by which electric energy is obtained. During the operation of a wind turbine, the generator will produce a huge amount of heat mainly in its windings and various internal wastes of iron core, primarily comprised of iron loss, copper loss, excitation loss and mechanical loss [7]. Besides, the temperature rise of the generator also has a correlation with power, operational condition, and duration of runs [8]. Moreover, there is a tendency of the unit-capacity enlargement of wind turbine which can be implemented by magnifying winding factor or magnetic field intensity. Since adding electromagnetic load is unsatisfactory with the restriction of magnetic saturation, at present, a popular method for enlarging the unit capacity is to increase inductance coil load. However, by applying this method, copper loss of bar will rise, which results in high coil temperature, acceleration of insulation aging and, eventually, damage of the machine. Because of this, a proper cooling method should be applied to control the internal temperature of various components of the generator within a permissible range. Hence, it can be concluded that the enlargement of the unit capacity of wind turbine mainly depends on the improvement of the cooling technology [9, 10].

2.3 Control system

As the wind speed and direction are changing all the time in the operation of wind turbine, auxiliary apparatus should be installed to adjust the operating status promptly to ensure the secure and stable operation of the wind turbine. The common system auxiliary apparatuses include: anemoscope, wind vane, yawing system, mechanical brake and thermometer. The anemoscope and the wind vane are used to detect immediate wind status; and the thermal sensor is responsible for monitoring the temperature changes in the generator and gearbox. When the operating status changes, the anemoscope, the wind vane and the thermal sensor will feed back the detected signal to the control system in the nacelle, then the input signal is diagnosed and processed by the control system and finally output to the yawing system and the mechanical brake, which changes the operating status of the wind turbine. Meanwhile, the control system has functions of displaying and recording parameters such as instantaneous mean wind speed, mean wind direction and mean power and other operating parameters. In addition, frequency converter is equipped in the control system, which aims at converting the unstable frequency of wind turbine signal to suffice to the demands of parallel operation. Therefore, the control system is also called control converter. In the operation, as a core component for the failure-free operation of wind turbine, the control system will produce a large amount of heat, which needs to be taken away timely.

3 Current wind turbine cooling systems

As has been mentioned above, in the operation of wind turbine, the gearbox, generator and control system will produce a large amount of heat [11]. In order to ensure the secure and stable operation of wind turbine, effective cooling measure has to be implemented to these components. Since the early wind turbines had lower power capacity and correspondingly lower heat production, the natural air cooling method was sufficient to meet the cooling requirement. As the power capacity increases, merely natural air cooling can no longer meet the requirement. The current wind turbines adopt forced air cooling and liquid cooling prevalently, among which, the wind generating set with power below 750 kW usually takes forced air cooling as a main cooling method. As to large- and medium-scale wind generating set with power beyond 750 kW, a liquid recirculation cooling method can be implemented to satisfy the cooling requirement [11].

3.1 Forced air cooling system

The forced air cooling system comes up where a znatural air cooling system cannot meet the cooling demands. When the air temperature in the wind turbine exceeds a certain prescribed value, to achieve the cooling objective, the control system will open the flap valve connecting internal and external environment of the nacelle and, meanwhile, fans installed in the wind turbine are switched on, which produce forced air blast to the components inside the nacelle. As the performance of air cooling ventilation system has a decisive influence on the cooling effect and operating performance of the wind turbine, the ventilation system should be well designed [9]. Thus, the design of the ventilation system is vital to an air cooling system project.

In the implementation of a forced air cooling system, different combinations are chosen according to the amount of system heat production and heat dissipation of various components. For a wind turbine with a power below 300 kW, since the heat dissipation of the generator and control the converter is relatively low, their heat is removed mainly by the cooling fans installed on the high-speed shaft, and the gearbox is cooled using a method of splash lubrication due to the rotation of the gear, where the heat of formation (or producing heat) is delivered through the gearbox and additional fins to the nacelle, and finally taken away by the fans. The cooling performance is mainly subject to the ventilating condition in nacelle [5]. By comparison, a wind turbine with power capacity beyond 300 kW possesses a comparatively larger heat production and, therefore, it is not sufficient for the gearbox to control the temperature rise only by the cooling fan installed on the high-speed shaft and the radiated rib on the box. The method of lubricating oil circulation can realize effective cooling. The basic operating procedure is described as follows: the gearbox is configured with an oil circulation supply system, driven by a pump and an external heat exchanger. The oil temperature can be adjusted under the permissible maximum value by regulating the oil delivery rate and the wind speed flowing through the heat exchanger according to the temperature rise status of the lubricating oil. This circulating lubrication cooling method is mature and secure in performance, while, on the other hand, it introduces a set of attachments which costs about 10% of the gearbox's manufacturing cost [5]. Considering the cooling for the increasing heat production in the generator and the converter, it can be implemented by enlarging the internal ventilation space and internal air passage of coil. Usually, the generator has both internal and external fans. And the radiating rib with an internal air passage is welded on the outer edge of the stator frame. Thereby, the internal circulating cooling air follows a circuit flowing through the terminal stator winding, iron core and the internal air passage of the radiating rib, while the external cooling air flows directly through the surface of the radiating ribs, as shown in Fig. 3 [12]. Theoretically, the more input air and the higher speed of the fan, the better the cooling effect. However, this will lead to increase flow resistance and power consumption, all of which result in a lower generator efficiency. Therefore the working condition of the generator fan should be designed rationally [13].

Comparing with other cooling method, the forced air cooling system has several advantages, such as simple structure, easy management and maintenance, and low initial and running cost. However, since the cooling air is from external environment, the cooling performance might become low because of the environment

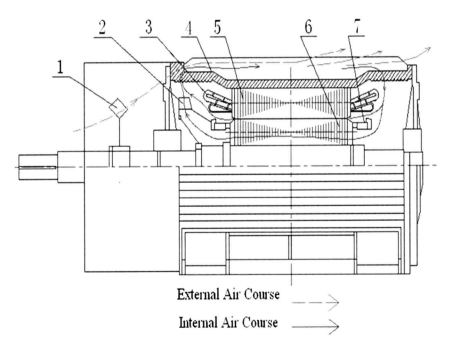

Figure 3: Forced air cooling method for generator: 1, external fan; 2, internal fan; 3, stator winding; 4, stator frame; 5, stator iron core; 6, rotor iron core; 7, rotor winding.

changes. Furthermore, during the ventilation of the nacelle, the severe corrosion on the set possibly caused by blown sand and rain goes against the long-term secure operation of the set. As the power capacity of the wind generating set keeps increasing, merely adopting forced air cooling method could not meet the cooling demands. Hence, liquid cooling systems are emerging.

3.2 Liquid cooling system

From the thermodynamics knowledge, the thermal equilibrium equation of a wind turbine cooling system can be described as $Q = q_m C_p (t_1-t_2)$, where Q is the total system heat, q_m is the mass flux of the cooling medium, C_p is the mean specific heat at constant pressure of the cooling medium between temperature t_1 and t_2. t_1 and t_2 are the inlet and outlet temperature of the cooling medium. As the liquid medium's concentration and specific heat capacity are much greater than that of the gaseous medium, the cooling system adopting liquid medium can obtain much larger cooling capability as well as a more compact system structure which can solve the problem of low cooling output and the enormous size of the air cooling system. The structure of the cooling system is shown in Fig. 4.

During the operation of a wind turbine, the cooling medium firstly flows through the oil cooler, exchanging heat with lubrication oil and taking away the heat produced by the gearbox. Then it flows into the heat exchanger fixed around the stator

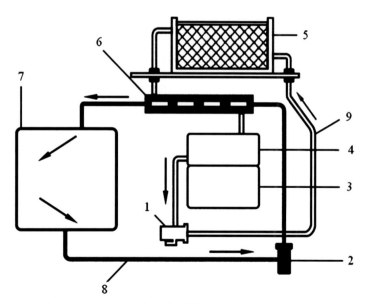

Figure 4: Cooling system adopting liquid cooling method [9]: 1, water pump; 2, oil pump, 3, generator; 4, generator heat exchanger; 5, external radiator; 6, oil cooler; 7, gearbox; 8, lubricating oil pipeline; 9, cooling medium pipeline.

winding, absorbing the heat produced by the generator. Finally, it will be pumped out and get cooled by an external radiator, by which the flow is prepared for the next cycle of heat exchange. In normal working condition, the cooling water pump always stays in working mode to deliver the internal heat to the external radiator through cooling medium. And the lubricating oil pump can be controlled by the temperature sensor in the gearbox. When the oil temperature exceeds the rated value, the pump switches on, delivering the oil to the oil cooler outside the gearbox; while the oil temperature falls below the rated value, the circuit is cut off to stop the cooling system. Besides, as the control converter in each wind generating set varies to each other, there will be difference in the amount of heat produced among these converters. When the heat production is relatively low, the forced air cooling generated by the fan fixed in the nacelle is sufficient for the control converter and other heat producing components; while if the heat production is comparatively large, a radiator outside the control converter can be installed to control its temperature rise through cooling medium taking away the heat in the same way of gearbox and generator.

With respect to the MW wind turbine with a larger power capacity, the gearbox, generator and control converter all produce comparatively large amount of heat. As shown in Fig. 5, cooling these components mentioned above usually needs two independent sets of cooling system – one shared by the generator and control converter and the other for the gearbox [14]. In an oil cooling system, the lubricating oil is pumped up to lubricate the gearbox; the heated oil is then to be delivered to the oil cooler on top of the central nacelle to be cooled by forced air. The cooled lubricating oil is then delivered back to the gearbox for use of the next cycle. A liquid cooling system is a closed-loop system containing an ethylene glycol aqueous solution-air heat exchanger, a water pump, valves, and control devices for temperature, pressure and flux. The cooling medium in the closed-loop system flows through the generator and the control converter to take away their produced heat.

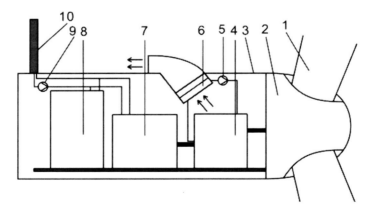

Figure 5: A cooling system for one MW wind turbine [14]: 1, blade; 2, hub; 3, nacelle; 4, gearbox; 5 and 9, hydraulic pump; 6, oil cooler; 7, generator; 8, converter; 10, heat exchanger.

Then it gets cooled in the external radiator on top of the rear of the nacelle, and finally runs back to the generator and the control converter to begin the next cooling cycle.

At present, the cooling mediums commonly used in the liquid cooling system are water and ethylene glycol aqueous solution. Comparing with water, ethylene glycol aqueous solution has better anti-freeze property. Table 1 shows freezing points of ethylene glycol aqueous solution in different densities. By adding a certain amount of stabilizers and preservatives, the minimum working temperature can extend to −50°C, but keeps its heat transfer performance equivalent to that of water [19].

Besides, in order to enhance the heat-exchange performance, the external heat exchanger adopts an effective and compact plate-fin structure, which is usually made of the light metal, aluminum. The heat exchanger exposed to the external environment is prone to be corroded, which will affect the durable, reliable operation of the heat exchanger. Therefore, necessary anti-corrosion treatments need to be implemented, like coating the aluminum flakes with anti-corrosive allyl resin coverings and employing hydrophilic membranes on its outer surface. Having been treated with this method, the acid rainproof of the aluminum fins and the anti-salt corrosion property can be 5–6 times as large as those of the ordinary ones. In the design of the heat exchanger, due to relatively large difference of the cooling system operating loads in winter and summer, the summer operating mode is adopted as the design condition, while the heat transfer efficiency can be controlled through a bypassing method in winter.

Comparing with the wind turbine adopting the air cooling method, the one adopting liquid cooling system has a more compact structure. Although it increases the cost of heat exchanger, cooling medium and corresponding laying of connecting pipelines, it extremely enhances the cooling performance for the wind generating

Table 1: Freezing points of ethylene glycol aqueous solution in different densities.

Density (%)	Freezing point (°C)
0	0
5	−2.0
10	−4.3
20	−9
30	−17
40	−26
50	−38
60	−50.1
70	−48.5
80	−41.8
85	−36
90	−26.8
100	−13

set, and thus facilitates the generating efficiency. Meanwhile, the design of the sealed nacelle prevents the invasion of wind, blown sand and rain, creating a good working surrounding for the wind turbine, which greatly extends the duration of the devices.

4 Design and optimization of a cooling system

As has been mentioned above, the increasing power capacity of wind turbines calls for a matching cooling system. With the widespread use of MW wind turbines, the liquid cooling system has been prevalently used in current wind turbines. Accordingly, the design and optimization of a liquid cooling system is briefly introduced in this section. Since currently very few researches are conducted on the heat dissipating regularity in wind turbine operation and experimental data are scarce, the following research is based on a steady working condition, where the heat production of the generating set is under a steady-state condition. According to the ambient conditions and technical requirements provided by wind turbine companies, the liquid cooling system is designed and analyzed under the maximum heat load. On this basis, the commercial software, MATLAB, is used for the purpose of optimal design, and the interaction and mechanism of action are investigated among parameters, such as wind speed, fin combinations, etc. These researches are somehow valuable to be referred to for the design and optimization of the MW wind turbine cooling system.

4.1 Design of the liquid cooling system

The cooling system of one certain MW wind turbine is shown in Fig. 5. This section proposes the design of the liquid cooling system for the generator and the control converter, which is shown as follows [14]. And as the designs of oil cooling system and liquid cooling system are basically the same, contents on those will be excluded due to restriction of the article length.

4.1.1 Given conditions

This MW wind turbine is located in the coastal area with a temperature ranging from −35 to 40°C. The start-up wind speed is 4 m/s, while the shutdown wind speed is 25 m/s. The relationship between the generated output P and wind speed $V_{c,in}$ is shown in Fig. 6. Other initial parameters are shown in Table 2. The objective is to design a liquid cooling system to meet the cooling demands of the wind turbine and to control its structural sizes to be most favorable for the durable operation of the wind turbine based on the giving ambient conditions and technical requirements from the wind turbine companies. Focusing on this objective, this section introduces how to select key components and explain the method of optimal computation to obtain the size of the ethylene glycol aqueous solution-air-typed heat exchanger.

4.1.2 Selection of the cooling medium

To meet the technical requirement of −35°C for the minimum ambient temperature in winter, the ethylene glycol aqueous solution with a concentration of 50% and a freezing point of −38°C is picked according to Cao [15] and Tan [16].

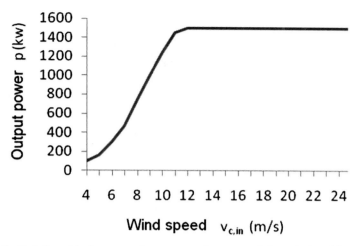

Figure 6: Relationship between the generated output of the wind turbine and the wind speed [14].

Table 2: Given parameters.

Items	Generator	Control converter	External radiator
Efficiency, η	97%	–	–
Heat dissipation (kW)	3% of the output	19	–
Maximum inlet water temperature (°C)	50	45	–
Flux (l/min)	50	60	–
Pressure loss (MPa)	0.08	0.1	≤0.01 liquid side
External dimensions (m × m × m)	–	–	$1.900 \times 0.820 \times 0.200$

4.1.3 Selection and design of the radiator

Normally, the operating performance of a cooling system mainly depends on the selection and the design of the heat exchanger. The heat exchanger in a practical operation should be, more or less, vibration-proof, because the vibration in the nacelle is driven by wind. In addition, if the wind turbine is located in a coastal area with comparatively high humidity, the heat exchanger should be corrosion-proof as well. Considering all the requirements mentioned above, the final choice for the radiator is an aluminum plate-fin heat exchanger with not only high heat transfer efficiency, but also a compact, light and firm structure [17, 18]. As shown in Fig. 7, where Channel A is air-flow passage, and B is the channel for ethylene glycol aqueous solution. The distribution of the channel is ABABABAB.... The detailed design of this cross-current plate-fin heat exchanger can be referred to Wang [17] and only necessary introduction is covered in this section due to space limitation.

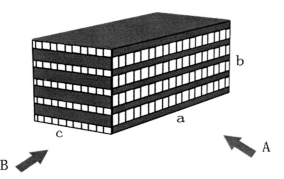

Figure 7: Core unit of a cross-current plate-fin heat exchanger [17]: (A) air-flow
passage; (B) ethylene glycol aqueous solution channel; (a) the width of
the core unit; (b) height of the core unit; (c) thickness of the core unit.

4.1.3.1 Selection of the fin unit and related dimension calculation

1. Calculation of the heat transfer area of air side and liquid side.
   Assuming that the density, thermal coefficient, constant-pressure specific heat
   capacity and kinetic viscosity of air and ethylene glycol aqueous solution stay
   constant in the heat transfer, their values are selected according to the inlet and
   outlet mean temperature.
2. Calculation of heat transfer temperature difference and heat transfer coefficient.
3. Fin efficiency and surface efficiency of the air side and liquid side.
4. Total heat transfer coefficient of the air side and liquid side.
5. Checking and calculating heat exchanger thickness.

After obtaining the heat transfer coefficient and logarithmic mean temperature
difference of both air side and liquid side, the real transfer area and heat exchanger
thickness can be calculated. If the actual calculated thickness c_{real} of heat exchanger
does not equal the given c, the value of c should be reassumed and calculated following
steps (1)–(5) of the flow path until the calculated c_{real} equals the default c.

4.1.3.2 Calculation to other parameters of the heat exchanger

1. Pressure loss on the liquid side.
   In order to meet the technological requirement and the pump selection requirement, the resistance of the heat exchanger should be checked in the design
   process. When the fluid is in a pump circulation in the plate-fin heat exchanger,
   the resistance calculation can be divided into three parts, i.e. inlet tube, outlet
   tube and central part of the heat exchanger [17].
2. Calculation of heat exchanger efficiency and weight.

4.1.3.3 Selection of the head plate for the plate-fin heat exchanger
According to Liu *et al.* [19] and Zhou *et al.* [20], staggered perforated plate header
is selected in order to obtain well-proportioned flux distribution and well-controlled
fluid friction loss.

4.1.3.4 Anti-corrosion measures

The heat exchanger exposed to the external environment is prone to be corroded, which will affect the durable, reliable operation of the heat exchanger. Therefore, necessary anti-corrosion treatments need to be implemented, like coating the aluminum flakes with anti-corrosive allyl resin coverings and employing hydrophilic membranes on its outer surface. Having been treated with this method, the acid rainproof of the aluminum fins and the anti-salt corrosion property can be 5–6 times as large as those of the ordinary ones. In the design of the heat exchanger, due to relatively large difference of the cooling system operating loads in winter and summer, the summer operating mode is adopted as the design condition, while the heat transfer efficiency can be controlled through a bypassing method in winter.

4.1.4 Flow resistance calculation of the liquid cooling system and pump selection

The liquid cooling pipeline system is comprised of a steel tube part and a pressure hose part. In view of the various factors, the following pipe diameters should be selected: steel tube and pressure hose diameter of the main trunk $D_1 = 48$ mm, branch steel tube and pressure hose's diameter $D_2 = 42$ mm. The on-way resistance and local resistance can be calculated based on the selected tube diameter, with which the circulating pump can be selected.

4.2 Optimization of the liquid cooling system

Based on the design method mentioned above, by utilizing MATLAB software, the optimization of the liquid cooling system is performed. Since the external radiator is the core component of the liquid cooling system, its structural dimension has an important impact on the cooling effect of the wind turbine and the weight of the nacelle. The subject of optimization in this section is the external radiator shown in Fig. 5. The constraint conditions are: the external radiator is fixed in the frame on top of the rear of the nacelle, with a limitation of frame size of 1.900 m × 0.820 m × 0.200 m; and the actual maximum size of the core unit of the external radiator is 1.800 m × 0.800 m × 0.200 m excluding the size of stream sheet and head. Under these conditions, the optimization procedure is shown as follows.

4.2.1 Derivation of the thickness of the heat exchanger core unit

The functional relation of the thickness of the heat exchanger can be derived from the heat transfer equation and the heat transfer coefficient equation and so forth as follows:

Total heat transfer:

$$Q = k_h \Delta t_m F_h \tag{1}$$

where Q is the heat transfer quantity of the heat exchanger, k_h is the total heat transfer coefficient on the liquid side, Δt_m is the heat transfer mean temperature difference, F_h is the total heat transfer area on the liquid side, given as

$$F_h = f_1(c, cc, ch) \tag{2}$$

where c is the thickness of the core unit of the heat exchanger, cc is the dimension of the fin unit on the airside, and ch is the fin unit dimension on the liquid side.

From eqns (1) and (2), the core unit thickness is obtained as

$$c = f_2(Q, k_h, \Delta t_m, cc, ch) \tag{3}$$

From the known condition:

$$Q = f_3(v_{c,in}) \tag{4}$$

Total heat transfer coefficient:

$$k_h = f_4(\alpha_c, \alpha_h, \eta_{0,c} \cdot \eta_{0,h}, F_c, F_h) \tag{5}$$

Heat transfer coefficient on the airside:

$$\alpha_c = f_5(v_{c,in}, cc) \tag{6}$$

Heat transfer coefficient on the liquid side:

$$\alpha_h = f_6(v_h, ch) \tag{7}$$

Flow velocity of the fluid:

$$v_h = f_7(c, cc, ch) \tag{8}$$

Fin efficiency on the air side:

$$\eta_{0,c} = f_8(cc, a_c) \tag{9}$$

Fin efficiency on the liquid side:

$$\eta_{0,h} = f_9(ch, a_h) \tag{10}$$

Total heat transfer area on the airside:

$$F_c = f_{10}(c, cc, ch) \tag{11}$$

From eqns (2), (5) and (11), the total heat transfer coefficient based on the total heat transfer area on liquid side can be expressed as

$$k_h = f_{11}(c, cc, ch, v_{c,in}) \tag{12}$$

Heat transfer mean temperature difference,

$$\Delta t_m = f_{12}(t_{c,in}, t_{c,out}, t_{h,in}, t_{h,out}) \tag{13}$$

where $t_{c,in}$ and $t_{c,out}$ represent the inlet and outlet temperature of the air, $t_{h,in}$ and $t_{h,out}$ are inlet and outlet temperatures of the ethylene glycol aqueous solution respectively, in which $t_{c,in}$ and $t_{h,out}$ are known quantities.

In addition that

$$t_{c,out} = f_{13}(cc, ch, v_{c,in}, Q) \tag{14}$$

$$t_{h,in} = f_{14}(Q) \tag{15}$$

From eqns (13) and (15):

$$\Delta t_m = f_{15}(c, cc, ch, v_{c,in}) \tag{16}$$

After substituting eqns (4), (12) and (16) into eqn (3), the functional relation of the heat exchanger's thickness can be simplified to:

$$c = f_{16}(cc, ch, v_{c,in}) \tag{17}$$

On the basis of the deduced relational expression of the heat exchanger's thickness, the thickness dimension is optimized with a method as follows.

Assume that when the wind turbine is running the wind speeds $v_{c,in}$ are under n different circumstances and, thus, there will be n pairs of generated output values and heat dissipation values corresponding to them. After choosing a dimension pair of the fin ('cc' and 'ch' in the equation), n different thicknesses of the heat exchanger core unit ('c') would be obtained, matching n circumstances, respectively, according to the above equations. On this basis, by changing Z types of fin pairs on the air and liquid sides, Z heat exchanger core unit thicknesses meeting design requirements (c_{max1}, c_{max2}, ..., c_{maxz}) can be obtained; therefore Z corresponding resistance on the liquid side and the heat exchanger weight can be obtained. The optimization computing task of the heat exchanger core unit is to find an air-and-liquid-side fin pair solution that not only can meet the cooling demands under various working condition, but also is able to minimize the system power consumption or the total weight of the system.

4.2.2 Optimization procedure of the heat exchanger core unit

1. As the wind turbine usually works under the condition that the wind speed exceeds 8 m/s, thus only the condition with a wind speed ranging from 8 to 25 m/s will be considered. Giving a state point every time by increasing speed of 1 m/s, the wind will be with 18 different velocities. The rated heat dissipating capacity of the radiator corresponding to various wind velocities can be obtained from the generator power graph, shown in Table 3 and Fig. 6.
2. Based on the overall consideration of the maximum rated inlet temperature required for the generator and the control converter as well as the temperature rise of fluid in the pipeline network, the radiator outlet ethylene glycol aqueous solution temperature can be selected as: $t_{h,out} = 43°C$. Other hypotheses are the same with the statement in Section 4.1.
3. Assuming that the airside fin and the liquid side fin are selected from one of the five types of straight fins and one of the five types of serrated fins, respectively, the collocation types for the air and liquid side fin pairs sums up to 25, with their specific parameters shown in Table 4.

Table 3: Relationship between inlet air velocity and heat dissipation of the heat exchanger [21].

$v_{c,in}$ (m/s)	Q (kW)
8	41.5
9	49
10	56.5
11	62.5
12	64
13	64
14	64
15	64
16	64
17	64
18	64
19	64
20	64
21	64
22	64
23	64
24	64
25	64

Table 4: Parameters of the air and liquid side fin pairs [21].

Parameter	Types of the air side fin pairs					Types of the liquid side fin pairs				
	cc1	cc2	cc3	cc4	cc5	ch1	ch2	ch3	ch4	ch5
Fin height, L_c (mm)	12	9.5	6.5	4.7	3.2	3.2	4.7	6.5	9.5	12
Fin height, δc (mm)	0.15	0.2	0.3	0.3	0.3	0.3	0.3	0.3	0.2	0.15
Fin interval, mc (mm)	1.4	1.7	1.7	2.0	4.2	4.2	2.0	1.4	1.7	1.4

The optimization procedure is shown in Fig. 8. The computational procedure is as follows. Firstly, choose an air and liquid side fin pair. Secondly, read the wind velocity and rated heat dissipating capacity of the heat exchanger and then calculate the heat exchanger thickness c to satisfy these conditions using an iterative method. Finally, calculate the weight of the heat exchanger core unit, pressure drop on the liquid side and other parameters like heat exchanger efficiency and so forth until all the calculation completes.

4.2.3 Interpretation of the optimization computing result

4.2.3.1 Wind condition numbers corresponding to the calculated heat exchanger thicknesses larger than 0.2 m based on various fin pair collocations

After choosing any air and liquid side fin pair, 18 heat exchanger thicknesses can be obtained corresponding to 18 wind conditions in order to match the cooling

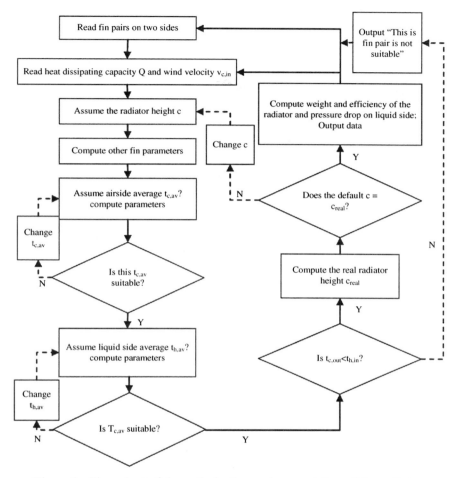

Figure 8: Flow chart of the optimization and computation of the radiator.

demands of the system. The calculated results by MATLAB forms are in Fig. 9. It can be concluded from the figure that, as the fin height on the airside declines, the fin height on the liquid side rises and the wind condition numbers with a heat exchanger thickness exceeding 0.2 m increases correspondingly, thus leading to an unreasonable collocation.

4.2.3.2 Selected heat exchanger thickness corresponding to different fin pairs and the weight of the heat exchanger

When the 18 values of c (heat exchanger thickness) are all less than 0.2 m corresponding to 18 wind conditions, it indicates that this fin pair can meet the system's cooling and dimension requirements simultaneously. Similarly, it can be found that when the fin pair is cc1 and ch1, indicating that the airside fin height reaches its maximum and that of the liquid side is of its minimum, the selected thickness c_{max} and weight W_{max} is the smallest, and thus the structure of the heat exchanger

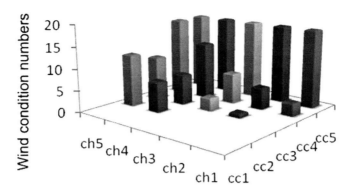

Fin pair collocations

Figure 9: Wind condition numbers corresponding to heat exchanger thickness exceeding 0.2 m based on various fin pair collocations [21].

is the most compact and lightest under this circumstance. Moreover, it can be found in the figure that if the airside fin size is changed and the liquid fin size is kept unchanged, there will be an obvious influence on the selected heat exchanger thickness and weight; while if the liquid side fin is changed instead of the airside one, the change will be comparatively smaller.

It can be found from the above computing results, the heat exchanger adopting fin pair cc1 and ch1 can reach its lightest and most compact structure, which is, thus, in favor of operating at a high altitude for the generating set. Considering that the liquid side pressure drop is far less than that of the cooling medium running through the generator and the control converter, this optimization focuses on the weight of the cooling system. From computing results, the fin pair cc1 and ch1 are adopted as the optimum fin pair. The results show that it would obtain a more effective fin function and comparatively higher fin efficiency by selecting low and thick fin on the liquid side with a large heat transfer coefficient and by selecting the high and thin fin on the small-heat-transfer-coefficient airside.

The relationship, shown in Fig. 10 [21], between the computed values of the parameters and the wind speeds under 18 wind conditions adopting fin pair cc1 and ch1 are computed by the commercial software, MATLAB. It can be concluded from the modeling result that as the wind speed rises up, the computed values of the heat exchanger thickness and the corresponding heat exchanger weight decline gradually, while the pressure drop on the liquid side and the heat exchanger's efficiency increase continuously. The maximum heat exchanger thickness is 116 mm, appearing at 10 m/s wind speed, and the pressure drop on liquid side is 341.5 Pa which meets the system pressure drop requirements.

The above optimization of liquid cooling systems is limited to the computed result of heat transfers and structural dimension and weight, and based on the hypothesis of steady working condition and thus no comprehensive conclusion can be drew for the dynamic property, power consumption and operating cost of

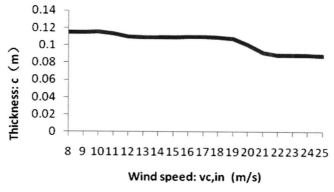

Figure 10: Relationship between wind speed and thickness of the heat exhanger adopting fin pair cc1 and ch1.

the wind turbine cooling system. In order to perform an exhaustive and objective optimization of the system, we will utilize tools such as computer-aided design, dynamic numerical simulation and so forth combining with the heat production data in the real-time wind turbine operation to launch a thorough and meticulous research. By doing these, we are able to render theoretical support to the future design of high-performance, reliable and economic wind turbines.

5 Future prospects on new type cooling system

With the wide application of wind energy at coastline area, the unit capacity of wind turbines increases greatly [22], which leads to a large rise of the heat production in gearboxes, generators and converters. The current liquid cooling system will no longer be sufficient to meet the cooling demands of wind turbines. Besides, the current cooling system has not taken full advantage of the ambient environment, such as high wind velocity, large solar energy density, etc. Therefore, there is a great potential to improve the operation economical efficiency. To solve the two problems above, some new cooling techniques for wind turbines are proposed in this section as follows.

5.1 Vapor-cycle cooling methods

From the theories of thermodynamics, all the current wind turbine cooling methods absorb heat by utilizing the specific heat capacity of the medium, while the vapor-cycle cooling uses the gasification latent heat of the boiling fluid to deliver the heat. As the gasification latent heat of the fluid is much larger than its specific heat capacity, the effect of the vapor-cycle cooling will be more obvious [24]. Therefore, the implementation of the vapor-cycle cooling technology is worthy of consideration. As shown in Fig. 11, its feature is that the system sets a vapor-cycle refrigerating engine outside the nacelle. During the operation, the cooling medium flows through the heat exchangers which is installed outside the gearbox, the generator, and the control converter successively,

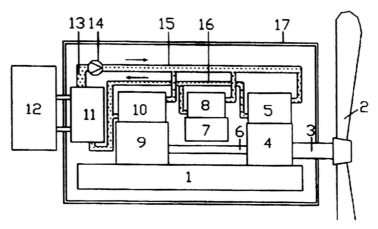

Figure 11: Structural diagram of the wind turbine adopting vapor-cycle cooling method: 1, saddle piece; 2, impeller blade; 3, low-speed shaft; 4, gearbox; 5, gearbox heat exchanger; 6, high-speed shaft; 7, control converter; 8, control converter heat exchanger; 9, generator; 10, generator heat exchanger; 11, cooling medium heat exchanger; 12, vapor-cycle refrigerating engine; 13, cooling medium; 14, circulating pump; 15, cooling medium inlet pipe; 16, cooling medium return pipe; 17, nacelle cover.

and take away the heat produced by these components. The heated cooling medium flows through an evaporator, getting cooled by exchanging heat with the refrigeration coolant circulating in the vapor-cycle refrigerating engine, and then perform the next round of cooling driven by circulating pump, which ensures that every component operates under a proper working condition. One thing worth mentioning is that the refrigerating engine can use bellows as connections among the cooling medium heat exchangers so that it will be more adaptable to the rotating working condition of the nacelle because of the change in wind direction during its running. The condenser of the refrigerating engine can be installed outside of the pylon in order to enhance convective heat transfer.

Comparing with the current forced air cooling and the liquid cooling method, the vapor-cycle cooling system introduces an extra cost of vapor-cycle refrigerating engine and additional power consumption for the cooling of the medium, however, it can adjust cooling capacity flexibly according to the cooling demands, and also can provide the optimal working condition for the wind turbine which paves the way for the next generation of high-power wind turbines.

5.2 Centralized cooling method

The current cooling solutions and the one discussed in Section 5.1 all aim at one single generating set. However, one with a complicated system is a prevalent issue, which leads to a huge difficulty for installation and maintenance, and, therefore, lowers down the operating reliability and the economical efficiency. Especially, with the increasing unit capacity of wind turbines, the heat production in the operation procedure will increase sharply and the disadvantage mentioned above of the traditional stand-alone cooling method will be more acute. Hence, the centralized cooling method is taken into consideration. As shown in Figs 12 and 13, the feature of a centralized cooling system for wind turbines is that it sets a cooling unit in the wind power plant, providing centralized cooling capacity to every wind turbine in it.

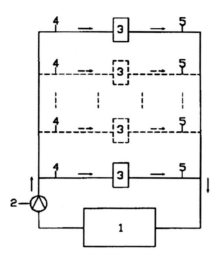

Figure 12: Schematic diagram of wind turbine centralized cooling system: 1, refrigerating engine; 2, circulating pump; 3, wind turbine; 4, cooling medium carrier pipe; 5, cooling medium return pipe.

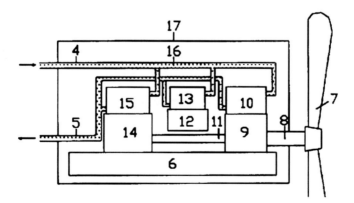

Figure 13: Structural diagram of wind turbine centralized cooling system: 6, saddle piece; 7, impeller blade; 8, low-speed shaft; 9, gearbox; 10, gearbox heat exchanger; 11, high-speed shaft; 12, control converter; 13, control converter heat exchanger; 14, generator; 15, generator heat exchanger; 16, cooling medium; 17, nacelle cover.

Similar with the wind turbine adopting vapor-cycle cooling method, the cooling medium runs sequentially through the heat exchangers outside the gearbox, the generator and the control converter, taking away the heat produced by these components. The heated cooling medium is then delivered to the refrigerating engine in the wind power plant through the cooling medium return pipe to get centralized refrigerated and once again delivered back to every wind turbine through the carrier pipe to perform the next round of cooling.

In a practical application, different types of air coolers or refrigeration units can be selected for the cooling unit according to the cooling demands of the generator. Compared with air coolers, refrigeration units are able to obtain lower cooling temperature. When the individual refrigeration unit cannot suffice to the cooling demands, multiple refrigeration units can be adopted; when the cooling demands are changed because of climate, season and other factors, they can be flexibly regulated by just changing the number of devices in operation. As to the wind power plant with a large-scale geographical area and a large number of generators, an alternative option is to set several centralized refrigeration units at different proper locations, according to the practical situation, to better satisfy the cooling demands and rationalize the distribution of the cooling medium delivery lines, aiming at obtaining high dependability and economy. With regard to wind turbines under low operating temperature, in order to reduce the cooling capacity wastage in the delivery of cooling medium, the system can adopt those pipe lines with good heat-insulating property and conduct some treatments of thermal insulation, such as coating the pipe with heat-insulating layers etc. The carrier and return pipes of the cooling medium can be laid out along with electric cables within the pylon so as to avoid corrosion of blown sand and rain to the pipeline. Besides, the joint of the cooling medium carrier and return pipe lines and the wind turbines can also be connected with bellows.

Compared with the forced air cooling and the liquid cooling systems, although the initial cost and the operating cost of the centralized cooled wind power generating system is higher, it has its own advantages, such as huge cooling capacity and flexible adaptability. Besides, the refrigeration units can aptly adopt various cooling manners according to the operating requirements of the wind power devices so as to suffice to the high-power cooling demands, which pave the way for the new generation of high-power wind turbine sets. Comparing to the vapor-cycle cooling method, this system simplifies the internal cooling devices of the wind turbine, and lowers the operating weight to make it suitable for the operation of wind turbines at a high altitude, which, thereby, eases the maintenance of the devices. In addition, when dealing with cooling demand fluctuation caused by changing of season, climate or other factors, no adjustments to individual wind turbines are required, but simply the adjustment to refrigeration units is needed.

5.3 Jet cooling system with solar power assistance

The cooling technologies and solutions mentioned above are all driven by electricity but the solar power as a clean energy from the atmosphere is omitted. This omission, to some extent, lowers the generating efficiency. Actually, the current heat-driven cooling technique is fairly mature. Because of these, the solar power jet cooling method can be implemented in the wind turbine cooling system.

As shown in Figs 14 and 15 [26], the features of a wind turbine adopting a cooling system with solar power jets include a solar thermal collector covering on the nacelle and a jet refrigerating engine with a secondary refrigerant circulation through heat exchangers of the gearbox, the control converter and the generator.

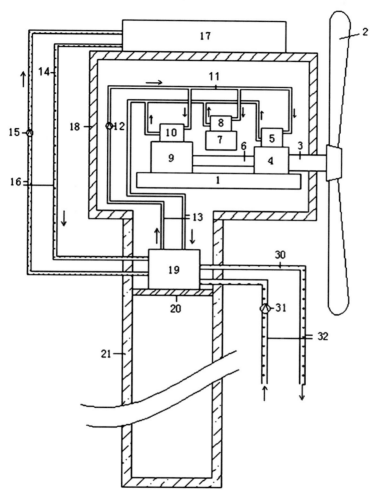

Figure 14: Schematic diagram of solar power jet cooling wind turbine: 1, saddle piece; 2, impeller blade; 3, low-speed shaft; 4, gearbox; 5, gearbox heat exchanger; 6, High-speed shaft; 7, control converter; 8, control converter heat exchanger; 9, generator; 10, generator heat exchanger; 11, secondary refrigerant; 12, circulating pump I; 13, carrier pipe of secondary refrigerant; 14, heat storage agent; 15, circulating pump II; 16, carrier pipe of heat storage agent; 17, solar thermal collector; 18, nacelle cover; 19, jet refrigerating engine; 20, platform of the pylon; 21, pylon; 30, cooling medium; 31, circulating pump III; 32, carrier pipe of the cooling medium.

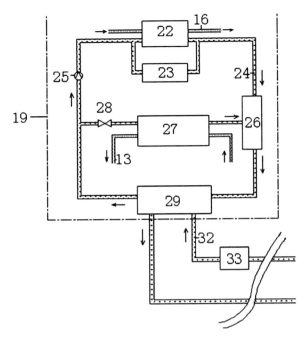

Figure 15: Schematic diagram of the solar power jet refrigerating engine: 22, heat accumulator; 23, auxiliary heater; 24, refrigerant; 25, circulating pump IV; 26, jet apparatus; 27, evaporator; 28, throttle; 29, condenser; 33, filter. The remaining symbols have been annotated above.

Similar to the wind turbine introduced in Section 5.1, during the operation, the secondary refrigerant sequentially flows through the heat exchanger outside the gearbox, the generator and the control converter, removing the heat produced by these components. The heated secondary refrigerant is then driven to the jet refrigerating engine by the circulating pump through the carrier pipes and performs heat exchange with liquid refrigerant in the evaporator. Finally the cooled secondary refrigerant is then delivered back to the above three heat exchangers to absorb heat. This circulation cycles to ensure the durable and secure operation of the wind turbine.

In the operation of the jet refrigerating engine, the solar thermal collector converts the solar energy into heat energy, leading to a rise of the heat storage agent in it. The heat is then delivered by the heat storage agent and stored in the heat accumulator. The heat-released heat storage agent flows back to the heat accumulator driven by a circulating pump, thus completes a circuit of solar energy conversion. Meanwhile, the refrigerant absorbs heat in the heat accumulator and gasifies and thus supercharges the pressure. The refrigerant is then ejected through the jet apparatus, leaving a low pressure close to vacuum at the jet tip. The low pressure steam refrigerant in the evaporator is thus drawn into the jet apparatus due to the pressure difference. The mixed steam refrigerant from the jet apparatus is then ejected into the condenser and performs heat exchange with the cooling medium flowing through the condenser. The cooled refrigerant is shunted into two parts by pipelines. One of the streams flows back to the heat accumulator to complete the power cycle while the other passes through the throttle to the evaporator, where it performs heat exchange with the secondary refrigerant and gasifies, turning into a low pressure steam. So far a refrigerating cycle is completed.

It is worth mentioning that sea water or underground water can be used as cooling medium in the condenser. To prevent blockage resulting in bad cooling effect, a filter needs to be installed on the cooling medium carrier pipe on the way to the condenser. In the area where water is hard to reach, the cooling medium can be the other liquid processed by the seawater heat exchanger or the underground heat exchanger. No matter which conditions, for those components and pipelines submerged in water, effective anti-corrosion treatments should be implemented. Those treatments include choosing the resistant material, protective coating technology and cathodic protection technology. In the night or the abnormal working condition of the heat accumulator, an auxiliary heater can be used to heat the refrigerant flowing through to offset the energy shortage of the heat accumulator.

Comparing with the current wind turbine cooling system, the jet cooling system with solar power assistant possesses many advantages, such as lower power consumption, better cooling performance and environment friendly. Only the circulating pump and auxiliary heater need electricity support in the cooling system, which enables low electricity consumption of the whole system; while the solar power jet cooling method can obtain temperature below the ambient temperature, which will ensure the temperature in the optimal working condition. And the refrigerant in this system can be chosen from the non- chlorofluorocarbons substances (NON-CFC) which will be of great help to the environment protection. Besides, since the jet refrigerating engine has a simpler and lighter structure than other ones in the current technology, this merit will enhance the total anti-fatigue property of the nacelle; also because it is usually installed in the nacelle, which will avoid the corrosion caused by blown sand and rain, the working performance and life span will be effectively secured. Therefore, it can satisfy the highly efficient and dependable working requirements of the high-power wind turbine.

5.4 Heat pipe cooling gearbox

The three solutions mentioned above mainly focus on the entire cooling effect of all the heat producing components. On the other hand, considering different structural conditions and cooling demands of various components and using corresponding combinatorial solutions will possibly further enhance the operating economical efficiency, which is also a future trend for wind turbine cooling systems. Since the gearbox has a simple structure and an ample space for installation, a cooling solution adopting gravity heat pipe is proposed in this section while the generator and the control converter remain adopting a traditional liquid cooling system.

As shown in Figs 16 and 17 [27], the feature of a wind turbine system adopting a heat pipe cooling gearbox is that it has several apparatuses, including a gravity heat pipe connected with the gearbox to refrigerate the lubricating oil in it, a

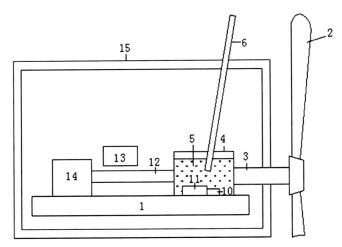

Figure 16: Schematic diagram of the wind turbine system adopting heat pipe cooling gearbox: 1, saddle piece; 2, impeller blade; 3, low-speed shaft; 4, gearbox; 5, lubricating oil; 6, gravity heat pipe; 10, temperature controller; 11, electric heater; 12, high-speed shaft; 13, control converter; 14, generator; 15, nacelle cover.

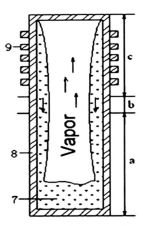

Figure 17: Schematic diagram of gravity heat pipe: 7, working substance; 8, shell of pipe; 9, radiating rib; a, evaporation zone; b, insulation zone; c, condensation zone.

temperature controller and an electric heater aiming at preventing extreme cold of the lubricating oil in winter. During the operation of the wind turbine, the heat produced by the gearbox is firstly absorbed by lubricating oil, and then delivered to the evaporation zone of the heat pipe. The working substance in the heat pipe evaporates because of heat absorption. The vapor ascends in the axial direction along the heat pipe, passing the thermal insulation zone, and then releases heat and condenses in the condensation zone outside the nacelle. Finally the liquor condensate returns to the evaporation zone with the assistance of gravity, which begins the next evaporation procedure and so forth to perform effective cooling on the gearbox. In winter, in order to avoid getting extremely cold lubricating oil caused by the large cooling capacity of heat pipes, which is common in the northern cold area, the gearbox is installed with a temperature controller, performing real-time monitoring of the gearbox lubricating oil temperature. In addition, the heat pipe in the system can adopt inclination arrangement to ensure the liquor condensate returning to the evaporation zone smoothly, by which to ensure continuity and validity of the cooling system. Besides, in the operation, the number and distribution of heat pipes can be selected according to the real heat transfer capacity.

Compared with the current cooling system with a wind turbine gearbox and the above three solutions, the wind turbine system adopting a heat pipe cooling gearbox has many advantages to satisfy the wind turbine cooling demands, such as simple structure, self-driven, high heat transfer efficiency, low cost, easy to install and repair, small need on lubricating oil and so forth. Besides, it can keep the gearbox operating under the condition of optimum working temperature by selecting proper working substance in the heat pipe.

The wind turbine cooling systems introduced above are only limited examples of the future possible systems. With the development of the wind turbine technology, various cooling solutions will emerge. However, only those cooling systems with better cooling effect, higher economical efficiency and lower power consumption will be widespread in the future development.

References

[1] Wang, C.X. & Zhang, Y., *Wind Power*, Beijing: China Electric Power Press, 2003.
[2] Yin, L. & Liu, W.Z., *Wind Power*, Beijing: China Electric Power Press, 2002.
[3] Ni, S.Y., Lecture on wind power, first lecture: types and configurations of wind turbine. *Solar Energy*, (2), pp. 6–10, 2000.
[4] Bryan d'Emil. Wind with miller, http://www.windpower.org/en/kids/index.htm?d=1,2006-12-19
[5] Wang, D.G., The calculation of thermal power of the increasing gear box for wind generator. *Wind Power*, (3), pp. 22–24, 2003.
[6] Ye, H.Y., *Control Technology on Wind Electricity Generation Unit*, Beijing: China Machine Press, pp. 51–52, 2002.
[7] Ding, S.N., *Heat and Cooling of Large Electric Machine*, Beijing: Science Press, 1992.
[8] Wen, Z.W., Analysis on temperature field of large synchronous machine based on numerical method, Ph.D. Dissertation, Graduate University of Chinese Academy of Sciences, Institute of Electrical Engineering, Xi'an, 2006.
[9] Ni, T.J., Major cooling methods and features of large generator. *Dongfang Electric Review*, **20(2)**, pp. 31–37, 2006.

[10] Tang, Y.Q. & Shi, N., *Electric Machinery*, Beijing: China Machine Press, pp. 271–280, 2001.

[11] Yuan, W.W. & Jiang, Y.L., Cooling systems in wind turbine. *World Sci-Tech R & D*, **29(2)**, pp. 80–85, 2007.

[12] Liao, Y., CFD analysis on the flow and temperature field of a 660kW wind turbine rotor, http://www.efluid.com.cn/software/show.asp?id=3116&softid=12&tradeid=-1

[13] Sun, M.L., Optimization design of a 600 kW/125 kW wind turbine. *Shanghai Medium and Lange Electrical Machines*, **(3)**, pp. 6–7, 2003.

[14] Yuan, W.W., Investigation on the wind turbine cooling system, Master Thesis, Nanjing University of Aeronautics Astronautics, Nanjing, 2008.

[15] Cao, J., Experimental study on viscidity change of alcohol aqueous solution at low temperature. *Petroleum Planning & Engineering*, **16(2)**, pp. 21, 2005.

[16] Tan, Z.C., Determination of the physicochemical property of glycol dibasicand and its aqueous solution. *Chemical Engineering (China)*, **(1)**, pp. 41–45, 1983.

[17] Wang, S.H., *Plate-Fin Heat Exchanger*, Beijing: Chemical Industry Press, 1984.

[18] Liu, W.H., Guo, X.M. & Huang, H., *New Technique and Development of Refrigeration and Air-Conditioning*, **8**, Beijing: China Machine Press, pp. 108–109, 2005.

[19] Liu, M.S., Li, N. & Dong, Q.W., Research progress of plate-fin heat exchanger in china. *Process Equipment & Piping*, **44(6)**, pp. 9–12, 2007.

[20] Zhou, A.M., Wen, J., Li, Y.Z. & Ma, Y.S., A study on the flow distribution and resistance of various heads of the plate-fin heat exchangers. *Chemical Engineering & Machinery*, **33(5)**, pp. 271, 2006.

[21] Jiang, Y.L., Yuan, W.W., Zhang, Q. & Wang, Z., Cooling System optimum design in MW wind turbines. *Journal of Nanjing University of Aeronautics & Astronautics*, **40(2)**, pp. 199–204, 2008.

[22] European Wind Energy Association (EWEA), A blueprint to achieve 12% of the world's electricity from wind power by 2020. *Wind Force*, **12**, pp. 46, 2004.

[23] Jiang, Y.L., Yuan, W.W., Li, H.L., *et al.*, Wind turbine cooling system adopting vapor-cycle cooling method, People's Republic of China. Patent of Invention. No. ZL200610039658.9, 2007.

[24] Ruan, L. & Gu, G.B., Development and current technical difficulties of vapor-cycle cooling technique. *Electric Age*, **(7)**, pp. 31–34, 2001.

[25] Jiang, Y.L., Yuan, W.W., Li, H.L., *et al.*, Wind turbine cooling system adopting centralized cooling method, People's Republic of China. Patent of Invention. No. 200610097464.4, 2006.

[26] Jiang, Y.L., Zhang, Q., Wang, Z., Han, Y.Q., *et al.*, Wind turbine adopting jet cooling system with solar power assistance, People's Republic of China. Patent of Invention. No. 200810195077.3, 2008.

[27] Jiang, Y.L., Zhang, Q., Wang, Z., et al., Wind turbine system adopting heat pipe cooling gearbox, People's Republic of China. Patent of Invention. No. 200810124198.9, 2008.

CHAPTER 20

Wind turbine noise measurements and abatement methods

Panagiota Pantazopoulou
BRE, Watford, UK.

This chapter presents an overview of the types, the measurements and the potential acoustic solutions for the noise emitted from wind turbines. It describes the frequency and the acoustic signature of the sound waves generated by the rotating blades and explains how they propagate in the atmosphere. In addition, the available noise measurement techniques are being presented with special focus on the technical challenges that may occur in the measurement process. Finally, a discussion on the noise treatment methods from wind turbines referring to a range of the existing abatement methods is being presented.

1 Introduction

Wind turbines generate two types of noise: aerodynamic and mechanical. A turbine's sound power is the combined power of both. Aerodynamic noise is generated by the blades passing through the air. The power of aerodynamic noise is related to the ratio of the blade tip speed to wind speed. The mechanical noise is associated with the relevant motion between the various parts inside the nacelle. The compartments move or rotate in order to convert kinetic energy to electricity with the expense of generating sound waves and vibration which is transmitted through the structural parts of the turbine.

Depending on the turbine model and the wind speed, the aerodynamic noise may seem like buzzing, whooshing, pulsing, and even sizzling. Downwind turbines with their blades downwind of the tower cause impulsive noise which can travel far and become very annoying for people. The low frequency noise generated from a wind turbine is primarily the result of the interaction of the aerodynamic lift on the blades and the atmospheric turbulence in the wind. High frequency noise is also generated due to the interaction of the air turbulence and the blades during their rotation, constantly changing angle of attack.

Depending on the rotational speed of the turbine, the size and the airflow wind turbines emit infrasound, low and high frequency sound waves. Large wind turbines produce infrasound of 8–12 Hz range. Small turbines can achieve higher blade tip velocities that can give low frequency noise of 20–20 kHz range.

To get a better feeling of the frequency ranges that are audible and are produced by wind turbines, we can see Table 1.

The levels of infrasound radiated by the large wind turbines are very low in comparison to other sources of acoustic energy in this frequency range. However, the annoyance is often connected with the periodic nature of the emitted sounds rather than the frequency of the acoustic energy. Because low frequencies travel farther than the high frequencies due to their long wavelengths, they become a cause of irritation for residents living not so close to wind farms.

Sound is a series of waves that travel through air in the form of disturbances and reaches our eardrums. Any natural or artificial obstacles such as hills and buildings play a shielding effect role, reflecting the sound, while most of the times absorbing some of the acoustic energy. Trees and ground vegetation also attenuate sound and change its directivity patterns. The distance and obstacles that are located between the source and the receiver have a significant impact on the acoustic 'line of sight'.

The ways to reduce noise from wind turbines are mostly focused on blade design optimisation and choice of wind farm location as it still a new field and new techniques are under development. A major challenge that needs to be addressed before abatement techniques are put in action; is to establish accurate measurement methods of wind turbine noise in order to define the parameters of the problem and find an appropriate acoustic solution for it. Noise measurements from wind turbines is a complex task because the background noise levels are comparable to the noise levels from wind turbine when it starts operating at certain wind speeds. This is the reason a commonly used approach that overcomes the above issue needs to be established.

Current acoustic treatment methods range from designing quieter wind turbine compartments to placing wind turbines as far as possible from residential areas. Wind turbine blades are aerodynamically shaped to avoid causing abrupt air flow disturbances and the turbine is placed upwind to eliminate interference of the flow between the tower and the wind turbine blades. Apart from aerodynamic solutions other treatments include sound proofing of the nacelle to reduce mechanical noise and careful selection of the wind turbine site to attenuate noise until it reaches the receiver.

Table 1: Important frequency ranges.

Description	Frequency range
Normal hearing	20 Hz to 20 kHz
Normal speech	100 Hz to 3 kHz
Low frequency	20–200 Hz
Infra sound	<16 Hz

It is common sense that the extensive use of wind turbines is included in the current and the future environmental plans of every country, fact which will certainly place a strong demand for dealing noise issues in the near future.

2 Noise types and patterns

2.1 Sources of wind turbine sound

The sources of sounds emitted from operating wind turbines can be divided into two categories: (1) mechanical sounds, from the interaction of turbine components, and (2) aerodynamic sounds, produced by the flow of air over the blades. A summary of each of these sound generation mechanisms follows, and a more detailed review is included in [1].

More specifically, wind turbines produce energy by the rotational motion of the blades due to the wind flow. Rotating blades are known to emit three different types of acoustic signature:

- tonal noise
- broadband noise
- mechanical noise

Tonal noise is characterised by discrete frequencies and it is generated by the periodical rotation of the turbine blades. An example of tonal noise history is shown in Fig. 1. Tonal noise is caused by the unsteady air velocity due to the blade rotation which disturbs the flow on the blade surface. It is directional as it is produced at the direction the blades meet the airflow and therefore is dependent on the observer position.

As the blades rotate the load on the blade surface changes periodically causing analogous changes in the unsteady pressure on the blade surface inducing sound waves. Depending on the position of the observer in relation to the turbine while

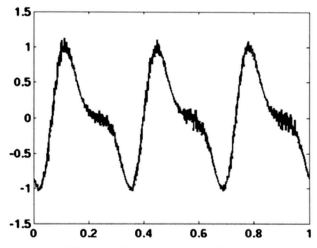

Figure 1: Tonal noise time history.

the blade is in operation the observer receives variable acoustic signals. Also the sound component in the direction of the observer varies with time and a sound wave is generated. Normally, the flow through the blades is distorted (non-uniform) and therefore the angle of attack of each blade varies continuously as the turbine rotates causing the sound generation to be highly directional and more frequent. This change in the angle of attack can be very abrupt especially when velocity discontinuities occur in the inflow profile resulting in rapid changes in the blade loading, flow disturbance and therefore generation of acoustic waves.

Such type of acoustic waves can be produced when the mounting tower of the wind turbine interferes with the flow passing through the wind turbine blades. This causes local velocity instabilities in the flow field which moves through the blades producing pulsing low frequency noise. A downwind design is shown in Fig. 2a.

It becomes self-explanatory that when the reverse order occurs, the disturbances ease as the blades encounter only the free field flow disturbances. An upwind turbine is shown in Fig. 2b.

The noise generated by downwind designs has low frequency in the range of 20–100 Hz and it is caused when the turbine blade encounters localised flow deficiencies due to the flow around a tower. It can also be impulsive described by short acoustic impulses or thumping sounds that vary in amplitude with time. It is caused by the interaction of wind turbine blades with disturbed air flow around the tower of a downwind machine.

2.2 Infrasound

A special category of the tonal noise released by wind turbines is infrasound. As we have already mentioned while low frequency ranges at the bottom of human perception (10–200 Hz), the *infrasound* is below the common limit of human perception. Sound with frequency below 20 Hz is generally considered infrasound, even though there may be some human perception in that range. A distinctive characteristic of infrasound is that it can travel very far because of its long wavelength that dissipates with a low rate and therefore it makes it easier to 'survive' and be present in our everyday life.

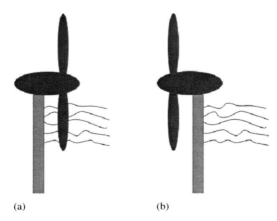

(a)   (b)

Figure 2: (a) Downwind turbine; (b) upwind turbine.

Considering the nature of infrasound, it becomes apparent that downwind wind turbines are more likely to produce significant levels of infrasound levels. This is because the tower–blade flow interaction creates air velocity instabilities that produce very low frequency acoustic waves called infrasound. Although downwind wind turbines have been used in the past, the modern technology has moved away from that noisy design to the upwind designs that give higher frequency noise levels and therefore are less irritating for humans.

One example of low frequency sound and infrasound from a modern turbine is shown in Fig. 3.

Broadband noise is dominated by high frequencies greater than 100 Hz and it is characterised by non-periodic signals that constitute an envelope that varies periodically. A typical broadband signal is shown in Fig. 4. Its main source of generation is the interaction of the wind turbine blades with atmospheric turbulence, and also described as a characteristic "swishing" or "whooshing" sound.

To determine the relative importance of tone noise and broadband noise we consider the narrow-band frequency spectrum of a signal. Figure 5 shows a typical spectrum of tonal and broadband noise. At higher frequencies the broadband random noise dominates the spectrum.

2.3 Mechanical generation of sound

A wind turbine consists of mechanical components that move or rotate in order to capture the motion of the turbine and convert it to energy. Sources of such sounds include:

1. Gearbox
2. Generator
3. Yaw drives
4. Cooling fans
5. Auxiliary equipment (e.g. hydraulics)

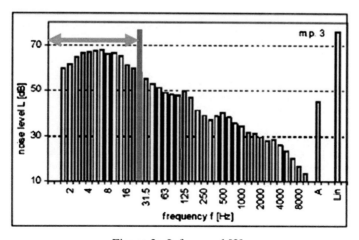

Figure 3: Infrasound [2].

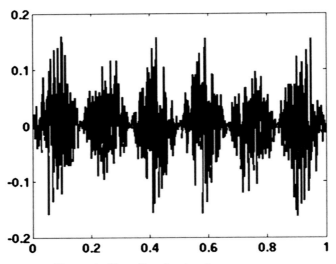

Figure 4: Broadband noise time spectrum.

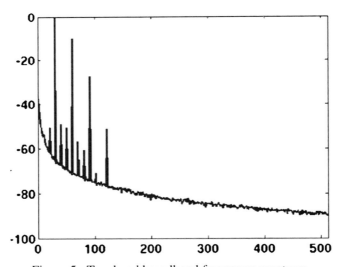

Figure 5: Tonal and broadband frequency spectrum.

Generally the mechanical sound is low frequency sound although it might have a broadband component that comes from the relative motion from each of the above parts. The turbine's metal parts come in contact with each other, such as the generator, the gearbox and the shafts and they emit noise and as they vibrate.

Because wind turbines can have different constructions they might have different sound emissions because of the way in which they operate. For instance they may have blades which are rigidly attached to the hub or may have blades that can be pitched (rotated around their long axis). Some have rotors that always turn at a constant or near-constant speed while other designs might change the rotor speed as the wind changes. Wind turbine rotors may be upwind or downwind of the tower.

It is worth mentioning that the hub, rotor, and tower may act as loudspeakers, transmitting the mechanical sound and radiating it. The transmission of noise from the mechanical parts of a wind turbine can take place in two ways:

- structure-borne
- air-borne

Structure-borne sound [3] is a sound that is propagated through structures as vibration and subsequently radiated as sound. The intensity and the frequency of structure-borne sound depend on many factors such as the rotational speed of the wind turbine, as well as the type and the material of the mechanical parts that vibrate. Air-borne [4] means that the sound is directly propagated from the component surface or interior into the air.

Structure-borne sound is transmitted along other structural components before it is radiated into the air. For example, Fig. 6 shows the type of transmission path and the sound power levels for the individual components for a 2 MW wind turbine. Note that the main source of mechanical sounds in this example is the gearbox, which radiates sounds from the nacelle surfaces and the machinery enclosure.

Utility scale turbines are usually insulated to prevent mechanical noise from proliferating outside the nacelle or tower. Small turbines are more likely to produce noticeable mechanical noise because of insufficient insulation.

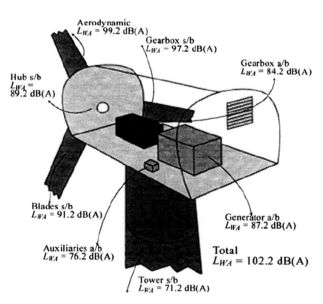

Figure 6: Sound power levels of wind turbine components [1].

3 Sound level

In order to understand how sound propagate we need to identify the nature of sound. Sound is a series of waves and it is characterised by two properties: amplitude (loudness) and frequency. Therefore there are sounds that consist of combinations of low and high magnitude with low and high frequency. The human ear can detect a very wide range of both sound levels and frequencies, but it is more sensitive to some frequencies than others.

Sound is generated by numerous mechanisms and is always associated with rapid small-scale pressure fluctuations, which produce sensations in the human ear. Sound waves are characterised in terms of their amplitude, wavelength (λ), frequency (f) and speed c, as follows:

$$c = \lambda f \tag{1}$$

The speed of sound is a function of the medium through which it travels, and it generally travels faster in more dense mediums. Sound propagates as a wave as shown in Fig. 7. In air it travels at a speed of 340 m/s and in water 1500 m/s. As sound travels it transports acoustic energy with it which attenuates the further it travels. As the sound propagates it disturbs the fluid from its mean state.

The pressure at a position $\mathbf{x}$ is $p = p_0 + p'(x, t)$ with $p'/p_0 \ll 1$.

Sound is measured in dB and the sound pressure level (SPL) is defined as

$$\text{SPL} = 20\log_{10}(p'_{\text{rms}} / 2 \times 10^{-5})\,\text{dB} \tag{2}$$

where p'_{rms} is the mean square level of fluctuation [5].

As sound energy travels through the air, it creates a sound wave that exerts pressure on receivers such as an ear drum or microphone and it makes our eardrums vibrate [6]. Human whisper releases an acoustic power of 10^{-10} W and a large jet transport at take off emits about 10 W.

The threshold of pain is between 130 and 140 dB. The threshold of hearing is around 0 dB. The sound *power* level from a single wind turbine is usually between 90 and 105 dB(A). Figure 8 shows a few examples of sound pressure levels from everyday life.

As described above, the decibel scale is logarithmic. A sound level measurement that combines all frequencies into a single weighted reading is defined as a

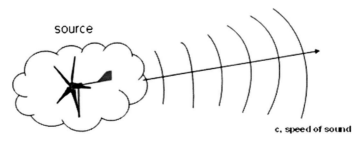

Figure 7: Sound propagation.

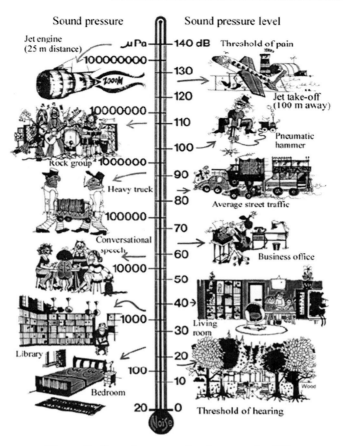

Figure 8: Sound pressure levels examples [9].

broadband sound level. For the determination of the human ear's response to changes in sound, sound level meters are generally equipped with filters that give less weight to the lower frequencies. There are a number of filters that accomplish this:

- A-weighting: This is the most common scale for assessing environmental and occupational noise. It approximates the response of the human ear to sounds of medium intensity.
- B-weighting: This weighting is not commonly used. It approximates the ear for medium-loud sounds, around 70 dB.
- C-weighting: Approximates response of human ear to loud sounds. It can be used for low frequency sound.
- G-weighting: Designed for infrasound.

Details of these scales are discussed by Beranek and Ver [7].

Once the A-weighted sound pressure is measured over a period of time, it is possible to determine a number of statistical descriptions of time-varying sound and to account for the greater community sensitivity to night-time sound levels [8]. Terms commonly used in describing environmental sound include:

- L10, L50, and L90: The A-weighted sound levels that are exceeded 10%, 50%, and 90% of the time, respectively.
- Leq: *Equivalent Sound Level:* The average A-weighted SPL which gives the same total energy as the varying sound level during the measurement period of time.
- Ldn: *Day-Night Level:* The average A-weighted sound level during a 24-h day, obtained after addition of 10 dB to levels measured in the night between 10 p.m. and 7 a.m.

4 Factors that affect wind turbine noise propagation

Propagation refers to how sound travels. Attenuation refers to how sound is reduced by various factors. Many factors contribute to how sound propagates and is attenuated, including air temperature, humidity, barriers, reflections, and ground surface.

The ability to hear a wind turbine also depends on the ambient sound level. When the background sounds and wind turbine sounds are of the same magnitude, the wind turbine sound gets lost in the background. The most important factors are:

- Source characteristics (directivity, height)
- Distance of the source from the observer
- Air absorption
- Ground effects (reflection and absorption on the ground)
- Weather effects (wind speed, temperature, humidity)
- Shape of the land – land topology

4.1 Source characteristics

The source characteristics such as height and directivity can affect the sound propagation path and its power or intensity. The higher a source is located, the higher the sound power loss rate is. This means that wind turbine that is mounted on a tower relatively high to a residential estate it has relatively low noise impact on the residents as the sound energy attenuates until it reaches the human ear. The directivity of an acoustic source has also a significant impact on the sound perceived by the human ear. For example when the sound is forced to follow a certain directional path determined by the geometrical shape it is placed in, such as conical speaker, the radiation field is concentrated towards a certain area leaving quite zones in the opposite direction.

In general, as sound propagates without obstruction from a point source, the initial sound energy decreases and it is being distributed over a larger and larger area as the distance from the source increases.

For example, in the case of spherical excitation or a monopole noise source the sound is radiated in all directions and the sound level is reduced by 6 dB for each doubling of distance from the source.

A moving train is a *line source*, and it emits equal sound power output per unit length of the train line. A line source produces cylindrical spreading, resulting in a sound level reduction of 3 dB per doubling of distance. The spherical propagation is associated with the three-dimensional propagation and the line source with the one-dimensional sound wave propagation, respectively. When two monopole sources of equal strength but opposite phase are put together at a short distance they produce a dipole, which is referred as two-dimensional sound wave propagation. Figure 9 shows the sound directivities of a monopole and a dipole.

4.2 Air absorption

Air absorption of sound is driven by two mechanisms: molecular relaxation and air viscosity. Molecular relaxation is the transition of a molecule from an excited energy level to another excited level of lower energy. High frequencies are absorbed more than low because they have short wave length and therefore the waves dissipate as they travel through the air molecules. The air absorption must be taken into account at high frequencies when calculating the reverberation time of a room. It is due to friction between air particles as the sound wave travels through the air. The amount of absorption depends on the temperature and humidity of the atmosphere [10].

4.3 Ground absorption

The ground can contribute to the sound attenuation by two mechanisms sound absorption and sound reflection. When the sound hits the ground the acoustic energy loss depends on the reflection coefficient of the surface. On hard surfaces attenuation occurs due to the acoustic energy losses on reflection while on porous surfaces, sound levels are being reduced due to the increased absorption of the ground. High frequencies are generally attenuated more than low frequencies.

The reflection coefficient depends on the impedance of the two media, in this case, air and ground, and represents the absorbency of the ground in a homogeneous

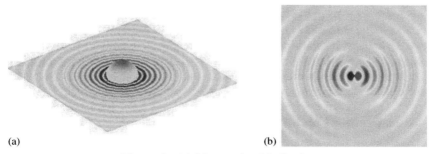

(a)   (b)

Figure 9: (a) Monopole; (b) dipole.

way, i.e. the entire infinite plane is assigned with the same reflection coefficient [11].

When the source and receiver are both close to the ground, the sound wave reflected from the ground may interfere destructively with the direct wave. This effect (called the *ground effect* [12]) is normally noticed over distances of several meters and more, and in the frequency range of 200–600 Hz.

4.4 Land topology

The topology and the shape of the land can significantly affect the magnitude and the direction of sound. For example, trees and high altitude vegetation can contribute to the sound attenuation. However, a long series of trees several hundreds of meters long is required in order to achieve significant attenuation.

Also significant attenuation can be achieved by the use of natural or artificial barriers or obstacles such as hills and buildings that exist on the ground. The level of impact on the sound reduction of an obstacle depends on whether it is high enough to obscure the 'line of sight' between the noise source and receiver.

Due to their short wavelength, high frequencies are trapped by the obstacles preventing them from travelling far, unlike the low frequencies.

Similarly to the ground reflection theory, the material of the barriers plays a dominant role in the sound propagation and this is the reason barriers are often used for noise treatment purposes [13]. A barrier is most effective when placed either very close to the source or to the receiver.

4.5 Weather effects, wind and temperature gradients

The wind and the temperature can affect the propagation of the sound in the atmosphere under certain weather conditions. The mean uniform wind flow determines the background noise levels and it alters the sound pressure downwind and upwind. When a wind is blowing there will always be a wind gradient. A wind gradient results in sound waves propagating upwind being 'bent' upwards and those propagating downwind being 'bent' downwards.

The temperature is another factor that affects sound radiation however; it becomes important only when the high temperature gradients occur. Such dramatic changes in the temperature profile are unlikely to happen in the atmosphere close to the ground but they can occur at high altitude layers. Any temperature differences in the atmosphere can cause local variations in the sound speed since the latter depends on the temperature of the gas. Higher temperatures produce higher speeds of sound. When sound waves are propagating through the atmosphere and meet a region of non-uniformity, some of their energy is re-directed into many other directions. This phenomenon is called refraction [14].

5 Measurement techniques and challenges

As we have already mentioned in previous sections, low frequency noise emissions from wind turbines have given rise to health effects to neighbours. Resident complaints about the irritating noise from the wind turbines has led scientists and engineers to invent ways to assess the levels of noise in the near and in the far field and eventually close to dwellings where residents live.

Measuring noise from wind turbines is not an easy task due to the fact that background noise levels increase as the wind speed increases. So, as the wind turbines start rotating the background noise levels are being intensified. Also when we think that the wind turbines are placed outdoors where trees, leaves and vegetation in general, are present then it becomes apparent that the background noise is comparable to the noise emissions from the turbines. This makes it extremely difficult to measure sound from wind turbines accurately. At wind speeds around 8 m/s and above, it generally becomes a quite abstruse issue to discuss sound emissions from modern wind turbines, since background noise will generally mask any turbine noise completely.

To assess the potential levels of infrasound and low frequency noise around a wind farm and at neighbouring locations of interest, the measurements are undertaken using a measuring system capable of capturing frequencies from 1 Hz to 20 kHz. Measurements are performed at internal and external locations placing the microphones at locations where noise is considered more audible when occurred. Assuming that the appropriate equipment is being used and the calibration procedures have been followed, the standard procedure to take noise measurements is the following:

- Ambient background noise levels evaluation .
- Spot measurements of noise levels inside dwellings subject to access or prediction if access is prohibited.
- Exterior acoustic measurements at neighbouring facades to assess annoyance.
- Spot measurements of wind speed.
- The evaluation and reporting of measurements made.

It is worth mentioning that there is a variety of sophisticated model available but they are all at developing stages. A typical time history of measured sound pressure levels is shown in Fig. 10.

5.1 For small wind turbines.

Our experience indicates that in practice, field measuring is a challenging task not only due to the difficulty to estimate background levels but also due to the complexity of the required experimental set up. One has to count for the directional noise emission depending on which side the wind is blowing and the fact that modern small wind turbines rotate around their vertical axis making the noise measurement techniques even more demanding.

Because of the importance of background noise in determining the acceptability of the overall noise level, it is crucial to measure the background ambient noise levels for all the wind conditions in which the wind turbine will be operating. Sound propagation is a function of the source sound characteristics (direction and

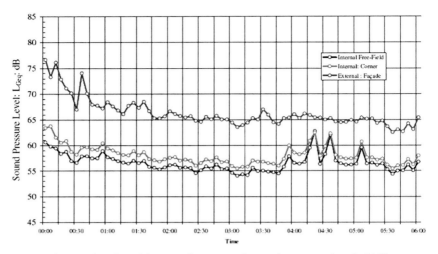

Figure 10: Time history of measured sound pressure levels [15].

height), distance, air absorption, reflection and absorption and weather effects such as changes of wind speed and temperature with height.

Given the current situation difficulties, a good idea to tackle with such a problem would be at a first stage to use the anechoic chamber to remove background noise and depending on the results, to employ measurement techniques and methods which enable characterisation of the noise emission from wind turbines at a receptor location. The use of a vertical board mounted with a microphone suggested in the document of IEA [16] recommended practices in case of such a problem. The document recommends the use of a large vertical board with a microphone at the designated position with its diaphragm flushed with the board surface or to suppress wind induced noise on a microphone. Figure 11 shows the background noise levels and the noise levels from a wind turbine in operation.

6 Abatement methods

In principle there are two ways to reduce noise: either make the source less noisy or make the receiver more sound proof. The same principle can be applied to the wind turbines when considered as noise sources.

Also if we think the factors that affect the noise propagation we can easily guess a few measures we can take in order to control noise from wind turbines. More specifically locating wind farms as far and as high as possible from residential areas and improving the sound insulation in houses we can significantly limit the sound levels received by the human ears. Natural obstacles and vegetation can also prevent noise from reaching people's homes.

A systematic and scientific consideration of all those factors along with real time measurements leads to a study known as environmental impact assessment. Such assessment is essential to evaluate the current environmental conditions at a

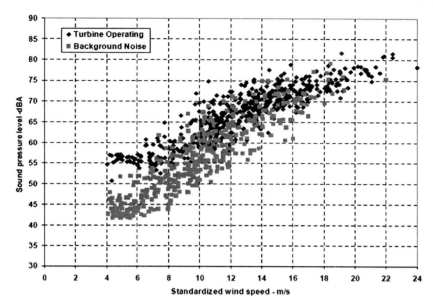

Figure 11: Background and operation noise measurements for a small wind turbine [17].

proposed erection of a wind farm. Among others, it assesses how the proposed wind farm will effect the environment and the public health, and whether any resultant changes in conditions are deemed acceptable or unacceptable. Best code guidelines and policy standards are being practised and followed in order to ensure environmental protection including noise and nuisance.

Recalling that the type of noise generated from a wind turbine can be mechanical and aerodynamic, we are able to investigate and find solutions for noise treatment.

Starting with the mechanical parts of a wind turbine that rotate or move in relation to each other such as gears that are creating structure-borne noise and vibration, it becomes clear about the need for designing gearboxes for quiet operation. Wind turbines use special gearboxes, in which the gear wheels are designed slightly flexible in order to reduce mechanical noise. One way of doing this is to ensure that the steel wheels of the gearbox have a semi-soft, flexible core, but a hard surface to ensure strength and long time wear.

In addition to designing quiet parts, insulating those parts seems to be another way to tackle that type of noise. Soundproofing and mounting equipment on sound dampening buffer pads helps to deal with this issue. In addition, special sound dampening buffer pads separate the gearboxes from the nacelle frame to minimise transmission of vibrations to the tower

There are limited acoustic solutions that can be applied to reduce noise from the mechanical movements due to the space restriction inside the nacelle as well as the necessity not to disturb the efficient functionality of the mechanical parts with added acoustic treatment. Although for large wind turbines mechanical noise is not much of an issue since those are located far from dwellings, for the small house, mounted wind turbines vibration can have significant impact on people's lives living in the house where the wind turbine is attached.

Moving on to the aerodynamic noise, there are various techniques or even technologies to decrease sound from the wind turbine blades. As somebody would expect most of those techniques originate from designing more aerodynamic blades [18] and adjusting the rotational speed of the turbine.

Again, revising section 'sources of wind turbine sound' we can remind ourselves what causes aerodynamic noise and therefore suggest potential solutions. Some of the noise causes we have discussed concern downwind designs, blade speed and shape and interaction of the airflow between the tower and the wind turbine.

Nowadays most rotors are upwind i.e. the rotor faces into the wind, reducing the risk of causing localised flow instabilities that are responsible for impulsive noise. Although there are still quite a few downwind turbines (where the rotor faces away from the wind) in use, new improved design features have been incorporated aiming at reducing impulsive noise such as increasing the distance between rotor and tower.

In addition to designing upwind turbines the shape of the tower and the nacelle are aerodynamically streamlined in order to reduce any noise that is created by the wind passing the turbine.

To limit the generation of aerodynamic sounds from wind turbines the rotor's rotational speed may be restricted in order to reduce the tip speeds. Large variable speed wind turbines often rotate at slower speeds in low winds, and in increased speeds in higher winds until the limiting rotor speed is reached. This results in much quieter operation in low winds than a comparable constant speed wind turbine. Many modern wind turbines have embedded special control programs that reduce the inflow angle and rpm of the rotor depending on the time of day or year, the wind speed and the wind direction. The noise can be significantly reduced at the expense of power output.

Wind turbine blades are constantly being redesigned to make them more efficient and less noisy. The broadband tip vortex noise caused by rotating wind turbines can be tackled by giving to the blade tip an aerodynamic shape that decreases generation of vorticity [19]. Forward sweeping into the direction of the incoming flow of the blade could result in quieter operation. Figure 12 shows graphically how design changes in the blade tip and shape can result in noise reduction i.e. how three different blade tip geometries can produce three different noise profiles [20].

When it comes to small wind turbines (under 30 kW) the ways to reduce noise are similar to those for large turbines. This means that they also have often variable-speed controls. The interesting fact is that small wind turbine designs may even have higher tip speeds in high winds than large wind turbines. This can result in greater sound generation than would be expected, compared to larger machines. Many modern microwind turbines rotate over their vertical axis regulating power in high winds by turning out of the wind. This additional functionality in operation

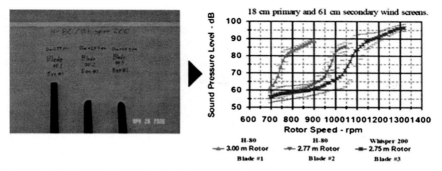

Figure 12: Three different blade tip geometries and the three corresponding noise levels.

can affect the nature of the sound generation from the wind turbine during power regulation. In general domestic wind turbines apart from noise effects can generate vibration signals which are transmitted through the walls causing annoyance to people inside the house. Ways to deal with those problems are associated with increased wall insulation and with locating the turbine at a distance from the residential areas at a height over 3 m from the ground and often being kept switched off during the night.

7 Noise standards

Currently, there are no common international noise standards or regulations for sound pressure levels from wind turbines. Every country, however, defines noise limits and regulations for human exposure depending on the time of the day.

A standard that is being internationally used is:

International Electrotechnical Commission IEC 61400-11 Standard: Wind turbine generator systems – Part 11: Acoustic noise measurement techniques [21].

The IEC 61400-11 standard defines:

- The quality, type and calibration of instrumentation to be used for sound and wind speed measurements.
- Locations and types of measurements to be made.
- Data reduction and reporting requirements.

The standard requires measurements of broad band sound, sound levels in one-third octave bands and in narrow-bands. These measurements are all used to determine the sound power level of the wind turbine.

8 Present and future

Noise from wind turbines is an issue that is gaining increasing concern for government bodies, regulators and the public. The pressure on governments to cut carbon emissions, forces them to make extensive use in their future environmental plans to install as many wind turbine as necessary. It becomes apparent that a number of improvements in standards and regulations is needed to ensure that communities can reliably anticipate noise from wind turbines and to ensure that the data are available to make those sound estimations.

Also research and development in establishing commonly used standards is essential for manufactures and planners to be able to conduct accurate measurements.

We have already mentioned in this chapter that the challenge in measuring noise from wind turbines is that the background noise levels increase as the wind speed increases making it difficult to accurately measure sound levels from the turbines. It is therefore essential that more investigation is being carried out into measuring background noise as a function of both time of day and wind speed and also taking into consideration the factors that affect propagation such as reflection and absorption, ground topology and weather phenomena. This further research would help in establishing accurate and practical noise standards that consequently would inform and guide manufacturers and installers to make comprehensive sound power level measurements, based on new standards, available to the public.

Standards are also needed for the measurement of noise from small wind turbines. These should encounter not only for noise but also for vibration effects since this seems to be a major problem for house mounted wind turbines. These would also inform planners and installers on their decision to select type of wind turbine, location as well as hours of operation.

References

[1] Wagner, S., Bareis, R. & Guidati, G., *Wind Turbine Noise*, Springer-Verlag: Berlin, 1996.
[2] Rogers, L.A., Manwell, F.J. & Wright, S., *Wind Turbine Acoustic Noise*, Renewable Energy Research Laboratory, 2006.
[3] Junger, C. Miguel & Feit David, *Sound, Structures and their Interaction*, Acoustical Society of America, 1993.
[4] Morse, P.M. & Ingard, K.U., *Theoretical Acoustics*, McGraw-Hill: New York, 1968.
[5] Pierce, D. Allan, *Acoustics: An Introduction to Its Physical Principles and Applications*, Acoustical Society of America, 1989.
[6] Dowling, P.A. & Ffowcs Williams, E.J., *Sound and Sources of Sound*, Ellis Horwood, 1983.
[7] Beranek, L.L. & Ver, I.L., *Noise and Vibration Control Engineering: Principles and. Applications*, Wiley: New York, 1992.
[8] Planning Policy Guidance Note 24 'Planning and Noise' (PPG24), Department of the Environment, 1994.
[9] Bruel and Kjaer Instruments, http://www.bkhome.com/
[10] Harris, C., Absorption of sound in air versus humidity and temperature. *Journal of the Acoustical Society of America*, **40**, p. 148, 1966.

[11] Herbert S. Ribner, Reflection, transmission, and amplification of sound by a moving medium. *Journal of the Acoustical Society of America*, **29(4)**, pp. 435–441, April 1957.

[12] Attenborough, K., Sound propagation close to the ground. *Annual Review of Fluid Mechanics*, **34**, pp. 51–82, 2002.

[13] Aylor, D., Noise reduction by vegetation and ground. *Journal of the Acoustical Society of America*, **51**, p. 197, 1971.

[14] Ingard, U., A Review of the influence of meteorological conditions on sound propagation. *Journal of the Acoustical Society of America*, **25**, p. 405, 1953.

[15] The measurement of low frequency noise at three UK wind farms, Department for Trade and Industry, 2006.

[16] Yoshinori Nii, Hikaru Matsumiya & Tetsuya Kogaki, Acoustic performances of a vertical board for wind turbine noise emission measurements. *Acoustical Science and Technology*, **24(2)**, pp. 83–89, 2002.

[17] Migliore, P., van Dam, J. & Huskey, A., *Acoustic Tests of Small Wind Turbines*, NREL SR-500-34601. National Renewable Energy Laboratory: Golden, CO.

[18] Howe, S.M., *Acoustics of Fluid-Structure Interactions*, 2nd ed, Cambridge University Press, 1998.

[19] Arakawa, C., Fleigl, O., Iidal, M. & Shimookal, M., Numerical approach for noise reduction of wind turbine blade tip with earth simulator. *Journal of the Earth Simulator*, **2**, pp. 11–33, 2005.

[20] Brian D. Vick & Clark, R. Nolan, Affect of new blades o noise reduction of small wind turbine water systems, *USDA-Agricultural Research Service*.

[21] International Electrotechnical Commission, International Standard IEC 61400-11, Wind Turbine Generator Systems – Part 11: Acoustic Noise Measurement Techniques.

CHAPTER 21

Wind energy storage technologies

Martin J. Leahy, David Connolly & Denis N. Buckley
The Charles Parsons Initiative, University of Limerick, Ireland.

Energy storage is widely recognised as a key enabling technology for renewable energy and particularly for wind and photovoltaics. Distributed generation could also help, but the location of wind resources and consumption are almost mutually exclusive. The main thrust of the US DoE Energy Storage Programme ($615 M) is in the direction of batteries and CAES. Advanced battery storage (electric vehicles (EV), flow batteries, second-life EV batteries) has the potential to reduce the need for grid infrastructure as it is not topographically, geologically and environmentally limited. This chapter includes a brief examination into the energy storage techniques currently available to assist large-scale wind penetrations. These are pumped-hydroelectric energy storage (PHES), underground pumped-hydroelectric energy storage (UPHES), compressed air energy storage (CAES), battery energy storage (BES), flow battery energy storage (FBES), flywheel energy storage (FES), supercapacitor energy storage (SCES), superconducting magnetic energy storage (SMES), hydrogen energy storage system (HESS), thermal energy storage system (TESS) and finally, EVs. The objective was to identify the following for each: how it works; advantages; applications; cost; disadvantages and future potential.

A brief comparison was then completed to indicate the broad range of operating characteristics available for energy storage technologies. It was concluded that PHES/UPHES, FBES, HESS, TESS and EVs are the most promising techniques to undergo further research. The remaining technologies will be used for their current applications in the future, but further development is unlikely.

1 Introduction

Traditional electricity networks are supplied by large, centralised and highly predictable generating stations. An inherent characteristic of such networks is that supply must meet demand at all times. Typically matching supply with demand on a network requires backup sources of power, such as an open-cycle gas turbine, or by a storage system. Wind energy is inherently intermittent, variable and non-dispatchable (cannot be switched on 'on demand'). Consequently, the need for such backup sources of power increases as the proportion of wind generation on the system increases. To reduce fuel demands, it is desirable that the backup source is a storage facility rather than further primary generation. In this chapter we will discuss the various options for electricity storage including both large-scale centralised storage and smaller-scale distributed storage.

Storage systems such as PHES have been in use since 1929 [1], primarily to level the daily load on the electricity network between night and day. As the electricity sector is undergoing a lot of change, energy storage is becoming a realistic option [2] for: restructuring the electricity market; integrating renewable resources; improving power quality; aiding the shift towards distributed energy; and helping the network operate under more stringent environmental requirements. In addition, energy storage can optimise the existing generation and transmission infrastructures whilst also preventing expensive upgrades. Power fluctuations from renewable resources will prevent their large-scale penetration into the network. However energy storage devices can manage these irregularities and thus aid the amalgamation of renewable technologies. In relation to conventional power production, energy storage devices can improve overall power quality and reliability, which is becoming more important for modern commercial applications. Finally, energy storage devices can reduce emissions by aiding the transition to newer, cleaner technologies such as renewable resources and the hydrogen economy. Therefore, Kyoto obligations can be met (and penalties avoided). A number of obstacles have hampered the commercialization of energy storage devices including: a lack of experience – a number of demonstration projects will be required to increase customer's confidence; inconclusive benefits – consumers do not understand what exactly are the benefits of energy storage in terms of savings and also power quality; high capital costs – this is clearly an issue when the first two disadvantages are considered; responsibility for cost – developers view storage as 'grid infrastructure' whereas the Transmission System Operator (TSO) views it as part of the renewable energy plant.

However, as renewable resources and power quality become increasingly important, costs and concerns regarding energy storage technologies are expected to decline. This chapter identifies the numerous different types of energy storage devices currently available. The parameters used to describe an energy storage device and the applications they fulfil are explored first. This is followed by an analysis of each energy storage technology currently available indicating their: operation and the advantages; applications; cost; disadvantages; future; and finally, a brief comparison of the various technologies is provided.

2 Parameters of an energy storage device

Below is a list of parameters used to describe an energy storage device. These will be used throughout the chapter:

* *Power capacity* is the maximum instantaneous output that an energy storage device can provide, usually measured in kilowatts (kW) or megawatts (MW).

- *Energy storage capacity* is the amount of electrical energy the device can store, usually measured in kilowatt-hours (kWh) or megawatt-hours (MWh).
- *Efficiency* indicates the quantity of electricity that can be recovered as a percentage of the electricity used to charge the device.
- *Response time* is the length of time it takes the storage device to start releasing power.
- *Round-trip efficiency* indicates the quantity of electricity which can be recovered as a percentage of the electricity used to charge and discharge the device.

In addition to the above, the following parameters are also used in describing rechargeable batteries and flow batteries:

- *Charge-to-discharge ratio* is the ratio of the time it takes to charge the device relative to the time it takes to discharge the device. For example, if a device takes 5 times longer to charge than to discharge, it has a charge-to-discharge ratio of 5:1.
- *Depth of discharge* (DoD) is the percentage of the battery capacity that is discharged during a cycle.
- *Memory effect*. If certain batteries are never fully discharged they 'remember' this and lose some of their capacity.

3 Energy storage plant components

Every energy storage facility is comprised of three primary components: storage medium, power conversion system (PCS), and balance of plant (BOP).

3.1 Storage medium

The storage medium is the 'energy reservoir' that retains the potential energy within a storage device. Storage media range from mechanical (PHES), chemical (BES) and electrical (SMES) potential energy.

3.2 Power conversion system

It is necessary to convert from alternating current (AC) to direct current (DC) and vice versa, for all storage devices except mechanical storage devices, e.g. PHES and CAES [3]. Consequently, a PCS is required that acts as a rectifier while the energy device is charged (AC to DC) and as an inverter when the device is discharged (DC to AC). The PCS also conditions the power during conversion to ensure that no damage is done to the storage device.

The customization of the PCS for individual storage systems has been identified as one of the primary sources of improvement for energy storage facilities, as each storage device operates differently during charging, standing and discharging [3].

The PCS usually costs from 33 to 50% of the entire storage facility. Development of PCSs has been slow due to the limited growth in distributed energy resources e.g. small-scale power generation technologies ranging from 3 to 10,000 kW [4].

3.3 Balance of plant

BOP comprises all additional works and ancillary components required to

* house the equipment
* control the environment of the storage facility
* provide the electrical connection between the PCS and the power grid

It is the most variable cost component within an energy storage device due to the various requirements for each facility. The BOP "typically includes electrical interconnections, surge protection devices, a support rack for the storage medium, the facility shelter and environmental control systems" [3]. The BOP may also include foundations, roadways, access, security equipment, electrical switchgear and metering equipment. Development activities including all paperwork, design, planning, safety, training and their costs are often included here.

4 Energy storage technologies

Energy storage devices by their nature are typically suitable for a very particular set of applications. This is primarily due to the potential power and storage capacities that can be obtained from the various devices. Therefore, in order to provide a fair comparison between the various energy storage technologies, they have been grouped together based on the size of power and storage capacity that can be obtained. Four categories have been created: devices large power (>50 MW) and storage (>100 MWh) capacities; devices with medium power (1–50 MW) and storage capacities (5–100 MWh); devices with medium power or storage capacities but not both; and finally, a section on energy storage systems.

The following energy storage technologies will be discussed under the respective groups: PHES, UPHES and CAES will be discussed together as they all have the potential for large power and storage capacities; BES and FBES will be discussed together as they have the potential for medium power and storage capacities, while SCES, FES and SMES will be grouped together as they all have either medium-scale power or storage capacities, and finally HESS, TESS and EVs will be discussed together as these are generally smaller energy storage systems. Before commencing it is worth noting which category each technology falls into. Only the technologies common by category will be compared against each other after they have been analysed. HESS, TESS and EVs have unique characteristics as these are energy systems, requiring a number of different technologies, which can be controlled differently. As energy systems transform from a fossil fuel production based on centralised production, to a renewable energy system, based on intermittent decentralised production, it is imperative that system flexibility is maximised. An ideal option to achieve this is by integrating the three primary sectors within any energy system: the electricity, heat and transport sectors. HESS, TESS and EVs provide unique opportunities to integrate these three sectors and hence increase the renewable energy penetrations feasible. As a result it is difficult to compare HESS, TESS and EVs to the other energy storage technologies directly as energy storage is only part of the purpose of those systems.

4.1 Pumped-hydroelectric energy storage

PHES is the most mature and largest capacity storage technique available. A pump and turbine have been combined in a single device optimised for this purpose. PHES consists of two large reservoirs located at different elevations and a number of pump turbine units (see Fig. 1). During off-peak electrical demand, water is pumped from the lower reservoir to the higher reservoir where it is stored until it is needed. When required (i.e. during peak electrical production) the water in the upper reservoir is released through the turbines, which are connected to generators that produce electricity. Therefore, during production a PHES facility operates similarly to a conventional hydroelectric system.

The efficiency of operational pumped storage facilities is in the region of 50–85% with more modern units at the upper end. However, variable speed machines are now being used to improve this. The efficiency is limited by the efficiency of the pump/turbine unit used in the facilities [2]. Until recently, PHES units have always used fresh water as the storage medium. However, in 1999 a PHES facility using seawater as the storage medium was constructed [6] (see Fig. 2); corrosion was

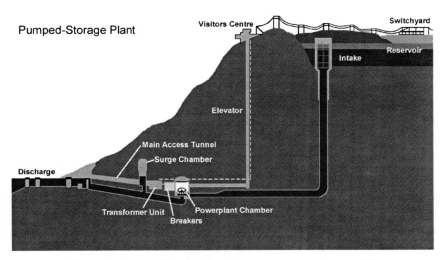

Figure 1: Pumped-hydroelectric energy storage layout [5].

Figure 2: Pumped-hydroelectric storage facility using seawater at Okinawa, Japan [6].

prevented by using paint and cathodic protection. A typical PHES facility has 300 m of hydraulic head (the vertical distance between the upper and lower reservoir). The power capacity (kW) is a function of the flow rate and the hydraulic head, while the energy stored (kWh) is a function of the reservoir volume and hydraulic head. To calculate the mass power output of a PHES facility, the following relationship can be used [7]:

$$P_C = \rho g Q H n \tag{1}$$

where P_C is the power capacity in W, ρ the mass density of water in kg/m^3, g the acceleration due to gravity in m/s^2, Q the discharge through the turbines in m^3/s, H the effective head in m and n is the efficiency.

Also, the storage capacity of the PHES may be evaluated with the following equation [8]:

$$S_C = V \rho g H n \tag{2}$$

where S_C is the storage capacity in megawatt-hours (MWh) and V is the volume of water that is drained and filled each day in m^3. It is evident that the power and storage capacities are both dependent on the head and the volume of the reservoirs. However, facilities should be designed with the greatest hydraulic head possible rather than largest upper reservoir possible. It is much cheaper to construct a facility with a large hydraulic head and small reservoirs, than to construct a facility of equal capacity with a small hydraulic head and large reservoirs because: less material needs to be removed to create the reservoirs required; smaller piping is necessary, hence, smaller boreholes during drilling; the pump turbine is physically smaller. Currently, there is over 90 GW in more than 240 PHES facilities in the world – roughly 3% of the world's global generating capacity. Individual facilities can store up to 15 GWh of electrical energy from in plant with power ratings from 30 to 4000 MW [2].

4.1.1 Applications of PHES

Similar to large storage capacities, PHES also has a fast reaction time, making it ideal for load levelling applications, where the plant can vary its effective load on the system from the full name plate rating in the positive direction (pumping) to the full name plate rating in the negative direction (generating) (see Fig. 3). Facilities can have a reaction time as short as 10 min or less from complete shutdown (or from full reversal of operation) to full power [3]. In addition, if kept on standby, full power can even be reached within 10–30 s.

Also, with the recent introduction of variable speed machines, PHES systems can now be used for frequency regulation in both pumping and generation modes (this has always been available in generating mode). This allows PHES units to absorb power in a more cost-effective manner that not only makes the facility more useful, but also improves the efficiency by approximately 3% [3] and extends the life of the facility. PHES can also be used for peak generation and black starts (start generating without access to a main frequency set by other units on the grid) due to its large power capacity and sufficient discharge time. Finally, PHES provides a load for base-load generating facilities during off-peak production, hence, cycling these units can be avoided which improves their lifetime as well as their efficiency.

4.1.2 Cost of PHES

Cost ranges from $600/kW [2] to upwards of $2000/kW [3], depending on a number of factors such as size, location and connection to the power grid.

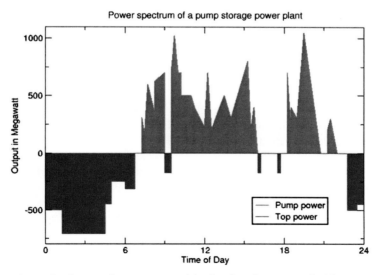

Figure 3: Output from a pumped-hydroelectric storage facility [5].

4.1.3 Disadvantages of PHES

A major disadvantage of PHES facility is its dependence on specific geological formations, because two large reservoirs with a sufficient hydraulic head between them must be located within close proximity to build a PHES system. As well as being rare, these geological formations normally exist in remote locations such as mountains, where construction is difficult and the power grid is not present, although large wind farm sites may provide a useful modern alternative. Finally, in order to make a PHES plant viable it must be constructed on a large scale. Although the cost per kWh of storage is relatively economical in comparison to other techniques, this results in a very high initial construction cost for the facility, therefore deterring investment in PHES, e.g. Bath County storage facility in the United States which has a power capacity of 2100 MW cost $1.7 billion in 1985.

4.1.4 Future of PHES

Currently, a lot of work is being carried out to upgrade old PHES facilities with new equipment such as variable speed devices which can increase capacity by 15–20% and efficiency by 5–10%. This is more desirable as energy storage capacity can be added without the high initial construction costs. Prospects of building new facilities are limited due to the "high development costs, long lead times and design limitations" [3]. However, a new concept that is showing a lot of theoretical potential is UPHES, discussed in the next section.

4.2 Underground pumped-hydroelectric energy storage

An UPHES facility has the same operating principle as PHES system: two reservoirs with a large hydraulic head between them. The only major difference between the two designs is the locations of their respective reservoirs. In conventional PHES, suitable geological formations must be identified to build the facility, as discussed earlier (see Section 5.1). However, UPHES facilities have been designed with the upper reservoir at ground level and the lower reservoir deep below the earth's surface. The depth depends on the amount of hydraulic head required for the specific application (see Fig. 4).

4.2.1 Applications of UPHES

UPHES can provide the same services as PHES: load levelling, frequency regulation, and peak generation. However, as UPHES does not need to be built at a suitable geological formation, it can be constructed anywhere with an area large enough for the upper reservoir. Consequently, it can be placed in ideal locations to function with wind farms, the power grid, specific areas of electrical irregularities, etc. The flexibility of UPHES makes it a more attractive option for energy storage than conventional PHES, but its technical immaturity needs to be addressed.

4.2.2 Cost of UPHES

The capital cost of UPHES is the deciding factor for its future. As it operates in the same way as PHES, it is a very reliable and cost-effective storage technique with

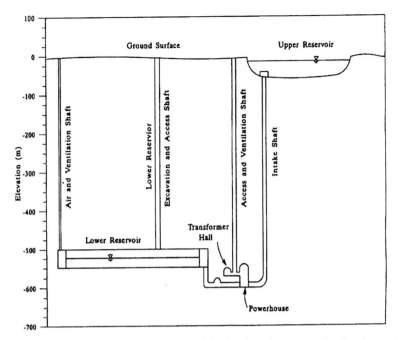

Figure 4: Proposed underground pumped-hydroelectric storage facility layout [7].

low maintenance costs. However, depending on the large capital costs involved, UPHES might not be a viable option as other technologies begin to develop larger storage capacities, e.g. flow batteries. Currently, no costs have been identified for UPHES, primarily due to the lack of facilities constructed. A number of possible cost-saving ideas have been put forward such as using old mines for the lower reservoir of the facility [7, 9]. Also, if something valuable can be removed to make the lower reservoir, it can be sold to make back some of the cost.

4.2.3 Disadvantages of UPHES

UPHES incorporates the same disadvantages as PHES (large-scale required, high capital costs, etc.), with one major exception. As stated previously (see Section 5.1), the most significant problem with PHES is its geological dependence. As the lower reservoir is obtained by drilling into the ground and the upper reservoir is at ground level, UPHES does not have such stringent geological dependences. The major disadvantage for UPHES is its commercial youth. To date there is very few, if any, UPHES facilities in operation. Therefore, it is very difficult to analyse and to trust the performance of this technology.

4.2.4 Future of UPHES

UPHES has a very bright future if cost-effective excavation techniques can be identified for its construction. Its relatively large-scale storage capacities, combined with its location independence, provide a storage technique with unique characteristics. However, as well as cost, a number of areas need to be investigated further in this area such as its design, power and storage capacities and environmental impact to prove it is a viable option.

4.3 Compressed air energy storage

A CAES facility consists of a power train motor that drives a compressor (to compress the air into the cavern), high pressure turbine (HPT), a low pressure turbine (LPT), and a generator (see Fig. 5).

In conventional gas turbines (GTs), 66% of the gas used is required to compress the air at the time of generation. Therefore, CAES pre-compresses the air using off-peak electrical power which is taken from the grid to drive a motor (rather than using gas from the GT plant) and stores it in large storage reservoirs. When the GT is producing electricity during peak hours, the compressed air is released from the storage facility and used in the GT cycle. As a result, instead of using expensive gas to compress the air, cheaper off-peak base-load electricity is used. However, when the air is released from the cavern it must be mixed with a small amount of gas before entering the turbine. If there was no gas added, the temperature and pressure of the air would be problematic. If the pressure using air alone was high enough to achieve a significant power output, the temperature of the air would be far too low for the materials and connections to tolerate [1]. The amount of gas

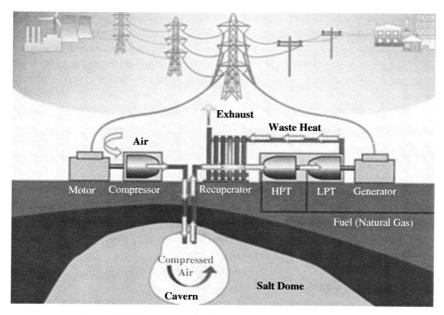

Figure 5: Compressed air energy storage facility [10].

required is so small that a GT working simultaneously with CAES can produce three times more electricity than a GT operating on its own, using the same amount of natural gas.

The reservoir can be man-made but this is expensive so CAES locations are usually decided by identifying natural geological formations that suit these facilities. These include salt-caverns, hard-rock caverns, depleted gas fields or an aquifer. Salt-caverns can be designed to suit specific requirements. Fresh water is pumped into the cavern and left until the salt dissolves and saturates the fresh water. The water is then returned to the surface and the process is repeated until the required volume cavern is created. This process is expensive and can take up to 2 years. The costs associated with hard-rock caverns are likely to be more than 50% higher. Finally, aquifers cannot store the air at high pressures and therefore have a relatively lower energy capacity.

CAES uses both electrical energy and natural gas so its efficiency is difficult to predict. It is estimated that the efficiency of the entire cycle is in the region of 64% [11] to 75% [3]. Typical plant capacities for CAES are in the region of 50–300 MW. The life of these facilities is proving to be far longer than existing GTs and the charge/discharge ratio is dependent on the size of the compressor used, as well as the size and pressure of the reservoir.

4.3.1 Applications of CAES

CAES is the only very large-scale storage technique other than PHES. CAES has a fast reaction time with plants usually able to go from 0 to 100% in less than 10 min, 10 to 100% in approximately 4 min and 50 to 100% in less than 15 s [2]. As a result, it is ideal for acting as a large sink for bulk energy supply and demand and also, it is able to undertake frequent start-ups and shutdowns. Furthermore, traditional GT suffer a 10% efficiency reduction from a 5°C rise in ambient temperatures due to a reduction in the air density. CAES use compressed air so they do not suffer from this effect. Also, traditional GTs suffer from excessive heat when operating on partial load, while CAES facilities do not. These flexibilities mean that CAES can be used for ancillary services such as frequency regulation, load following, and voltage control [3]. As a result, CAES has become a serious contender in the wind power energy storage market. A number of possibilities are being considered such as integrating a CAES facility with a number of wind farms within the same region. The excess off-peak power from these wind farms could be used to compress air for a CAES facility. *Iowa Association of Municipal Utilities* is currently planning a project of this nature [12].

4.3.2 Cost of CAES

The cost of CAES facilities are $425/kW [2] to $450/kW [3]. Maintenance is estimated between $3/kWh [13] and $10/kWh [14]. Costs are largely dependent on the reservoir construction. Overall, CAES facilities expect to have costs similar to or greater than conventional GT facilities. However, the energy cost is much lower for CAES systems.

4.3.3 Disadvantages of CAES

The major disadvantage of CAES facilities is their dependence on geographical location. It is difficult to identify underground reservoirs where a power plant can be constructed, is close to the electric grid, is able to retain compressed air and is large enough for the specific application. As a result, capital costs are generally very high for CAES systems. Also, CAES still uses a fossil fuel (gas) to generate electricity. Consequently, the emissions and safety regulations are similar to conventional GTs. Finally, only two CAES facilities currently exist, meaning it is still a technology of potential not experience.

4.3.4 Future of CAES

Reservoir developments are expected in the near future due to the increased use of natural gas storage facilities. The US and Europe are more likely to investigate this technology further as they possess acceptable geology for an underground reservoir (specifically salt domes). Due to the limited operational experience, CAES has been considered too risky by many utilities [14].

A number of CAES storage facilities have been planned for the future including:

- 25 MW CAES research facility with aquifer reservoir in Italy
- 3 × 100 MW CAES plant in Israel
- Norton Energy Storage LLC in America is planning a CAES with a limestone mine acting as the reservoir. The first of four phases is expected to produce between 200 and 480 MW at a cost of $50 to $480 million. The final plant output is planned to be 2500 MW.

Finally, proposals have also been put forward for a number of similar technologies such as micro-CAES and thermal and compressed air storage (TACAS). However, both are in the early stages of development and their future impact is not decisive. Although Joe Pinkerton, CEO of *Active Power*, declared that TACAS "is the first true minute-for-minute alternative to batteries for UPS industry" [3].

4.4 Battery energy storage

There are three important types of large-scale BES. These are: lead-acid (LA); nickel-cadmium (NiCd); sodium-sulphur (NaS). These operate in the same way as conventional batteries, except on a large scale, i.e. two electrodes are immersed in an electrolyte, which allows a chemical reaction to take place so current can be produced when required.

4.4.1 LA battery

This is the most common energy storage device in use at present. Its success is due to its maturity (research has been ongoing for an estimated 140 years), relatively low cost, long lifespan, fast response, and low self-discharge rate. These batteries are can be used for both short-term applications (seconds) and long-term applications (up to 8 h). There are two types of LA batteries; flooded lead-acid (FLA) and valve-regulated lead-acid (VRLA). FLA batteries are made up of two electrodes that are constructed using lead plates which are immersed in a mixture of water (65%) and sulphuric acid (35%) (see Fig. 6). VRLA batteries have the same operating principle as FLA batteries, but they are sealed with a pressure-regulating valve. This eliminates air from entering the cells and also prevents venting of the hydrogen. VRLA batteries have lower maintenance costs, weigh less and occupy less space. However, these advantages are coupled with higher initial costs and shorter lifetime.

Both the power and energy capacities of LA batteries are based on the size and geometry of the electrodes. The power capacity can be increased by increasing the surface area for each electrode, which means greater quantities of thinner electrode plates in the battery. However, to increase the storage capacity of the battery, the mass of each electrode must be increased, which means fewer and thicker plates. Consequently, a compromise must be met for each application. LA batteries can respond within milliseconds at full power. The average DC–DC efficiency of a LA battery is 75–85% during normal operation, with a life of approximately 5 years or 250–1000 charge/discharge cycles, depending on the depth of discharge [3].

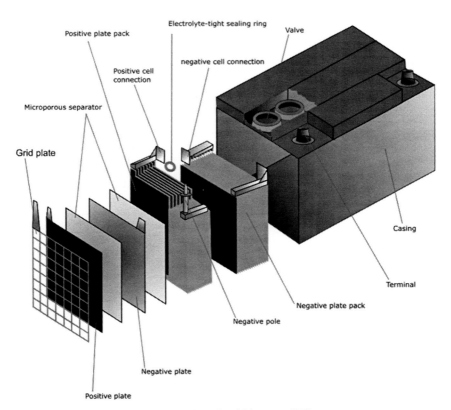

Figure 6: Lead acid battery [15].

4.4.1.1 Applications of LA battery

FLA batteries have two primary applications [3]:

1. Starting and ignition, short bursts of strong power e.g. car engine batteries
2. Deep cycle, low steady power over a long time

VRLA batteries are very popular for backup power, standby power supplies in telecommunications and also for UPS systems. A number of LA storage facilities are in operation today as can be seen in Table 1.

4.4.1.2 Cost of LA battery

Costs for LA battery technology have been stated as $200/kW to $300/kW [2], but also in the region of $580/kW [3]. Looking at Table 1, the cost variation is evident.

4.4.1.3 Disadvantages of LA battery

LA batteries are extremely sensitive to their environments. The typical operating temperature for a LA battery is roughly 27°C, but a change in temperature of 5°C or more can cut the life of the battery by 50%. However, if the DoD exceeds this, the cycle life of the battery will also be reduced. Finally, a typical charge-to-discharge ratio of a LA battery is 5:1. At faster rates of charge, the cell will be damaged.

Table 1: Largest LA and VRLA batteries installed worldwide [2].

Plant	Year of installation	Rated Energy (MWh)	Rate Power (MW)	Battery system alone		Total cost of the storage system*	
				Cost in $1995 ($/kWh)	Cost in $1995 ($/kWh)	Cost in $1995 ($/kW)	Cost in $1995 ($/kW)
CHINO California	1988	40	10	201	805	456	1823
HELCO Hawaii (VRLA)	1993	15	10	304	456	777	1166
PREPA Puerto Rico	1994	14	20	341	239	1574	1102
BEWAG Germany	1986	8.5	8.5	707	707	n/a	n/a
VERNON California (VRLA)	1995	4.5	3	305	458	944	1416

*Includes power conditioning system and balance-of-payment.

4.4.1.4 Future of LA battery

Due to the low cost and maturity of the LA battery it will probably always be useful for specific applications. The international *Advanced Lead-Acid Battery Consortium* is also developing a technique to significantly improve storage capacity and also recharge the battery in only a few minutes, instead of the current hours [2]. However, the requirements of new large-scale storage devices would significantly limit the life of a LA battery. Consequently, a lot of research has been directed towards other areas. Therefore, it is unlikely that LA batteries will be competing for future large-scale multi-MW applications.

4.4.2 NiCd battery

A NiCd battery is made up of a positive with nickel oxyhydroxide as the active material and a negative electrode composed of metallic cadmium. These are separated by a nylon divider (see Fig. 7). The electrolyte, which undergoes no significant changes during operation, is aqueous potassium hydroxide. During discharge, the nickel oxyhydroxide combines with water and produces nickel hydroxide and a hydroxide ion. Cadmium hydroxide is produced at the negative electrode. To charge the battery the process can be reversed. However, during charging, oxygen can be produced at the positive electrode and hydrogen can be produced at the negative electrode. As a result some venting and water addition is required, but much less than required for a LA battery.

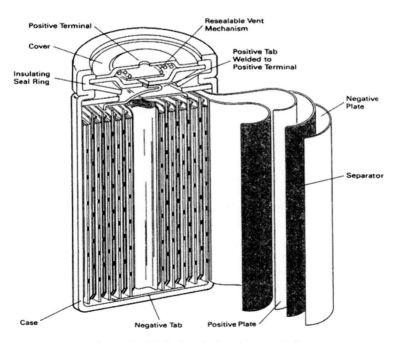

Figure 7: Nickel-cadmium battery [16].

There are two NiCd battery designs: vented and sealed. Sealed NiCd batteries are the common, everyday rechargeable batteries used in a remote control, lamp, etc. No gases are released from these batteries, unless a fault occurs. Vented NiCd batteries have the same operating principles as sealed ones, but gas is released if overcharging or rapid discharging occurs. The oxygen and hydrogen are released through a low pressure release valve making the battery safer, lighter, more economical, and more robust than sealed NiCd batteries.

The DC–DC efficiency of a NiCd battery is 60–70% during normal operation although the life of these batteries is relatively high at 10–15 years, depending on the application. NiCd batteries with a pocket-plate design have a life of 1000 charge/discharge cycles, and batteries with sintered electrodes have a life of 3500 charge/discharge cycles. NiCd batteries can respond at full power within milliseconds. At small DoD rates (approximately 10%) NiCd batteries have a much longer cycle life (50,000 cycles) than other batteries such as LA batteries. They can also operate over a much wider temperature range than LA batteries, with some able to withstand occasional temperatures as high as 50°C.

4.4.2.1 Applications of NiCd battery

Sealed NiCd batteries are used commonly in commercial electronic products such as a remote control, where light weight, portability, and rechargeable power are important. Vented NiCd batteries are used in aircraft and diesel engine starters, where large energy per weight and volume are critical [3]. NiCd batteries are ideal for protecting power quality against voltage sags and providing standby power in harsh conditions. Recently, NiCd batteries have become popular as storage for solar generation because they can withstand high temperatures. However, they do not perform well during peak shaving applications, and consequently are generally avoided for energy management systems.

4.4.2.2 Cost of NiCd battery

NiCd batteries cost more than LA batteries at $600/kW [3]. However, despite the slightly higher initial cost, NiCd batteries have much lower maintenance costs due to their environmental tolerance.

4.4.2.3 Disadvantages of NiCd battery

Like LA batteries, the life of NiCd batteries can be greatly reduced due to the DoD and rapid charge/discharge cycles. However, NiCd batteries suffer from 'memory' effects and also lose more energy during due to self-discharge standby than LA batteries, with an estimated 2–5% of their charge lost per month at room temperature in comparison to 1% per month for LA batteries [3]. Also, the environmental effects of NiCd batteries have become a widespread concern in recent years as cadmium is a toxic material. This creates a number of problems for disposing of the batteries.

4.4.2.4 Future of NiCd battery

It is predicted that NiCd batteries will remain popular within their current market areas, but like LA batteries, it is unlikely that they will be used for future large-scale projects. Although just to note, a 40 MW NiCd storage facility was constructed in Alaska; comprising of 13,760 cells at a cost of $35M [2]. The cold temperatures experienced were the primary driving force behind the use NiCd as a storage medium. NiCd will probably remain more expensive than LA batteries, but they do provide better power delivery. However, due to the toxicity of cadmium, standards and regulations for NiCd batteries will continue to rise.

4.4.3 NaS battery

NaS batteries have three times the energy density of LA, a longer life span, and lower maintenance. These batteries are made up of a cylindrical electrochemical cell that contains a molten-sodium negative electrode and a molten-sulphur positive electrode. The electrolyte used is solid β-alumina. During discharging, sodium ions pass through the β-alumina electrolyte where they react at the positive electrode with the sulphur to form sodium polysulphide (see Fig. 8). During charging, the reaction is reversed so that the sodium polysulphide decomposes, and the sodium ions are converted to sodium at the positive electrode. In order to keep the sodium and sulphur molten in the battery, and to obtain adequate conductivity in the electrolyte, they are housed in a thermally insulated enclosure that must keep it above 270°C, usually at 320–340°C.

A typical NaS module is 50 kW at 360 kWh or 50 kW at 430 kWh. The average round-trip energy efficiency of a NaS battery is 86% [2] to 89% [3]. The cycle life is much better than for LA or NiCd batteries. At 100% DoD, the NaS batteries can last approximately 2500 cycles. As with other batteries, this increases as the DoD decreases; at 90% DoD the unit can cycle 4500 times and at 20% DoD 40,000 times [3].

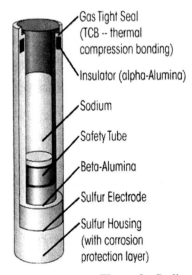

- **Low resistance, high efficiency due to**
  - Beta Alumina tube
  - Sulfur electrode design

- **High durability due to**
  - Corrosion protection layer
  - Sulfur electrode design

- **High energy density due to**
  - Cell properties and design

- **Intrinsic safety due to**
  - Incorporation of safety tube

Figure 8: Sodium-sulphur battery [3].

One of the greatest characteristics of NaS batteries is its ability to provide power in a single, continuous discharge or else in shorter larger pulses (up to five times higher than the continuous rating). It is also capable of pulsing in the middle of a long-term discharge. This flexibility makes it very advantageous for numerous applications such as energy management and power quality. NaS batteries have also been used for deferring transmission upgrades. Currently, NaS batteries cost $810/kW, but it is only a recently commercialised product. This cost is likely to be reduced as production increases, with some predicting reductions upwards of 33% [3]. The major disadvantage of NaS batteries is retaining the device at elevated temperatures above 270°C. It is not only energy consuming, but it also brings with it problems such as thermal management and safety regulations [17]. Also, due to harsh chemical environments, the insulators can be a problem as they slowly become conducting and self-discharge the battery.

A 6 MW, 8 h unit has been built by *Tokyo Electric Power Company* (TEPCO) and *NGK Insulators, Ltd.* (NGK), in Tokyo, Japan with an overall plant efficiency of 75% and is thus far proving to be a success (see Fig. 9). The materials required to create a NaS battery are inexpensive and abundant, and 99% of the battery is recyclable. The NaS battery has the potential to be used on a MW scale by combining modules. Combining this with its functionality to mitigate power disturbances, NaS batteries could be a viable option for smoothing the output from wind turbines into the power grid [3]. *American Electric Power* is planning to incorporate a 6 MW NaS battery with a wind farm for a 2-year demonstration [18, 19]. The size of the wind farm has yet to be announced but the results from this will be vital for the future of the NaS battery.

4.5 Flow battery energy storage

There are three primary types of FBES: vanadium redox (VR); polysulphide bromide (PSB); zinc-bromine (ZnBr). They all operate in a similar fashion: two charged electrolytes are pumped to the cell stack where a chemical reaction occurs, allowing current to be obtained from the device when required. The operation of each will be discussed in more detail during the analysis.

Figure 9: 6 MW, 8 h Sodium-sulphur energy storage facility in Tokyo, Japan [2].

4.5.1 VR flow battery

A VR battery is made up of a cell stack, electrolyte tank system, control system and a PCS (see Fig. 10). These batteries store energy by interconnecting two forms of vanadium ions in a sulphuric acid electrolyte at each electrode; with V^{2+}/V^{3+} in the negative electrode, and V^{4+}/V^{5+} in the positive electrode. The size of the cell stack determines the power capacity (kW) whereas the volume of electrolyte (size of tanks) indicates the energy capacity (kWh) of the battery.

As the battery discharges, the two electrolytes flow from their separate tanks to the cell stack where H^+ ions are passed between the two electrolytes through the permeable membrane. This process induces self-separation within the solution thus changing the ionic form of the vanadium as the potential energy is converted to electrical energy. During recharge this process is reversed. VR batteries operate at normal temperature with an efficiency as high as 85% [2, 3]. As the same chemical reaction occurs for charging and discharging, the charge/discharge ratio is 1:1. The VR battery has a fast response, from charge to discharge in 0.001 s and also a high overload capacity with some claiming it can reach twice its rated capacity for several minutes [2]. VR batteries can operate for 10,000 cycles giving them an estimated life of 7–15 years depending on the application. Unlike conventional batteries they can be fully discharged without any decline in performance [21]. At the end of its life (10,000 cycles), only the cell stack needs to be replaced as the electrolyte has an indefinite life and thus can be reused. VR batteries have been designed as modules so they can be constructed on-site.

4.5.1.1 Applications of VR flow battery

As the power and energy capacities are decoupled, the VR flow battery is a very versatile device in terms of energy storage. It can be used for every energy storage requirement including UPS, load levelling, peak shaving, telecommunications, electric utilities and integrating renewable resources. Although the versatility of flow batteries makes it extremely useful for a lot of applications, there are a number of competing devices within each area that perform better for their specific application. Consequently, although capable of performing for numerous applications,

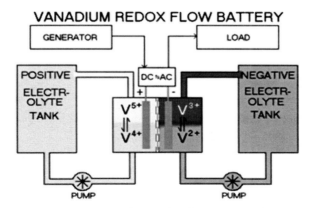

Figure 10: Vanadium redox flow battery [20].

VR batteries are only considered where versatility is important, such as the integration of renewable resources.

4.5.1.2 Cost of VR flow battery
There are two costs associated with flow batteries: the power cost (kW), and the energy cost (kWh), as they are independent of each other. The power cost for VR batteries is $1828/kW, and the energy cost is $300/kWh to $1000/kWh, depending on system design [3].

4.5.1.3 Disadvantages of VR flow battery
VR batteries have the lowest power density and require the most cells (each cell has a voltage of 1.2 V) in order to obtain the same power output as other flow batteries. For smaller-scale energy applications, VR batteries are very complicated in relation to conventional batteries, as they require much more parts (such as pumps, sensors, control units) while providing similar characteristics. Consequently, when deciding between a flow battery and a conventional battery, a decision must be made between a simple but constrained device (conventional battery), and a complex but versatile device (flow battery).

4.5.1.4 Future of VR flow battery
VR batteries have a lot of potential due to their unique versatility, specifically their MW power and storage capacity potential. However, the commercial immaturity of VR batteries needs to be changed to prove it is a viable option in the future.

4.5.2 PSB flow battery
PSB batteries operate very similarly to VR batteries. The unit is made up of the same components; a cell stack, electrolyte tank system, control system and a PCS (see Fig. 11). The electrolytes used within PSB flow batteries are sodium bromide as the positive electrolyte, and sodium polysulphide as the negative electrolyte.

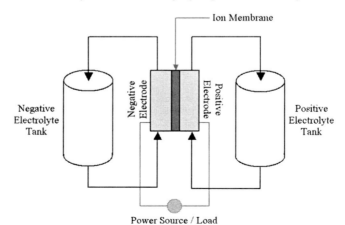

Figure 11: Polysulphide bromide flow battery [3].

During discharge, the two electrolytes flow from their tanks to the cell where the reaction takes place at a polymer membrane that allows sodium ions to pass through. Like VR batteries, self-separation occurs during the discharge process and as before, to recharge the battery this process is simply reversed. The voltage across each cell is approximately 1.5 V.

PSB batteries operate between 20 and 40°C, but a wider range can be used if a plate cooler is used in the system. The efficiency of PSB flow batteries approaches 75% according to [2, 3]. As with VR batteries, the discharge ratio is 1:1, since the same chemical reaction is taking place during charging and discharging. The life expectancy is estimated at 2000 cycles but once again, this is very dependent on the application. As with VR batteries the power and energy capacities are decoupled in PSB batteries.

4.5.2.1 Applications of PSB flow battery
PSB flow batteries can be used for all energy storage requirements including load levelling, peak shaving, and integration of renewable resources. However, PSB batteries have a very fast response time; it can react within 20 ms if electrolyte is retained charged in the stacks (of cells). Under normal conditions, PSB batteries can charge or discharge power within 0.1 s [2]. Therefore, PSB batteries are particularly useful for frequency response and voltage control.

4.5.2.2 Cost PSB flow battery
The power capacity cost of PSB batteries is $1094/kW and the energy capacity cost is $185/kWh [3].

4.5.2.3 Disadvantages PSB flow battery
During the chemical reaction small quantities of bromine, hydrogen, and sodium sulphate crystals are produced. Consequently, biweekly maintenance is required to remove the NaS by-products. Also, two companies designed and planned to build PSB flow batteries. *Innogy's Little Barford Power Station* in the UK wanted to use a 24,000 cell 15 MW 120 MWh PSB battery, to support a 680 MW combined-cycle gas turbine plant. Tennessee Valley Authority (TVA) in Columbus wanted a 12 MW, 120 MWh to avoid upgrading the network. However, both facilities have been cancelled with no known explanation.

4.5.2.4 Future PSB flow battery
Like the VR battery, PSB batteries can scale into the MW region and therefore must have a future within energy storage. However, until a commercial demonstration succeeds, the future of PSB batteries will remain doubtful.

4.5.3 Zinc-bromine (ZnBr) flow battery
These flow batteries are slightly different to VR and PSB flow batteries. Although they contain the same components: a cell stack, electrolyte tank system, control system, and a PCS (see Fig. 12) they do not operate in the same way.

During charging the electrolytes of zinc and bromine ions (that only differ in their concentration of elemental bromine) flow to the cell stack. The electrolytes

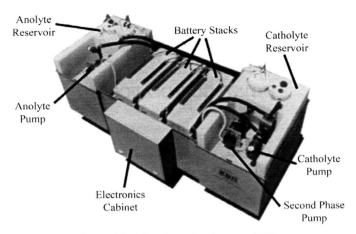

Figure 12: Zinc-bromine battery [22].

are separated by a microporous membrane. Unlike VR and PSB flow batteries, the electrodes in a ZnBr flow battery act as substrates to the reaction. As the reaction occurs, zinc is electroplated on the negative electrode and bromine is evolved at the positive electrode, which is somewhat similar to conventional battery operation. An agent is added to the electrolyte to reduce the reactivity of the elemental bromine. This reduces the self-discharge of the bromine and improves the safety of the entire system [22]. During discharge the reaction is reversed; zinc dissolves from the negative electrode and bromide is formed at the positive electrode. ZnBr batteries can operate in a temperature range of 20–50°C. Heat must be removed by a small chiller if necessary. No electrolyte is discharged from the facility during operation and hence the electrolyte has an indefinite life. The membrane however, suffers from slight degradation during the operation, giving the system a cycle life of approximately 2000 cycles. The ZnBr battery can be 100% discharged without any detrimental consequences and suffers from no memory effect. The efficiency of the system is about 75% [2] or 80% [3]. Once again, as the same reaction occurs during charging and discharging, the charge/discharge ratio is 1:1, although a slower rate is often used to increase efficiency [3]. Finally, the ZnBr flow battery has the highest energy density of all the flow batteries, with a cell voltage of 1.8 V.

4.5.3.1 Applications of ZnBr flow battery

The building block for ZnBr flow batteries is a 25 kW, 50 kWh module constructed from three 60-cell battery stacks in parallel, each with an active cell area of 2500 cm^2 [22]. ZnBr batteries also have a high energy density of 75–85 Wh/kg. As a result, the ZnBr batteries are relatively small and light in comparison to other conventional and flow batteries such as LA, VR and PSB. Consequently, ZnBr is currently aiming at the renewable energy backup market. It is capable of smoothing the output fluctuations from a wind farm [2], or a solar panel [22], as well as providing frequency control. Installations currently completed have used ZnBr flow batteries for UPS, load management and supporting microturbines, solar generators, substations and T&D grids [2].

4.5.3.2 Cost of ZnBr flow battery
The power capacity cost is $639/kW and the energy capacity cost is $400/kWh [3].

4.5.3.3 Disadvantages of ZnBr flow battery
It is difficult to increase the power and storage capacities into the large MW ranges as the modules cannot be linked hydraulically, hence the electrolyte is isolated within each module. Modules can be linked electrically though and plans indicate that systems up to 1.5 MW are possible. As stated, the membrane suffers from slight degradation during the reaction so it must be replaced at the end of the batteries life (2000 cycles).

4.5.3.4 Future of ZnBr flow battery
The future of ZnBr batteries is currently aimed at the renewable energy market. *Apollo Energy Corporation* plan to develop a 1.5 MW ZnBr battery to back up a 20 MW wind farm for several minutes. They hope to keep the wind farm operational for an additional 200+ h a year [2]. The results from this will be very decisive for the future of ZnBr flow batteries.

4.6 Flywheel energy storage

A FES device is made up of a central shaft that holds a rotor and a flywheel. This central shaft rotates on two magnetic bearings to reduce friction (see Fig. 13). These are all

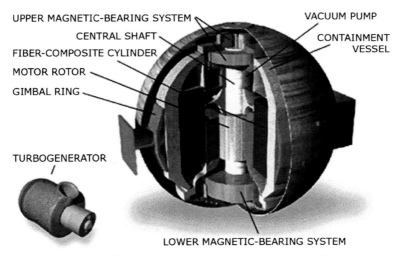

Figure 13: Flywheel energy storage device [23].

contained within a vacuum to reduce aerodynamic drag losses. Flywheels store energy by accelerating the rotor/flywheel to a very high speed and maintaining the energy in the system as kinetic energy. Flywheels release energy by reversing the charging process so that the motor is then used as a generator. As the flywheel discharges, the rotor/flywheel slows down until eventually coming to a complete stop.

The rotor dictates the amount of energy that the flywheel is capable of storing. Flywheels store power in direct relation to the mass of the rotor, but to the square of its surface speed. Consequently, the most efficient way to store energy in a flywheel is to make it spin faster, not by making it heavier. The energy density within a flywheel is defined as the energy per unit mass:

$$E_\mathrm{f} = \frac{I\omega^2}{2} = \frac{m_\mathrm{f}r^2\omega^2}{2} \tag{3}$$

where E_f is the total kinetic energy in J, I the moment of inertia in kg/m^2, ω the angular velocity of the flywheel in rad/s, m_f the mass of the flywheel in kg and r is the radius in m.

The power and energy capacities are decoupled in flywheels. In order to obtain the required power capacity, you must optimise the motor/generator and the power electronics. These systems, referred to as 'low speed flywheels', usually have relatively low rotational speeds, approximately 10,000 rpm and a heavy rotor made form steel. They can provide up to 1650 kW, but for a very short time, up to 120 s.

To optimise the storage capacities of a flywheel, the rotor speed must be increased. These systems, referred to as 'high speed flywheels', spin on a lighter rotor at much higher speeds, with some prototype composite flywheels claiming to reach speeds in excess of 100,000 rpm. However, the fastest flywheels commercially available spin at about 80,000 rpm. They can provide energy up to an hour, but with a maximum power of 750 kW.

Over the past number of years, the efficiency of flywheels has improved up to 80% [3], although some sources claim that it can be as high as 90% [1]. As it is a mechanical device, the charge-to-discharge ratio is 1:1.

4.6.1 Applications of FES

Flywheels have an extremely fast dynamic response, a long life, require little maintenance, and are environmentally friendly. They have a predicted lifetime of approximately 20 years or tens of thousands of cycles. As the storage medium used in flywheels is mechanical, the unit can be discharged repeatedly and fully without any damage to the device. Consequently, flywheels are used for power quality enhancements such as uninterruptible power supply (UPS), capturing waste energy that is very useful in EV applications and finally, to dampen frequency variation, making FES very useful to smooth the irregular electrical output from wind turbines.

4.6.2 Cost of FES

At present, FES systems cost between $200/kWh to $300/kWh for low speed flywheels, and $25,000/kWh for high speed flywheels [2]. The large cost for high speed flywheels is typical for a technology in the early stages of development.

Battery technology such as the LA battery is the main competitor for FES. These have similar characteristics to FES devices, and usually cost 33% less [3]. However, as mentioned previously (see Section 3.7.1), FES have a longer life span, require lower maintenance, have a faster charge/discharge, take up less space and have fewer environmental risks [2].

4.6.3 Disadvantages of FES

As flywheels are optimised for power or storage capacities, the needs of one application can often make the design poorly suited for the other. Consequently, low speed flywheels may be able to provide high power capacities but only for very short time period, and high speed flywheels the opposite. Also, as flywheels are kept in a vacuum during operation, it is difficult to transfer heat out of the system, so a cooling system is usually integrated with the FES device. Finally, FES devices also suffer from the idling losses: when flywheels are spinning on standby, energy is lost due to external forces such as friction or magnetic forces. As a result, flywheels need to be pushed to maintain its speed. However, these idling losses are usually less than 2%.

4.6.4 Future of FES

Low maintenance costs and the ability to survive in harsh conditions are the core strengths for the future of flywheels. Flywheels currently represent 20% of the $1-billion energy storage market for UPS. Due to its size and cycling capabilities, FES could establish even more within this market if consumers see beyond the larger initial investment. As flywheels require a preference between optimization of power or storage capacity, it is unlikely to be considered a viable option as a sole storage provider for power generation applications. Therefore, FES needs to extend into applications such as regenerative energy and frequency regulation where it is not currently fashionable if it is to have a future [3].

4.7 Supercapacitor energy storage

Capacitors consist of two parallel plates that are separated by a dielectric insulator (see Fig. 14). The plates hold opposite charges which induce an electric field, in which energy can be stored. The energy within a capacitor is given by:

$$E = \frac{CV}{2} \qquad (4)$$

where E is the energy stored within the capacitor (in J), V the voltage applied, and C is the capacitance found from [1]:

$$C = \frac{A\varepsilon_r\varepsilon_0}{d} \qquad (5)$$

where A is the area of the parallel plates, d the distance between the two plates, ε_r the relative permittivity or dielectric constant, and ε_0 is the permittivity of free

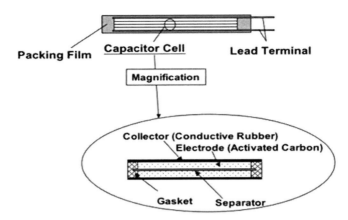

Figure 14: Supercapacitor energy storage device [24].

space (8.854×10^{-12} F/m). Therefore, to increase the energy stored within a capacitor, the voltage or capacitance must be increased. The voltage is limited by the maximum energy field strength (after this the dielectric breaks down and starts conducting), and the capacitance depends on the dielectric constant of the material used.

Supercapacitors are created by using thin film polymers for the dielectric layer and carbon nanotube electrodes. They use polarised liquid layers between conducting ionic electrolyte and a conducting electrode to increase the capacitance. They can be connected in series or in parallel. SCES systems usually have energy densities of 20–70 MJ/m^3, with an efficiency of 95% [2].

4.7.1 Applications of SCES

The main attraction of SCES is its fast charge and discharge, combined with its extremely long life of approximately 1×10^6 cycles. This makes it a very attractive replacement for a number of small-scale (<250 kW) power quality applications. In comparison to batteries, supercapacitors have a longer life, do not suffer from memory effect, show minimal degradation due to deep discharge, do not heat up, and produce no hazardous substances [1]. As a result, although the energy density is smaller, SCES is a very attractive option for some applications such as hybrid cars, cellular phones, and load levelling tasks. SCES is primarily used where pulsed power is needed in the millisecond to second time range, with discharge times up to 1 min [2].

4.7.2 Cost of SCES

SCES costs approximately $12,960/kWh [2] to $28,000/kWh [1]. Therefore, large-scale applications are not economical using SCES.

4.7.3 Disadvantages of SCES

SCES has a very low energy storage density leading to very high capital costs for large-scale applications. Also, they are heavier and bulkier than conventional batteries.

4.7.4 Future of SCES

Despite the small energy storage densities on offer, the exceptional life and cycling capabilities, fast response and good power capacity (up to 1 MW) of supercapacitors means that they will always be useful for specific applications. However, it is unlikely that SCES will be used as a sole energy storage device. One long-term possibility involves combining SCES with a battery-based storage system. SCES could smooth power fluctuations, and the battery provides the storage capacity necessary for longer interruptions. However, other technologies (such as flow batteries) are more likely to be developed for such applications. As a result, the future of SCES is likely to remain within specific areas that require a lot of power, very fast, for very short periods.

4.8 Superconducting magnetic energy storage

A SMES device is made up of a superconducting coil, a power conditioning system, a refrigerator and a vacuum to keep the coil at low temperature (see Fig. 15).

Energy is stored in the magnetic field created by the flow of DC in the coil wire. In general, when current is passed through a wire, energy is dissipated as heat due to the resistance of the wire. However, if the wire used is made from a superconducting material such as lead, mercury or vanadium, zero resistance occurs, so energy can be stored with practically no losses. In order to obtain this superconducting state within a material, it must be kept at a very low temperature. There are two types of superconductors: low-temperature superconductors that must be cooled to between 0 and 7.2 K, and high-temperature superconductors that have a temperature range of 10–150 K, but are usually in the 100 ± 10 K region. The energy stored within the coil (in J), E_C, can be obtained from [1]:

$$E = \frac{LI}{2} \qquad (6)$$

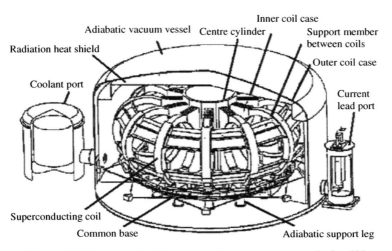

Figure 15: Superconducting magnetic energy storage device [2].

where L is the inductance of the coil, and I is the current passing through it. Therefore, material properties are extremely important as temperature, magnetic field, and current density are pivotal factors in the design of SMES. The overall efficiency of SMES is in the region of 90% [13] to 99% [3]. SMES has very fast discharge times, but only for very short periods of time, usually taking less than 1 min for a full discharge. Discharging is possible in milliseconds if it is economical to have a PCS that is capable of supporting this. Storage capacities for SMES can be anything up to 2 MW, although its cycling capability is its main attraction. SMES devices can run for thousands of charge/discharge cycles without any degradation to the magnet, giving it a life of 20+ years.

4.8.1 Applications of SMES

Due to the high power capacity and instantaneous discharge rates of SMES, it is ideal for the industrial power quality market. It protects equipment from rapid momentary voltage sags, and it stabilises fluctuations within the entire network caused by sudden changes in consumer demand levels, lightening strikes or operation switches. As a result, SMES is a very useful network upgrade solution with some sources claiming that it can improve the capacity of a local network by up to 15% [3]. However, due to high energy consumption of the refrigeration system, SMES is unsuitable for daily cycling applications such as peak reduction, renewable applications, and generation and transmission deferral [2].

4.8.2 Cost of SMES

SMES cost approximately $300/kW [2] to $509/kW [3]. It is worth noting that it is difficult to compare the cost of SMES to other storage devices due to its scales and purpose. In practical terms SMES should be compared to other network upgrade solutions where it is often very competitive or even less costly. Finally, the cost of storing electricity within a superconductor is expected to decline by almost 30% which could make SMES an even more attractive option for network improvements [3].

4.8.3 Disadvantages of SMES

The most significant drawback of SMES is its sensitivity to temperature. As discussed the coil must be maintained at an extremely low temperature in order to behave like a superconductor. However, a very small change in temperature can cause the coil to become unstable and lose energy. Also, the refrigeration can cause parasitic losses within the system. Finally, although the rapid discharge rates provide some unique applications for SMES, it also limits its applications significantly. As a result, other multifunctional storage devices such as batteries are usually more attractive.

4.8.4 Future of SMES

Immediate focus will be in developing small SMES devices in the range of 1–10 MW for the power quality market which has foreseeable commercial potential. A lot of work is being carried out to reduce the capital and operating costs of high-temperature SMES devices, as it is expected to be the commercial superconductor of choice once manufacturing processes are more mature, primarily due to cheaper cooling. There is a lot of market potential for SMES due to its unique application characteristics, primarily in transmission upgrades and industrial power quality [3]. However, one of the greatest concerns for SMES is its reliability over a long period of time.

4.9 Hydrogen energy storage system

HESS is the first of the three energy storage systems discussed in this chapter. HESS is the one of the most immature but also one of the most promising energy storage techniques available. As an energy storage system, HESS acts as a bridge between all three major sectors of an energy system: the electricity, heat and transport sectors. It is the only energy storage system that allows this level of interaction between these sectors and hence it is becoming a very attractive option for integrating large quantities of intermittent wind energy. There are three stages in HESS: create hydrogen; store hydrogen; use hydrogen (for required application).

4.9.1 Hydrogen production

There are three primary techniques to produce hydrogen: extraction from fossil fuels; reacting steam with methane; electricity (electrolysis). However, as producing hydrogen from fossil fuels is four times more expensive than using the fuel itself, and reacting steam with methane produces pollutants, electrolysis has become the most promising technique for hydrogen production going forward.

An electrolyser uses electrolysis to breakdown water into hydrogen and oxygen. The oxygen is dissipated into the atmosphere and the hydrogen is stored so it can be used for future generation. Due to the high cost of electrical production, only a small proportion of the current hydrogen production originates from electrolysis. Therefore, the most attractive option for future production is integrating electrolyser units with renewable resources such as wind or solar. In order to achieve this, an electrolyser must be capable of operating: with high efficiency; under good dynamic response; over a wide input range; under frequently changing conditions [2].

Recently a number of advancements have been made including higher efficiencies of 85%, wider input power capabilities, and more variable inputs. A new proton exchange membrane (PEM) has been developed instead of the preceding alkaline membranes. This can operate with more impure hydrogen, faster dynamic response, lower maintenance, and increased suitability for pressurization [2]. However, a PEM unit has lower efficiency (40–60%) so some development is still required. Electrolysers are modular devices so the capacity of a device is proportional to the number of cells that make up a stack. The largest commercial systems available can produce 485 Nm^3/h, corresponding to an input power of 2.5 MW. The lifetime of an electrolyser is proving difficult to predict due to its limited experience. However, research has indicated that the electrolyser unit will have the shortest lifespan within HESS. Some have predicted a lifespan in the region of 5–10 years but this is only an estimate [2].

4.9.1.1 Cost of hydrogen production

The estimated costs to produce power using an electrolyser are extremely varied. Predictions are as low as €300/kW [25] up to €1100/kW [2]. *ITM Power* in the UK claim to have produced an electrolyser that can operate with renewable sources, at a cost of $164/kW, and are currently planning to begin mass production in 2008 [26]. Maintenance costs are expected to be 3% of the capital cost [2].

4.9.1.2 Future of hydrogen production

Immediate developments are investigating the possibility of producing an electrolyser that can pressurise the hydrogen during electrolysis, as compressing the hydrogen after production is expensive and unreliable. Like all areas of HESS, the electrolyser needs a lot more development as well as technical maturity.

4.9.2 Hydrogen storage

A number of different options are currently available to store hydrogen:

1. *Compression*: The hydrogen can be compressed into containers or underground reservoirs. The cost of storing hydrogen in pressure vessels is $11/kWh to $15/kWh [2]. However, for underground reservoirs it is only $2/kWh [27]. This is a relatively simple technology, but the energy density and efficiency (65–70%) are low. Also, problems have occurred with the mechanical compression. However, this is at present the most common form of hydrogen storage for the transport industry, with the hydrogen compressed to approximately 700 bar (the higher the storage pressure, the higher the energy density, see Fig. 16). Although the energy required for the compression is a major drawback.
2. *Liquefied hydrogen*: The hydrogen can be liquefied by pressurising and cooling. Although the energy density is improved, it is still four times less than conventional petrol. Also, keeping the hydrogen liquefied is very energy intensive, as it must be kept below 20.27 K [28].
3. *Metal hydrides*: Certain materials such as nanostructured carbons and clathrate hydrate absorb molecular hydrogen. By absorbing the hydrogen in these materials, it can be easily transported and stored. Once required, the hydrogen is removed from the parent material. The energy density is similar to that obtained for liquefied hydrogen [28]. The extra material required to store the hydrogen is a major problem with this technique as it creates extra costs and mass. This is still a relatively new technology, so with extra development it could be a viable option; especially if the mass of material is reduced. Carbon-based absorption can achieve higher energy densities, but it has higher costs and even less demonstrations [2]. Both metal hydride or carbon-based absorption use thermal energy. This thermal heat could come from the waste heat of other processes with HESS, such as the electrolyser or fuel cell (FC), to improve overall efficiency.

Each storage technique is in the early stages of development and hence there is no optimum method at present with research being carried out in each area.

4.9.3 Hydrogen usage

There are two superior ways of using hydrogen:

1. internal combustion engine (ICE)
2. fuel cell

It is expected that the ICE will act as a transition technology while FCs are improving, because the modifications required to convert an ICE to operate on hydrogen are not very significant. However, the FC, due to its virtually emission-free, efficient and reliable characteristics, is expected to be the generator of choice for future hydrogen powered energy applications.

4.9.4 Fuel cell

A FC converts stored chemical energy, in this case hydrogen, directly into electrical energy. A FC consists of two electrodes that are separated by an electrolyte (see Fig. 17).

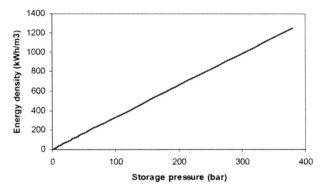

Figure 16: Energy density vs. pressure for hydrogen gas [2].

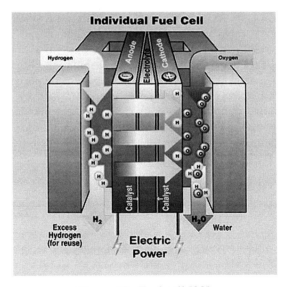

Figure 17: Fuel cell [29].

Hydrogen is passed over the anode (negative) and oxygen is passed over the cathode (positive), causing hydrogen ions and electrons to form at the anode. The electrons flow through an external circuit to produce electricity, whilst the hydrogen ions pass from the anode to the cathode. Here the hydrogen ions combine with oxygen to produce water. The energy produced by the various types of cells depends on the operation temperature, the type of FC, and the catalyst used (see Table 2). FCs do not produce any pollutants and have no moving parts. Therefore, theoretically it should be possible to obtain a reliability of 99.9999% in ideal conditions [30].

4.9.4.1 Cost of FC
All FCs cost between €500/kW and €8000/kW which is very high, but typical of an emerging technology [2]. These costs are expected to reduce as the technology ages and commercialization matures.

4.9.4.2 Future of FC
Immediate objectives for FCs include harnessing the waste heat more effectively to improve co-generation efficiency and also, combining FCs with electrolysers as a single unit. The advantage being lower capital costs although resulting in lower efficiency and increased corrosion [2]. FCs are a relatively new technology with high capital costs. However, with characteristics such as no moving parts, no emissions, lightweight, versatility and reliability, this is definitely a technology with a lot of future potential.

4.9.5 Disadvantages of HESS
The primary disadvantage with hydrogen is the huge losses due to the number of energy conversions required. Typically in a system that has high wind energy penetrations, by the time that hydrogen is actually being used for its final purpose it has gone through the following processes with corresponding efficiencies: (1) hydrogen is created by electrolysis – 85% efficient, (2) the hydrogen is stored – 65–70% efficient, (3) hydrogen is consumed in a FC car, power plant, or CHP unit – efficiency of 40–80%. This results in an overall efficiency ranging from 22 to 48%. In addition, this process assumes only one storage stage within the life of the hydrogen where as typically more than one storage stage would be necessary, i.e. stored when created, and stored at the location of use. Therefore, by implementing a "hydrogen economy", the efficiency of the system is very low that could result in very high energy costs and very poor utilization of limited resources such as wind or biomass. In summary, although the HESS offers huge flexibility, this flexibility is detrimental to the overall energy system efficiency.

4.9.6 Future of HESS
The use of hydrogen within the transport and electricity generation industries is expected to grow rapidly as electrolysis, storage techniques, and FCs become more commercially available. There are very ambitious hydrogen programs in the EU, USA and Japan, indicating increasing interest in hydrogen technology. Iceland is attempting to become the first 'hydrogen country' in the world by producing hydrogen from surplus renewable energy and converting its transport infrastructure from fossil fuels to hydrogen. In Norway, *Statkraft* plans to connect an electrolysis unit

Table 2: FC types [1].

Fuel cell	Electrolyte	Catalyst	Efficiency (%)	Operating temperature (°C)	Power output (kW)	Applications	Additional notes
Alkaline fuel cell (AFC)	Potassium hydroxide	Platinum	70	150–200	0.3–12	Widely used in the space industry (NASA)	Water produced by cell is drinkable. Can be easily poisoned by carbon dioxide (CO_2)
Polymer electrolyte membrane or proton exchange membrane (PEM)	Solid organic polymer	Platinum	45	80	50–250	Portable applications such as cars	Cell is sensitive to impurities so hydrogen used must be of good quality
Phosphoric acid fuel cell (PAFC)	Phosphoric acid	Platinum	40	150–200	200	Large stationary generation. Also co-generation (increases efficiency to 85%)	Can use impure hydrogen such as hydrogen from fossil fuels
Molten carbonate fuel cell (MCFC)	Potassium, sodium or lithium carbonate	Variety of non-precious metals	60	650	10–2000	Co-generation (increases efficiency to 85%)	High operating temperature and corrosive electrolyte result in short cell lifetime
Solid oxide fuel cells (SOFC)	Solid zirconium oxide	Variety of non-precious metals	60	1000	100	Utility applications. Prototype for co-generation exists (85% efficient)	High temperature causes slow start-up

to a large wind turbine and *Norsk Hydro* is continuing a project to provide Utsira Island with a wind hydrogen system. In Germany, *Siemens* and *P&T Technologies* are developing a wind hydrogen engine using an ICE. In the UK *Wind Hydrogen Limited* intend to develop large-scale wind hydrogen schemes. Finally, *HyGen* in California is developing a multi megawatt hydrogen generating and distributing network [2].

Car manufacturers are driving research in hydrogen for both the transport and infrastructure divisions. The automotive industry has engaged in setting up a strategy for the introduction of hydrogen to the transport sector with a number of single prototype projects advancing to fleet demonstrations [2]. Hydrogen is a serious contender for future energy storage due to its versatility. Once hydrogen can be produced effectively, it can be used for practically any application required. Consequently, producing hydrogen from renewable resources using electrolysis is currently the most desirable objective available. Primarily due to the versatility and potential of hydrogen to replace conventional fuel, "it is envisaged that the changeover to a hydrogen economy is less than 50 years from now" [2].

4.10 Thermal energy storage

Thermal energy storage (TES) involves storing energy in a thermal reservoir so that it can be recovered at a later time. A number of thermal applications are used instead of electricity to provide heating and cooling including aquifer thermal storage (ATS) and duct thermal storage (DTS). However, these are heat generation techniques rather than energy storage techniques and therefore will not be discussed in detail here. In terms of storing energy, there are two primary TES options. The first option is a technology which is used to supplement air conditioning in buildings and is displayed in Fig. 18. The second option is an energy storage system rather than a technology which will be discussed in more detail later.

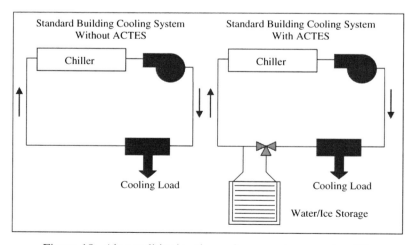

Figure 18: Air-conditioning thermal energy storage setup [3].

4.10.1 Air-conditioning thermal energy storage (ACTES)

The air-conditioning thermal energy storage (ACTES) units work with the air conditioning in buildings by using off-peak power to drive the chiller to create ice. During the day, this ice can be used to provide the cooling load for the air conditioner. This improves the overall efficiency of the cycle as chillers are much more efficient when operated at night time due to the lower external temperatures. Also, if ACTES units are used, the size of the chiller and ducts can be reduced. Chillers are designed to cope with the hottest part of the hottest day possible, all day. Therefore, they are nearly always operating below full capacity. If ACTES facilities are used, the chiller can be run at full capacity at night to make the ice and also at full capacity during the day; with the ice compensating for shortfalls in the chiller capacity. ACTES units lose approximately 1% of their energy during storage [3].

4.10.1.1 Cost of ACTES

If ACTES is installed in an existing building, it costs from $250 to $500 per peak kW shifted, and it has a payback period from 1 to 3 years. However, if installed during construction, the cost saved by using smaller ducts (20–40% smaller), chillers (40–60% smaller), fan motors, air handlers and water pumps will generally pay for the price of the ACTES unit. As well as this, the overall air conditioning cost is reduced by 20–60% [3].

4.10.1.2 Future of ACTES

Due to the number of successful installations that have already occurred, this technology is expected to grow significantly where air-conditioning is a necessity. It is however, dependent on the future market charges that apply, as this technology benefits significantly from cheaper off-peak power and demand charges. Finally, ACTES units will have to compete with other building upgrades such as lighting and windows, for funding in the overall energy saving strategies enforced [3].

4.10.2 Thermal energy storage system

The TESS can also be used very effectively to increase the flexibility within an energy system. As mentioned previously in this chapter, by integrating various sectors of an energy system, increased wind penetrations can be achieved due to the additional flexibility created. Unlike the HESS which enabled interactions between the electricity, heat and transport sectors, TES only combines the electricity and heat sectors with one another. By introducing district heating into an energy system, then electricity and heat can be provided from the same facility to the energy system using combined heat and power (CHP) plants. This brings additional flexibility to the system which enables larger penetrations of intermittent renewable energy sources. To illustrate the flexibility induced by TES on such a system, a snapshot of the power during different scenarios is presented below. The system in question contains a CHP plant, wind turbines, a thermal storage, a hot water demand, and an electrical demand as illustrated in Fig. 19.

During times of low wind power, a lot of electricity must be generated by the CHP plants to accommodate for the shortfall power production. As a result, a lot of hot water is also being produced from the CHP plant as seen in Fig. 19a. The high production of hot water means that production is now greater than demand, and consequently, hot water is sent to the thermal storage. Conversely, at times of high wind power, the CHP plants produce very little electricity and hot water. Therefore, there is now a shortage of hot water so the thermal storage is used to supply the shortfall, as seen in Fig. 19b.

4.10.2.1 Disadvantages of TESS

Similar to the HESS, the primary disadvantage with a TESS is the large investments required to build the initial infrastructure. However, the TESS has two primary advantages: (1) the energy system efficiency is improved with the implementation of a TESS. CHP production is approximately 85–90% efficient while conventional power plants are only 40% efficient, and (2) this technique has already been implemented in Denmark so it is a proven solution. On the negative side TES does not improve flexibility within the transport sector like the HESS, but this is inferior to the advantages it possesses. Therefore, in summary, the TESS does have disadvantages, but these are small in comparison to the advantages.

4.10.2.2 Future of TESS

Due to the efficiency improvements and maturity of this system, it is very likely that it will become more prominent throughout the world. Not only does it enable the utilization of more intermittent renewable energy (such as wind), but it also optimises the use of fuel within power plants, something that will become critical as biomass becomes more prominent. This system has been put into practice in Denmark which has the highest wind penetration in the world. In addition, Lund has

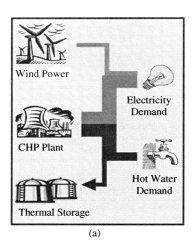 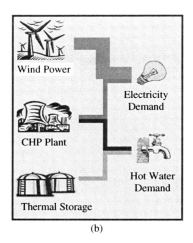

(a)   (b)

Figure 19: Energy system with district heating and thermal energy storage during: (a) a low wind scenario and (b) a high wind scenario.

outlined a roadmap for Denmark to use this setup in achieving a 100% renewable energy system at a lower cost than a conventional energy system [31]. Therefore, it is evident this technology can play a crucial role in future energy systems.

4.11 Electric vehicles

The final energy storage system that will be discussed in this chapter is the deployment of EVs. Once again, system flexibility and hence feasible wind penetrations are increased with the introduction of EVs into the transport sector. As illustrated in Fig. 20, EVs can feed directly from the power grid while stationary, at individual homes or at common recharging points, such as car parks or recharging stations. By implementing EVs, it is possible to make large-scale BES economical, combat the huge oil dependence within the transport sector and drastically increase system flexibility (by introducing the large-scale energy storage) [32]. Consequently, similar to the HESS and the TESS, EVs also provide a method of integrating existing energy systems more effectively.

4.11.1 Implementation of EV technology
EVs can be classified under three primary categories: (1) battery electric vehicles (BEV), (2) smart electric vehicles (SEVs) and (3) vehicle to grid (V2G).

4.11.2 Applications of EV technology
BEVs are plugged into the electric grid and act as additional load. In contrast, SEVs have the potential to communicate with the grid. For example, at times of high wind production, it is ideal to begin charging EVs to avoid ramping centralised production. In addition, at times of low wind production, charging vehicles should be avoided if possible until a later stage. V2G EVs operate in the same way

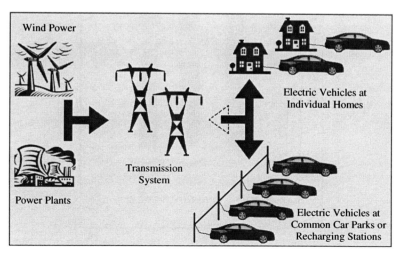

Figure 20: Schematic of electric vehicles and electric power grid.

as SEV, however, they have the added feature of being able to supply power back to the grid. This increases the level of flexibility within the system once again. All three types of EVs could be used to improve wind penetrations feasible on a conventional grid, with each advancement in technology increasing the wind penetrations feasible from approximately 30–65% [32] (from BEV to V2G).

4.11.3 Cost of EV technology

The costs associated with EVs are different to the costs quoted for other storage technologies. Consumers are not buying EVs to provide energy storage capacity for the grid, instead they are buying EVs as a mode of transport. Therefore, it is difficult to compare the costs of EVs under the conventional $/kW and $/kWh that other storage systems are compared with. As a result, below is a comparison between the price of EVs and conventional vehicles, as this comparison is more relevant when considering the uptake of EVs. Figure 21 illustrates the cost of owning a BEV and a conventional EV over a 105,000 km lifetime, with 25% of its life in urban areas. It is evident from Fig. 21 that BEVs are approximately 20% more expensive than conventional vehicles: while SEVs and V2G would be even more expensive, but these are still at the development stage. As SEVs and V2G EVs will enable significantly larger wind penetrations on the power grid than BEVs [32],

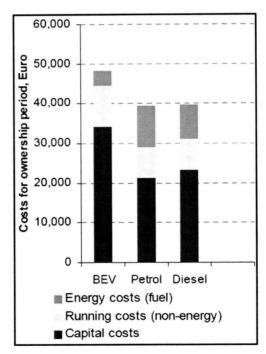

Figure 21: Cost of battery electric vehicles and conventional vehicles for a lifetime of 105,000 km (25% of which is urban driving) [33].

it is likely that economic incentives will be necessary to attract consumers to purchase SEV and V2G vehicles.

4.11.4 Disadvantages of EV technology

The primary disadvantage with EVs is the initial investment to establish the required infrastructure. Transmission lines will need to be upgraded to allow for high power capacities to and (in the case of V2G) from the electric cars, battery banks or charging stations will be required to replace conventional refuelling stations, and maintenance services will need to be established as we transfer from conventional ICEs to electric motors. In addition, travelling habits may need to be altered due to the alternative limitations associated with EVs instead of conventional vehicles, such as driving styles and time required for refuelling. Finally, the remaining issue with EVs is the driving range that can be obtained. Currently, hydrogen vehicles have a much larger range than EVs, although hydrogen vehicles are much less efficient. Therefore, depending on which of these factors is more important for different energy systems will most likely decide which of these technologies is preferred.

4.11.5 Future of EV technology

Electric vehicles are most likely going to be a key component in a number of future energy systems with large penetrations of intermittent renewable energy. This is primarily due to the two advantages mentioned in the introduction to this section: they reduce oil dependence and provide affordable large-scale energy storage. However, as mentioned already, alternative options such as hydrogen vehicles may reduce the attraction to EVs within energy systems which prioritise range over energy efficiency. However, steps are being taken in Europe, Japan and elsewhere to improve battery technology to the point where a 500 km range is standard. Key developments, e.g. in nanostructured lithium batteries may triple the energy density which can be achieved if this technology can be proven and manufactured cost effectively.

5 Energy storage applications

Energy storage devices can accommodate a number of network requirements. These include as follows [2].

5.1 Load management

There are two different aspects to load management:

1. *Load levelling*: Using off-peak power to charge the energy storage device and subsequently allowing it to discharge during peak demand. As a result, the overall power production requirements becomes flatter and thus cheaper baseload power production can be increased.
2. *Load following*: Energy storage device acts as a sink when power required falls below production levels and acts as a source when power required is above production levels.

Energy devices required for load management must be in the 1–100+ MW range as well as possessing fast response characteristics.

5.2 Spinning reserve

Spinning reserve is classified under two categories:

1. *Extremely fast response spinning reserve*: power capacity that is kept in the state of 'hot-stand-by'. As a result it is capable of responding to network abnormalities quickly.
2. *Conventional spinning reserve*: power capacity that requires a slower response.

Energy storage devices used for spinning reserve usually require power ratings of 10–400 MW and are required between 20 and 50 times per year.

5.3 Transmission and distribution stabilization

Energy storage devices are required to stabilise the system after a fault occurs on the network by absorbing or delivering power to generators when needed to keep them turning at the same speed. These faults induce phase angle, voltage and frequency irregularities that are corrected by the storage device. Consequently, fast response and very high power ratings (1–10 MW) are essential.

5.4 Transmission upgrade deferral

Transmission line upgrades are usually separated by decades and must be built to accommodate likely load and generating expansions. Consequently, energy storage devices are used instead of upgrading the transmission line until such time that it becomes economical to do so. Typically, transmission lines must be built to handle the maximum load required and hence it is only partially loaded for the majority of each day. Therefore, by installing a storage device, the power across the transmission line can be maintained constant even during periods of variable supply and demand. Then when demand increases, the storage device is discharged preventing the need for extra capacity on the transmission line to supply the required power. Therefore, upgrades in transmission line capacities can be postponed. Storage devices for this application must have a power capacity of kW to several hundreds of megawatts and a storage capacity of 1–3 h. Currently the most common alternative is portable generators; with diesel and fossil fuel power generators as long-term solutions and biodiesel generators as a short-term solution.

5.5 Peak generation

Energy storage devices can be charged during off-peak hours and then used to provide electricity when it is the most expensive, during short peak production periods.

5.6 Renewable energy integration

In order to aid the integration of renewable resources, energy storage can be used to:

1. Store renewable energy during off-peak time periods for use during peak hours (diurnal time-scale)
2. Match the output from renewable resources to the load required in the electricity market time-scale (quarter hour)
3. Smooth output fluctuations from a renewable resource in the millisecond to second time-scale
4. Significantly enhance the capacity factor (output which can be considered reliable) and associated payments attributable to wind
5. Facilitate the maximum inclusion of renewable electricity, by storing electricity from conventional generation for periods when total generating capacity cannot supply demand

A storage system used with renewable technology must have a power capacity of 10 kW to 100 MW, have fast response times (in some cases less than a second), excellent cycling characteristics (100–1000 cycles per year) and a good lifespan.

5.7 End-use applications

A survey in the US estimated that losses due to end-use and UPS applications were between \$119 billion and \$189 billion [2]. The most common end-use application is power quality which primarily consists of voltage and frequency control. Transit and end-use ride-through are applications requiring short power durations and fast response times, in order to level fluctuations, prevent voltage irregularities and provide frequency regulation. This is primarily used on sensitive processing equipment.

5.8 Emergency backup

This is a type of UPS except the units must have longer energy storage capacities. The energy storage device must be able to provide power while generation is cut altogether. Power ratings of 1 MW for durations up to 1 day are most common. For outages of several hours, days or weeks, diesel generators are more cost effective.

5.9 Demand side management

Demand side management (DSM) involves actions that encourage end-users to modify their level and pattern of energy usage. Energy storage can be used to provide a suitable sink or source in order to facilitate the integration of DSM. Conversely, DSM can be used to reduce the amount of energy storage capacity required in order to improve the network. Ultimately, the goal is to match consumption with generation by attributing some intelligence and control ability to the consumer, first at the industrial-scale and then domestic with tariff and other incentives for peak avoidance.

6 Comparison of energy storage technologies

To begin a brief comparison of the storage technologies within each category is discussed. These include: large power and storage capacities (PHES, UPHES, CAES); medium power and storage capacities (BES, FBES); large power or storage capacities (SCES, FES, SMES) energy storage systems (HESS, TESS, EVs). This is followed by an overall comparison of all the storage technologies.

6.1 Large power and energy capacities

The only devices capable of large power (>50 MW) and energy capacities (>100 MWh) are PHES, UPHES and CAES.

New PHES facilities are unlikely to be built as upgrades continue to prove successful. Once upgrades have been completed on existing PHES facilities, like all large-scale energy storage technologies the potential for PHES will depend heavily on the availability of suitable sites. It is widely believed that there are a limited number of suitable sites available for PHES. However, recent studies completed have illustrated the potential for seawater PHES [6, 34] as well as the potential for many more freshwater PHES sites than originally anticipated [35]. Therefore, if results continue in this fashion, PHES may only be constrained by economics and not technical feasibility, indicating that it could become a very important technology as fuel prices continue to rise in the future.

In theory UPHES could be a major contender for the future as it operates under the same operating principals as PHES: therefore, almost all of the technology required to construct such a facility is already available and mature. In addition, sites for UPHES will not be located in mountainous, isolated regions where construction is difficult and expensive. However, UPHES will still have unique site constraints of its own as it will require a suitable underground reservoir. Until such time that an extensive investigation is completed analysing the availability of such reservoirs, the future of UPHES will be uncertain.

Finally, the attractiveness of CAES depends on your opinion regarding the availability of gas and once again, the potential for suitable locations. It is a flexible, reliable, and efficient technology but it still needs gas to operate and an underground storage reservoir for the compressed air. Consequently, like PHES and UPHES, the potential for CAES will depend heavily on the availability of suitable locations. However, in addition the future of CAES may be decided based on the future availability of gas within an energy system. CAES by its nature is capital intensive and hence a long-term commitment. Therefore, if the energy system considering CAES has long-term ambitions to eliminate a dependence on gas, then this should be accounted for when analysing the feasibility of CAES.

In conclusion, it is evident that large-scale energy storage facilities all share one key issue: the availability of suitable locations. However, based on recent studies, suitable sites for PHES may be more prominent than originally anticipated. Therefore, until such time that the other large-scale storage technologies can display a similar potential for new facilities, it is likely that PHES will continue to lead the way in this category.

6.2 Medium power and energy capacities

This section includes BES and FBES. The only major contender from the BES storage technologies for future large-scale projects is the NaS battery. LA and NiCd will always be used for their existing applications but further breakthroughs are unlikely. FBES technologies (including VR, PSB and ZnBr) are all currently competing for the renewable energy market. Demonstration results for these batteries will be decisive for their future. It is worth noting that flow batteries are much more complex than conventional batteries. This is the reason conventional batteries will always be required. Conventional batteries are simple but constrained (power and storage capacities are coupled) while flow batteries are flexible but complex (power and storage capacities are independent but a number of extra parts required). The other key issue for this category will be the development of EVs. If technological advancements continue within EVs, then large-scale BES will most probably play an important role in future energy systems but not as stand-alone systems. Therefore, the future of this sector is very uncertain as various technologies continue to develop.

6.3 Large power or storage capacities

FES must be optimised to solve storage capacity issues (high speed flywheels, max 750 kW for 1 h) or power capacity issues (low speed flywheels, max 1650 kW for 120 s). SCES and SMES are only useful for power capacity issues. Therefore, FES, SCES and SMES are all used for power issues where a lot of power is required very fast. These technologies only differ in power capacity. FES is ideal for small power (up to 750 kW), SCES for medium power (up to 1 MW) and SMES for large power issues (up to 10 MW). The optimum technology depends on the power required for each specific application. Due to this very specific characteristic, these technologies are likely to be used for their specific purposes well into the future such as uninterruptable power supply.

6.4 Overall comparison of energy storage technologies

It is very difficult to compare the various types of energy storage techniques to one another as they are individually ideal for certain applications but no technology is perfect for everything. Consequently, for the purposes of this chapter, a number of illustrations are provided indicating the capabilities of each energy storage technology in relation to one another (see Figs. 22–26). This is followed by a table outlining the detailed characteristics of each storage technology (see Table 3) and a table indicating the cost of each technology (see Table 4). Finally, there is a table specifying the applications that each storage technology is suitable for (see Table 5).

6.5 Energy storage systems

The three energy storage systems that have been discussed above are HESS, TESS and EVs. The HESS provides an excellent level of flexibility within an energy system, by enabling the electricity, heat and transport sectors to interact

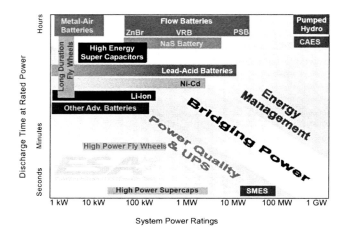

Figure 22: Discharge time vs. power ratings for each storage technology [37].

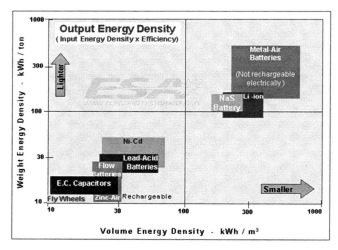

Figure 23: Weight energy density vs. volume energy density for each technology [37].

with one another. However, the primary disadvantage is the poor efficiencies due to the number of conversion required between creating hydrogen and using hydrogen. In contrast, the TESS increases the efficiency of the overall energy system. However a thermal energy system does not incorporate the transport sector. As a result, EVs (the third energy system discussed) are often combined with the TESS. Connolly *et al.* compared such a system with a hydrogen system and found that the TESS/EV energy system only needs 85% of the fuel that a HESS requires [37]. In addition, the TESS has already been implemented in Denmark and thus is a much more mature solution that a hydrogen economy.

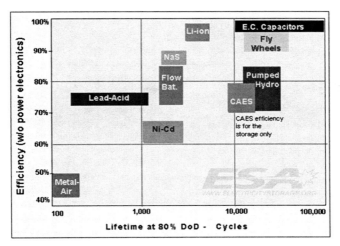

Figure 24: Efficiency and lifetime at 80% depth of discharge for each technology [37].

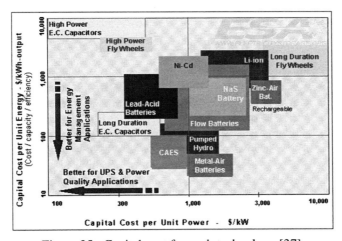

Figure 25: Capital cost for each technology [37].

However, if energy production is cheap or baseload renewable energy (i.e. biomass) is limited, the inefficiencies of the hydrogen energy system may be attractive. Therefore, a lot of potential exists but more research is required to truly quantify the benefits and drawbacks of each system.

Finally, it is evident from this chapter, that energy storage systems provide a much more promising solution to the integration of intermittent renewable energy than individual technologies. Energy storage technologies will most likely improve the penetrations of renewable energy on the electricity network but disregard the heat and transport sectors. Consequently, energy storage systems could be the key to finally replacing the need for fossil fuel with renewable energy.

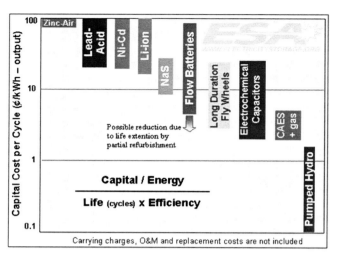

Figure 26: Cost per cycle for each technology [37].

7 Energy storage in Ireland and Denmark

Ireland and Denmark are very interesting countries as case studies for the deployment of energy storage systems. They have relatively small national electricity networks and large wind resources. They differ in that while Denmark is deeply interconnected to the electricity grids of continental Europe, Ireland as an island nation is not. Neither have substantial fossil fuel resources, but Ireland in particular is among the most dependent (92.6% in 2008) countries worldwide on imported fossil fuels for its electricity generation. In order to reduce greenhouse gases, Ireland's primary objective is to produce at least 40% of its electricity from renewable resources by 2020. Currently, Ireland's wind generating capacity has just surpassed 1000 MW, approximately 13% of the total Irish network capacity (see Table 6). However, previous reports had indicated that grid stability can be affected once wind capacity passed 800 MW [3]. As a result, Ireland will need to address the effects of grid intermittency in the immediate future.

Energy storage for an electric grid provides all the benefits of conventional generation such as, enhanced grid stability, optimised transmission infrastructure, high power quality, excellent renewable energy penetration and increased wind farm capacity. The main downsides of energy storage are the additional cost and the round-trip efficiency loss which may be as much as 30%. However, the operation of energy storage systems to store renewable generated electricity produces no carbon emissions and they do not rely on imported fossil fuels. As a result, energy storage is a very attractive option for increasing wind penetration onto the electric grid when it is needed.

However, currently Ireland's solution to the grid problems associated with the intermittency of wind generation is grid interconnection. EirGrid is in the process

Table 3: Characteristics of storage technologies [4].

	Power rating	Discharge duration	Response time	Efficiency	Parasitic losses	Lifetime	Maturity
Pumped hydro	100 – 4000 MW	4 – 12 h	sec – min	0.7 – 0.85	evaporation	30 y	commercial
CAES (in reservoirs)	100 – 300 MW	6 – 20 h	sec – min	0.64	-	30 y	commercial
CAES (in vessels)	50 – 100 MW	1 – 4 h	sec – min	0.57	-	30 y	concept
Flywheels (low speed)	<1650 kW	3 – 120 s	<1 cycle	0.9	~1%	20 y	commercial products
Flywheels (high speed)	<750 kW	<1 h	<1 cycle	0.93	~3%	20 y	prototypes in testing
Super-capacitors	<100 kW	<1 m	<1/4 cycle	0.95	-	10,000 cycles	some commercial products
SMES (Micro)	10 kW – 10 MW	1 s – 1 m	<1/4 cycle	0.95	~4%	30 y	commercial
SMES	10 – 10 MW	1 – 30 m	<1/4 cycle	0.95	~1%	30 y	design concept
Lead-acid battery	<50 MW	1 m – 8 h	<1/4 cycle	0.85	small	5 – 10 y	commercial
NaS battery	<10 MW	<8 h	n/a	0.75 – 0.86	5kW/kWh	5 y	in development
ZnBr flow battery	<1 MW	<4 h	<1/4 cycle	0.75*	small	2,000 cycles	in test/commercial units
V redox flow battery	<3 MW	<10 h	n/a	70 – 85%*	n/a	10 y	in test
Polysulphide Br flow battery	<15 MW	<20 h	n/a	60 – 75%*	n/a	2,000 cycles	in test
Hydrogen (Fuel Cell)	<250 kW	as needed	<1/4 cycle	0.34 – 0.40*	n/a	10 – 20 y	in test
Hydrogen (Engine)	<2 MW	as needed	seconds	0.29 – 0.33*	n/a	10 – 20 y	available for demonstration

*AC-AC efficiency.
**Discharge device. An independent charging device (electrolyser) is required.

Table 4: Cost characteristics of storage technologies [2].

	Capital cost			O&M cost		Cost certainty	Environmental issues	Safety issues
	Power-related cost ($/kW)	Energy-related cost ($/kWh)	BOP ($/kWh)	Fixed ($/kW-y)	Variable (c$/kWh)			
Pumped hydro	600	0 – 20	included	3.8	0.38		reservoir	exclusion area
CAES (in reservoirs)	425 – 480	3 – 10	50	1.42	0.01		gas emissions	none
CAES (in vessels)	517	50	40	3.77	0.27		gas emissions	pressure vessels
Flywheels (low speed)	300	200 – 300	~80				-	containment
Flywheels (high speed)	350	500 – 25,000	~1000	7.5	0.4		-	containment
Super-capacitors	300	82,000	10,000	5.55	0.5		-	-
SMES (Micro)	300	72,000	~10,000	26	2		-	magnetic field
SMES	300	2,000	~1,500	8	0.5		-	magnetic field
Lead-acid battery	200 – 300	175 – 250	~50	1.55	1.0		lead disposal	lead disposal, H2
NaS battery	259	245	~40	n/a	n/a		chemical handling	thermal reaction
ZnBr flow battery	1,500	200	included	n/a	n/a		chemical handling	chemical handling
V redox flow battery	n/a	175 –190	n/a	n/a	n/a		chemical handling	chemical handling
Polysulphide Br flow battery	1,200	175 –190	n/a	n/a	n/a		chemical handling	chemical handling
Hydrogen (Fuel Cell)	1100 – 2600	2 – 15	n/a	10.0	1.0		-	-
Hydrogen (Engine)	950 – 1850	2 – 15	n/a	0.7	0.77		emissions	-

Price list available　Price quotes available　Cost determined each project　Costs estimated

Table 5: Technical suitability of storage technologies to different applications [2].

Storage Application	Pumped hydro	Compressed air	Flywheel	Supercapacitors	Superconducting magnets	Lead-acid batteries	Advanced batteries	Flow batteries	Hydrogen fuel cell	Hydrogen engine
Storage Application			X	X	X	X		X	X	
Transit and end-use ride-through			X	X	X	X		X	X	
T&D stabilisation and regulation					X					
Peak generation	X	X	X			X	X	X	X	X
Fast response spinning reserve			X			X	X	X		
Conventional spinning reserve	X	X	X			X	X	X	X	X
Uninterruptible power supply			X	X	X	X	X	X	X	
Renewable integration			X	X	X	X	X	X		
Load levelling	X	X				X	X	X	X	X
Load following		X				X	X	X	X	X
Emergency back-up		X				X	X	X	X	X
Renewables back-up	X	X				X	X	X	X	X

Table 6: Network capacity for Ireland and Northern Ireland (correct as of May 2009).

Item	Republic of Ireland (MW)	Northern Ireland (MW)	All-Island (MW)
Total conventional capacity (MW)	6336.3	1968[*]	8902.3[♦]
Total wind capacity (MW)	1077	182[**]	1421.5[♦]
Total	7413.3	2150	10,323.8[♦]

Will increase to [*]2566 MW and [**]408 MW by August 2009.
[♦]Predicted values for August 2009.

of constructing a 500 MW interconnector to Wales that will allow for importing and exporting of electricity to and from Great Britain. Effectively, Great Britain will be Ireland's 'storage' device: excess electricity can be sold when the wind is blowing and electricity can be imported when it is not. A similar approach to improve grid stability was carried out in Denmark who installed large interconnectors to neighbouring Germany, Norway and Sweden (see Table 7).

However, Denmark discovered that they were only using approximately 500 MW of their wind generation at any time (see Fig. 27). The rest was being exported to Germany, Norway and Sweden.

Although this makes it possible for Denmark to implement a large amount of wind generation, Denmark is not only exporting wind power, but also its benefits. Firstly, Denmark exports its wind power cheaper than it buys power back. When excess wind power is available, Denmark needs to get rid of it, so Norway and Sweden pause their hydrogenerating facilities and buy cheaper wind power from Denmark. When wind production is low, Norway and Sweden turn back on their hydrogenerators, which have now stored large amounts of water, and sell power back to Denmark at a higher tariff. By using Great Britain as a power sink/source to accommodate wind power, Ireland too could face similar financial losses whilst exporting wind power.

Also, Germany uses the wind power generated in Denmark to reduce its own CO_2 emissions. Although Denmark is generating green power it is not profiting in terms of CO_2 reductions (green value). Consequently, if interconnection is continued to be used in Ireland to integrate wind power onto the grid, Ireland's green power could be used to reduce the CO_2 emissions of Great Britain rather than Ireland. By using energy storage technologies instead of interconnection, Ireland

Table 7: Grid interconnection in and out of Denmark.

Country	Interconnection from Denmark (MW)	Interconnection to Denmark (MW)
Germany	1200	800
Norway	950	1000
Sweden	610	580
Total	2760	2380

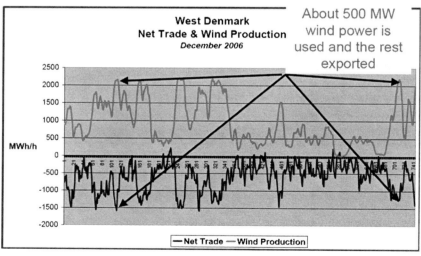

Figure 27: Wind production and interconnection trading for Denmark in December 2006 [4].

can develop an independent, stable and green electric grid. As renewable electricity deployment continues to expand worldwide at a rate of 30% p.a. for both wind and solar, these same issues will arise in larger countries. When the proportion of renewables on any country's electricity is large enough to require significant export and import activity, issues of interconnection, storage and green value will need to be addressed.

8 Conclusions

No one technology has all the ideal characteristics required for optimal grid integration of renewables. It is most likely that PHES (depending on topography, geology and feasibility), and FBES, will be the most attractive options for integrating renewable energy in the future because:

- PHES is a mature, proven, large-scale storage system that can act as an energy reservoir during times of excess electricity-production or as a producer during shortages in energy supply. In addition, recent studies have indicated that the lack of suitable sites may not be as severe as originally anticipated. Therefore, it is anticipated that PHES still has a role to play when integrating amounts of renewable energy.
- FBES facilities can act as the 'middle-man' between the grid and large-scale PHES facilities. As FBES are site-specific they can built where PHES is not an option and flexibility can compensate for the disadvantages of PHES.

In addition to the technologies identified above, all three energy storage systems discussed in this chapter have a huge potential in improve renewable energy penetrations in the future.

- The HESS is establishing itself as a serious contender for future power production more and more especially in the transport sector. Therefore, even if the HESS is not utilised for converting the hydrogen back to electricity, it is evident that hydrogen will probably be required for other applications such as heating or transport in the future. Therefore, it is an area that has a lot of future potential even though it can be an inefficient process.
- The TESS is not only capable of increasing the wind penetrations feasible within an energy system, but it also increases the overall efficiency of the energy system. Even more importantly, this technology has already been proven within the Danish energy system and hence does not carry the same risks as other options. However, the primary drawback of the TESS in comparison to the HESS is the transport sector: TESS does not account for the transport sector. However, this can be counteracted by combining the TESS with EVs.
- EVs are more efficient than both hydrogen and conventional vehicles. They also have the potential to make large-scale BES economical and hence vastly improve the flexibility within energy system. By combining EVs with the TESS huge reductions in fuel demands can be achieved as well as drastic increases in the potential to integrate renewable energy. Also, Lund has shown that this

technique can be extended further to a 100% renewable energy system [31]. As a result, this combination is one of the most promising solutions in the transition from a fossil fuel to a renewable energy system.

- All three energy storage systems can drastically improve the integration of wind energy. However, due to the very large initial investment required to implement any of these energy systems, it is very likely that either the HESS or the TESS/ EVs combination scenario will begin to compete with one another as the primary solution to integrating renewable energy. To decide which of these is the most beneficial will depend on whole range of issues including: (1) wind, wave, solar and tidal resource available; (2) biomass potential; (3) electricity, heating and transport demands; (4) energy system infrastructure already in place and so on.

In relation to the other technologies discussed in this chapter BES, FES, SMES, SCES and ACTES are always going to be used within the power sector but future operational breakthroughs are unlikely. Finally, although CAES reduces the amount of gas required it still uses gas for electricity production and therefore is likely to be a transition technology rather than a long-term solution depending on future availability of gas within the energy system.

After considering all the technologies, it is clear that integrating the electricity, heat and transport sectors of an energy system is probably the most effective method of increasing renewable energy usage. The additional flexibility that occurs by integrating these systems enables the grid operator to utilise the intermittent renewable resources more effectively. It is only then, after all these systems have been joined together, that individual energy storage technologies should be added as additional flexibility, especially if under current economic constraints.

References

[1] Cheung, K.Y., Cheung, S.T., Navin, De Silva, R.G., Juvonen, M.P., Singh, R. & Woo, J.J., *Large-Scale Energy Storage Systems*, Imperial College London, 2003.

[2] Gonzalez, A., Ó'Gallachóir, B., McKeogh, E. & Lynch, K., Study of electricity storage technologies and their potential to address wind energy Intermittency in Ireland. Sustainable Energy Ireland, 2004 http://www.sei.ie/Grants/Renewable_Energy_RD_D/Projects_funded_to_date/Wind/Study_of_Elec_Storage_Technologies_their_Potential_to_Address_Wind_Energy_Intermittency_in_Irl/.

[3] Baxter, R., Energy storage – a nontechnical guide, Oklahoma, PennWell Corporation, 2006.

[4] Sanidia National Laboratories, Energy Storage Systems, 2003. http://www.sandia.gov/ess/Technology/technology.html

[5] Wikipedia, Pumped-storage hydroelectricity, 2007 http://en.wikipedia.org/wiki/Pumped-storage_hydroelectricity

[6] Fulihara, T., Imano, H. & Oshima, K., Development of pump turbine for seawater pumped-storage power plant. *Hitachi Review*, **47(5)**, pp. 199–202, 1998.

[7] Wong, I.H., An underground pumped storage scheme in the Bukit Timah Granit of Singapore. *Tunnelling and Underground Space Technology*, **11(4)**, pp. 485–489, 1996.

[8] Figueirdo, F.C. & Flynn, P.C., Using diurnal power to configure pumped storage. *IEEE Transactions on Energy Conversion*, **21(3)**, pp. 804–809, 2006.

[9] Uddin, N., Preliminary design of an underground reservoir for pumped storage. *Geotechnical and Geological Engineering*, **21**, pp. 331–355, 2003.

[10] Argonne National Laboratory, Compressed Air Energy Storage (CAES) in Salt Caverns, 2009 http://web.ead.anl.gov/saltcaverns/uses/compair/index.htm

[11] Herr, M., *Economics of Integrated Renewables and Hydrogen Storage Systems in Distributed Generation*, University of London: Imperial College of Science, 2002.

[12] Energy Services, Wind plus compressed air equals efficient energy storage in Iowa proposal, 2003, http://www.wapa.gov/es/pubs/ESB/2003/03Aug/esb084.htm

[13] Schoenung, S., Characteristics and Technologies for Long vs. Short-Term Energy Storage, Sandia National Laboratories, 2001.

[14] Gordon, S. & Falcone, P., The emerging roles of energy storage in a competitive power market: summary of a DOE workshop, Sandia National Lab, 1995.

[15] University of Cambridge, Lead/acid batteries, 2007, http://www.doitpoms.ac.uk/tlplib/batteries/batteries_lead_acid.php

[16] RadioShack, Nickel-Metal Hydride Batteries, 2004, http://support.radioshack.com/support_tutorials/batteries/bt-nimh-main.htm

[17] Mpower, High Temperature Batteries, 2005, http://www.mpoweruk.com/high_temp.htm

[18] American Electric Power, AEP Commissions First U.S. Demonstration of the NAS Battery, 2005, http://aeptechcentral.com/nasbattery.htm

[19] Wald, M.L., Utility will use batteries to store wind power, The New York Times, Sept. 11, 2007, http://www.nytimes.com/2007/09/11/business/11battery.html?pagewanted=all

[20] PowerPidea, Vanadium redox batteries, 2006, http://peswiki.com/index.php/PowerPedia:Vanadium_redox_batteries

[21] Menictas, C., Hong, D., Yan, Z., Wilson, J., Kazacos, M. & Skyllas-Kazacos, M., Status of the vanadium Redox battery development program, *Proc. of Electrical Engineering Congress '94*, Sydney, 1994.

[22] Ball, G.J., Lex, P., Norris, B.L. & Scaini, V., Grid-connected solar energy storage using the zinc-bromine flow battery, ZBB Energy Corporation, 2002.

[23] Sheppard, G., Flywheels – a look to the future, http://www.upei.ca/~physics/p261/projects/flywheel2/flywheel2.htm

[24] NEC-TOKIN, Realizing Low ESR, 2004, http://www.nec-tokin.com/english/product/supercapacitor/feature.html

[25] Krom, L., *Renewable Hydrogen for Transportation Study*, L&S Associates, 1998.

[26] ITM Power Plc, Technology, 2007, http://www.itm-power.com/technology.html

[27] Padro, C. & Putsche, V., Survey of the economics of hydrogen technologies, National Renewable Energy Laboratory, 1999.

[28] Wikipedia, Liquid hydrogen, 2007, http://en.wikipedia.org/wiki/Liquid_hydrogen

[29] Naval Facilities Engineering Service Center, Impacts of the energy policy act of 2005 on federal facility operations, 2005, http://p2library.nfesc.navy.mil/issues/emergeoct2005/index.html

[30] Wikipedia, Fuel cell, 2007, http://en.wikipedia.org/wiki/Fuel_cell

[31] Lund, H. & Mathiesen, B.V., Energy system analysis of 100% renewable energy systems – the case of Denmark in years 2030 and 2050. *Energy*, **34(5)**, pp. 524–531, 2009.

[32] Lund, H. & Kempton, W., Integration of renewable energy into the transport and electricity sectors through V2G. *Energy Policy*, **36(9)**, pp. 3578–3587, 2008.

[33] Sustainable Energy Ireland. Hybrid electric and battery electric vehicles: technology, costs and benefits, 2007, http://www.sei.ie/News_Events/Press_Releases/Costs_and_benefits.pdf

[34] Spirit of Ireland, Working together for Ireland, http://www.spiritofireland.org/index.php

[35] Connolly, D. & Leahy, M., Development of a computer program to locate potential sites for pumped-hydroelectric energy-storage. *Energy*, 2009 (submitted).

[36] Connolly, D., Lund, H., Mathiesen, B.V. & Leahy, M., Ireland's pathway towards a 100% renewable energy-system: the first step, *Proc. of 5th Dubrovnik Conf. for Sustainable Development of Energy, Water and Environment Systems*, Dubrovnik, Croatia, 2009.

[37] Electricity Storage Association, Technology and Comparisons, 2003 http://electricitystorage.org/tech/technologies_comparisons.htm

Index

Management of Natural Resources, Sustainable Development and Ecological Hazards II

*Edited by: **C.A. BREBBIA**, Wessex Institute of Technology, UK, and **E. TIEZZI**, University of Siena, Italy.*

The first Conference was very well attended by a substantial group of scientists from all over the world and helped to underline the concern of the international community regarding the state of the planet. The basic premise of the meeting was the need to determine urgent solutions before we reach a point of irreversibility.

Our current civilisation has fallen into a self-destructive process by which natural resources are consumed at an increasing rate. This process has now spread across the planet in search of further sources of energy and materials. The aggressiveness of this quest is such that the future of our planet lies in the balance. The process is compounded by the pernicious effects of the resulting pollution.

On topics presented at the Second International Conference on Management of Natural Resources, Sustainable Devlopment and Ecological Hazards include the following headings: Greenhouse Issues; Ecosystems Modelling; Mathematical and System Modelling; Natural Resources Management; Environmental Indicators; Sustainability Development Studies; Recovery of Damaged Areas; Energy and the Environment; Socio Economic Factors; Soil Contamination; Waste Management; Water Resources; Environmental Management.

WIT Transactions on Ecology and the Environment, Vol 127
ISBN: 978-1-84564-204-4   **eISBN: 978-1-84564-381-2**
2009   464pp   £176.00

WIT PRESS ...for scientists by scientists

Energy and Sustainability II

Edited by: A.A. MAMMOLI, The University of New Mexico, and C.A. BREBBIA Wessex Institute of Technology, UK

The way in which our society exists, operates and develops is strongly influenced by the way in which energy is produced and consumed. No process in Industry can be performed without sufficient supply of energy, and without Industry there can be no production of commodities on which the existence of modern Society depends.

The energy systems evolved over a long period and more rapidly over the last two centuries, as a response to the requirements of Industry and Society, starting from combustion of fuels to exploiting nuclear energy and renewable resources. It is clear that the evolution of the energy systems is a continuous process, which involves constant technological development and innovation.

The presentation of the Second International Conference includes: Renewable Energy Technologies; Energy Management; Energy Policies; Energy and the Environment; Energy Analysis; Energy Efficiency; Energy Storage and Management.

WIT Transactions on Ecology and the Environment, Vol 121
ISBN: 978-1-84564-191-7   **eISBN: 978-1-84564-368-3**
2009   560pp   £213.00

WIT eLibrary

Home of the Transactions of the Wessex Institute, the WIT electronic-library provides the international scientific community with immediate and permanent access to individual papers presented at WIT conferences. Visitors to the WIT eLibrary can freely browse and search abstracts of all papers in the collection before progressing to download their full text.

Visit the WIT eLibrary at
http://library.witpress.com

WITPRESS   *...for scientists by scientists*

Tribology and Design

Edited by: **M. HADFIEL**, *Bournemouth University, UK,* **C.A. BREBBIA**, *Wessex Institute of Technology, UK and* **J. SEABRA**, *University of Porto, Portugal*

Tribology and Design 2010 is the 3rd International Conference in a series that originated with two meetings held at Bournemouth University, UK in 2005 and 2007. The Tribology and Design Conference explores the role of technology and design in the broader sense. It brings together colleagues from different disciplines interested in problems of surface interaction and design. The applications covered range from geomechanics to nano problems and from sustainability issues to advanced materials. It has never been so important for the designer to consider product and system durability in relation to reliability and sustainability issues. The topics for discussion also cover studies of tribology in nature and how the resulting lessons can be applied by the designers. Another important theme is the application of tribology in biomechanics, a field in which surface mechanics in general is of fundamental importance.

This book contains the papers presented at the Third International Conference, arranged into the following subject areas: Tribology in Space Applications; Reliability in Product Design; Nano-Tribology and Design; Tribology Under Extreme Conditions; Tribology in Geo-Mechanics; Energy Applications; Surface Measurements; Tribology in Biomechanics; Life-Oriented Design; Tribology and Nature; Design Tools; Surface Engineering; Lubricant Design; Test Methods; Advanced Materials; Analytical Studies; Sustainability and Tribology; Product Reliability; Corrosion Problems

WIT Transactions on Engineering Sciences, Vol 66
ISBN: 978-1-84564-440-6   eISBN: 978-1-84564-441-3
2010   304pp   apx £115.00

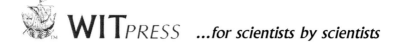

Structures Under Shock and Impact XI

Edited by: N. JONES, The University of Liverpool, UK , C.A. BREBBIA, Wessex Institute of Technology, UK and Ü. MANDER, University of Tartu, Estonia

The shock and impact behaviour of structures presents challenges to researchers, not only because of its obvious time-dependent aspects, but also because of the difficulties in specifying the external dynamic loading characteristics and in obtaining the full dynamic properties of materials. Thus it is important to recognise and fully utilise the contributions and understanding emerging from the theoretical, numerical and experimental studies, as well as investigations into material properties under dynamic loading conditions.

Of interest to engineers from civil, military, nuclear, offshore, aeronautical, transportation and other backgrounds, the topics covered include: Impact and Blast Loading Characteristics; Protection of Structures from Blast Loads; Energy Absorbing Issues; Structural Crashworthiness; Hazard Mitigation and Assessment; Behaviour of Steel Structures; Behaviour of Structural Concrete; Material Response to High Rate Loading; Seismic Engineering Applications; Interaction Between Computational and Experimental Results; Innovative Materials and Material Systems; Fluid Structure Interaction.

WIT Transactions on The Built Environment, Vol 113
ISBN: 978-1-84564-466-6   eISBN: 978-1-84564-467-3
Forthcoming   apx400pp   apx£152.00

Find us at
http://www.witpress.com